RETINAL DEVELOPMENT

This advanced text takes a developmental approach to the presentation of our current understanding of how vertebrates construct a retina. The book starts by examining how a patch of ectoderm becomes committed to make the eyes. It proceeds through the generation of the retinal neurons and how they connect up, culminating in the emergence of the first light responses. Written by experts in the field, each of the 17 chapters covers a specific step in this process, focusing on the underlying molecular, cellular and physiological mechanisms. There is also a special section on emerging technologies including genomics, zebrafish genetics and stem cell biology that are starting to yield important new insights into retinal development. Primarily aimed at professionals, both biologists and clinicians working with the retina, this book provides a concise and up-to-date view of what is known about vertebrate retinal development. Since the retina is 'an approachable part of the brain', this book is also attractive to all neuroscientists interested in development, as processes required to build this exquisitely organized system are ultimately relevant to all other parts of the central nervous system.

EVELYNE SERNAGOR is a neurophysiologist studying the role of early experience in guiding the development of retinal circuitry. She is a senior lecturer in Developmental Neuroscience at Newcastle University Medical School in the School of Neurology, Neurobiology and Psychiatry.

STEPHEN EGLEN is a lecturer in computational biology at the Department of Applied Mathematics and Theoretical Physics, University of Cambridge. He uses theoretical modelling techniques to understand and predict mechanisms of neural development.

BILL HARRIS has worked in the field of retinal development for over 30 years, studying problems of eye field specification, proliferation, cell determination and neural connectivity. He is a professor at Cambridge University in the Department of Physiology, Development and Neuroscience.

RACHEL WONG is Professor of Biological Structure at the University of Washington in Seattle, USA. She uses live-imaging techniques and electrophysiological approaches to study the assembly of neural circuits in the vertebrate retina.

RETINAL DEVELOPMENT

Edited by

Evelyne Sernagor
University of Newcastle upon Tyne, UK

Stephen Eglen
University of Cambridge, UK

Bill Harris
University of Cambridge, UK

Rachel Wong
University of Washington, Seattle, USA

CAMBRIDGE UNIVERSITY PRESS
Cambridge, New York, Melbourne, Madrid, Cape Town,
Singapore, São Paulo, Delhi, Mexico City

Cambridge University Press
The Edinburgh Building, Cambridge CB2 8RU, UK

Published in the United States of America by Cambridge University Press, New York

www.cambridge.org
Information on this title: www.cambridge.org/9781107411661

© Cambridge University Press 2006

This publication is in copyright. Subject to statutory exception
and to the provisions of relevant collective licensing agreements,
no reproduction of any part may take place without the written
permission of Cambridge University Press.

First published 2006
First paperback edition 2012

A catalogue record for this publication is available from the British Library

ISBN 978-0-521-83798-9 Hardback
ISBN 978-1-107-41166-1 Paperback

Cambridge University Press has no responsibility for the persistence or
accuracy of URLs for external or third-party internet websites referred to in
this publication, and does not guarantee that any content on such websites is,
or will remain, accurate or appropriate.

'This book is dedicated to the prevention of blindness.'

Contents

	List of contributors	page ix
	Foreword	xi
	Alan Bird	
	Preface	xv
	Acknowledgements	xv
1	Introduction – from eye field to eyesight	1
	Rachel O. L. Wong	
2	Formation of the eye field	8
	Michael E. Zuber and William A. Harris	
3	Retinal neurogenesis	30
	David H. Rapaport	
4	Cell migration	59
	Leanne Godinho and Brian Link	
5	Cell determination	75
	Michalis Agathocleous and William A. Harris	
6	Neurotransmitters and neurotrophins	99
	Rachael A. Pearson	
7	Comparison of development of the primate *fovea centralis* with peripheral retina	126
	Anita Hendrickson and Jan Provis	
8	Optic nerve formation	150
	David W. Sretavan	
9	Glial cells in the developing retina	172
	Kathleen Zahs and Manuel Esguerra	
10	Retinal mosaics	193
	Stephen J. Eglen and Lucia Galli-Resta	
11	Programmed cell death	208
	Rafael Linden and Benjamin E. Reese	
12	Dendritic growth	242
	Jeff Mumm and Christian Lohmann	
13	Synaptogenesis and early neural activity	265

Evelyne Sernagor

14 Emergence of light responses   288
    Evelyne Sernagor and Leo M. Chalupa

New perspectives

15 Regeneration: transdifferentiation and stem cells   307
    Jennie Leigh Close and Thomas A. Reh

16 Genomics   325
    Seth Blackshaw

17 Zebrafish models of retinal development and disease   342
    James M. Fadool and John E. Dowling

    Index   371

Colour plate section between pp. 304 and 305.

Contributors

Michalis Agathocleous, Department of Anatomy and Physiology, University of Cambridge, Downing Street, Cambridge CB2 3DY UK

Seth Blackshaw, Department of Neuroscience and Center for High-Throughput Biology, Johns Hopkins University School of Medicine, BRB 329, 773 N. Broadway Avenue, Baltimore, MD 21287, USA

Leo M. Chalupa, Distinguished Professor of Ophthalmology and Neurobiology, Chair, Section of Neurobiology, Physiology and Behavior, Division of Biological Sciences, UC Davis, One Shields Avenue, Davis, CA 95616, USA

Jennie Leigh Close, Neurobiology and Behavior Program, 357420 Health Sciences Center, University of Washington, School of Medicine, Seattle, WA 98195, USA

John E. Dowling, Department of Molecular and Cellular Biology, The Biological Laboratories, Harvard University, 16 Divinity Avenue, Cambridge, MA 02138, USA

Stephen J. Eglen, Department of Applied Mathematics and Theoretical Physics, Centre for Mathematical Sciences, Wilberforce Road, Cambridge CB3 0WA, UK

Manuel Esguerra, University of Minnesota, Department of Neuroscience, 6-145 Jackson Hall, 321 Church St SE, Minneapolis, MN 55455, USA

James M. Fadool, Department of Biological Science, Florida State University, 235 Biomedical Research Facility, Tallahassee, FL 32306-4340, USA

Lucia Galli-Resta, Istituto di Neuroscienze CNR, 56100 Pisa, Italy

Leanne Godinho, Department of Molecular and Cellular Biology, Harvard University, 16 Divinity Avenue, Cambridge, MA 02138, USA

William A. Harris, Department of Physiology Development and Neuroscience, University of Cambridge, Downing Street, Cambridge CB2 3DY, UK

Anita Hendrickson, Biological Structure, Box 357420, University of Washington, Seattle, WA 98195, USA

Rafael Linden, Instituto de Biofísica da UFRJ, CCS, bloco G, Cidade Universitaria, 21949-900, Rio de Janeiro, Brazil

Brian Link, Department of Cell Biology, Neurobiology and Anatomy, Medical College of Wisconsin, Milwaukee, Wisconsin 53226, USA

Christian Lohmann, Max-Planck Institute of Neurobiology, Am Klopferspitz 18, 82152 Planegg-Martinsried, Germany

Jeff Mumm, Luminomics, 1508 South Grand Blvd., St. Louis, MO 63104, USA

Rachael A. Pearson, Developmental Biology Unit, Institute of Child Health, University College London, 30 Guilford Street, London WC1N 1EH, UK

Jan M. Provis, Research School of Biological Sciences, The Australian National University, GPO Box 475, Canberra, ACT 2601, Australia

David H. Rapaport, Division of Anatomy, Department of Surgery, University of California, San Diego, School of Medicine, 9500 Gilman Drive, La Jolla, California 92093-0604, USA

Benjamin E. Reese, Neuroscience Research Institute and Department of Psychology, University of California at Santa Barbara, Santa Barbara, CA 93106-5060, USA

Thomas A. Reh, Neurobiology and Behavior, 357420 Health Sciences Center, University of Washington, School of Medicine, Seattle, WA 98195, USA

Evelyne Sernagor, School of Neurology, Neurobiology and Psychiatry, Medical Sciences, University of Newcastle upon Tyne, Framlington Place, Newcastle upon Tyne NE2 4HH, UK

David W. Sretavan, Department of Ophthalmology, University of California, San Francisco, CA 94143, USA

Rachel O. L. Wong, Department of Biological Structure, University of Washington, HSB G514, Box 357420, Seattle, WA 98195-7420, USA

Kathleen Zahs, University of Minnesota, Department of Physiology, 6-125 Jackson Hall, 321 Church Street SE, Minneapolis, MN 55455, USA

Michael E. Zuber, Department of Ophthalmology, SUNY Upstate Medical University, 750 East Adams Street, Syracuse, NY 13210, USA.

Foreword

The editors have assembled an impressive authorship to produce this book on development of the retina. There are several reasons why this is timely. Over the last decade there have been rapid advances in our understanding of the mechanisms involved in formation of the eye and determination of the fate of cells. This has been driven by an explosion of laboratory techniques that have allowed the study of gene expression and characterization of cell and tissue behaviour.

As a consequence there is increasing knowledge of what determines cell function, and of the behavioural relationship between cells. This has resulted in an understanding of genetically determined disease in humans. Many genes' products have been identified during development because they are highly expressed and mutations in these genes have been identified as being responsible for developmental abnormalities in man. Some of these genes express at low levels in adult life fulfilling a house-keeping function, and mutations in these have also been identified as giving rise to progressive retinal degeneration.

Findings from studies of development are of crucial importance to the current attempts to devise biological treatment of retinal diseases. There is ample evidence that growth factors delay cell death due to apoptosis in genetically determined retinal dystrophies in animals, and therapeutic trials in man have been initiated. There is still some doubt as to which agent may be the most appropriate to achieve suppression of apoptosis. Our knowledge of the mechanisms of programmed cell death is derived largely from studies of development and alternative therapeutic approaches may become evident as this work progresses.

There are also efforts to explore the possible role of cell transplantation. This is a major development in medicine in general, and the potential of treating retinal disease has been explored for some years. This was initiated by attempts to replace photoreceptor cells in retinal dystrophies. Many of the early efforts were disappointing but success has been achieved. Cell transplantation may also be applicable to other retinal diseases. Replacement of retinal pigment epithelium would be important in treatment of age-related macular disease, and of endothelial and pericytes would be appropriate in retinal vascular disease such as diabetic retinopathy. Many questions need to be addressed to accomplish success with this approach. What is the most appropriate source of the cells capable of assuming the functional characteristics of retinal cells? What environmental conditions would induce these pluri potential cells to form neurons, retinal pigment epithelium, glial cells and vascular cells, and how could these cells be induced to assume appropriate functional relationships with neighbouring cells?

Studies of development are likely to provide answers to these questions. The process of regeneration of the eye in amphibia has been intriguing since it was first observed nearly three centuries ago. With modern techniques it is possible to identify the biology of the phenomenon. The constant enlargement of the retina from its anterior edge in fish throughout life allows investigation of the mechanisms of cell and tissue generation. The relevance of this observation to mammals is illustrated by the observation that pluripotential cells can be retrieved from human donor eyes from the posterior ciliary body.

Thus this book is of great interest both to biologists and to those involved in the study of, and developing treatment for, retinal disease in humans. There are questions that can be addressed only by the study of the embryonic retina, and others that can most easily be answered by the development biologist. This book gives an invaluable account of the biology of the developing retina that demonstrates the value of such studies. Above all it illustrates well the value of research from one discipline to those in another.

Professor Alan Bird
Moorfields Eye Hospital, London

Preface

Vision is undoubtedly our most 'cherished' sense, and blindness the most tragic loss in perceiving the world around us. Visual perception begins in the eye, of which the retina is the most important component for interpreting visual signals, including colour, shape and movement. The retina is an ocular extension of the brain specialized in receiving and processing light and images. Although it is merely a few 100 micrometres thick and contains only seven cell types, the retina performs very sophisticated visual processing. Ultimately, it sends ALL information about the outside world to visual centres of the brain via the optic nerve in the form of coded electrical impulses. Understanding how the retina is organized and how it functions is thus of fundamental importance for understanding the entire visual system. It is therefore not surprising that the retina has been the focus of attention of many scientists since the late nineteenth century, when Cajal, in 1893, provided the first account of the anatomical organization of the vertebrate retina.

Although our knowledge of how the retina is organized and functions in adult organisms is absolutely essential, understanding how it is assembled during development is no less important. Indeed, when normal development is impaired, irreversible damage can result, in some cases even blindness. Moreover, understanding how the retina develops is attractive not only to developmental neuroscientists interested in vision, but to all neuroscientists interested in development, because the retina is 'an approachable part of the brain', and developmental processes required to build this exquisitely organized system, with well-defined layers and a limited number of cell types, are ultimately relevant to all other parts of the central nervous system.

In the last 10 to 15 years, the advent of powerful new techniques in genetics, molecular biology, imaging and electrophysiology have led to a huge leap forward in our understanding of how the retina develops. The goal of this book is to review all these new advances, while placing them in a chronological context of developmental events, from cell proliferation to the building of neural circuits involved in visual processing. Our intent is to deliver a well-illustrated source of up-to-date information for scientists interested in retinal research, retinal development or development of other parts of the vertebrate brain. We hope that the information gathered will provide deeper insights to all students and researchers aiming to achieve a better understanding of this fascinating part of the brain. We have also deliberately highlighted many open questions in every chapter, in the hope that they will inspire other scientists.

Reference

Cajal, S. R. (1893). *The Structure of the Retina. La Retine des Vertébrés*. Springfield IL: Thomas Springfield.

Acknowledgements

We wrote this book because it seemed to us there was none available on the development of the vertebrate retina and that such a book would be valuable to both developmental biologists and to those doing translational work, especially in developmental aspects of retinal disease. Our plan was to cover the development of the retina ontogenetically like a developmental story.

We each do research on different aspects of retinal development and were able to write on these aspects but knew that this story would not be complete unless it covered stages and approaches that were the domain of others. Part of the fun of this book for us has been working with these scientists who have tried very hard to oblige us by conforming to a standard chapter style, which we thought was important to the cohesion and uniformity of a book that has been the effort of many. We therefore take this opportunity to thank all the authors for their excellent contributions and for putting up with our prescriptive demands.

Finally just a quick word on why, in spite of our desire for uniformity between chapters, in some chapters the figures show the pigment epithelium at the top while in other chapters the pigment epithelium is shown at the bottom. We tried to get everyone to do it the same way, but this just stirred a large debate that prompted one of us to write this poem in frustration.

> The eye is a globe when looked at whole
> It has both a dorsal and ventral pole.
>
> Clinical explorers that chart passages North
> Claim the PE's on top. "We took photos. We've got'em"
> But Dev Neuro types who to the South sally forth
> Take photos that show it's at the bottom.

Finally, we decided that, as long as the authors labelled their figures, we would allow both views.

Besides the authors, we would like to thank Katrina Halliday at Cambridge University Press who helped us put this enterprise together, as well as Clare Georgy and Jo Bottrill who have helped in the final production of the book. We would also like to thank Anne Cowell for her relentless, scrutinizing editing work. We also thank baby Samuel for lending us his beautiful eyes for the cover of the book.

Finally, we would like to thank Professor Alan Bird who agreed to write a foreword, and our many colleagues who, since the time of Ramon y Cajal, have contributed to our current understanding of retinal development.

1
Introduction – from eye field to eyesight

Rachel O. L. Wong

University of Washington, Seattle, USA

Vision begins at the retina, a light-sensitive tissue at the back of the eye that comprises highly organized, laminated networks of nerve cells. Investigating the mechanisms of retinal development is fundamentally important to gaining a basic knowledge of how vision is established. In this book, we present the sequence of developmental events and the mechanisms involved in shaping the structure and function of the vertebrate retina.

1.1 Formation of the eye

The eye is derived from three types of tissue during embryogenesis: the neural ectoderm gives rise to the retina and the retinal pigment epithelium (RPE), the mesoderm produces the cornea and sclera, and the lens originates from the surface ectoderm (epithelium). During embryogenesis (Figure 1.1), the eyes develop as a consequence of interactions between the surface ectoderm and the optic vesicles, evaginations of the diencephalon (forebrain). These optic vesicles are connected to the developing central nervous system by a stalk that later becomes the optic nerve. When the optic vesicles contact the ectoderm, inductive events take place to cause the epithelium to form a lens placode. The lens placode then invaginates, pinches off eventually and becomes the lens. During these events, the optic vesicle folds inwards and forms a bilayered cup, the optic cup. The outer layer of the optic cup differentiates into the RPE whereas the inner layer differentiates into the retina. The iris and ciliary body develop from the peripheral edges of the retina. The sclera is derived from mesenchymal cells of neural crest origin, which also migrate to form the cornea and trabecular meshwork of the anterior chamber of the eye. During early development, the hyaloid artery and vein provide the major blood supply to the eye; these structures is later disassembled, leaving behind the ophthalmic artery and veins. In humans, eye development begins at around 22 days of development and is not completed until several months after birth (Mann, 1964).

1.2 Basic organization of the mature vertebrate retina

Since the early investigations of Cajal in the past century (see Cajal, 1972), it is well established that the vertebrate retina comprises five major classes of nerve cells or neurons

Retinal Development, ed. Evelyne Sernagor, Stephen Eglen, Bill Harris and Rachel Wong.
Published by Cambridge University Press. © Cambridge University Press 2006.

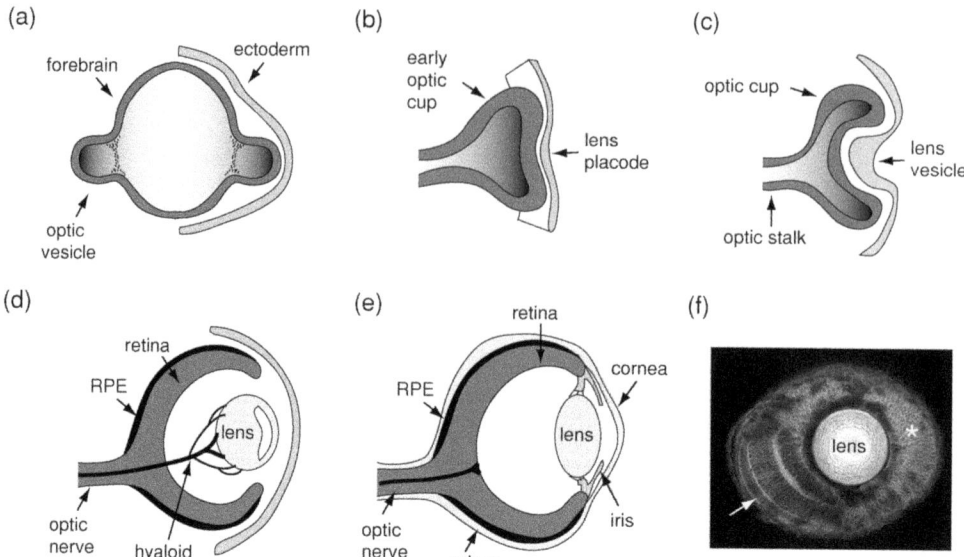

Figure 1.1 Development of the eye. (a) Optic vesicles that form from the neural tube give rise to the two eyes. (b) Contact between the optic vesicles and the surface ectoderm produces a lens placode. (c) The lens placode pushes into the optic vesicle, resulting in the formation of an optic cup and a lens vesicle. (d) The outer surface of the optic cup becomes the retinal pigment epithelium (RPE), and the inner surface becomes the retina. (e) Location of the retina within the mature eye. (f) Example of a developing vertebrate eye (zebrafish), showing the developing lens and the retina. All cell membranes are labelled here by expression of fluorescent protein in a transgenic animal, and imaged in the live animal. The region of the retina deeper within the eye (arrow) shows its characteristic lamination pattern whereas peripheral retina (example, asterisk) is last to differentiate and laminate.

(Figure 1.2; see Wässle, 2004, for review). Rod and cone photoreceptors convert light information to chemical and electrical signals that are relayed to interneurons in the outer retina. Bipolar interneurons are contacted by photoreceptors and convey signals from the outer retina to the inner retina. Transmission from photoreceptors is modulated by horizontal cells that also contact the bipolar cells. In the inner retina, bipolar cells form chemical synapses with their major targets, the retinal ganglion cells and amacrine interneurons. Amacrine cells not only modulate signals from the bipolar cells by providing inhibition directly onto ganglion cells, but also modulate transmitter release from the bipolar cells. Light information leaves the retina via axons of the retinal ganglion cells that collectively form the optic nerve.

The cell bodies and connections of retinal neurons are arranged in layers (Figure 1.2). This laminar organization of the retina is stereotypic across species. Connections are restricted to two major laminae, the outer plexiform layer (OPL) and the inner plexiform layer (IPL). Müller glial cells, which span the depth of the retina, provide important structural and functional support for the retinal neurons. Embedded within this basic organization of the vertebrate retina are many specialized subcircuits, working together in parallel to process

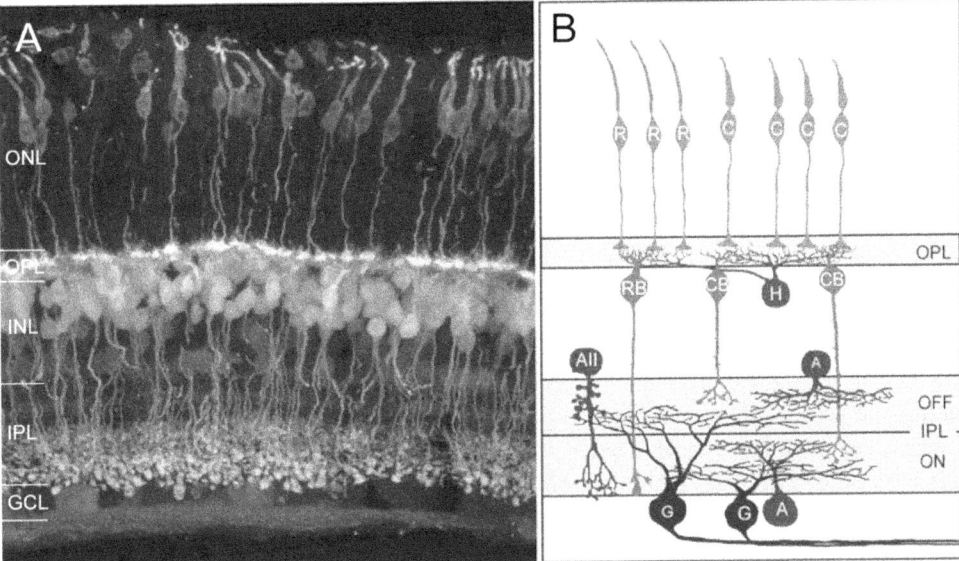

Figure 1.2 Basic circuitry of the vertebrate retina. (A) Cross-section of a mouse retina showing the laminated distribution of cell bodies and neuronal processes visualized by cell-staining methods. ONL, outer nuclear layer comprising photoreceptors; OPL, outer plexiform layer; INL, inner nuclear layer; IPL, inner plexiform layer; GCL, ganglion cell layer. Image provided by J. Morgan. (B) Schematic representation of the basic wiring diagram of the retina. R, rods; C, cones; RB, rod bipolar cell; CB, cone bipolar cell; H, horizontal cell; AII, AII-type amacrine cell in the rod pathway; A, amacrine cell; G, ganglion cell.

different features of the image. Rods are sensitive to low light levels and a rod-driven circuit exists for visualizing objects under dim light conditions. In vertebrates, this circuit involves connections between rod photoreceptors, rod bipolar cells and a special type of amacrine cell, the AII amacrine cell (Figure 1.2B). Increments (ON) and decrements (OFF) in light intensity are detected and processed along two vertical pathways. Cone photoreceptors contact a variety of cone bipolar cells, some of which are depolarized (ON) and others, hyperpolarized (OFF) by increased illumination. ON- and OFF-cone bipolar cells contact retinal ganglion cells, which respond to changes in illumination according to their bipolar input. Together, the ON and OFF pathways provide contrast information. In addition to these basic features, the retina also has specialized circuits that can compute other features of the visual scene, such as the direction of motion or orientation of edges.

Work to date has identified components of several subcircuits of the retina, demonstrating a high degree of correlation between structure and function. For example, connections involving ON and OFF components are largely confined to distinct sublaminae within the IPL (Figure 1.2). The relationship between structure and function of circuits in the mature retina has facilitated studies aimed at understanding the mechanisms essential to its development, and also studies that wish in general to determine what factors are essential for the development of the central nervous system.

1.3 Development of the vertebrate retina

Vision, of course, relies on the proper development of the retina. Much progress has been made towards our understanding of the cellular and molecular mechanisms underlying retinal development. We will present the major areas of retinal research, beginning at the stage when the eyes form to when the retina performs its mature task, processing the visual scene.

1.3.1 Specification of the eye field and building blocks of the retina

Retinal development begins with specification of the eye primordia during early stages of embryonic life. One major area of investigation thus encompasses studies aimed at elucidating the genes and molecular signalling pathways required for the formation of the two eyes (Chapter 2). Once the eye fields are defined, the next step concerns the generation of the appropriate cell types, their numbers and distributions (Chapter 3). Here, considerations have been given to cell-intrinsic (genetic) and extrinsic (environmental) signals that act in concert to specify cell fate. Decisions to become one or another type of retinal cell appear to depend on many factors, including the time of cell genesis (Chapter 5). Cell division occurs at the retinal surface abutting the pigment epithelium. From this location, postmitotic cells migrate to their final locations within the retina. Although cellular mechanisms underlying neuronal migration are well studied in the central nervous system in general, our understanding of this process in the vertebrate retina is only in its infancy (Chapter 4). One thing that is evident, however, is that, like other parts of the nervous system, there is an overproduction of retinal neurons during development, many of which die before eye opening. The mechanisms regulating naturally occurring cell death in the retina are important for controlling cell number and distribution (Chapter 11). Also, understanding what evokes cell survival or death in the retina is likely to have implications for the regenerative capacity of the vertebrate retina. To date, mammalian retinas show a limited ability to regenerate whereas in other vertebrates, such as the zebrafish, retinas have a tremendous capability to regenerate. Studies comparing the development of different vertebrate species are thus important for the discovery of genes and cellular interactions that support regeneration of the vertebrate retina (Chapter 15).

1.3.2 Wiring cell components of the retina

Following the generation of each cell type, the major sequence of developmental events in the retina pertains to the formation and maintenance of connections between its cellular components, and between the retina and its brain targets. For the latter, the formation of the optic nerve is of primary importance in order to wire the eye to the brain (Chapter 8). Within the retina, organization of its networks occurs progressively and with precision. First, the various cell types need to express their appropriate neurotransmitters for intercellular communication. These transmitters, as well as neurotrophic molecules, play essential roles in the survival and differentiation of the retina (Chapter 6). Second, to communicate with their neighbours, retinal neurons need to extend processes. The dendritic processes of retinal neurons, their input surface, are contacted by presynaptic cells. Conversely, the axons of retinal neurons, their output processes, synapse onto their target cells. It should be noted, however, that the processes of amacrine cells are both pre- and postsynaptic in nature. One important requirement for dendritic outgrowth of retinal neurons, studied most widely in retinal ganglion cells, is that their arbors overlap by defined amounts, leading to tiling and complete coverage of the retinal surface. Different cell types show different amounts of overlap. How these mosaics of cell territories are established during development is fascinating and important to study because they relate to spatial processing by each cell population (Chapter 10). In fact, retinal ganglion cells that can sample at high acuity have small dendritic arbors that hardly overlap whereas those that detect motion primarily show greater overlap. One idea is that early contact between neighbouring cells of the same type regulates their spacing via adhesion-based signalling. However, there is also evidence for intrinsic factors limiting how large a dendritic arbor retinal ganglion cells, and perhaps other retinal neurons, can grow. Our knowledge of the factors that control the growth of retinal neurons is only just beginning to deepen.

Another essential wiring pattern in the retina is that the processes of ON and OFF bipolar, amacrine and ganglion cells stratify within their appropriate sublaminae. Much work has been focused on determining the role of intrinsic factors as well as cell–cell interactions in shaping the stratification of these cell types. Indeed, the use of state-of-the-art live-imaging techniques, transgenic mice and mutants lacking specific cell types or molecular interactions is beginning to help unravel the mechanisms that regulate neurite patterning in the retina (Chapter 12).

Accurate processing of visual information not only necessitates that the axons and dendrites of retinal neurons target their correct synaptic partners, but importantly, that they form the appropriate balance of excitatory and inhibitory connections. Synapse formation has been studied for many decades in a variety of animals. Traditionally, this developmental event has been investigated using electron microscopy (EM) methods that enable synapses to be visualized at the ultrastructural level (Chapter 13). Photoreceptors and bipolar cells form ribbon synapses, which can be recognized under EM by an electron-dense ribbon-like structure flanked by synaptic vesicles containing neurotransmitter. Amacrine cells form conventional synapses whereby pre- and postsynaptic densities and synaptic vesicles are observed at the contact site, but ribbons are absent. At the EM level, then, it is possible to distinguish photoreceptor (outer retina), bipolar and amacrine synapses. At present, the study of synaptogenesis in the vertebrate retina is restricted to fixed tissue, but modern methods of live-cell labelling using fluorescently tagged synaptic proteins (Morgan et al., 2005) are likely to help us gain a dynamic view of this fundamental developmental process in live tissue.

1.3.3 Properties of early circuits in the retina and the emergence of light sensitivity

It is perhaps surprising that early circuits of the retina are functional and able to generate electrical activity before the retina is sensitive to light. Amacrine cells and the ganglion cells form the first synaptic circuit in the retina. Photoreceptors develop much later and bipolar cells needed to connect the outer retina to the inner retina form their connections after the retina is wired to visual targets in the brain. The activity produced by the early amacrine–ganglion cell network demonstrates unique spatiotemporal patterns, which is characteristic of many vertebrates studied thus far. These patterns, and their potential function in synaptic wiring, will be discussed in detail (Chapter 13).

Light responses emerge shortly before eye opening in mammals, and in the embryo of turtles and zebrafish. Few studies to date have examined the nature of these responses, and in particular how the region of space encoded by retinal neurons becomes defined is not well understood. With improvements in electrophysiological techniques that allow detailed assessment of the physiological properties of retinal neurons and their early and mature responses to light stimuli, this gap in our knowledge is beginning to fill (Chapter 14).

1.4 Concluding remarks

Although a large part of this book is dedicated to the architecture, connectivity and function of retinal neurons, the maturation of glial cells and their role in retinal development is also considered (Chapter 9). Recent studies certainly demonstrate that glial cells are integral and essential components of the retina, and that they play a significant role in maturation of retinal neurons and their connectivity.

A common theme throughout the book concerns reference to different vertebrates, ranging from zebrafish to primate. Such diversity in the study of vertebrate retinal development has led to the discovery of developmental mechanisms that are unique or common across vertebrates. Moreover, each vertebrate has features that offer investigation of specific developmental events. For example, the rapid development and relative transparency of the embryonic zebrafish eye permits visualization of retinal development in vivo (Chapter 17). In particular, this has enabled cell division and migration in the retina to be followed. The presence of a fovea in monkeys allows us to study the mechanisms underlying the development of this specialized region of the retina, which is necessary for high-acuity vision in human (Chapter 7). Thus, in the future, studies based on different vertebrates are likely to continue to yield basic information of how the retina develops.

A major goal of this book is not only to present the current knowledge of how the vertebrate retina develops, but also to convey a sense of progress in our understanding of the mechanisms involved. This progress has largely been fuelled by advances in several technical areas. First, molecular methods now enable the identification of gene products expressed in specific retinal cell types, and at distinct periods of development. This knowledge should help us better understand the molecular pathways that specify cell identity (Chapter 16). Second, it is now possible to visualize and track retinal cells in live tissue by expression of fluorescent proteins in transgenic fish or mice (Chapter 12). Third, new ways to record physiological responses from not only one, but dozens of retinal neurons simultaneously, has led to the discovery of patterned activity during development (Chapter 13). Such methods can also be used to study light responses from a multitude of retinal ganglion cells during development. Together with conventional approaches used successfully over the decades, such technological advances will push the frontiers ahead in our quest to understand how the vertebrate retina attains its structure and function.

References

Cajal, S. R. (1972). *The Structure of the Retina*, ed. S. A. Thorpe and M. Glickstein. Springfield, Illinois: Charles Thomas.

Mann, I. (1964). *The Development of the Human Eye*. New York: Grune and Stralton.

Morgan, J., Huckfeldt, R. and Wong, R. O. (2005). Imaging techniques in retinal research. *Exp. Eye Res.*, **80**, 297–306.

Wässle, H. (2004). Parallel processing in the mammalian retina. *Nat. Rev. Neurosci.*, **5**, 747–57.

2

Formation of the eye field

Michael E. Zuber
SUNY Upstate Medical University, Syracuse NY, USA

William A. Harris
University of Cambridge, Cambridge, UK

2.1 Introduction

Vertebrate eyes originate from a single field of neuroectodermal cells in the anterior region of the neural plate called the eye field (sometimes referred to as the eye anlage, eye primordia or presumptive eye). The origins of the eye field can be traced back to the 32-cell-stage blastula in which a subset of blastomeres is competent, but not yet committed, to form retina. This chapter begins with a discussion of retinal competence and the maternal molecules and cell–cell interactions that take place in and bias early blastomeres toward a retinal fate. Transplantation experiments have shown that the entire presumptive neural plate of midgastrula embryos can form retina, demonstrating the remarkable coordination of neural development with eye formation. Neural induction and the neural patterning events critical for defining where the eye field forms in the developing nervous system will be addressed. Cultured amphibian anterior neural plates form eyes demonstrating that the eye field is specified (committed to form the eye) by the neural plate stage. A conserved set of transcription factors collectively referred to as eye field transcription factors are required for normal eye formation and are expressed in the eye field of the neural plate stage embryo. These genes and their functional interactions, which are required for and under some circumstances sufficient to drive eye field and eye formation will be described. A description of how the single vertebrate eye field separates to form the eye primordia that eventually give rise to the two eyes concludes this chapter.

2.2 Retinal competence

The eye field is not specified until early neurula stages (see below). However, even at early cleavage stages only a subset of blastomeres is competent to contribute to the eye field. Fate-mapping, transplantation and ablation experiments have shown that nine dorsal animal (retinogenic) blastomeres of the 32-cell-stage *Xenopus laevis* embryo normally contribute progeny to each eye (Figure 2.1 and Moody, 1987). A certain level of plasticity is observed during normal development. The proportion of retinal cells derived from any given

Retinal Development, ed. Evelyne Sernagor, Stephen Eglen, Bill Harris and Rachel Wong.
Published by Cambridge University Press. © Cambridge University Press 2006.

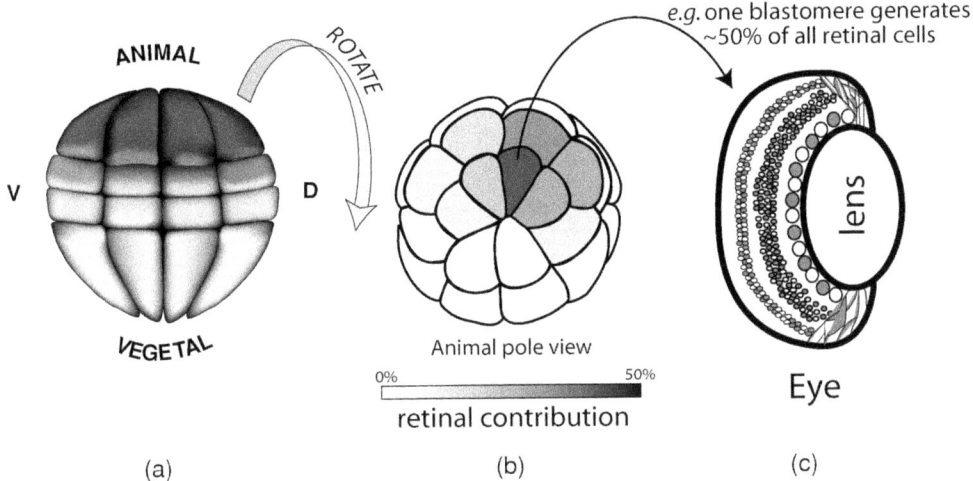

Figure 2.1 Multiple blastomeres generate eye-forming progeny. (a) Three-dimensional schematic diagram of a 32-cell-stage *Xenopus* embryo. (b) Animal pole view shows the subset of blastomeres that normally produce retinal progeny in the left eye. The shaded blastomeres produce fewer than 1% (lightly shaded) to as many as 50% (darkly shaded) of the retinal cells. (c) The progeny of retinogenic blastomeres are distributed throughout the retina (Huang and Moody, 1993).

retinogenic blastomere varies from animal to animal and no significant spatial segregation of their progeny is observed, that is, clones derived from the different blastomeres are found intermixed in the eye (Figure 2.1 and Moody, 1987; Huang and Moody, 1993).

Normally, only dorsal animal blastomeres deposit progeny in the eyes, however, all animal blastomeres are competent to contribute to the eyes. If equatorial or ventral animal blastomeres are transplanted to the retinogenic zone they are reprogrammed in response to interactions with their new neighbours and generate normal sized eyes (Figure 2.2a and Huang and Moody, 1993). If dorsal animal blastomeres in the centre of the retinogenic zone are killed nearby dorsal blastomeres compensate, generating more retinal progeny, resulting in tadpoles with normal eyes (Figure 2.2b). Therefore, cell–cell interactions are important in both determining the location and regulating the size of the retinogenic zone.

Although dorsal animal blastomeres are biased, they are not committed to a retinal lineage at the 32-cell stage. When transplanted to ventral vegetal locations, they retain their neural fate but don't make retina (Figure 2.2c and Gallagher *et al.*, 1991). Conversely, ventral vegetal blastomeres transplanted to retinogenic locations never contribute progeny to the retina (Figure 2.2d and Huang and Moody, 1993). Because zygotic transcription does not begin until later in development, these results show that inherited maternal determinants restrict the location of the retinogenic zone to the animal side of the embryo.

What are the maternal and cell–cell signals that determine whether a given animal blastomere will contribute progeny to the eye field and retina? How do animal blastomeres differ from their vegetal cousins? Suppression of bone morphogenetic protein (BMP) signalling appears to be necessary for animal blastomeres to generate retinal progeny (Moore and

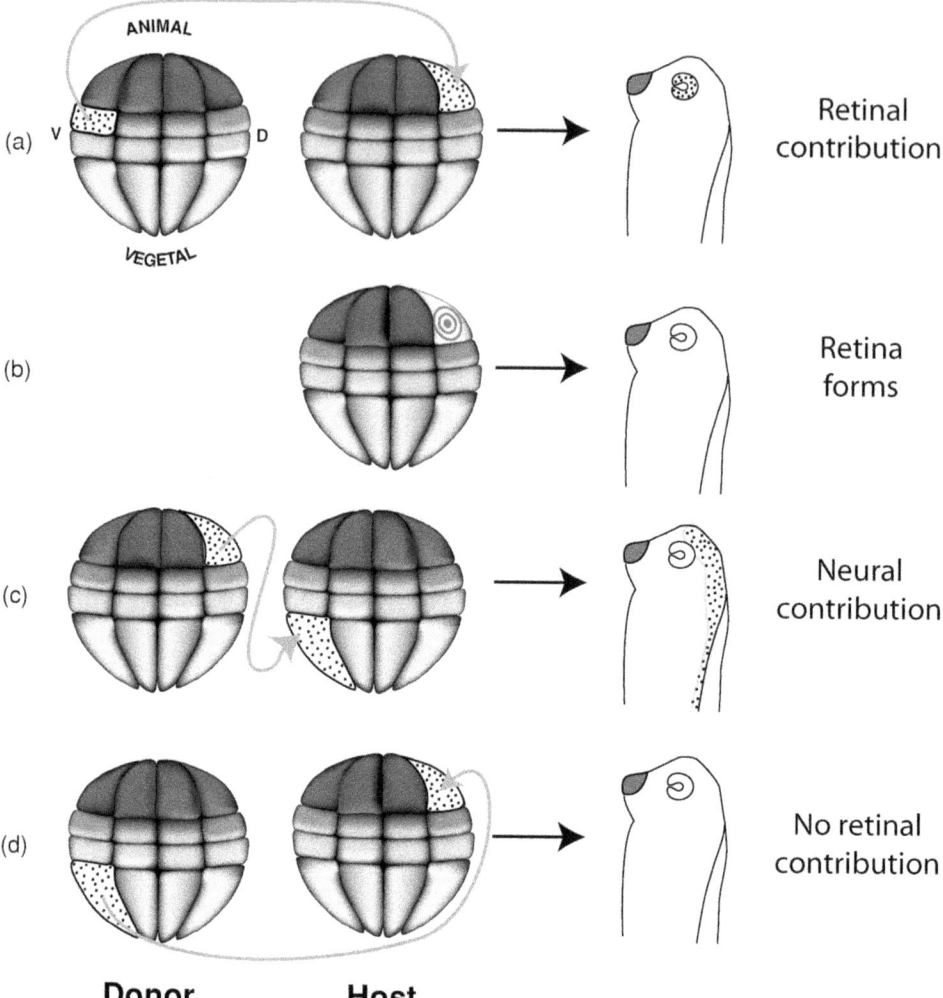

Figure 2.2 Transplantation and ablation experiments highlight the plasticity of blastomeres that contribute to retinal formation. Bull's eye indicates the position of ablated blastomere. See main text for a detailed description.

Moody, 1999). When BMP signalling is activated in retinogenic animal blastomeres, they fail to generate retinal progeny. Blocking BMP signalling in animal blastomeres that normally generate epidermis (by using BMP4 antagonists or dominant negative forms of BMP receptor) alters their fate and they generate retinal progeny. These results are consistent with a model in which eye formation is tightly coupled with neural induction, as inhibition of BMP signalling is required for neural induction (see below). Consistent with this model, the neural inducer noggin can also promote the progeny of animal blastomeres to the retinal lineage.

Interestingly, neither BMP inhibition nor neural induction by noggin can change the fate of vegetal blastomeres – even after they are transplanted to the retinogenic zone (Moore and Moody, 1999). Misexpression of *cerberus*, a member of the cysteine knot (DAN) family of genes, however, can redirect vegetal progeny into the retina (Bouwmeester *et al.*, 1996; Hsu *et al.*, 1998; Moore and Moody, 1999). Cerberus is secreted into the extracellular space, binds to Nodal, BMP and Wingless-Int (Wnt) proteins via independent sites and antagonizes the signalling cascades of all three of these molecules, an activity that appears to be required for *cerberus*' ability to induce heads (Bouwmeester *et al.*, 1996; Piccolo *et al.*, 1999; Silva *et al.*, 2003). Simultaneous repression of BMP and Wnt signalling in vegetal blastomeres does not result in retinal progeny (Moore and Moody, 1999). It may be that all three signalling cascades must be inhibited to drive vegetal progeny to a retinogenic fate. Alternatively, inhibition of Nodal signalling (or modulation of an unidentified Cerberus-regulated signalling cascade) might be sufficient to relieve the inhibition of maternal determinants present in the vegetal-half of the 32-cell stage embryo and reprogramme vegetal blastomeres to the retinal lineage.

In the 32-cell-stage embryo maternal determinants and cell–cell interactions, bias a subset of blastomeres toward a retinal cell lineage. The descendants of nine dorsal animal blastomeres are determined, but not necessarily destined, to contribute to the eye field and eyes. Inhibition of BMP signalling is important in keeping progeny in the retinogenic lineage. Considerable regulation takes place in the blastula, consequently the eventual fate of retinogenic cells is not established until the neural plate stage.

2.3 Neural induction and patterning

Our eyes are an externally visible extension of the veiled central nervous system (CNS). The physical separation of the eyes from the brain conceals the fact that the embryonic development of these two apparently distinct organs is tightly coordinated. Early evidence came in the 1960s when Nieuwkoop showed in amphibians that dissociated ectodermal cells form anterior neural structures including eyes when they are re-associated in culture (Nieuwkoop, 1963). These results supported the two-signal model of activation/transformation proposed to regulate neural patterning (Nieuwkoop *et al.*, 1952; Nieuwkoop and Nigtevecht, 1954). In this model, neural induction converts the entire dorsal ectoderm into anterior neuroectoderm. Although the exact timing and specific signalling systems involved in neural induction are still controversial, BMP inhibitors, fibroblast growth factors (FGFs) and Wnts play major roles as early as the blastula stage (Streit *et al.*, 2000; Wilson *et al.*, 2000; Wilson and Edlund, 2001; Stern, 2002; Bertrand *et al.*, 2003; Kuroda *et al.*, 2004). During the transformation step, more caudal regions of the presumptive neural plate are patterned to form the mid- and hindbrain. In the absence of caudalizing signals, the neural plate maintains an anterior neural character. Consistent with this model, experiments using transplantation and region-specific molecular markers show that in midgastrula embryos all regions of the presumptive neural plate have the capacity to form eyes. However, at later developmental stages, eye formation is restricted to the most anterior neural plate, specifically, the presumptive forebrain (Saha and Grainger, 1992; Li et al., 1997).

The signalling systems regulating early neural patterning include Wnts, FGFs, BMPs, Nodals and retinoic acid (RA) – the first three of which are also involved in neural induction. These signalling systems and their inhibitors also regulate forebrain development, and thereby eye field and eye formation. Early on, these pathways generate a simple anterior-posterior patterning to the neural plate and induce secondary signalling centres. The secondary signalling centres positioned near and within the neural plate act locally to specify the presumptive forebrain into regions that will eventually generate the anteriodorsal telencephlon and eyes, the ventral hypothalamus and the caudal diencephalon. The five signalling systems mentioned above are all indirectly required for normal eye field and eye formation since they are all involved in early patterning of the neural plate and/or forebrain prior to eye field formation. The focus here, will be on the signalling systems most directly linked to the formation of the eye field. Current evidence suggests that regulation of Wnt signalling is critical for patterning the eye field within the forebrain.

Wnt proteins bind to the Frizzled•low density lipoprotein receptor-related protein (Fz•LRP) complex and transduce a signal to the intracellular protein Dishevelled (Dsh). Activated Dsh inactivates the glycogen synthase kinase-3β•Adenomatous Polyposis Coli•Axin (GSK3β•APC•Axin) protein complex. Active GSK3β•APC•Axin degrades the transcriptional regulator β-catenin. Therefore, in the presence of Wnt ligands, GSK3β•APC•Axin is inactivated, β-catenin levels increase in the cytoplasm and eventually the nucleus. In the nucleus, β-catenin interacts with transcription factors including T cell-specific transcription factor (TCF) or lymphoid enhancer binding factor 1 (LEF1) to control the transcription of Wnt target genes (reviewed in Logan and Nusse, 2004).

In zebrafish embryos mutant for the *masterblind* (*mbl*) gene, the eyes and telencephalon are reduced in size or absent while the diencephalon is expanded. $mbl^{-/-}$ is the result of a mutation in Axin, which abolishes its ability to complex with GSK3β and degrade β-catenin. Consequently, Wnt signalling is overactive in $mbl^{-/-}$ embryos. Overexpression of wild-type Axin rescues the $mbl^{-/-}$ phenotype (Heisenberg et al., 2001; van de Water et al., 2001). These results suggest that patterning of the forebrain is dependent on the level of Wnt activity. High Wnt activity promotes diencephalic fates, while lower or no Wnt activity results in telencephalon and eye fates. Consistent with this hypothesis, transplantation of cells expressing Wnt1 or Wnt8b into the presumptive forebrain of wild-type embryos mimicked the $mbl^{-/-}$ phenotype (Houart et al., 2002).

The effects of Wnt antagonists also suggest reduced Wnt signalling is necessary for eye field formation. The inhibitor dickkopf1 (dkk1) binds to the Wnt receptor complex and down-regulates receptor expression on the cell surface (Mao et al., 2001). Overexpression of *dkk1* results in an increase in the size of the eyes and telencephalon with a corresponding reduction in more caudal tissues including the diencephalon (Shinya et al., 2000). Secreted Frizzled related proteins (SFRPs) are soluble proteins structurally similar to Wnt receptors that bind Wnt ligands and inhibit signalling through Wnt receptor (Uren et al., 2000). Secreted Frizzled related protein 1 is expressed in the anterior neural plate and its overexpression expands the expression domain of the eye field markers *Rx3*, *Pax6* and *Six3*. Conversely, interfering with *SFRP1* expression reduces the size of the eye field (Esteve *et al.*, 2004). *tlc* is closely related to SFRP1 and SFRP5 and is expressed first in the anterior neural plate and later in the anterior neural border (ANB) – a signalling centre required for normal forebrain patterning. Removal of ANB cells blocks telencephalon gene expression and results in forebrain degeneration (Shimamura and Rubenstein, 1997; Houart *et al.*, 1998). *tlc* can restore telencephalon gene expression in ANB-ablated embryos and induce ectopic telencephalon gene expression when expressed in presumptive diencephalon (Houart *et al.*, 2002).

Although each of the Wnt inhibitors described above generate similar gross effects (overexpression favours more anterior forebrain fates), a closer look reveals they affect forebrain patterning differently. *Dkk1* induces both telencephalon and eye fates at the expense of diencephalon (Shinya *et al.*, 2000). Embryos overexpressing *SFRP1* coexpress telencephalon and eye field markers throughout the most anterior neural plate with no effect on diencephalon markers, while *tlc* induces telencephalon at the expense of both the eye field and diencephalon (Houart *et al.*, 2002; Esteve *et al.*, 2004). One explanation for these differences could be that these inhibitors block Wnt signalling with differing efficiency and/or that Wnt signalling may pattern forebrain regions in a dose-dependent manner. This idea is supported by the phenotypes observed in LiCl treated embryos. Lithium mimics Wnt signalling by inhibiting GSK3β (Klein and Melton, 1996). Lithium dosage has a graded effect on forebrain patterning. Embryos exposed to low lithium doses have small eyes and a loss of eye-specific *Pax6* expression. At higher doses, however, the entire forebrain is lost (van de Water *et al.*, 2001). A dosage-based model also helps to explain other previously conflicting results. While most experiments show that Wnt activity is detrimental to eye formation, overexpression of the Wnt receptor Frizzled 3 (XFz3) in *Xenopus* induces ectopic eye field markers (*Pax6* and *Rx1*) and eyes (Rasmussen *et al.*, 2001). Insulin-like growth factors (IGFs) can also induce ectopic eyes in *Xenopus*. However, IGFs induce eyes by antagonizing the activity of the Wnt signal transduction pathway (Richard-Parpaillon *et al.*, 2002). Together, these results indicate that an intermediate level of Wnt signalling is necessary to pattern the eye field.

In addition to the classical (or canonical) Wnt/β-catenin signalling pathway described above, Wnt ligands can also signal through non-canonical, β-catenin-independent pathways. Evidence suggests that these different Wnt signalling pathways antagonize each other's activities with respect to eye field formation. For example, in zebrafish, Wnt/β-catenin signalling via the Wnt8b ligand and Fz8a receptor inhibits eye field specification. In direct contrast β-catenin-independent signalling through Wnt11 and Fz5 promotes eye field formation, at least in part via antagonism of canonical Wnt signalling (Cavodeassi *et al.*, 2005). Moreover, in *Xenopus*, β-catenin-independent signalling through Wnt4 and Fz3 is required for eye field formation (Maurus *et al.*, 2005). These results appear to explain the apparent conflicting effects of Wnts and Wnt inhibitors on eye field formation. Forebrain patterning and specification of the eye field may be dependent on precise coordination of canonical and non-canonical Wnt signalling.

In summary, neural induction generates an early neural plate with an anterior neural fate bias. The presumptive neural plate is then patterned under the influence of caudalizing signals to generate different brain regions (fore-, mid- and hindbrain). Precise control of Wnt signalling is critical for patterning the forebrain. Disruptions of Wnt activity via components of the signalling cascades or the activity of Wnt inhibitors can have dramatic effects on eye formation. The evidence suggests that a gradient, or possibly distinct levels (low, intermediate and relatively high), of Wnt signalling modulated via both canonical and non-canonical signalling systems are required to pattern the telencephalon, eye field and diencephalon within the presumptive forebrain.

2.4 Eye field specification

Classical transplantation and fate-mapping experiments have determined the timing and location of eye field formation. A group of eye field transcription factors are synchronously expressed in the eye field during its specification. Each of these genes is required for normal eye formation and together they can induce the formation of additional eye fields and functional eyes. Although early models suggested a single signalling cascade, recent experiments point to a more complex model in which these genes act synergistically in a self-regulating feedback network to convert a region of the anterior neural plate into the eye field.

Transplantation experiments from the beginning of the last century demonstrated that the amphibian eye field is located in the neural plate. Removal of the anterior neural plate results in eyeless animals (Spemann, 1901; Lewis, 1907; Adelmann, 1929a). If this same region is transplanted to the belly wall, or simply grown in culture, a histologically normal eye will form (Adelmann, 1929b; Lopashov and Stroeva, 1964; Li et al., 1997).

Fate-mapping experiments have more precisely resolved the location and shape of the eye field. In amphibians and fish, the eye field is a single crescent of cells spanning the breadth of the early anterior neural plate (Figure 2.3 and Brun, 1981; Eagleson and Harris, 1990; Kimmel et al., 1990; Eagleson et al., 1995; Woo and Fraser, 1995). In contrast, at the earliest stages analysed, chick fate maps indicate the presence of two distinct eye fields separated by the embryonic midline (Fernandez-Garre et al., 2002). These differences could result from species-specific developmental processes. However, in amphibians, fish and mice the single eye field also resolves into two bilaterally symmetric eye primordia and several genes critical for eye field specification (see below) are expressed across the embryonic chick midline. These results suggest that a single eye field analogous to that observed in amphibians and fish may also exist in chick.

The location and timing of eye field specification in the anterior neural plate is synchronized with the coordinated expression of a group of eye field transcription factors or EFTFs (Table 2.1). The expression pattern of the *Xenopus* EFTFs *ET*, *Rx*, *Pax6*, *Six3*, *Lhx2*, *tll* and *Optx2* overlap in the presumptive eye field during and immediately following its specification (Figure 2.3 and Zuber et al., 2003). The EFTFs of other species have a similar pattern

Table 2.1. *The eye field transcription factors (EFTFs) (Xenopus homologues) including the origin of their abbreviated names and the transcription factor family to which they belong. Alternative names are shown in parentheses*

EFTF	Long name	Transcription factor family
ET (Tbx3)	<u>E</u>ye <u>T</u>-box	T-box
Rx1(Rax)	<u>R</u>etina homeobo<u>x</u>-1	Paired-like homeobox
Pax6	<u>P</u>aired homeobo<u>x</u>-6	Paired homeobox
Six3	<u>Si</u>ne oculis-related homeobo<u>x</u>-3	Six family of homeobox
Lhx2	<u>L</u>IM <u>h</u>omeobo<u>x</u>-2	LIM (<u>L</u>in 11, <u>I</u>sl-1, <u>M</u>ec-3) homeobox
tll (Tlx)	<u>T</u>ai<u>ll</u>ess	Nuclear receptor-type
Optx2 (Six6)	<u>Op</u>tic <u>Six</u> gene <u>2</u>	Six family of homeobox

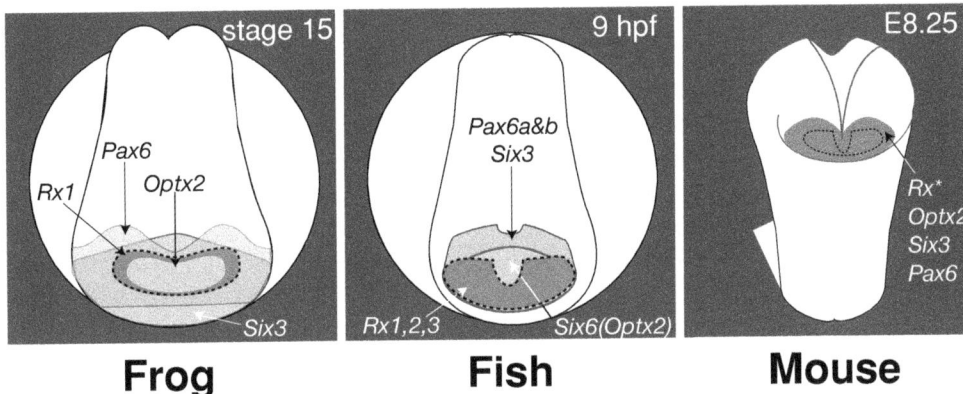

Figure 2.3 Eye field transcription factors have overlapping expression patterns during eye field formation. Anterior, frontal views of neural plate-staged frog (*Xenopus laevis*), fish (zebrafish and medakafish), and mouse embryos show the expression domains of *Rx*, *Pax6*, *Six3* and *Optx2* homologues (Oliver *et al.*, 1995; Mathers *et al.*, 1997; Seo *et al.*, 1998; Toy and Sundin, 1999; Inoue *et al.*, 2000; Chuang and Raymond, 2002; Zuber *et al.*, 2003; Bailey *et al.*, 2004). Two *Pax6* (*Pax6a* and *Pax6b*) and three *Rx* (*Rx1*, *Rx2* and *Rx3*) homologues have been identified in fish. The expression pattern of mouse *Rx* (*Rax*) at embryonic day (E)8.25 was estimated from its published expression patterns at E7.5 and E8.5. The location of the presumptive eye field is indicated with a black dashed line. hpf, hours post-fertilization. For colour version, see Plate 1.

of coordinated expression. For example, at neural plate stages *Pax6*, *Rx*, *Six3* and *Optx2* are also observed in a single band of expression in the chick, zebrafish and mouse embryo (Figure 2.3 and Walther and Gruss, 1991; Li *et al.*, 1994; Oliver *et al.*, 1995; Mathers *et al.*, 1997; Bovolenta *et al.*, 1998; Toy *et al.*, 1998; Ohuchi *et al.*, 1999; Toy and Sundin, 1999; Chuang and Raymond, 2002).

Eye field transcription factors have been highly conserved through evolution and genetic evidence from multiple species demonstrates they are required for vertebrate eye formation.

Functional inactivation of *Pax6*, *Rx*, *Lhx2*, *tll*, *Six3* and *Optx2* results in frogs, fish, rodents and/or humans with abnormal or no eyes (Hill *et al.*, 1991; Mathers *et al.*, 1997; Porter *et al.*, 1997; Hollemann *et al.*, 1998; Chow *et al.*, 1999; Isaacs *et al.*, 1999; Zuber *et al.*, 1999; Wawersik and Maas, 2000; Yu *et al.*, 2000; Loosli *et al.*, 2001, 2003; Tucker *et al.*, 2001; Carl *et al.*, 2002; Li *et al.*, 2002; Andreazzoli *et al.*, 2003; Lagutin *et al.*, 2003; Voronina *et al.*, 2004). Eye field transcription factors are not only necessary for eye formation; in some contexts, they are also sufficient. Overexpression of *Pax6*, *Six3*, *Rx* and *Optx2* can expand or induce eye tissues within the vertebrate nervous system (Oliver *et al.*, 1996; Mathers *et al.*, 1997; Andreazzoli *et al.*, 1999; Chow *et al.*, 1999; Loosli *et al.*, 1999; Zuber *et al.*, 1999, 2003; Bernier *et al.*, 2000; Chuang and Raymond, 2001). Together these results demonstrate a critical role for EFTFs in early eye formation.

Many vertebrate EFTFs were originally identified as homologues of *Drosophila* genes required for fly eye formation. *Pax6*, for example, is a homologue of *Drosophila eyeless* (*ey*) (Quiring *et al.*, 1994). Based on its remarkable evolutionary conservation, requirement for normal eye formation in multiple species and the ability of mammalian *Pax6* orthologues to induce ectopic fly eyes, *ey*/*Pax6* was hailed as a potential master regulator of eye formation (Halder *et al.*, 1995; Callaerts *et al.*, 1997). In initial models, *ey* was placed atop a hierarchy of genes that drove eye formation in the eye portion of the eye-antennal imaginal disc complex. However, a much more complex set of interactions soon emerged. The genes *twin-of-eyeless* (*toy*), *sine oculis* (*so*), *optix*, *eyes absent* (*eya*), *dachshund* (*dac*) and *eye gone* (*eyg*) are all required for fly eye formation. Individually or in combinations, these genes can also induce ectopic eyes, sometimes in the absence of *ey*. For instance, both *toy* and *optix* can induce ectopic eyes via an *ey*-independent mechanism (Czerny *et al.*, 1999; Seimiya and Gehring, 2000; Punzo *et al.*, 2004). In *Drosophila*, a model evolved in which *toy*, *ey*, *so*, *optix*, *eya*, *dac* and *eyg* behave as a network with hierarchical components as well as regulatory feedback loops including protein•protein interactions (Chen *et al.*, 1997; Pignoni *et al.*, 1997; Kumar and Moses, 2001).

An analogous transcription factor network is at work during vertebrate eye formation. *Pax6*, *Six3*, *Rx* and *Optx2* activate each other's expression, while inactivation of each can reduce the expression of the others (Andreazzoli *et al.*, 1999; Chow *et al.*, 1999; Loosli *et al.*, 1999; Zuber *et al.*, 1999, 2003; Bernier *et al.*, 2000; Chuang and Raymond, 2001; Lagutin *et al.*, 2001, 2003; Wargelius *et al.*, 2003). The evidence suggests that the network is conserved among species. For example, vertebrate *Pax6* and *Six3* and their fly homologues (*toy*, *ey*, *so* and *optix*) cross-regulate each other's expression in flies, frogs, fish and mice (Pignoni *et al.*, 1997; Halder *et al.*, 1998; Seimiya and Gehring, 2000; Carl *et al.*, 2002; Goudreau *et al.*, 2002; Zuber *et al.*, 2003). Eye field transcription factors can act synergistically and, as in the fly, functional interactions among the vertebrate EFTFs involve protein•protein complexes and multiple levels of regulation (Zuber *et al.*, 1999, 2003; Mikkola *et al.*, 2001; Li *et al.*, 2002; Stenman *et al.*, 2003).

Based on their coordinated expression and the extensive analysis demonstrating the presence of a genetic network, Kumar and Moses proposed that *Drosophila* eye field specification might be driven by the coordinated expression of the fly EFTFs (Kumar and

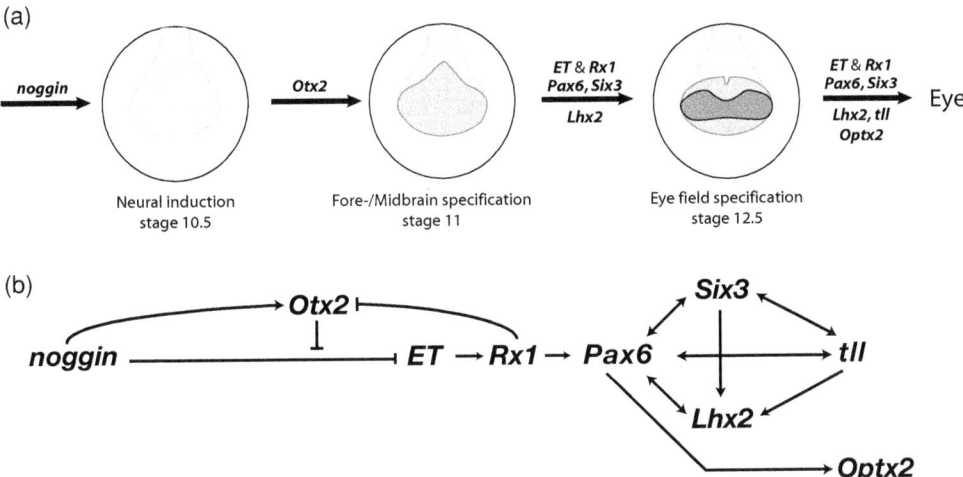

Figure 2.4 A summary model illustrates the developmental events (a) and genetic interactions (b) during eye field specification. The neural plate forms in response to neural inducers such as noggin. The initial neural plate is then patterned, generating fate-restricted regions. *Otx2* is required for forebrain and midbrain specification. The early-expressed EFTFs *ET*, *Rx1*, *Pax6*, *Six3* and *Lhx2* act coordinately to specify the eye field. Although not required for the initial specification of the eye field, *tll* and *Optx2* play later roles in eye formation. In (b) bars and arrows indicate the repression and induction of target gene expression, respectively. For example, noggin repressed *ET* expression, while ET induces *Rx1* expression.

Moses, 2001). This model was subsequently tested in vertebrates by coexpressing the EFTFs in developing *Xenopus* embryos. The coordinated expression of the EFTFs *ET*, *Rx*, *Pax6*, *Six3*, *Lhx2*, *tll* and *Optx2* with the anterior neural patterning gene *Otx2* is sufficient to induce ectopic eye fields and eyes (Zuber *et al.*, 2003). In contrast to ectopic retinal tissues induced by individual EFTFs, cocktail-induced eyes are generated at a higher frequency, are larger and develop outside the nervous system at various locations on the body including the belly. Most remarkable, when exposed to a brief flash of light, these eyes generated electroretinograms (ERGs) that are typical of normal eyes, demonstrating that the combined expression of the EFTFs can generate functional third eyes with the retinal cell types and neural circuits required for sight (E. Solessio, personal communication).

Eye field transcription factor cocktail subsets and inductive analysis were used to characterize the network of interactions between vertebrate EFTFs in frogs and generate a model for the interactions required for eye field specification (Figure 2.4). In the model, coordinated expression of the early EFTFs, *ET*, *Rx*, *Pax6*, *Six3* and *Lhx2* specify the eye field within the presumptive forebrain. Although not required for its initial specification, the later-expressed EFTFs *Optx2* and *tll* either cement eye field formation or are required at later developmental stages for normal eye formation (Figure 2.4).

To a large extent the mechanisms of eye field specification appear to be conserved among vertebrate species. For example, the morphological defects and EFTF expression patterns in $Rx^{-/-}$, $Pax6^{-/-}$, $Lhx2^{-/-}$, $Six3^{-/-}$, $Six6^{-/-}$ and $tll^{-/-}$ mice are consistent with the hierarchical aspects of the frog model. In frog, $Rx1$ function is predicted to be required prior to $Pax6$ and $Lhx2$. Consistent with this order, $Rx^{-/-}$ mice lack optic sulci, vesicles and cups, while $Pax6^{-/-}$ and $Lhx2^{-/-}$ mice develop optic vesicles and even rudimentary optic cups (Grindley et al., 1995; Porter et al., 1997; Zhang et al., 2000). $Rx^{-/-}$ mice lack normal $Pax6$ expression in the optic primordia, while $Rx1$ expression is unaffected in the $Pax6^{-/-}$ mouse (Zhang et al., 2000). Neither $Optx2$ nor tll are required for the initial steps of eye field specification in *Xenopus*. $Six6^{-/-}$ ($Optx2^{-/-}$) mice have normal (although small) eyes, while $tll^{-/-}$ mice do not develop retinal defects until three weeks after birth (Yu et al., 2000; Li et al., 2002). In mouse as well as frog, altering $Optx2$ levels have no effect on $Pax6$, $Six3$ or Rx expression (Li et al., 2002; Zuber et al., 2003).

In spite of the similarities outlined above, inconsistencies and potential differences still exist between the working models of vertebrate eye field specification. For instance, in frog ET activates $Rx1$ expression and both genes are proposed to play an early role in eye field specification. Although *Xenopus ET* is most homologous to $Tbx3$, $Tbx3$ null mice have no reported eye phenotype (Papaioannou, 2001; Davenport et al., 2003). $Tbx3$ is highly related to $Tbx2$ and both genes are members of the same subfamily of T-box containing genes (Papaioannou, 2001). The eyes of $Tbx2$ null mice do develop abnormally, however the expression of *Xenopus Tbx2* is not detected in the early neural plate suggesting that it is not involved in initiating eye field specification (Takabatake et al., 2002; Harrelson et al., 2004). The fish $Rx3$ is required for normal eye formation. In $Rx3$ mutants, both $Tbx2$ and $Tbx3$ expression is lost in the retina, suggesting that $Rx3$ is genetically upstream of these genes (Mathers et al., 1997; Loosli et al., 2001, 2003). This directly contradicts the frog model in which $Rx1$ is downstream of ET ($Tbx3$) (Figure 2.4 and Zuber et al., 2003).

The discrepancies highlighted above may be a consequence of the different techniques used to identify functional interactions between gene products, for example, inductive analysis in frog and mutant analysis in mice. In addition, some genomes carry duplicate copies or highly homologous EFTFs. For example, distinct Rx homologues with similar expression patterns have been reported in *Xenopus* (2), medakafish (2) and zebrafish (3) suggesting functional redundancy and complicating the identification of true functional orthologues (Casarosa et al., 1997; Mathers et al., 1997; Chuang et al., 1999; Winkler et al., 2000; Loosli et al., 2001). Clearly, a significant amount of feedback and cross-regulation is built into the system. Although there are clear advantages to the developing organism, it complicates analysis, making additional investigations necessary to more clearly define the genetic interactions necessary and required for vertebrate eye field specification.

The coordinated expression of the EFTFs is sufficient to generate ectopic eye fields and functional eyes. However, it is currently unclear how this coordinated expression is established. Recent evidence suggests that Wnt signalling, discussed in the previous section, directly controls expression of at least one EFTF. In the zebrafish anterior neural plate, Wnt11-expressing cells induce $rx3$ expression in neighbouring cells via a non-canonical Wnt signalling pathway (Cavodeassi et al., 2005). In *Xenopus*, non-canonical Wnt4 signalling activates expression of EAF2 a component of the RNA polymerase II elongation factor complex. Wnt4 and EAF2 are both required for eye formation and EAF2 regulates *Rx* expression in vitro (Maurus *et al.*, 2005). Wnt signalling is unlikely to induce every EFTF, however, since their expression patterns are not identical (Figure 2.3). Instead, distinct, as yet unidentified signalling systems coordinate EFTF expression.

In summary, transplantation and fate-mapping experiments have defined the location of the eye field. Eye field specification is synchronized with the coordinated expression of a group of EFTFs, most of which are required for normal eye development. When expressed individually, some EFTFs can induce ectopic, eye-like structures. The EFTFs form a self-regulating feedback network. Ectopic coexpression of these factors mimics the endogenous eye field, inducing ectopic, functional eyes. The signalling systems that coordinate their expression and the functional interactions among the EFTFs that are required for eye field specification and eye development remain largely unknown.

2.5 Separating the eye field into two eye primordia

All vertebrate embryos have a pair of eyes organized symmetrically across the body midline. As described above, the eye field forms as a single domain spanning the early anterior neural plate. In order to form two separate eyes, the single eye field must be split into two lateral eye primordia. Failure to do so results in cyclopic animals with one large midline eye. This is because the entire eye field is competent to form eye tissue. Midline as well as lateral anterior neural plate can form an eye (Adelmann, 1936).

The mechanism by which the eye field separates appears to be species specific. In amphibians, signals originating from the prechordal mesoderm underlying the anterior neural plate are responsible for separation of the eye field. Experiments in the 1930s by Mangold and Adelmann demonstrated that removal of the amphibian prechordal mesoderm results in cyclopic animals (Adelmann, 1930, 1934; Mangold, 1931).

Alterations in the expression patterns of EFTFs and fate-mapping experiments in *Xenopus* supports a mechanism in which prechordal mesoderm represses retinal fate in the midline of the anterior neural plate. *Xenopus ET, Rx1, Pax6, Six3, Lhx2* and *Optx2* are all expressed in the anterior neural plate. Their expression patterns are variable in size and extend to other regions of the neural plate, but all cover the eye field and each is expressed as a single band at stage 15 (Zuber *et al.*, 2003). Without exception, the expression of each EFTF is repressed in the midline and resolves into the two eye primordia by stage 18 (Figure 2.5 and Casarosa *et al.*, 1997; Hirsch and Harris, 1997; Li *et al.*, 1997; Mathers *et al.*, 1997; Zuber *et al.*, 1999; Zhou *et al.*, 2000; Viczian *et al.*, 2006). If midline cells within the single *Xenopus* eye field are labelled with tracking dyes their progeny are found, not in the eyes, but at the midline in the ventral hypothalamus and optic stalk suggesting that they have been reprogrammed to ventral diencephalic fates (Figure 2.5 and Eagleson *et al.*, 1995; Li *et al.*, 1997). Furthermore, transplanted chick prechordal plate represses *Pax6* expression in the anterior neural plate (Li *et al.*, 1997). In the absence of the prechordal plate, *Pax6* expression remains strong in the midline and chick embryos develop a single medial optic

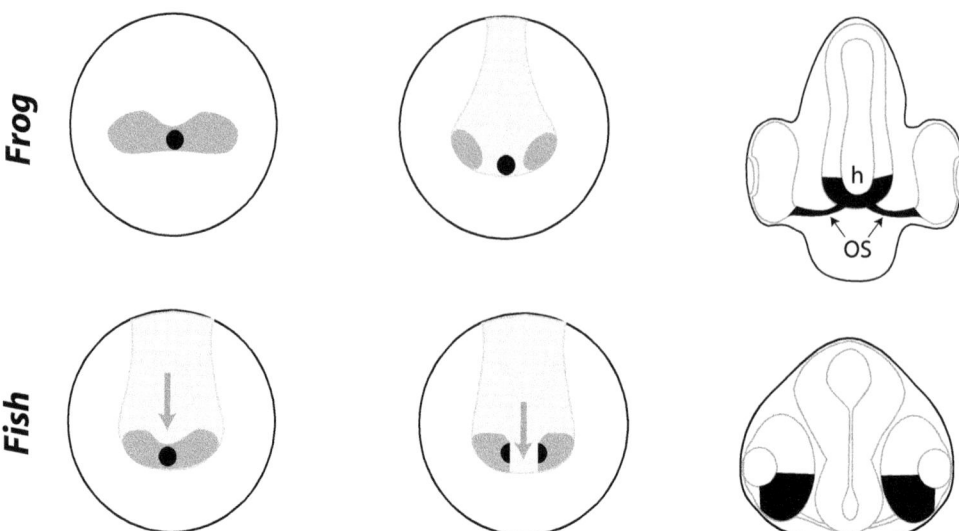

Figure 2.5 Two mechanisms drive eye field separation. In the frog *Xenopus laevis* midline eye field cells (labelled black) change fate and generate progeny that eventually form ventral diencephalic structures including the hypothalamus (h) and optic stalk (OS). In contrast, midline eye field cells of the fish are pushed into the right and left eye primordial by migrating diencephalic precursor cells. The result: midline eye field cells generate ventral retina.

vesicle (Pera and Kessel, 1997). In summary, evidence suggests that in amphibians and chick signals from the prechordal mesoderm represses retinal fate in the midline resulting in formation of the two eye primordia.

In zebrafish, fate mapping indicates that eye field separation involves the movement of cells within the neural plate. As in frogs, the fish eye field first forms as a continuous region spanning the anterior neural plate and can be molecularly identified by its expression of *Pax6* and members of the *Six* and *Rx* transcription factor families (Chuang and Raymond, 2002). At the end of gastrulation, the single eye field also expresses the zinc-finger containing *odd paired-like* (*opl*) gene, while more posterior diencephalic precursor cells express the *forkhead* gene, *foxb1.2* (Moens *et al.*, 1996; Grinblat *et al.*, 1998; Odenthal and Nusslein-Volhard, 1998). In direct contrast to results from the frog, fish midline eye field cells always contribute to the eyes (Figure 2.5 and Varga *et al.*, 1999). Zebrafish midline eye field cells don't change fate, rather they are pushed to either the right or left eye primordia as posterior, *foxb1.2*-expressing, diencephalic precursor cells move their way anteriorly along the midline (Figure 2.5 and Woo and Fraser, 1995; Varga *et al.*, 1999). Genetic evidence also supports this model since zebrafish mutations that effect the raustral migration of CNS cells (including *silberblick*, *knypek* and *trilobite*) are cyclopic (Hammerschmidt *et al.*, 1996; Heisenberg *et al.*, 1996; Heisenberg and Nusslein-Volhard, 1997; Marlow *et al.*, 1998).

In fish, mutants have been invaluable in identifying the signalling systems regulating eye field separation. *cyclops* (*cyc*) is a Nodal-related member of the transforming growth factor-β family of secreted signalling molecules (Feldman *et al.*, 1998; Rebagliati *et al.*, 1998; Sampath *et al.*, 1998). In *cyc* mutants posterior diencephalic precursors express *opl* instead of *foxb1.2* and fail to move anteriorly resulting in cyclopic embryos lacking ventral forebrain (Hatta *et al.*, 1991; Varga *et al.*, 1999). Zebrafish mutations in Nodals (*cyc* and *squint*), Nodal cofactors (*one-eyed pinhead*; *oep*) and Nodal signal transducing transcription factors (*schmalspur*) all result in cyclopia (Hammerschmidt *et al.*, 1996; Feldman *et al.*, 1998; Pogoda *et al.*, 2000). Activation of the Nodal signalling cascade (by overexpression of Nodal receptor ActRIIB or the transcription factor Smad2) rescues the *cyc* phenotype (Gritsman *et al.*, 1999).

In mice, null mutations effecting Nodal signalling are early embryonic lethal, however genetic combinations of heterozygous Nodal (*Nodal$^{+/-}$*) with heterozygous *Nodal receptor* or *Smad2* mutations cause cyclopia (Nomura and Li, 1998; Song *et al.*, 1999). Interestingly, ablation studies show that in order to induce cyclopia in wild-type fish embryos, it is necessary to remove prechordal plate (which also migrates anteriorly during gastrulation) as well as medial ventral diencephalic precursors (Varga *et al.*, 1999). *cyc* and *oep* are both expressed in the prechordal mesoderm, have prechordal plate defects and are cyclopic (Hammerschmidt *et al.*, 1996; Feldman *et al.*, 1998; Sampath *et al.*, 1998). These results suggest that the mechanisms driving fish and frog eye field separation are perhaps not quite so dissimilar as they initially appear and that Nodal signalling in the prechordal mesoderm is a general requirement for vertebrate eye field separation.

Nodal signalling modulates its effects on the eye field, at least in part, through the hedgehog (Hh) family of secreted morphogens. The transcriptional regulator of the Nodal signalling pathway, Smad2, directly modulates the *Sonic hedgehog* (*Shh*) promoter in zebrafish and chick neural tissue (Muller *et al.*, 2000). *Sonic hedgehog* expression is absent from the neuroectoderm of early stage *cyc* mutants (Krauss *et al.*, 1993). *Smad2* overexpression induces *Shh* expression and rescues the *cyc* eye phenotype in fish (Muller *et al.*, 2000). A conserved requirement for Hh in vertebrates is also demonstrated by the fact that humans, mice and fish lacking functional *Hh* signalling are cyclopic (Chiang *et al.*, 1996; Roessler *et al.*, 1996; Nasevicius and Ekker, 2000). Hedgehog overexpression expands the ventral forebrain (including optic stalk) at the expense of the retina. In fish, frogs and mice, ectopic Hh represses the EFTF *Pax6* while activating the expression of the optic stalk marker *Pax2* and other ventral forebrain markers (Barth and Wilson, 1995; Ekker *et al.*, 1995; Macdonald *et al.*, 1995; Shimamura *et al.*, 1995; Shimamura and Rubenstein, 1997; Perron *et al.*, 2003).

In summary, separation of the single eye field into the two eye primordia is required for normal eye formation. Fate-mapping studies suggest that at least two different mechanisms are used – physical displacement of eye field cells and/or reprogramming of midline eye field cells to ventral forebrain fates. In spite of these distinct mechanisms, misexpression and genetic analysis demonstrate that eye field separation is controlled through the Nodal and Hh signalling systems in vertebrates.

2.6 Concluding remarks

The embryonic location of the vertebrate eye field was first determined over a century ago. It has only been in the last decade, however, that the genes required for eye field and eye formation have been identified. Consequently, many unanswered questions remain regarding how the eye field is specified within the anterior neural plate.

A current model of vertebrate eye field formation, based on experiments in frogs, suggests that the combined expression of a group of EFTFs is sufficient to specify the eye field in the anterior neural plate. Is this mechanism conserved among vertebrate species? Are all these genes required for eye field formation? What are the upstream regulators that coordinate their expression? What are the functional interactions among the EFTFs necessary for eye field specification? Do the EFTFs act coordinately to regulate later aspects of eye formation? How are the dorso-ventral and proximo-distal axes of the eye field established?

The answers to these questions, and others, will provide a better understanding of how the cells of the early anterior neural plate are specified to form the eye field and then generate all the cells of the mature retina.

Acknowledgements

Thanks to Andrea Viczian for constructing the chapter figures and table. The authors acknowledge support from Research to Prevent Blindness through a Career Development Award to MZ.

References

Adelmann, H. B. (1929a). Experimental studies on the development of the eye. I. The effect of the removal of the median and lateral areas of the anterior end of the urodelan neural plate on the development of the eyes (*Triton teniatus* and *Amblystoma punctatum*). *J. Exp. Zool.*, **54**, 249–90.

Adelmann, H. B. (1929b). Experimental studies on the development of the eye. II. The eye-forming potencies of the urodelan neural plate (*Triton teniatus* and *Amblystoma punctatum*). *J. Exp. Zool.*, **54**, 291–317.

Adelmann, H. B. (1930). Experimental studies on the development of the eye. III. The effect of the ('Unterlagerung') on the heterotopic development of median and lateral strips of the anterior end of the neural plate of Amblystoma. *J. Exp. Zool.*, **57**, 223–81.

Adelmann, H. B. (1934). A study of cyclopia in *Amblystoma punctatum*, with special reference to the mesoderm. *J. Exp. Zool.*, **67**, 217–81.

Adelmann, H. B. (1936). The problem of cyclopia. *Q. Rev. Biol.*, **11**, 161–82.

Andreazzoli, M., Gestri, G., Angeloni, D., Menna, E. and Barsacchi, G. (1999). Role of Xrx1 in Xenopus eye and anterior brain development. *Development*, **126**, 2451–60.

Andreazzoli, M., Gestri, G., Cremisi, F. *et al.* (2003). Xrx1 controls proliferation and neurogenesis in Xenopus anterior neural plate. *Development*, **130**, 5143–55.

Bailey, T. J., El-Hodiri, H., Zhang, L. *et al.* (2004). Regulation of vertebrate eye development by *Rx* genes. *Int. J. Dev. Biol.*, **48**, 761–70.

Barth, K. A. and Wilson, S. W. (1995). Expression of zebrafish nk2.2 is influenced by sonic hedgehog/vertebrate hedgehog-1 and demarcates a zone of neuronal differentiation in the embryonic forebrain. *Development*, **121**, 1755–68.

Bernier, G., Panitz, F., Zhou, X. et al. (2000). Expanded retina territory by midbrain transformation upon over-expression of *Six6* (*Optx2*) in Xenopus embryos. *Mech. Dev.*, **93**, 59–69.

Bertrand, V., Hudson, C., Caillol, D., Popovici, C. and Lemaire, P. (2003). Neural tissue in ascidian embryos is induced by FGF9/16/20, acting via a combination of maternal GATA and Ets transcription factors. *Cell*, **115**, 615–27.

Bouwmeester, T., Kim, S., Sasai, Y., Lu, B. and De Robertis, E. M. (1996). Cerberus is a head-inducing secreted factor expressed in the anterior endoderm of Spemann's organizer. *Nature*, **382**, 595–601.

Bovolenta, P., Mallamaci, A., Puelles, L. and Boncinelli, E. (1998). Expression pattern of cSix3, a member of the Six/sine oculis family of transcription factors. *Mech. Dev.*, **70**, 201–3.

Brun, R. B. (1981). The movement of the prospective eye vesicles from the neural plate into the neural fold in *Ambystoma mexicanum* and *Xenopus laevis*. *Dev. Biol.*, **88**, 192–9.

Callaerts, P., Halder, G. and Gehring, W. J. (1997). PAX-6 in development and evolution. *Annu. Rev. Neurosci.*, **20**, 483–532.

Carl, M., Loosli, F. and Wittbrodt, J. (2002). Six3 inactivation reveals its essential role for the formation and patterning of the vertebrate eye. *Development*, **129**, 4057–63.

Casarosa, S., Andreazzoli, M., Simeone, A. and Barsacchi, G. (1997). Xrx1, a novel Xenopus homeobox gene expressed during eye and pineal gland development. *Mech. Dev.*, **61**, 187–98.

Cavodeassi, F., Carreira-Barbosa, F. Young, R. M. et al. (2005). Early stages of zebrafish eye formation require the coordinated activity of Wnt11, Fz5, and the Wnt/beta-catenin pathway. *Neuron*, **47**, 43–56.

Chen, R., Amoui, M., Zhang, Z. and Mardon, G. (1997). Dachshund and eyes absent proteins form a complex and function synergistically to induce ectopic eye development in Drosophila. *Cell*, **91**, 893–903.

Chiang, C., Litingtung, Y., Lee, E. et al. (1996). Cyclopia and defective axial patterning in mice lacking *Sonic hedgehog* gene function. *Nature*, **383**, 407–13.

Chow, R. L., Altmann, C. R., Lang, R. A. and Hemmati-Brivanlou, A. (1999). *Pax6* induces ectopic eyes in a vertebrate. *Development*, **126**, 4213–22.

Chuang, J. C. and Raymond, P. A. (2001). Zebrafish genes *rx1* and *rx2* help define the region of forebrain that gives rise to retina. *Dev. Biol.*, **231**, 13–30.

Chuang, J. C. and Raymond, P. A. (2002). Embryonic origin of the eyes in teleost fish. *BioEssays*, **24**, 519–29.

Chuang, J. C., Mathers, P. H. and Raymond, P. A. (1999). Expression of three Rx homeobox genes in embryonic and adult zebrafish. *Mech. Dev.*, **84**, 195–8.

Czerny, T., Halder, G., Kloter, U. et al. (1999). *Twin of eyeless*, a second *Pax-6* gene of Drosophila, acts upstream of eyeless in the control of eye development. *Mol. Cell*, **3**, 297–307.

Davenport, T. G., Jerome-Majewska, L. A. and Papaioannou, V. E. (2003). Mammary gland, limb and yolk sac defects in mice lacking *Tbx3*, the gene mutated in human ulnar mammary syndrome. *Development*, **130**, 2263–73.

Eagleson, G. W. and Harris, W. A. (1990). Mapping of the presumptive brain regions in the neural plate of *Xenopus laevis*. *J. Neurobiol.*, **21**, 427–40.

Eagleson, G., Ferreiro, B. and Harris, W. A. (1995). Fate of the anterior neural ridge and the morphogenesis of the Xenopus forebrain. *J. Neurobiol.*, **28**, 146–58.

Ekker, S. C., Ungar, A. R., Greenstein, P. *et al.* (1995). Patterning activities of vertebrate hedgehog proteins in the developing eye and brain. *Curr. Biol.*, **5**, 944–55.

Esteve, P., Lopez-Rios, J. and Bovolenta, P. (2004). SFRP1 is required for the proper establishment of the eye field in the medaka fish. *Mech. Dev.*, **121**, 687–701.

Feldman, B., Gates, M. A., Egan, E. S. *et al.* (1998). Zebrafish organizer development and germ-layer formation require nodal-related signals. *Nature*, **395**, 181–5.

Fernandez-Garre, P., Rodriguez-Gallardo, L., Gallego-Diaz, V., Alvarez, I. S. and Puelles, L. (2002). Fate map of the chicken neural plate at stage 4. *Development*, **129**, 2807–22.

Gallagher, B. C., Hainski, A. M. and Moody, S. A. (1991). Autonomous differentiation of dorsal axial structures from an animal cap cleavage stage blastomere in Xenopus. *Development*, **112**, 1103–14.

Goudreau, G., Petrou, P., Reneker, L. W. *et al.* (2002). Mutually regulated expression of *Pax6* and *Six3* and its implications for the *Pax6* haploinsufficient lens phenotype. *Proc. Natl. Acad. Sci. U. S. A.*, **99**, 8719–24.

Grinblat, Y., Gamse, J., Patel, M. and Sive, H. (1998). Determination of the zebrafish forebrain: induction and patterning. *Development*, **125**, 4403–16.

Grindley, J. C., Davidson, D. R. and Hill, R. E. (1995). The role of *Pax-6* in eye and nasal development. *Development*, **121**, 1433–42.

Gritsman, K., Zhang, J., Cheng, S. *et al.* (1999). The EGF-CFC protein one-eyed pinhead is essential for nodal signaling. *Cell*, **97**, 121–32.

Halder, G., Callaerts, P. and Gehring, W. J. (1995). Induction of ectopic eyes by targeted expression of the *eyeless* gene in Drosophila. *Science*, **267**, 1788–92.

Halder, G., Callaerts, P., Flister, S. *et al.* (1998). *Eyeless* initiates the expression of both *sine oculis* and *eyes absent* during Drosophila compound eye development. *Development*, **125**, 2181–91.

Hammerschmidt, M., Pelegri, F., Mullins, M. C. *et al.* (1996). Mutations affecting morphogenesis during gastrulation and tail formation in the zebrafish, *Danio rerio*. *Development*, **123**, 143–51.

Harrelson, Z., Kelly, R. G., Goldin, S. N. *et al.* (2004). Tbx2 is essential for patterning the atrioventricular canal and for morphogenesis of the outflow tract during heart development. *Development*, **131**, 5041–52.

Hatta, K., Kimmel, C. B., Ho, R. K. and Walker, C. (1991). The cyclops mutation blocks specification of the floor plate of the zebrafish central nervous system. *Nature*, **350**, 339–41.

Heisenberg, C. P. and Nusslein-Volhard, C. (1997). The function of silberblick in the positioning of the eye anlage in the zebrafish embryo. *Dev. Biol.*, **184**, 85–94.

Heisenberg, C. P., Brand, M., Jiang, Y. J. *et al.* (1996). Genes involved in forebrain development in the zebrafish, *Danio rerio*. *Development*, **123**, 191–203.

Heisenberg, C. P., Houart, C., Take-Uchi, M. *et al.* (2001). A mutation in the Gsk3-binding domain of zebrafish Masterblind/Axin1 leads to a fate transformation of telencephalon and eyes to diencephalon. *Genes Dev.*, **15**, 1427–34.

Hill, R. E., Favor, J., Hogan, B. L. *et al.* (1991). Mouse small eye results from mutations in a paired-like homeobox-containing gene. *Nature*, **354**, 522–5.

Hirsch, N. and Harris, W. A. (1997). Xenopus *Pax-6* and retinal development. *J. Neurobiol.*, **32**, 45–61.

Hollemann, T., Bellefroid, E. and Pieler, T. (1998). The Xenopus homologue of the Drosophila gene *tailless* has a function in early eye development. *Development*, **125**, 2425–32.

Houart, C., Westerfield, M. and Wilson, S. W. (1998). A small population of anterior cells patterns the forebrain during zebrafish gastrulation. *Nature*, **391**, 788–92.

Houart, C., Caneparo, L., Heisenberg, C. *et al.* (2002). Establishment of the telencephalon during gastrulation by local antagonism of Wnt signaling. *Neuron*, **35**, 255–65.

Hsu, D. R., Economides, A. N., Wang, X., Eimon, P. M. and Harland, R. M. (1998). The Xenopus dorsalizing factor Gremlin identifies a novel family of secreted proteins that antagonize BMP activities. *Mol. Cell*, **1**, 673–83.

Huang, S. and Moody, S. A. (1993). The retinal fate of Xenopus cleavage stage progenitors is dependent upon blastomere position and competence: studies of normal and regulated clones. *J. Neurosci.*, **13**, 3193–210.

Inoue, T., Nakamura, S. and Osumi, N. (2000). Fate mapping of the mouse prosencephalic neural plate. *Dev. Biol.*, **219**, 373–83.

Isaacs, H. V., Andreazzoli, M. and Slack, J. M. (1999). Anteroposterior patterning by mutual repression of orthodenticle and caudal-type transcription factors. *Evol. Dev.*, **1**, 143–52.

Kimmel, C. B., Warga, R. M. and Schilling, T. F. (1990). Origin and organization of the zebrafish fate map. *Development*, **108**, 581–94.

Klein, P. S. and Melton, D. A. (1996). A molecular mechanism for the effect of lithium on development. *Proc. Natl. Acad. Sci. U. S. A.*, **93**, 8455–9.

Krauss, S., Concordet, J. P. and Ingham, P. W. (1993). A functionally conserved homolog of the Drosophila segment polarity gene *hh* is expressed in tissues with polarizing activity in zebrafish embryos. *Cell*, **75**, 1431–44.

Kumar, J. P. and Moses, K. (2001). EGF receptor and Notch signaling act upstream of *Eyeless/Pax6* to control eye specification. *Cell*, **104**, 687–97.

Kuroda, H., Wessely, O. and Robertis, E. M. (2004). Neural induction in Xenopus: requirement for ectodermal and endomesodermal signals via Chordin, Noggin, beta-Catenin, and Cerberus. *PLoS Biol.*, **2**, E92.

Lagutin, O., Zhu, C. C., Furuta, Y. *et al.* (2001). *Six3* promotes the formation of ectopic optic vesicle-like structures in mouse embryos. *Dev. Dyn.*, **221**, 342–9.

Lagutin, O. V., Zhu, C. C., Kobayashi, D. *et al.* (2003). *Six3* repression of Wnt signaling in the anterior neuroectoderm is essential for vertebrate forebrain development. *Genes Dev.*, **17**, 368–79.

Lewis, W. H. (1907). Experiments on the origin and differentiation of the optic vesicle in amphibia. *Am. J. Anat.*, **7**, 259–76.

Li, H. S., Yang, J. M., Jacobson, R. D., Pasko, D. and Sundin, O. (1994). *Pax-6* is first expressed in a region of ectoderm anterior to the early neural plate: implications for stepwise determination of the lens. *Dev. Biol.*, **162**, 181–94.

Li, H., Tierney, C., Wen, L., Wu, J. Y. and Rao, Y. (1997). A single morphogenetic field gives rise to two retina primordia under the influence of the prechordal plate. *Development*, **124**, 603–15.

Li, X., Perissi, V., Liu, F., Rose, D. W. and Rosenfeld, M. G. (2002). Tissue-specific regulation of retinal and pituitary precursor cell proliferation. *Science*, **297**, 1180–3.

Logan, C. Y. and Nusse, R. (2004). The Wnt signaling pathway in development and disease. *Annu. Rev. Cell Dev. Biol.*, **20**, 781–810.

Loosli, F., Winkler, S. and Wittbrodt, J. (1999). *Six3* over-expression initiates the formation of ectopic retina. *Genes Dev.*, **13**, 649–54.

Loosli, F., Winkler, S., Burgtorf, C. *et al.* (2001). Medaka eyeless is the key factor linking retinal determination and eye growth. *Development*, **128**, 4035–44.

Loosli, F., Staub, W., Finger-Baier, K. C. *et al.* (2003). Loss of eyes in zebrafish caused by mutation of chokh/rx3. *EMBO Rep.*, **4**, 894–9.

Lopashov, G. V. and Stroeva, O. G. (1964). *Development of the Eye; Experimental Studies*. New York, NY: D. Davey.

Macdonald, R., Barth, K. A., Xu, Q. *et al.* (1995). Midline signaling is required for *Pax* gene regulation and patterning of the eyes. *Development*, **121**, 3267–78.

Mangold, O. (1931). Das Determinationsproblem. III. Das Wirbeltierauge in der Entwicklung und Regeneration. *Ergeb Biol.*, **7**, 196–403.

Mao, B., Wu, W., Li, Y. *et al.* (2001). LDL-receptor-related protein 6 is a receptor for Dickkopf proteins. *Nature*, **411**, 321–5.

Marlow, F., Zwartkruis, F., Malicki, J. *et al.* (1998). Functional interactions of genes mediating convergent extension, knypek and trilobite, during the partitioning of the eye primordium in zebrafish. *Dev. Biol.*, **203**, 382–99.

Mathers, P. H., Grinberg, A., Mahon, K. A. and Jamrich, M. (1997). The *Rx* homeobox gene is essential for vertebrate eye development. *Nature*, **387**, 603–7.

Maurus, D., Heligon, C., Burger-Schwarzler, A., Brandli, A. W. and Kuhl, M. (2005). Noncanonical Wnt-4 signaling and EAF2 are required for eye development in *Xenopus laevis*. *EMBO J.*, **24**, 1181–91.

Mikkola, I., Bruun, J. A., Holm, T. and Johansen, T. (2001). Superactivation of *Pax6*-mediated transactivation from paired domain-binding sites by DNA-independent recruitment of different homeodomain proteins. *J. Biol. Chem.*, **276**, 4109–18.

Moens, C. B., Yan, Y. L., Appel, B., Force, A. G. and Kimmel, C. B. (1996). *valentino*: a zebrafish gene required for normal hindbrain segmentation. *Development*, **122**, 3981–90.

Moody, S. A. (1987). Fates of the blastomeres of the 32-cell-stage Xenopus embryo. *Dev. Biol.*, **122**, 300–19.

Moore, K. B. and Moody, S. A. (1999). Animal-vegetal asymmetries influence the earliest steps in retina fate commitment in Xenopus. *Dev. Biol.*, **212**, 25–41.

Muller, F., Albert, S., Blader, P. *et al.* (2000). Direct action of the nodal-related signal cyclops in induction of sonic hedgehog in the ventral midline of the CNS. *Development*, **127**, 3889–97.

Nasevicius, A. and Ekker, S. C. (2000). Effective targeted gene 'knockdown' in zebrafish. *Nat. Genet.*, **26**, 216–20.

Nieuwkoop, P. D. (1963). Pattern formation in artificially activated ectoderm (*Rana pipiens* and *Ambystoma punctatum*). *Dev. Biol.*, **7**, 255–79.

Nieuwkoop, P. D. and Nigtevecht, G. V. (1954). Neural activation and transformation in explants of competent ectoderm under the influence of fragments of anterior notochord in urodeles. *J. Embryol. Exp. Morphol.*, **2**, 175–93.

Nieuwkoop, P. D., Botterenbrood, E. C., Kremer, A. *et al.* (1952). Activation and organization of the central nervous system in Amphibians. *J. Exp. Zool.*, **120**, 1–108.

Nomura, M. and Li, E. (1998). Smad2 role in mesoderm formation, left-right patterning and craniofacial development. *Nature*, **393**, 786–90.

Odenthal, J. and Nusslein-Volhard, C. (1998). Fork head domain genes in zebrafish. *Dev. Genes Evol.*, **208**, 245–58.

Ohuchi, H., Tomonari, S., Itoh, H., Mikawa, T. and Noji, S. (1999). Identification of chick rax/rx genes with overlapping patterns of expression during early eye and brain development. *Mech. Dev.*, **85**, 193–5.

Oliver, G., Mailhos, A., Wehr, R. *et al.* (1995). *Six3*, a murine homologue of the *sine oculis* gene, demarcates the most anterior border of the developing neural plate and is expressed during eye development. *Development*, **121**, 4045–55.

Oliver, G., Loosli, F., Koster, R., Wittbrodt, J. and Gruss, P. (1996). Ectopic lens induction in fish in response to the murine homeobox gene *Six3*. *Mech. Dev.*, **60**, 233–9.

Papaioannou, V. E. (2001). T-box genes in development: from hydra to humans. *Int. Rev. Cytol.*, **207**, 1–70.

Pera, E. M. and Kessel, M. (1997). Patterning of the chick forebrain anlage by the prechordal plate. *Development*, **124**, 4153–62.

Perron, M., Boy, S., Amato, M. A. *et al.* (2003). A novel function for Hedgehog signaling in retinal pigment epithelium differentiation. *Development*, **130**, 1565–77.

Piccolo, S., Agius, E., Leyns, L. *et al.* (1999). The head inducer Cerberus is a multifunctional antagonist of Nodal, BMP and Wnt signals. *Nature*, **397**, 707–10.

Pignoni, F., Hu, B., Zavitz, K. H. *et al.* (1997). The eye-specification proteins So and Eya form a complex and regulate multiple steps in Drosophila eye development. *Cell*, **91**, 881–91.

Pogoda, H. M., Solnica-Krezel, L., Driever, W. and Meyer, D. (2000). The zebrafish forkhead transcription factor FoxH1/Fast1 is a modulator of nodal signaling required for organizer formation. *Curr. Biol.*, **10**, 1041–9.

Porter, F. D., Drago, J., Xu, Y. *et al.* (1997). *Lhx2*, a LIM homeobox gene, is required for eye, forebrain, and definitive erythrocyte development. *Development*, **124**, 2935–44.

Punzo, C., Plaza, S., Seimiya, M. *et al.* (2004). Functional divergence between eyeless and twin of eyeless in Drosophila melanogaster. *Development*, **131**, 3943–53. Erratum in *Development* 2004 Sep, **131**, 4635.

Quiring, R., Walldorf, U., Kloter, U. and Gehring, W. J. (1994). Homology of the *eyeless* gene of Drosophila to the *Small eye* gene in mice and Aniridia in humans. *Science*, **265**, 785–9.

Rasmussen, J. T., Deardorff, M. A., Tan, C. *et al.* (2001). Regulation of eye development by frizzled signaling in Xenopus. *Proc. Natl. Acad. Sci. U. S. A.*, **98**, 3861–6.

Rebagliati, M. R., Toyama, R., Haffter, P. and Dawid, I. B. (1998). Cyclops encodes a nodal-related factor involved in midline signaling. *Proc. Natl. Acad. Sci. U. S. A.*, **95**, 9932–7.

Richard-Parpaillon, L., Heligon, C., Chesnel, F., Boujard, D. and Philpott, A. (2002). The IGF pathway regulates head formation by inhibiting Wnt signaling in Xenopus. *Dev. Biol.*, **244**, 407–17.

Roessler, E., Belloni, E., Gaudenz, K. *et al.* (1996). Mutations in the human *Sonic Hedgehog* gene cause holoprosencephaly. *Nat. Genet.*, **14**, 357–60.

Saha, M. S. and Grainger, R. M. (1992). A labile period in the determination of the anterior-posterior axis during early neural development in Xenopus. *Neuron*, **8**, 1003–14.

Sampath, K., Rubinstein, A. L., Cheng, A. M. *et al.* (1998). Induction of the zebrafish ventral brain and floorplate requires cyclops/nodal signaling. *Nature*, **395**, 185–9.

Seimiya, M. and Gehring, W. J. (2000). The Drosophila homeobox gene *optix* is capable of inducing ectopic eyes by an eyeless-independent mechanism. *Development*, **127**, 1879–86.

Seo, H. C., Drivenes, O. Ellingsen, S. and Fjose, A. (1998). Expression of two zebrafish homologs of the murine *Six3* gene demarcates the initial eye primordia. *Mech. Dev.*, **73**, 45–57.

Shimamura, K. and Rubenstein, J. L. (1997). Inductive interactions direct early regionalization of the mouse forebrain. *Development*, **124**, 2709–18.

Shimamura, K., Hartigan, D. J., Martinez, S., Puelles, L. and Rubenstein, J. L. (1995). Longitudinal organization of the anterior neural plate and neural tube. *Development*, **121**, 3923–33.

Shinya, M., Eschbach, C., Clark, M., Lehrach, H. and Furutani-Seiki, M. (2000). Zebrafish Dkk1, induced by the pre-MBT Wnt signaling, is secreted from the prechordal plate and patterns the anterior neural plate. *Mech. Dev.*, **98**, 3–17.

Silva, A. C., Filipe, M., Kuerner, K. M., Steinbeisser, H. and Belo, J. A. (2003). Endogenous Cerberus activity is required for anterior head specification in Xenopus. *Development*, **130**, 4943–53.

Song, J., Oh, S. P., Schrewe, H. *et al.* (1999). The type II activin receptors are essential for egg cylinder growth, gastrulation, and rostral head development in mice. *Dev. Biol.*, **213**, 157–169.

Spemann, H. (1901). Uber Correlationen in der Entwickelung des Auges. *Verh. Anat. Ges.*, **15**, 61–79.

Stenman, J., Yu, R. T., Evans, R. M. and Campbell, K. (2003). Tlx and Pax6 co-operate genetically to establish the pallio-subpallial boundary in the embryonic mouse telencephalon. *Development*, **130**, 1113–22.

Stern, C. D. (2002). Induction and initial patterning of the nervous system – the chick embryo enters the scene. *Curr. Opin. Genet. Dev.*, **12**, 447–51.

Streit, A., Berliner, A. J., Papanayotou, C., Sirulnik, A. and Stern, C. D. (2000). Initiation of neural induction by FGF signaling before gastrulation. *Nature*, **406**, 74–8.

Takabatake, Y., Takabatake, T., Sasagawa, S. and Takeshima, K. (2002). Conserved expression control and shared activity between cognate T-box genes *Tbx2* and *Tbx3* in connection with Sonic hedgehog signaling during Xenopus eye development. *Dev. Growth Differ.*, **44**, 257–71.

Toy, J. and Sundin, O. H. (1999). Expression of the *optx2* homeobox gene during mouse development. *Mech. Dev.*, **83**, 183–6.

Toy, J., Yang, J. M., Leppert, G. S. and Sundin, O. H. (1998). The *optx2* homeobox gene is expressed in early precursors of the eye and activates retina-specific genes. *Proc. Natl. Acad. Sci. U. S. A.*, **95**, 10643–8.

Tucker, P., Laemle, L., Munson, A. *et al.* (2001). The eyeless mouse mutation (ey1) removes an alternative start codon from the *Rx/rax* homeobox gene. *Genesis*, **31**, 43–53.

Uren, A., Reichsman, F., Anest, V. *et al.* (2000). Secreted frizzled-related protein-1 binds directly to Wingless and is a biphasic modulator of Wnt signaling. *J. Biol. Chem.*, **275**, 4374–82.

van de Water, S., van de Wetering, M. Joore, J. *et al.* (2001). Ectopic Wnt signal determines the eyeless phenotype of zebrafish masterblind mutant. *Development*, **128**, 3877–88.

Varga, Z. M., Wegner, J. and Westerfield, M. (1999). Anterior movement of ventral diencephalic precursors separates the primordial eye field in the neural plate and requires cyclops. *Development*, **126**, 5533–46.

Viczian, A. S., Bang, A. G., Harris, W. A. and Zuber, M. E. (2006). Expression of *Xenopus laevis Lhx2* during eye development and evidence for divergent expression among vertebrates. *Dev. Dyn.*, **235**, 1133–41.

Voronina, V. A., Kozhemyakina, E. A., O'Kernick, C. M. *et al.* (2004). Mutations in the human *RAX* homeobox gene in a patient with anophthalmia and sclerocornea. *Hum. Mol. Genet.*, **13**, 315–22.

Walther, C. and Gruss, P. (1991). *Pax-6*, a murine paired box gene, is expressed in the developing CNS. *Development*, **113**, 1435–49.

Wargelius, A., Seo, H. C., Austbo, L. and Fjose, A. (2003). Retinal expression of zebrafish six3.1 and its regulation by Pax6. *Biochem. Biophys. Res. Commun.*, **309**, 475–81.

Wawersik, S. and Maas, R. L. (2000). Vertebrate eye development as modeled in Drosophila. *Hum. Mol. Genet.*, **9**, 917–25.

Wilson, S. I. and Edlund, T. (2001). Neural induction: toward a unifying mechanism. *Nat. Neurosci.*, **4** *(Suppl.)*, 1161–8.

Wilson, S. I., Graziano, E., Harland, R., Jessell, T. M. and Edlund, T. (2000). An early requirement for FGF signaling in the acquisition of neural cell fate in the chick embryo. *Curr. Biol.*, **10**, 421–9.

Winkler, S., Loosli, F., Henrich, T., Wakamatsu, Y. and Wittbrodt, J. (2000). The conditional medaka mutation eyeless uncouples patterning and morphogenesis of the eye. *Development*, **127**, 1911–19.

Woo, K. and Fraser, S. E. (1995). Order and coherence in the fate map of the zebrafish nervous system. *Development*, **121**, 2595–609.

Yu, R. T., Chiang, M. Y., Tanabe, T. *et al.* (2000). The orphan nuclear receptor Tlx regulates Pax2 and is essential for vision. *Proc. Natl. Acad. Sci. U. S. A.*, **97**, 2621–5.

Zhang, L., Mathers, P. H. and Jamrich, M. (2000). Function of *Rx*, but not *Pax6*, is essential for the formation of retinal progenitor cells in mice. *Genesis*, **28**, 135–42.

Zhou, X., Hollemann, T., Pieler, T. and Gruss, P. (2000). Cloning and expression of *xSix3*, the Xenopus homologue of murine *Six3*. *Mech. Dev.*, **91**, 327–30.

Zuber, M. E., Perron, M., Philpott, A., Bang, A. and Harris, W. A. (1999). Giant eyes in *Xenopus laevis* by over-expression of XOptx2. *Cell*, **98**, 341–52.

Zuber, M. E., Gestri, G., Viczian, A. S., Barsacchi, G. and Harris, W. A. (2003). Specification of the vertebrate eye by a network of eye field transcription factors. *Development*, **130**, 5155–67.

3

Retinal neurogenesis

David H. Rapaport

*University of California, San Diego
School of Medicine, USA*

3.1 Introduction

In the past half-century the field of biology has witnessed a burgeoning of understanding of the biochemistry, molecular and cell biology of cell signalling. More recently, a significant effort was made to focus the techniques and concepts of biology to a mechanistic understanding of the nervous system. Within the area of development, perhaps the cardinal question has been how to signal immature cells to form the diverse organs, tissues and differentiated cells of the body – a particularly challenging question in the nervous system given the great diversity of cell types to be made. Because of its combination of diverse cell types within a highly structured tissue the vertebrate retina has served as an important model tissue in pursuit of answers to such questions. Specifically, the retina displays a laminar cytoarchitecture, and seven cell types that are largely confined to one of three laminae. These include receptors (rod and cone photoreceptors), short and long projection neurons (bipolar and retinal ganglion cells, respectively), local circuit neurons (horizontal and amacrine cells) and glia (Müller cells). The constancy of retinal structure and cell types across vertebrates allows cross-species comparisons to be readily made. Further, almost all retinal cell types exhibit multiple levels of differentiation. For example, there are several subtypes of ganglion cells or amacrine cells based on morphology, transmitter content, synaptic connectivity, etc. Thus, explanation of determination and differentiation can be sought at multiple levels of specificity.

Initial studies of cell fate acquisition in the retina focused on nature versus nurture issues. However, as has been concluded in most arenas in which this debate has raged, it now appears that some combination best approximates how retinal cells decide what to mature into. The ultimate aim of retinal development is to produce the right cell types in the right proportions at the right stages in development to allow formation of functional circuits. This is achieved not just by turning on the genetic programmes to make the products that characterize different cell types at the right times, but in concert with the production of new cells. Indeed, ongoing studies are suggesting that the 'targets' of signals for cell fate determination are the progenitors rather than, presumably naïve, postmitotic cells. Thus, a

Retinal Development, ed. Evelyne Sernagor, Stephen Eglen, Bill Harris and Rachel Wong.
Published by Cambridge University Press. © Cambridge University Press 2006.

better understanding of normal retinal development requires working out the mechanism(s) promoting progression through the cell cycle and maintenance and departure from the cell cycle. This chapter focuses on these issues – how retinal progenitors proceed through the cell cycle and how and when they produce postmitotic daughters. Much of this occurs relatively early in development, primarily from late optic vesicle and optic cup stages. Finally, the significance of sequentially ordered cell production will be explored in the context of the mechanism(s) of cell fate determination.

3.2 The cell cycle in the retina

The neuroepithelium of the optic vesicle and cup that generates the retina proceeds through the various stages of the cell cycle as in any other tissue in the body (Figure 3.1a). Chromosomes are duplicated during a DNA synthesis period called 'S-phase', and the cells become tetraploid. Subsequently they undergo a division whereby the chromosome pairs are sorted, aligned, segregated and separated to two daughters during mitotic or 'M-phase' of the cell cycle. In M-phase the chromosome complement is reduced to the normal two of a diploid cell. Between M- and S-phase are two temporal gaps, of varying length, called interphase (Figure 3.1). The first gap, G1 interphase, is between M- and S-phases, and G2 interphase is between S- and M-phases. Therefore cells are diploid during G1 and tetraploid during G2. Finally, a cell leaves the cell cycle after M-phase, sometime during G1. It may or may not retain the potential to re-enter the cycle and is said to be in a stage referred to as G0 (Figure 3.1b).

Early investigators made two seminal observations about cell division in the CNS. First, dyes that bind nucleic acids readily demonstrated the chromatin that aggregates during M-phase of the cell cycle. Such profiles are called mitotic figures, and the great majority line the lumen of the neural tube (Figure 3.1b, for review see Sidman, 1970; Jacobson, 1978; Rakic, 1981). In the retina this corresponds to the outermost (or scleral) surface, which, upon formation of the optic cup, is adjacent to the retinal pigment epithelium (Figures 3.1b, 3.2A–C, H–K). Second, using spectrophotometry to measure DNA content, it was early noted that profiles farthest away from the mitotic figures have approximately double the concentration of DNA, and are tetraploid (Figure 3.2D–G) (Sauer and Chittenden, 1959). Among the hypotheses advanced to account for these observations was that DNA replication, which occurs during S-phase, is spatially separate from cytokinesis, occurring in M-phase, with the nuclei migrating between the inner and outer surfaces of the neuroepithelium during interphase periods (Figure 3.1b).

Understanding of cell cycle and cell production was significantly advanced by the development of a new tool that relied on the incorporation of the radiolabelled nucleotide precursor of thymine, thymidine. It was important that the nucleotide be thymine since it is the one unique to DNA. Tritiated-thymidine (^{3}H-TdR) is taken up by dividing cells and incorporated into DNA during S-phase. Radiolabelled cells are demonstrated by autoradiography of tissue sections (Figure 3.2D/E, H/I). Any cells postmitotic before the ^{3}H-TdR was administered are not labelled. A larger cohort (eventually all) of the dividing cells is

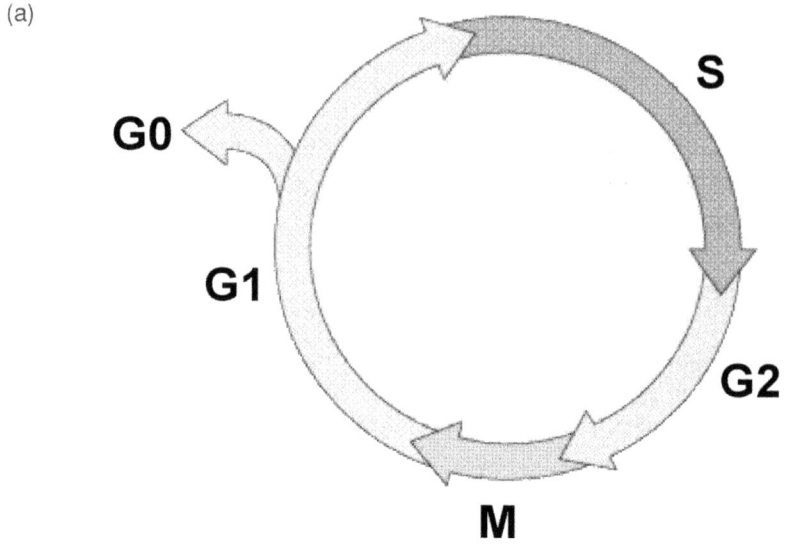

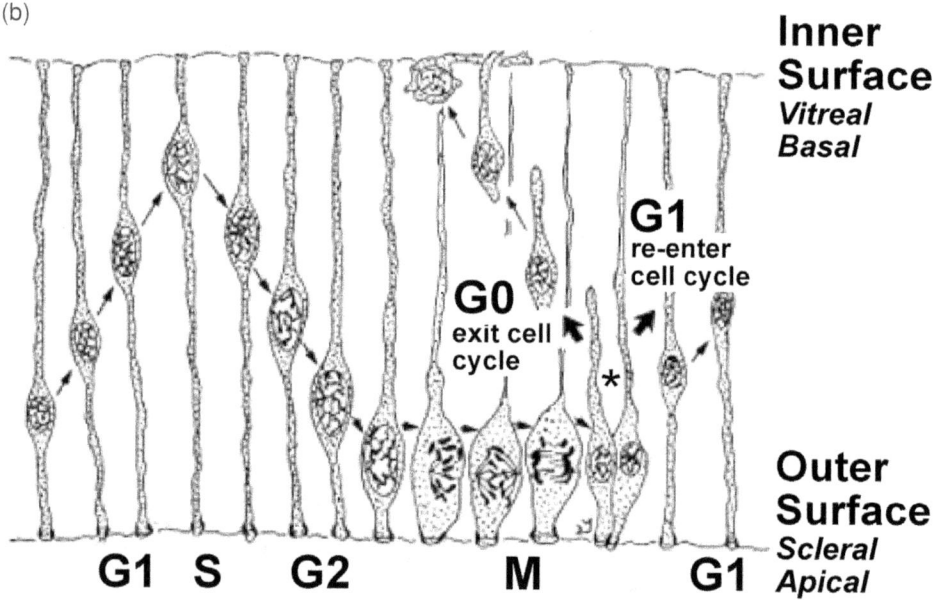

Figure 3.1 (a) Diagram showing the stages of the cell cycle. Chromosomal replication and DNA synthesis occurs in S-phase. Progenitors then spend a variable time tetraploid during G2 interphase whereupon they enter the mitotic phase (M-phase) and undergo cytokinesis to return to a diploid condition. At this point in time the cells face the decision of exiting the cell cycle (enter G0) or returning to S-phase by passing through G1 interphase. (b) This diagram fits the cell cycle into the context of retinal structure. At this age the presumptive retina is a pseudostratified epithelium forming the inner layer of the optic cup. This layer is apposed at its outer (scleral) surface to the retinal pigment

labelled as the interval following administration increases, limited only by the availability of ^{3}H-TdR. Eventually the ^{3}H-TdR is dispersed, incorporated or metabolized, and the cells that continue to divide lose their signal (by approximately 50% each round). Recently other thymine precursors such as 5'-Bromo-2'-deoxy-uridine (BrdU) have been administered, ones for which antibodies are available and allow immunohistochemistry to demonstrate their presence (Figure 3.2F/G, J/K). Regardless of the label, use of short post-injection intervals allows the localization of S-phase nuclei and tracking their position through the cell cycle (Figure 3.1b, 3.2D–G).

One of the first tissues to be studied with the new techniques for tracking cell cycle and cell genesis was the vertebrate retina (Sidman, 1961). Confirming the earlier hypothesis, S-phase occurs at a site distant from the luminal surface of the neural tube (Figure 3.1b), as this is where radio- or immunolabelled profiles are seen one hour following ^{3}H-TdR (Figure 3.2D/E) or BrdU administration (Figure 3.2F/G). A thin cell-sparse marginal zone separates the labelled cells from the innermost (vitreal) surface of the retina. Several hours later the radiolabelled profiles will have migrated to the outer retinal surface where they can be observed as mitotic figures (Figure 3.2H/I). Following mitosis the progeny either return to the inner zone to again replicate DNA, or leave the cell cycle, migrate away from the neural tube lumen and take up adult laminar positions (Figure 3.1b). The transit of nuclei during the cell cycle has come to be known as 'interkinetic nuclear migration', and the neuroepithelium spanning the migratory pathway forms the 'ventricular zone'.

3.3 The rate of progression through the cell cycle

The cell cycle is, by definition, a dynamic process studied by static neuroanatomical techniques only through time-consuming and resource-intensive experiments. However, a number of ways have evolved to derive data on cell cycle timing. The most straightforward is to simply count the number of cells in the retina at two time-points during development. More widely applied techniques for determining cell cycle timing involve short survivals following injection of a DNA synthesis marker (Figure 3.3). One can measure the proportion of labelled mitotic figures (Figure 3.3a) or count the labelled progenitors at successive intervals until the maximum number is reached (Figure 3.3b).

Zebrafish retinal progenitors rapidly increase in absolute numbers at stages prior to 16 hours post-fertilization (hpf) and after 24 hpf. However, between 16 hpf and 24 hpf the increase slows considerably. Computing cell cycle timing from these data gave

---

Figure 3.1 (*cont.*) epithelium but remains separated by the potential space of the obliterated neural tube lumen (optic vesicle). The nuclei of progenitors undergo S-phase distal to the neural tube lumen but enter M-phase at the luminal surface. A vitreally extending cell process retracts as cells round up for M-phase and again extends outward. During the interphase periods the cell nuclei migrate out (G2) and in (G1) within this process, in what is known as interkinetic nuclear migration. Following M-phase daughter cells are faced with the decision of exiting the cell cycle, migrating to the proper laminar position, extending axons and dendrites and differentiating, or re-entering S-phase to once again replicate DNA.

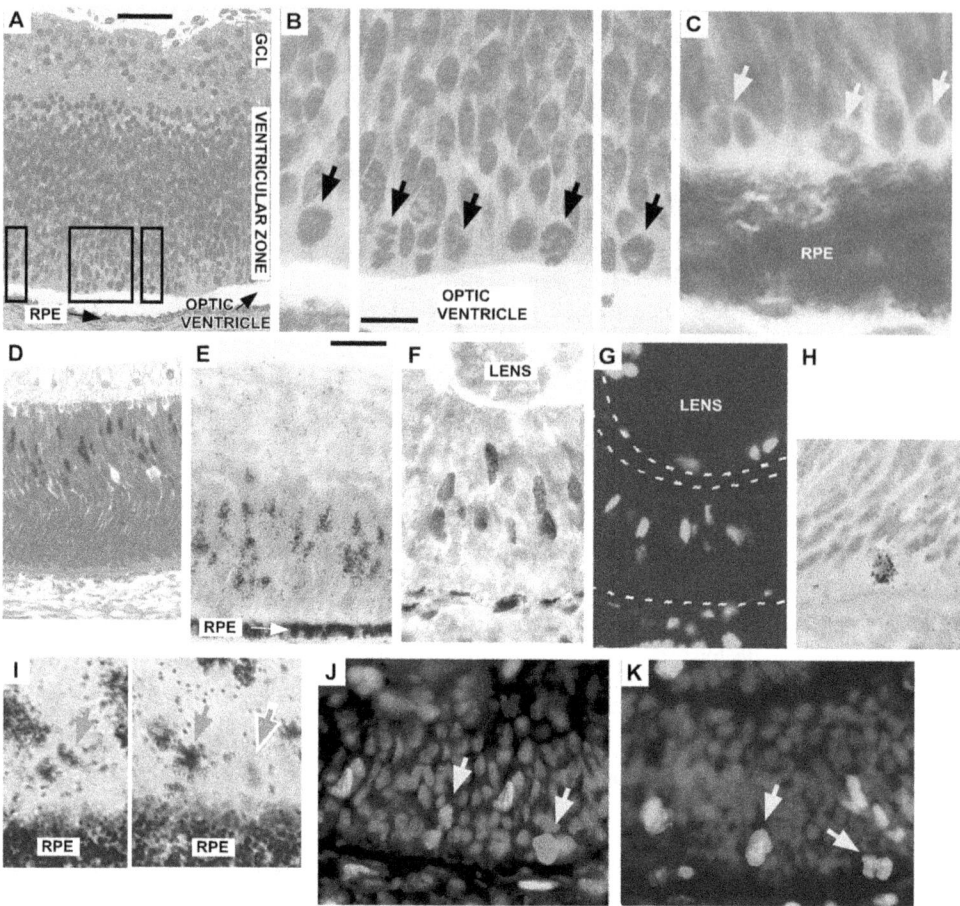

Figure 3.2 (A–C) Mitotic figures are present at the outer surface of the retina, against the lumen of the optic ventricle, a space eventually obliterated as the neurosensory retina and the retinal pigment epithelium (RPE) become apposed. Panel A is a low-power micrograph showing the retinal neuroepithelium of a postnatal day (P)0 mouse. The ganglion cell layer (GCL) is separated from the ventricular zone by a cell sparse region that will become the inner plexiform layer (IPL). During histological processing the neurosensory retina has separated from the RPE revealing the potential space that is the optic ventricle. This is not normally visible at this stage of development. Rectangles indicate areas shown at high power in panel B where arrows indicate mitotic figures. Panel C illustrates the same region of a P0 cat retina. Scale bar in A is 50 μm; in B is 10 μm. Scale bar in B applies to C as well. (D–G) The pattern of labelling in the developing retina shortly after administration of a marker of DNA synthesis. In all cases the labelled S-phase nuclei are distal to the outer surface of the retina, just below the IPL. Panels D and E are autoradiographs of sections through a P1 cat and an embryonic day (E)58 monkey retina, respectively, one hour following administration of ^{3}H-TdR. Labelled cells have dense silver grains in the photographic emulsion over their nuclei. Panels F and G are sections through a Stage 33/34 *Xenopus* retina 30 minutes following administration of BrdU. Labelled cells are demonstrated immunohistochemically using a horseradish peroxidase (brown, Panel F) and a fluorescein isothiocyanate- (green, Panel G) conjugated secondary antibody. In Panel G dotted white lines indicate the margins of the

lengths of 8 to 10 hours before 16 hpf and after 24 hpf, slowing to 32 to 49 hours between 16 to 24 hpf (Table 3.1) (Li *et al.*, 2000). During these stages the zebrafish retina develops from an optic primordium into a well-defined optic vesicle, eventually a cup. At a comparable stage *Xenopus* retinal cells replicate at a rate of 8.6 h/div (Table 3.1, Rapaport *et al.*, unpublished). Chick retinal progenitor cells (RPCs) complete a cycle in approximately 5 hours in the one-day-optic-vesicle-stage embryo, and in 10 hours in the 6-day embryo (Table 3.1) (Fujita, 1962). Most cell cycle data from mammals come from retinas at significantly later stages of development. The cell cycle in the retina of a newborn mouse takes 28 to 30 hours (Gloor *et al.*, 1985; Young, 1985a), and 28 hours in the postnatal day (P2) rat (Denham, 1967). Another study in the rat, using multiple techniques, indicates that the cell cycle is a bit longer – 32 to 39 hours between birth (P0) and P2 (Table 3.1) (Alexiades and Cepko, 1996).

All studies of cell cycle timing in the retina agree that it slows during development. In the developing frog retina it lengthens from approximately 8 hours at Stage 27 to as long as 56 hours at Stage 37/38 (Rapaport *et al.*, unpublished). Interestingly, over the same developmental interval in the rat, though much longer in terms of hours, cell cycle lengthens similarly, from 14 hours at embryonic day (E)14 to as long as 56 hours at P8 (Table 3.1) (Alexiades and Cepko, 1996). The apparent steady increase would not seem to support a model where the genesis of different cell types can be modulated by changing cell cycle rate. The only vertebrate retina studied so far to show modulated cycle timing is that of the zebrafish where it is relatively rapid (8 to 10 hours) before 16 hpf and after 24 hpf, but much slower in the intervening 8-hour period (Li *et al.*, 2000). Mitotic deceleration occurs, as the optic vesicle becomes a cup, earlier than in other studies of RPC cycle timing. It is unclear what the function of the modulation is. It may be important in scaling the output of progenitors to the rate of increase in the size of the retina. Further data need to be collected to determine both the significance and generality of modulated cell cycle timing.

3.4 Mode of cell division: symmetrical versus asymmetrical

Another mechanism affecting the production of cells is the mode of division, of which three can be described. Early RPCs go through a period of symmetrical divisions, each daughter returning to the cell cycle. This mode allows the pool of progenitors to expand

←

Figure 3.2 (*cont.*) retina and the lens. Scale bar in A applies to D. Scale bar in E is 25 µm and applies to F and G as well. (H–K) Mitotic figures labelled by a DNA synthesis marker after an intermediate survival period. By now cells in S-phase have moved to the ventricular surface and entered M-phase. Panel H shows a labelled prophase nucleus (arrowed) of a P0 cat retina 2.5 hours after administration of ^{3}H-TdR. Panel I shows labelled anaphase and prophase nuclei (arrowed), as well as an unlabelled metaphase figure (arrow with white border) in an E58 monkey retina one hour after administration of ^{3}H-TdR. In Panels J and K BrdU-positive (S-phase) cells in Stage 33/34 *Xenopus* retina appear green following indirect fluorescence immunohistochemistry. Cells in M-phase have been demonstrated with an antibody to phosphohistone H3, in this case demonstrated with a tetramethyl rhodamine isothiocyanate- (red) conjugated secondary antibody. No mitotic figures are BrdU-positive 30 minutes after injection (J), but they are after a one hour interval (K). Scale bar in B applies to H and I, and scale bar in E applies to J and K. For colour version, see Plate 2.

(a) Per cent labelled mitotic figure method

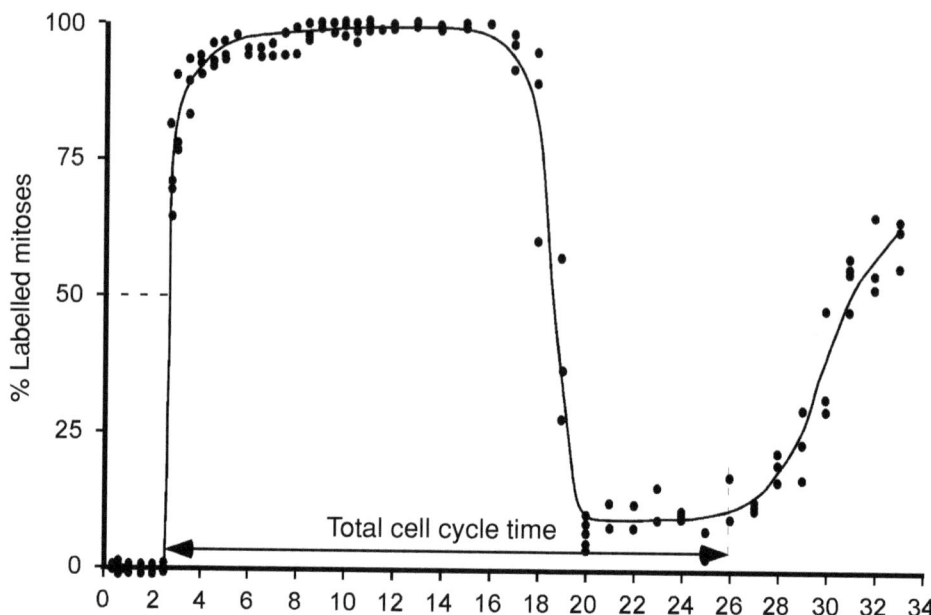

(b) Cumulative labelling method

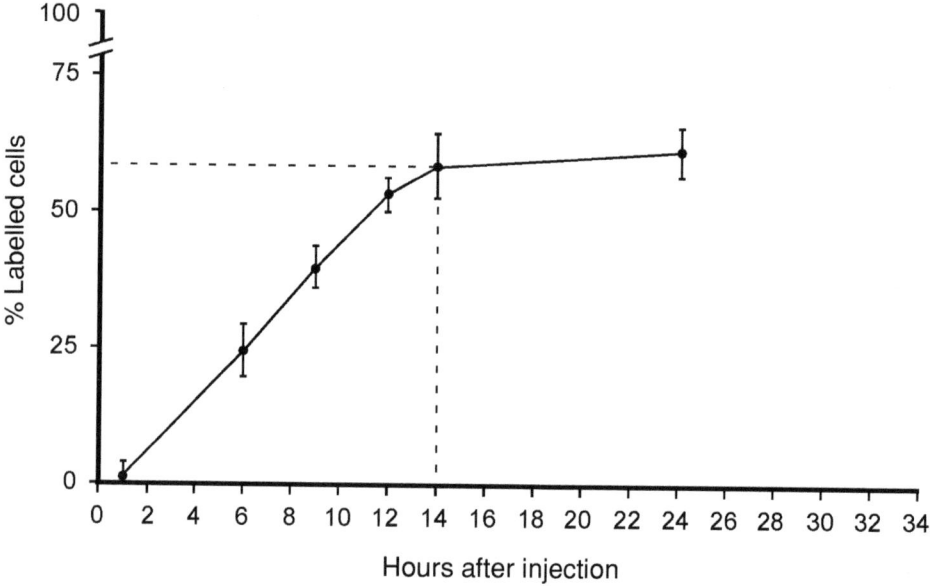

Hours after injection

Figure 3.3 Two ways of determining cell cycle timing using short survival periods following injection of a marker incorporated during DNA synthesis (S-phase). (a) A pulse of the DNA synthesis marker is given and the percent of labelled mitotic figures is counted at varying intervals. The amount

Table 3.1. *Data on cell cycle timing in the retina of various vertebrates*

Experiment (reference)	Species	Method	Cell cycle length	Cell cycle length	Cell cycle length
Rapaport et al. (unpub.)	Xenopus	Window labelling	*Stg. 27*[a] **8.6 h**	*Stg. 33/34* **36 h**	*Stg. 37/38* **56 h**
Li et al. (2000)	Zebrafish	Counting cell number	*16 hpf*[b] **8–10 h**	*16–24 hpf* **32–49 h**	*24 hpf* **8–10 h**
Fujita (1962)	Chick	Per cent labelled mitotic figures	*Stg. 7*[c] **5 h**	*Stg. 29*[c] **10 h**	
Young (1985a)	Mouse	Per cent labelled mitotic figures	*P1* (central retina) **30 h**	*P1* (periph. retina) **28.5 h**	
Gloor et al. (1985)	Mouse	Cumulative labelling	*P1* **28 h**		
Denham (1967)	Rat	Per cent labelled mitotic figures	*P2* (central retina) **28 h**	*P5* (central retina) **28 h**	
Alexiades and Cepko (1996)	Rat	Cumulative labelling	*E14* **14 h**	*P1* **36 h**	*P8* **55 h**
Alexiades and Cepko (1996)	Rat	Window labelling	*E14* **15 h**	*P1* **32 h**	

[a] *Xenopus* stages from Nieuwkoop and Faber (1956).
[b] Italics indicate stage of development at which observations were made.
[c] Chick stages from Hamburger and Hamilton (1951).
Abbreviations: hpf, hours post-fertilization, E, embryonic day (day post-conception), P, postnatal day (day since birth).

exponentially. Later, RPCs divide asymmetrically, one daughter returning to the cell cycle, the other exiting, migrating and differentiating (Figure 3.1b). This mode provides a steady source of new cells at a relatively constant rate. At some late stage, progenitors go through a final, symmetrical division in which both daughters become postmitotic. To distinguish

Figure 3.3 (*cont.*) injected is predetermined to be available for a finite period, significantly less than the cell cycle. The time between the injection until the first labelled mitotic figures is a measure of G2, and the whole cell cycle is traversed during the interval between this point, through the disappearance of labelled mitotic figures (as the concentration of the label available for uptake decreases) to the point when labelled mitotic figures appear again. The reappearance of labelled mitotic figures indicates the cells that do not exit the cell cycle, but return to this point. This graph is modified from Young, 1985a. (b) With the DNA synthesis marker continuously available the population of retinal progenitors become labelled as they move through S-phase. Eventually a maximum number of labelled cells is reached, representing all the cells in the retina that are dividing at that point in time (growth fraction). The time to reach the growth fraction is a direct measure of $G2 + M + G1$. The length of the entire cell cycle can be calculated by deriving the proportional duration of S-phase from the number of labeled cells after a short interval as a proportion of the time to reach the growth fraction. This graph is modified from Alexiades and Cepko (1996).

early and late modes of symmetrical division this one is called 'terminal'; it ultimately signals the end of proliferation. The traditional view is that RPCs move through these steps sequentially and in one direction. It is equally possible that progenitors vary their mode of division to suit proliferative demands at any particular point in development. For example, zebrafish retina produces all ganglion cells during a short, rapid, early window of genesis. Subsequently the tissue undergoes a phase in which no postmitotic cells are produced (Hu and Easter, 1999; Li *et al.*, 2000). This could reflect a slowing of cell cycle (as discussed) or a return to a symmetric mode of division. The latter allows expansion of the proliferative pool before the genesis of other retinal cell types. Similarly, cell birth in the marsupial wallaby begins just before it enters the pouch (P0), stops on P30 and resumes again some 20 days later to finish after P85 (Harman and Beazley, 1989). Besides the sheer magnitude of the interval of quiescence of cell genesis, the major difference between the zebrafish and wallaby is that, for the latter, several cell types, not just ganglion cells, are generated during the early phase, and some of these are also produced by the late phase. Though not as dramatic, similar decreases in cell production are seen in the developing retina of all vertebrates that have been examined including the monkey (LaVail *et al.*, 1991) and the rat (Rapaport *et al.*, 2004). In the rat, steady state kinetics is a feature of the early phase of development, changing to exponential for the late phase (Rapaport *et al.*, 2004). A period of symmetric divisions between the early and late phases of cell production would allow for expansion of the pool of RPCs in anticipation of a final high output. Indeed, late-phase cells, rods, bipolar cells and Müller cells are the most numerous in the retina.

3.5 Mitotic spindle orientation and the mode of cell division

There has been long-standing interest in whether the plane of cleavage of dividing cells is indicative of mode of cell division. Cleavage perpendicular to the plane of the ventricular surface allows both daughters to maintain a connection with the ventricular surface (Figure 3.4a), while parallel cleavage 'releases' the non-ventricular daughter (Figure 3.4c). Orientation of cleavage has also been shown to be a mechanism for distributing cellular contents between daughters. In one mode the cytoplasmic contents are equally divided. In the other, the cues terminating cell division, controlling cell migration and differentiation can be sequestered to a single daughter (Figure 3.4d/e).

Most RPCs divide with their spindle oriented parallel (horizontal) to the ventricular (luminal) surface (Figure 3.4a), although the presence of some with obliquely, even perpendicularly, oriented spindles has been noted (Figure 3.4b) (Rapaport and Stone, 1983; Silva *et al.*, 2002; Cayouette and Raff, 2003; Das *et al.*, 2003; Tibber *et al.*, 2004). If cleavage away from the plane of the ventricular surface favours asymmetric mitosis, then there should be a greater frequency of them with age. In the fetal cat retina, neither the number of obliquely oriented mitoses, nor the mean angle of mitotic spindle orientation varied with development (Figure 3.4f) (Rapaport and Stone, 1983). Time-lapse imaging of dividing cells in zebrafish retina failed to demonstrate any mitotic profiles with one daughter unattached to the outer

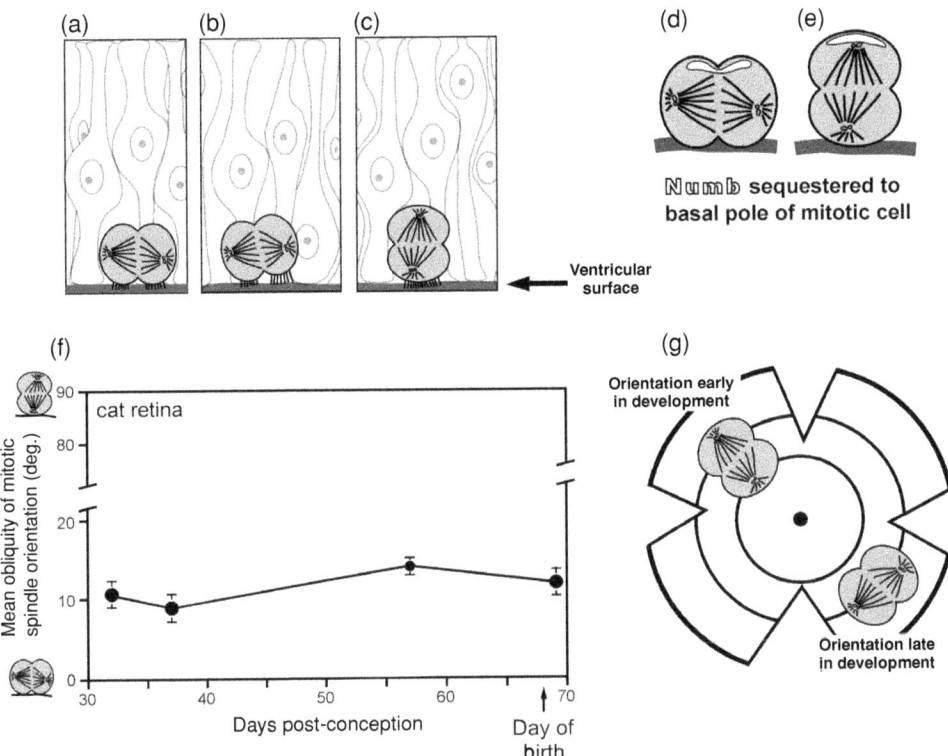

Figure 3.4 (a) The great majority of mitotic figures in the developing retina have their spindles oriented parallel to the ventricular surface and the plane of cleavage orthogonal. Both daughters maintain junctional complexes with the outer surface of the retina. (b) Spindle orientation typically varies about the horizontal plane by up to 10 to 15 degrees. Both daughters still maintain connection with the ventricular surface. (c) In rare cases spindle orientation can be as much as 90 degrees off the horizontal plane, now perpendicular to the ventricular surface, and the plane of cleavage parallel. The distal (basal) daughter may no longer maintain connection to the outer margin of the retina. (d) Under the conditions shown in (a) Numb and other proteins sequestered to the basal pole of progenitor cells are symmetrically distributed to both daughters. (e) Cleavage perpendicular to the ventricular surface distributes basally sequestered proteins such as Numb so that only the distal daughter inherits it. (f) The mean orientation (±S.E.M.) of cleavage of a sample of mitotic figures at the ventricular surface of the developing cat retina during prenatal development (data from Rapaport and Stone, 1983). Obliquity of orientation of cleavage is measured as degrees off the axis of the ventricular surface at the site of the mitotic figure. The mean cleavage angle is very close to horizontal (0 degrees) varying between 9 and 14 degrees throughout development. There is very little variation despite the fact that at the early age the progeny of a minority of progenitors are leaving the cell cycle, while at the oldest age most progenitors are producing postmitotic daughters. (g) A unique pattern of cleavage recently shown in zebrafish retina. Throughout development the cleavage plane is parallel to the ventricular surface. However, at early stages cells in mitosis separate along a radial axis, generating a central and a peripheral daughter. Late in retinogenesis the cleavage axis rotates so as to be tangential to the edge of the retina.

surface of the retinal neuroepithelium (Das *et al.*, 2003). On the other hand, a previously unobserved feature of spindle orientation did emerge (Figure 3.4g). Early in zebrafish and rat retinal development mitotic cleavage is oriented radially along an axis that extends from central to peripheral like the spokes of a wheel. Later the plane rotates 90 degrees so that it is then tangential to concentric circles emanating from the centre. It is suggested that this change promotes asymmetric divisions, but no direct data are available. It could be significant that two mutant fish strains, *sonic youth* and *lakritz*, known to exhibit delayed retinal differentiation (Neumann and Nuesslein-Volhard, 2000; Kay *et al.*, 2001) fail to demonstrate this shift.

Renewed interest in the relationship of spindle orientation to mitotic fate grew from the observation that it can asymmetrically distribute cytoplasmic constituents between progeny. For example, Notch1 and Numb are sequestered basally in RPCs, and parallel cleavage segregates all of the proteins to the non-ventricular daughter (Figure 3.4e) (Zhong *et al.*, 1996; Cayouette *et al.*, 2001; Silva *et al.*, 2002; Cayouette and Raff, 2003). Notch has been shown to be involved in keeping RPCs in an undifferentiated state (Austin *et al.*, 1995; Dorsky *et al.*, 1995, 1997) while Numb can antagonize this function (Wakamatsu *et al.*, 1999; French *et al.*, 2002). However, the data do not appear to support the hypothesis that Numb sequestration leads the non-ventricular daughter to exit the cell cycle and differentiate. In the chick retina, Numb expression does not correlate with expression of differentiation markers expected to indicate asymmetric division (Silva *et al.*, 2002). Further, at a late stage of rat retinal development, when many cells are being born, Numb distribution was not predictive of mode of division. However, it did appear to relate to cell fate. Specifically, when cleavage is orthogonal to the ventricular surface, and Numb is distributed equally between daughters (Figure 3.4d), both differentiated as the same cell type, typically rods. The progeny of parallel cleavage (Figure 3.4e), either or both of which could become postmitotic, adopted different phenotypes (Cayouette and Raff, 2003).

3.6 The order of cell birth in the retina

Long-term survival following administration of a DNA synthesis marker leads to a smaller cohort of labelled cells than short-term. The nucleoside is rapidly dispersed throughout the body, concentration halving within 5 to 20 minutes depending on the species and the route of administration (Blenkinsopp, 1968; Nowakowski and Rakic, 1974; Hickey *et al.*, 1983). Further, with each division the concentration of the label incorporated into a cell's DNA halves, and with no available pool for replenishment, decreases below the threshold of detection within two to three cell divisions. The cells that remain labelled weeks, months and even years following labelling are those that underwent their terminal cell division within one or two cell cycles of the one during which the label was incorporated. In other words these cells underwent their 'birthday' at or near the time of injection.

A complete study of retinal cell birthdays must satisfy a number of criteria. The age of injection of DNA synthesis markers must be early and late enough to include periods when no labelled cells are present. In this way the timing of initiation and cessation of cell genesis is determined. In addition, the tissue must be prepared in such a way that the features defining each of the retinal cell types can be resolved. The potential phenotype(s) can be narrowed considerably based on the laminar location of cell bodies. However, this can be unreliable, particularly when attempting to distinguish between ganglion cells and amacrine cells in the ganglion cell layer and the variety of neuron types and glia in the inner nuclear layer (INL). Most studies rely on thin-sectioned tissue in which one can resolve details of cytoplasmic and nuclear morphology characteristic of each cell type (Blanks and Bok, 1977; LaVail et al., 1991; Rapaport and Vietri, 1991; Strettoi and Masland, 1995; Rapaport et al., 2004). This still leaves a problem distinguishing amacrine cells from ganglion cells in the retinal ganglion cell layer, since these share similar morphology. Labelling with a cell-specific marker or retrogradely transported dye can allow more reliable identification of these and other cell phenotypes.

Studies that successfully characterize the complete sequence of cell genesis are few. To date they have been published for the chick (Prada et al., 1991), monkey (LaVail et al., 1991), wallaby (Harman and Beazley, 1989) and rat (Rapaport et al., 2004). The data from these diverse species indicate that the sequence of cell genesis in the vertebrate retina is highly conserved. But for minor details (to be discussed), in all species studied cell birth proceeds in the following order (Figure 3.5): retinal ganglion cells, horizontal cells, cone photoreceptors, amacrine cells, bipolar cells, rod photoreceptors and Müller glia. The first three cell types generated are located in the ganglion cell layer, the INL and the outer nuclear layer (ONL) respectively, and this supports an outside-in pattern (relative to the ventricular zone). However, subsequent cell types are found in many other layers and indicate that, in the retina, cell birthdate is related more to phenotype than to laminar location. The first cells to be generated, ganglion cells and horizontal cells, are the largest in the retina, supporting the proposition that large cells are generated before small. Likewise, the last cells to be born are the Müller glial cells, supporting a trend in the CNS for glia to be generated late.

Another feature of the genesis of retinal cells that appears conserved is a separation into distinct phases. The early phase, consisting of ganglion cells, horizontal cells and cones is separated by a varying length of time from a later phase, when the remaining cell types are generated (Figure 3.5). The interval between phases can last from 2 to 3 weeks, during which all cell birth ceases entirely such as in the wallaby (Harman and Beazley, 1989), to 4 to 7 days and 10 to 20 days of lower density cell birth in the rat and monkey, respectively (LaVail et al., 1991; Rapaport et al., 2004). The two phases of genesis differ also in their kinetics with phase 1 genesis ramping up and down much more quickly than phase 2, during which cell birth begins more gradually (Figure 3.5). The cells of the early phase are relatively few in number and, in the adult, exhibit strong central-to-peripheral gradients of density. It is likely that the two patterns of cell production allow for the interplay between the length of the developmental period, rate of production and expansion of the retinal area, the end result being the adult patterns of cell distribution across the retina.

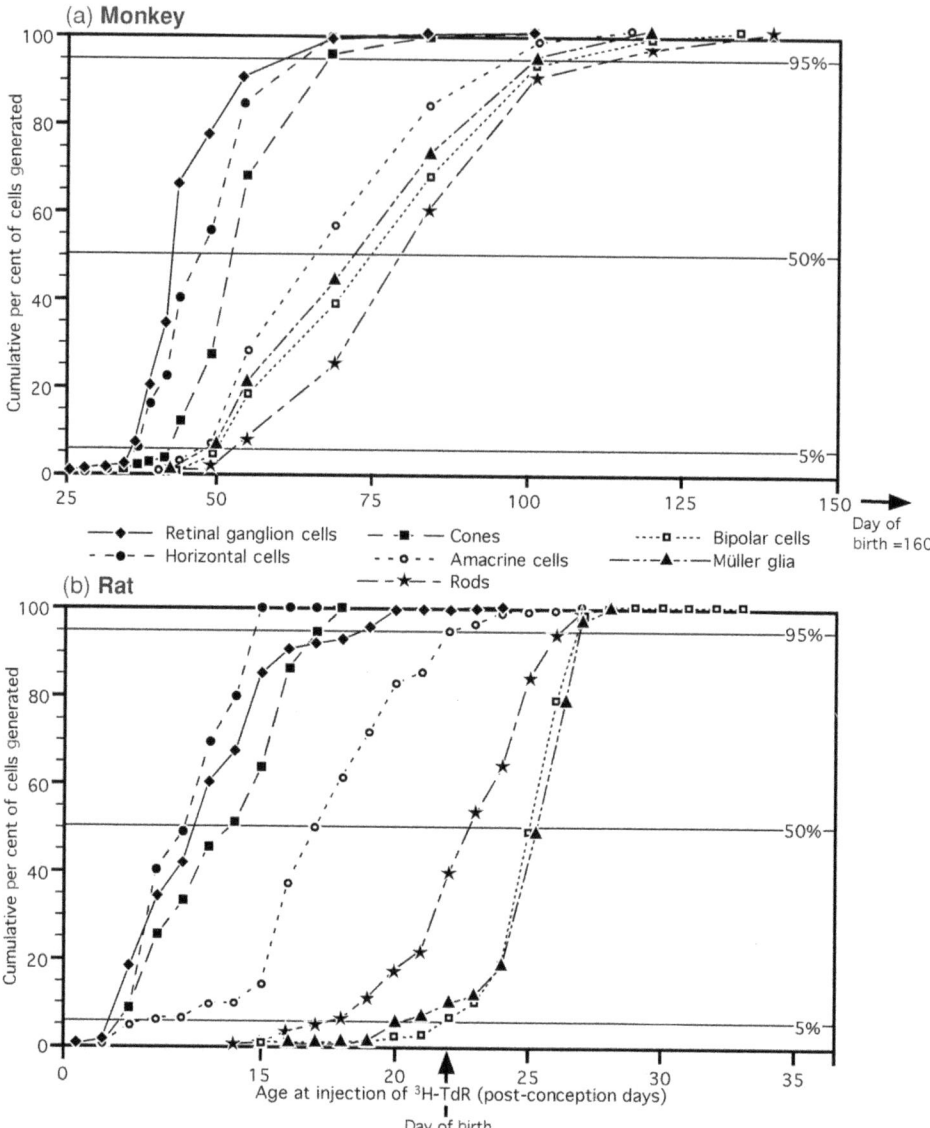

Figure 3.5 The timing and sequence of retinal cell genesis in the Rhesus macaque monkey (a) and laboratory rat (b) based on pulse labelling with ^{3}H-TdR at varying stages of development. Heavily labelled cells are said to have undergone their birthday on the day of injection. The cumulative proportion of labelled cells best shows the interrelationships between cell types. Though there is significant overlap in the timing of genesis of different cell types there is a definite order. Each cell type reaches the landmarks of 5%, 50% and 95% of cells generated in a defined sequence. Changes in the ordinal position of cell types are very rare. Though the time over which their retinas are generated differs significantly, the sequence is nearly identical for monkey and rat. In both species two phases of cell production can be recognized differing in kinetics of proliferation and time of genesis. Phase 1, consisting of retinal ganglion cells, horizontal cells and cones, is early, and separated by a gap from phase 2. In phase 1 cell genesis is particularly rapid as shown by the steeper slopes of the graphs. Both species also demonstrate a late, prolonged cessation of ganglion cell genesis so that horizontal cells and cones bypass them. This is presumed to result from the late addition of displaced amacrine cells to the ganglion cell layer.

3.6.1 Retinal ganglion cells

Many studies have looked at the time of genesis of retinal ganglion cells, and all support their primacy in the sequence in the retina. This is particularly striking given the wide variety of species studied – fish (Hollyfield, 1972; Sharma and Ungar, 1980; Hu and Easter, 1999), amphibia (Jacobson, 1976; Holt *et al.*, 1988; Stiemke and Hollyfield, 1995), birds (Fujita and Horii, 1963; Kahn, 1974; Prada *et al.*, 1991) and particularly mammals (*gerbil* (Wikler *et al.*, 1989); *hamster* (Sengelaub *et al.*, 1986); *mouse* (Dräger, 1985); *cat* (Zimmerman *et al.*, 1988); *rat* (Reese and Colello, 1992; Galli-Resta and Ensini, 1996; Rapaport *et al.*, 2004); *ferret* (Reese *et al.*, 1994); *monkey* (LaVail *et al.*, 1991; Rapaport *et al.*, 1992)). Indeed, the suggestion has been made that ganglion cells are the default phenotype of retinal progenitors (Reh, 1989). Within the population of retinal ganglion cells several functional types have been distinguished based on an array of anatomical, physiological and behavioural criteria (for review see Stone, 1983). These also appear to differ in the timing of their genesis. In cat, monkey and rat small retinal ganglion cells are generated early and increasingly larger cells are generated at later developmental stages (Walsh *et al.*, 1983; Walsh and Polley, 1985; Rapaport *et al.*, 1992, 2004). Unequivocal identification of ganglion cell type relies on considerably more features than soma size, but to some extent these data suggest the small (W-, γ-, or konio-) cells are generated early, X-, β-, midget or parvo- cells during an intermediate period and Y-, α-, parasol, or magno- cells are generated late. In the thick, perifoveal retinal ganglion cell layer of the monkey the cells labelled by the earliest ^{3}H-TdR injections are located against the vitreal surface. There is a clear trend for subsequent injections to label cells deeper in the layer (Rapaport *et al.*, 1996). Taken with data from retrograde labelling studies indicating that retinal ganglion cells projecting to parvocellular layers of the dorsal lateral geniculate nucleus (LGNd) are located vitreal to those terminating in the magnocellular layers (Perry and Silveira, 1988) the conclusion that X-cells are generated before Y-cells (or the same cell type by any other name) is supported. It is intriguing that, while ganglion cells as a class fit the general 'large cells generated early' paradigm relative to other retinal cell types, within this class the relationship of size to birthday is reversed.

Depending on the species, placement of the eyes and consequent degree of binocular vision, a varying proportion of ganglion cells in temporal retina, ranging from small (mouse, rat, rabbit) to about half (cat, ferret) to all (macaque monkey, human) project ipsilaterally at the optic chiasm (Stone, 1966; Stone *et al.*, 1973; Provis and Watson, 1981; Jeffery, 1985; Fukuda *et al.*, 1989). These cells also exhibit differences in their time of genesis. Retinal ganglion cell birth in mouse begins around E11. Data from injection of a retrograde tracer into the visual targets on one side of the brain demonstrate that the first of these project ipsilaterally. Contralateral projecting ganglion cells tend to be generated after the ipsilateral ones (Dräger, 1985). Similar studies in species in which development is more protracted and in which the population of ipsilaterally projecting ganglion cells is more substantial paint a clearer picture – the non-crossing ganglion cells in temporal retina are born significantly (up to four days in the cat) before the ones that cross (Reese *et al.*, 1992).

3.6.2 Amacrine cells

The smallest cells in the retinal ganglion cell layer are the last to be generated (Walsh et al., 1983; Walsh and Polley, 1985; Rapaport et al., 1992; Reese et al., 1994). However, the evidence suggests that these are not ganglion cells with small somata, but amacrine cells that reside in the retinal ganglion cell layer. Because small ganglion cells and displaced amacrine cells overlap in soma size, determining their separate birthdays requires the use of a marker specific to one or the other. The most robust technique at this point in time is to inject a dye into the visual centres of the brain, or apply it to the optic nerve, and allow retrograde transport back to the bodies of ganglion cells. This is particularly effective since the prime feature distinguishing the two phenotypes is the presence of an axon within the optic nerve, terminating in brain nuclei. There are also a number of markers for amacrine cells, including many related to the diversity of neurotransmitters they express. Nevertheless, problems remain – how can one be sure that a labelling protocol fills all retinal ganglion cells, especially the smallest ones, or that a cell marker labels all individuals and is specific to only that one type? In the rat, ganglion cell genesis begins at least two days before the genesis of displaced amacrine cells, and displaced amacrine cells are the last neurons in this layer to be generated (Reese and Colello, 1992). If not for the presence of the IPL, 'displaced' and 'normally placed' amacrine cells would be contiguous, as would be the timing of their birth. The hypothesis was early advanced that displaced amacrine cells are the first of this type to be born, and that upon migrating outward they find the IPL incompletely formed and continue beyond it into the ganglion cell layer. However, the 'displaced amacrine cell as developmental error' hypothesis cannot readily account for the fact that some types form highly organized mosaics in the ganglion cell layer and INL, which mirror each other, cell for cell. For example, acetylcholine-containing amacrine cells have highly stratified dendritic fields in the ON and OFF sublayers of the IPL. For each cholinergic amacrine cell in the ganglion cell layer ramifying in the ON sublamina there is a single, matching cell in the INL that ramifies in the OFF sublamina (Vaney et al., 1981; Masland and Tauchi, 1986; Rodieck and Marshak, 1992). This arrangement is too regular to result from an error and indicates that there are developmental signals directing a subset of amacrine cells to be generated early and to take up residence in the ganglion cell layer. In this way amacrine cell bodies are adjacent to where they extend dendrites, whether displaced or not. However, some amacrine cells in the ganglion cell layer may indeed be ectopic. Thus, a distinction has been suggested between 'displaced' and 'misplaced' amacrine cells (Vaney, 1990).

As hinted, amacrine cells are an extremely diverse class with many subtypes distinguished by morphology and, in particular, neurotransmitter phenotype (for review see Vaney, 1990; Masland, 2001). Though a majority of amacrine cells express glutamate or γ-aminobutyric acid (GABA) the diversity of neurotransmitter phenotypes is greatly increased by the many that express, often in concert with GABA, other transmitters such as 5-hydroxytryptamine, dopamine, acetylcholine, and including many peptides (substance P, VIP, somatostatin, etc.). The GABAergic amacrine cells appear to be born within the first cohort of amacrine cells, from E13 in the rat (Lee et al., 1999), while acetylcholine, dopamine and corticotrophin-releasing factor (CRF) expressing amacrine cells are not generated in this species until two to three days later (Evans and Battelle, 1987; Zhang and Yeh, 1990; Reese and Colello, 1992). Similarly, amacrine cell genesis in the chick extends from approximately E3 to E9 (Prada *et al.*, 1991); however the subset of dopaminergic (tyrosine hydroxylase immunoreactive) amacrines ceases its genesis by E7 (Gardino *et al.*, 1993). These results suggest that the timing of genesis may be a factor in the determination of the specific amacrine phenotype adopted. However, only a few of the many subtypes of amacrine cells have been studied, and more data are needed. Eventually interest should focus on whether amacrine phenotype is determined intrinsically, before or at the time of birth, or by local cues present when the cells arrive at the inner margin of the INL, the place that they will eventually settle as the retina achieves adult cytoarchitecture.

3.6.3 Horizontal cells

One usually thinks of cell birth as the first step of a progression that proceeds uninterrupted from genesis through migration and differentiation, but an initially surprising, though since shown to be robust, finding is that there can be a substantial interval from the birth of a cell to its differentiation. Horizontal cells are among the first cells to be born in the retina. They migrate to an appropriate position in the neuroepithelium, settling at the future position of the outer plexiform layer (OPL). They remain here, for up to three weeks in the cat (Zimmerman *et al.*, 1988), obvious as an interrupted band of large, round, but otherwise immature, somata (Figure 3.2D). When late-generated bipolar cells appear synaptogenesis between bipolar, horizontal cells and photoreceptors proceeds, and the OPL emerges. Prior to that, horizontal cells are transiently GABAergic, exert a trophic influence on cones (which express $GABA_A$ receptors) and facilitate the establishment of synapses (Redburn and Madtes, 1986; Versaux-Botteri *et al.*, 1989) and their own maturation. Expression of GABA is down-regulated when cone synaptogenesis is complete, at which time there is significant dendritic growth and remodelling to achieve an adult horizontal cell phenotype (Messersmith and Redburn, 1993; Mitchell *et al.*, 1995, 1999; Huang *et al.*, 2000).

3.6.4 Photoreceptors

There is great inter-species variability in the proportion of photoreceptors that are rods and cones. Chick retinas and the macular regions of macaque and human retinas are described as cone dominated or pure cone; the frog, *Xenopus laevis*, has approximately equal numbers of cones and rods (Chang and Harris, 1998), and mouse, cat or owl monkey are rod dominated or pure rod. The two types of photoreceptors may share developmental history, as there is evidence for a generic 'photoreceptor' stage of determination before rod or cone fate is adopted (Harris and Messersmith, 1992). However, in all species studied there are distinct temporal windows for the birth of cones and rods, with cone genesis generally beginning before rod genesis and ending before the end of rod genesis (Carter-Dawson and LaVail, 1979; Stenkamp *et al.*, 1997). Although inter-species variation in the ordinal relations of the birth of retinal cell types is rare, it is most frequent for rod and cone photoreceptors. In the macaque rods are the last cell type to be labelled by ^{3}H-TdR (LaVail et al., 1991) (Figure 3.5a), while in the rat rod genesis begins early in the late phase of cell genesis, sequentially between amacrine cells and bipolar and Müller cells (Rapaport et al., 2004). However, the analysis performed in the monkey included the fovea centralis, and therefore the samples over-represent cones. When analysis was confined to the peripheral retina, where rods are the predominant photoreceptor, their ordinal position shifted earlier, between amacrine cells and bipolar cells (LaVail et al., 1991). This is where they are found in the rat (Figure 3.5b) (Rapaport et al., 2004). In other words, when comparing retinal regions containing a similar complement of rods and cones (in this case rod dominated) the ordinal positions of rods and cones in the sequence of retinal cell birth correspond. This suggests that timing of genesis may reflect, at least partly, the number of cells produced. Other data support this hypothesis. In the chick, which has a pure cone retina, photoreceptors and amacrine cells switch positions in the sequence of cell genesis (Kahn, 1974; Prada et al., 1991). This gives more time for the genesis of both cell types, moving the peak of genesis for cones to a later age and the peak for amacrines to an earlier one. Similarly, rods and amacrine cells swap positions in the sequence of cell genesis in *Xenopus* (see Section 3.7) where amacrine cells are, perhaps, proportionately fewer.

Most vertebrates have multiple cone types with different spectral sensitivities due to expressing various opsins, the light-sensitive portion of a photopigment molecule. A species may have several opsins; for example, zebrafish express four, one each with maximal sensitivity in the red (L or long-wavelength), green (M or medium-wavelength), blue (S or short-wavelength) and ultraviolet (UV) parts of the spectrum (Nawrocki et al., 1985; Vihtelic et al., 1999; Chinen et al., 2003). Therefore zebrafish has proved a particularly valuable model of cone development in much the same way that rod development has been studied in rod-dominated rodents. Within the population of zebrafish cones those expressing different opsins are generated in a specific order. Double-labelling experiments identifying opsin phenotype and cell birth have demonstrated that the first postmitotic zebrafish cones are the red-sensitive ones, followed sequentially by green, blue and UV (Larison and Bremiller, 1990; Stenkamp et al., 1997), all of which are generated before rods. It is unclear at this point if this order results from cell-autonomous or non-autonomous influences; however, the fact that cones, as a class, can be stained with specific markers before their opsin is expressed (Carter-Dawson et al., 1986; Gonzalez-Fernandez and Healy, 1990; Harris and Messersmith, 1992; Hauswirth et al., 1992; Johnson et al., 2001) suggests that the cone differentiation path includes a generic step before the specific opsin phenotype is determined.

Similar to horizontal cells, photoreceptors, particularly cones, are slow to mature, at least in terms of their function in phototransduction. Growth of photoreceptor outer segments is one of the last events of retinal development (Olney, 1968; Weidman and Kuwabara, 1968). Further, cones become postmitotic in zebrafish about a month before their opsins are expressed (Stenkamp et al., 1996, 1997), and a similar delay is imposed on the monkey (genesis – E43, opsin expression – E75, for foveal cones) (Bumsted et al., 1997). Indeed, cone opsin expression is detected only after rod opsin (both mRNA and protein) despite the fact that cones are generated early and rods late. Perhaps the cell types that are born significantly prior to their terminal differentiation play other roles in retinal development during the intervening time. In this context it may be relevant that photoreceptors initially extend a process all the way out to the IPL, into two sublayers reminiscent of the ON and OFF substrata of the adult (Johnson *et al.*, 1999). By the time bipolar cells are born these processes retract and synaptogenesis in the OPL proceeds. This intriguing result suggests that photoreceptors, a component of the outer retina, may play a role in setting up the functional organization of the inner retina.

3.6.5 Müller glia

The timing of Müller cell genesis has long been controversial. Ultrastructural studies of the developing retina, at the time that cells are first becoming postmitotic, demonstrate profiles with radial processes (Bhattacharjee and Sanyal, 1975; Meller and Tetzlaff, 1976) and the presence of inner and outer limiting membranes (Uga and Smelser, 1973; Kuwabara and Weidman, 1974). Since adult Müller cells are radially aligned, and their endfeet form the limiting membranes, the assumption was made that the early presence of these features indicates that the cells were formed early. However, any study that has directly determined the birth of Müller cells using traditional labelling with ^{3}H-TdR or BrdU has found them to be heavily labelled only after injections late in development. Indeed, most studies show Müller glia as the last cell type in the retina to be born (Blanks and Bok, 1977; Young, 1985b; Harman and Beazley, 1989; LaVail *et al.*, 1991; Prada *et al.*, 1991). Birthdating glia is tricky since they tend to proliferate slowly and maintain the ability to divide in the mature CNS. The conflicting data could be explained by a model in which Müller cells are determined early, transiently assume an adult-like morphology, and even continue slowly dividing. Analysis of postnatal clones in the rodent retina show that, if there is a Müller glial cell in a clone, there is at least one other cell that is not a Müller glial cell. Thus, if Müller glia are dividing in the postnatal retina, they are probably acting like multipotent progenitors (Turner and Cepko, 1987). At some late point, as adult retinal architecture is achieved, they could go through a burst of rapid division and then settle into a quiescent state. On the other hand, a number of cell types, including RPCs themselves, are radially aligned during development and could form inner and outer endfeet. In the developing cerebral cortex progenitors have been shown to act as radial guides (Noctor *et al.*, 2001, 2002, 2004). Similarly, RPCs may assume a Müller-cell-like morphology and function in guiding the migration of their daughters. It is significant that a myriad of Müller-specific markers such as glial fibrillary acidic protein, glutamine synthetase, and cellular retinaldehyde-binding protein are expressed only late in retinal development (De Leeuw *et al.*, 1990; Sarthy *et al.*, 1991; Peterson *et al.*, 2001; Kuzmanovic *et al.*, 2003), after the birth of Müller glia, as defined traditionally.

3.7 Cell lineage and cell genesis

Nearly simultaneously three papers were published tracing the lineage of RPCs, showing they could produce clones of any combination of phenotypes (Turner and Cepko, 1987;

Holt *et al.*, 1988; Wetts and Fraser, 1988). They all interpreted the absence of restricted progenitors, even ones for glia, as evidence that cell fate is determined by extrinsic, non-autonomous factors. This contrasts with the conclusion of cell birthday studies in the monkey, an animal that allows high temporal resolution (LaVail *et al.*, 1991), and of slightly later published chimera studies showing sharp borders and normal composition of retinal clones labelled significantly earlier in development (Williams and Goldowitz, 1992). Eventually it was understood that, despite being multi-potential, RPCs do not generate all fates at all times. Recent models suggest that they pass through periods of competence to make different cell types (Rapaport and Dorsky, 1998; Livesey and Cepko, 2001) so that early in development a restricted repertoire of fates is available, and this changes so that different fates are available at different stages of development. The mechanism and limits of competence are not well studied for the retina. Progenitors may pass through temporally separate periods of competence for each cell type, strictly reflecting the order that they are born. Less strict periods of competence may allow more than one type to be generated simultaneously. The ubiquitous early and late phases of cell genesis may partly reflect passage from competence to make a cohort of early phenotypes to a later one. Alternatively, competence could be loose and stochastic, such that all fates are possible at all times although the probability of adopting any particular one could change. Either of the later two hypotheses can account for the overlap in the schedules of production of different cell types. In these models, phenotypes that are neighbours in the sequence of cell genesis and/or in the same phase (i.e. bipolar cells then Müller cells, Figure 3.5) might often switch birth order (i.e. there would be nearly as high a probability of generating a Müller cell before a bipolar cell as vice versa). However, there would be a much lower probability of cell types distant from each other in the sequence, perhaps in different phases, to switch ordinal positions (i.e. rod before a cone). As the interval of retinal cell birth shortens such as between macaque (100 days), rat (20 days) and frog (2 days) the most parsimonious model changes from strict competence windows to stochastic.

Recent studies have been testing the limits of competence by examining how strict the order of cell genesis is for individual progenitor cells. This analysis was performed on *Xenopus laevis* because it develops entirely externally, embryos are available at all stages and protocols have been developed to transfect DNA at these stages (Holt *et al.*, 1990; Ohnuma *et al.*, 2002). The techniques used to trace the lineage of RPCs from just before their first postmitotic division were adapted to this task. The utility of *Xenopus* as a developmental model led to numerous previous attempts to determine retinal cell birthdays (Jacobson, 1976; Holt *et al.*, 1988; Stiemke and Hollyfield, 1995). However, the ability to resolve a sequence is poor. Lineage tracing was combined with labelling of DNA synthesis to examine the order of cell birth at the resolution of individual clones. By using *Xenopus*, in which no apparent order of genesis is seen at the population level, the sequence of cell genesis, and the role of competence, was put to its most rigorous test.

Stage 20 *Xenopus* embryos were transfected with green fluorescent protein (GFP) cDNA by lipofection. Small amounts of GFP-DNA were injected to achieve low-efficiency labelling. Only 60% of the eyes of mature tadpoles (Stage 41) had labelled clones, and of these more than 82% had only one or two clones (Figure 3.6a). At various times following GFP labelling, BrdU was injected into the embryos. In an embryo this small the BrdU remains available for more than 24 hours, so the paradigm provides cumulative, not pulse labelling. For each clone the phenotype of the members that were BrdU-positive, and those BrdU-negative, was determined (Figure 3.6b). The data were analysed in two ways – one was visual and intuitive, the second attempted to quantify the labelling. Both methods led to the same, strong conclusion. The 'graphical' data are illustrated in Figure 3.6c. A random selection of 25 of the 61 clones that contained a mixture of BrdU-positive and BrdU-negative profiles are represented as rows. The rows are arranged from top to bottom to make the smoothest transition from all white circles (all cells BrdU-positive) to all black ones (Figure 3.6c, all BrdU-negative). If there is significant variation in the sequence of cell genesis within clones no ordering will produce a smooth transition. However, if the individual clones tell part of a larger cell birth sequence a pattern will emerge because clones containing early generated cell types that are BrdU-positive will be 'higher' on the stack (Figure 3.6c) than ones with later generated cells BrdU-positive. With cell types ordered horizontally according to the hypothesized sequence of cell birth an imaginary diagonal line can be drawn across the population of clones, more or less separating BrdU-positive from BrdU-negative cells (Figure 3.6c). Only one clone exhibited an order that was inconsistent with the proposed sequence of genesis. In this clone (arrows in Figure 3.6c), a horizontal cell is BrdU-positive while a cone is BrdU-negative, indicating that the cone became postmitotic before the horizontal cell. This clone might represent a rarely occurring cell birth order, or could be due to experimental error. While either possibility cannot be ruled out, anomalies occurred at very low frequency, well within the limit of experimental error.

To verify the impression conveyed by the visual ordering of cell genesis a quantitative analysis was undertaken. First, the subpopulation of clones that contained a mixture of BrdU-labelled and -unlabelled members, and in which all cases of one or more individual type was BrdU-negative, was defined. For these the per cent of the other cell types that were BrdU labelled (Figure 3.6d) was determined. If cell birth followed a specific sequence the order of phenotypes should be apparent according to their BrdU labelling. For example, some clones contained retinal ganglion cells, all of which were BrdU-negative. Determining the proportion of other cell types that were BrdU-negative and -positive in these subsets of clones allows rank ordering by percentage. The exercise is repeated for each cell type. Eventually the results will show that if ganglion cells are the first to be generated, clones with ganglion cells still dividing (BrdU-positive) must have any other cell types BrdU-positive. Further, if cones follow ganglion cells, clones in which all cones are BrdU-negative must also have all ganglion cells BrdU-negative, while any other cell type can be BrdU-positive. Finally, if Müller cells are the last cell type to be generated, clones in which all Müller cells are BrdU-negative should have all other cell types BrdU-negative. The data shown in Figure 3.6d bears out the sequence of cell genesis in *Xenopus*, one that is very similar to that seen in other vertebrates.

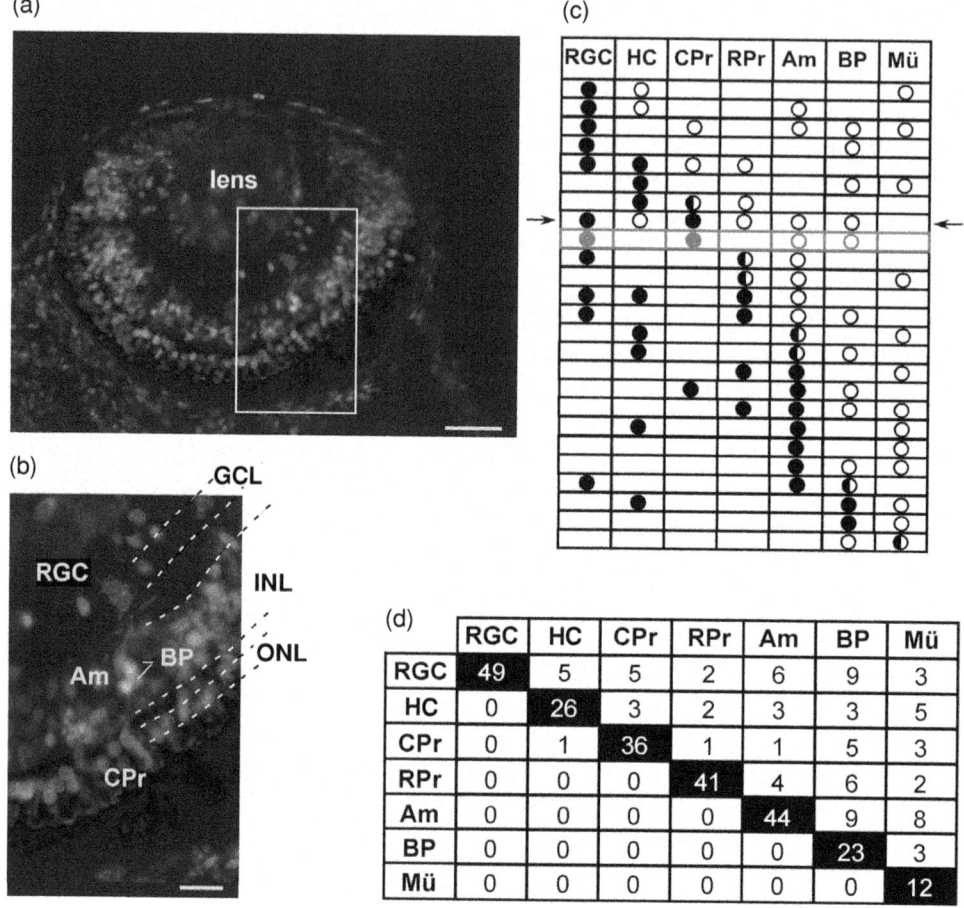

Figure 3.6 (a–b) Low- (a) and high- (b) power photomicrograph of a section through a Stage 41 *Xenopus* retina that was injected with DOTAP liposomal transfection reagent/GFP-cDNA (red immunofluoresence) at Stage 19 and with BrdU at Stage 27 (green immunofluoresence). The blue immunofluoresence demonstrates rhodopsin, a specific marker of rods. This retina had a single transfected clone consisting of five radially aligned cells. The cell in the ONL is negative for rhodopsin and is therefore a cone; it is also negative for BrdU. In the INL there are two, overlapping, bipolar cells (BP) and an amacrine cell (Am). All are BrdU-positive. The retinal ganglion cell (RGC) is BrdU-negative. It is concluded that at the time of BrdU administration the ganglion cell and cone were postmitotic and the amacrine cell and bipolars were still dividing. GCL, ganglion cell layer. Scale bar in (a) equals 50 μm, in (b) equals 25 μm. (c) A graphical representation of 25 of 61 clones containing a mix of BrdU-positive and BrdU-negative cells. Each horizontal line represents a clone and each box a cell type. The composition of clones is indicated by the presence of a circle. If the circle is white then all instances of that cell type are BrdU-positive, if black then all are BrdU-negative, and if half-and-half the BrdU labelling was heterogeneous. The clone in (a/b) is indicated by colored text. This sample shows that there is a smooth transition from all cells BrdU-positive to all cells BrdU-negative, if the cell types are arranged in the proper sequence. Of all clones analysed only one, indicated by flanking arrows, had a reversal in the order of genesis, in this instance the cone was postmitotic while the horizontal cell (HC) was still

Analysis of cell birth within clones provides data on the order of cell genesis that bear strongly on the mechanism of cell fate determination in the retina, specifically the role and limits of cell competence. The most novel and interesting result is that the order of cell genesis is rigidly fixed. Though a progenitor may or may not generate a cell of a particular phenotype during its productive life, if it does, its ordinal position relative to other types is inviolate. As a consequence, retinal ganglion cells, if present in a clone, are always born before any other cell type; cones, always before other cell types except ganglion cells, and so on through the sequence. Such precise temporal patterning is consistent with a model involving an intrinsic mechanism controlling competence to become a specific phenotype, one type at a time. The validity of the model is strengthened by being demonstrated in an animal that fails to demonstrate a sequence of cell genesis at the population level. The fact that, even though retinal cell types display a well-defined sequence of genesis when one looks at a particular developmental landmark (i.e. 5%, 50% or 95% of cells of a type generated), many, sometimes all, types are generated contemporaneously (Figure 3.5) can now be interpreted as the result of the individual progenitors being at different stages of development relative to each other. The longer the period of retinal cell birth the less variance between progenitor stages, and the more separate the curves of cell genesis (Figure 3.5). There is no gradually shifting bias to become any of a number of cell types with development, and no stochastic element to cell birth order. Further, RPCs do not become competent to produce a subset of possible types; rather progenitors are competent to make only one cell type at a time, and this competence shifts in only one direction.

3.8 Concluding remarks

Evidence demonstrates that numerous inductive cues can influence cell fate but that the target must be competent to respond to them. Originally it was assumed that postmitotic daughters were unbiased as they left the cell cycle and were directed by inducers to migrate and settle into a certain layer and adopt an appropriate phenotype. However, this does not

---

Figure 3.6 (*cont.*) dividing. CPr, cone photoreceptor, Mü, Müller glia; RPr, rod photoreceptor. (d) Table illustrating quantitative analysis of order of cell birth within individual clones in *Xenopus* retina. The diagonal line of black boxes denotes what are called the 'baseline clones' – one each for the seven cell types in the retina, in which all occurrences of that cell type are postmitotic (BrdU-negative). For example, 49 clones contain one or more ganglion cells, all of which are BrdU-negative, and 26 clones have only BrdU-negative horizontal cells. All the open boxes are called 'comparison clones'. They represent the other cell types in the baseline clones and indicate the number of that phenotype that are still dividing (i.e. BrdU-positive). For example, of the 41 rod baseline clones there are none containing BrdU-positive ganglion cells, horizontal cells or cones, and 4, 6, and 2 containing BrdU-positive amacrine cells, bipolar cells and Müller glia. When the phenotypes are ordered horizontally and vertically in the sequence of cell genesis all comparison clones to the left of the baseline diagonal, but one, contain no BrdU-positive cells. The one exception represents the clone indicated by the arrow in (c). All the comparisons to the right of the diagonal contain at least one BrdU-positive cell and they tend to be ordered from low to high. This is not strict since not all clones contain all cell types, but within clones the order is strict. For colour version, see Plate 3.

accord well with observations showing the expression of cell-specific markers immediately upon exiting M-phase, perhaps even before (Barnstable *et al.*, 1985; Waid and McLoon, 1995). Transplanted cortical progenitors can change their fate, but only if they undergo S-phase in the host environment (McConnell, 1988; McConnell and Kaznowski, 1991). The implication is that inductive cues act, not on naïve postmitotic cells, but on the progenitors themselves, such that at the time a daughter leaves the cell cycle its fate is determined. The discovery that retinal progenitors are generated in a rigid sequence suggests that there are competence mechanisms for each cell type.

As models of cell fate determination move the site of action from postmitotic daughters to progenitors there is increasing interest in interactions between genes that control the cell cycle and those influencing cell fate (Ohnuma *et al.*, 2001; Ohnuma and Harris, 2003). Experiments show Notch–Delta signalling to be part of the competence mechanism of retinal progenitors (Austin *et al.*, 1995, Dorsky *et al.*, 1995, 1997) and also to be involved in signalling Müller glial cell fate (Bao and Cepko, 1997; Ohnuma *et al.*, 1999; Furukawa *et al.*, 2000). More generally it has been suggested (Vetter and Moore, 2001) that signals terminating proliferation of a neural progenitor also signal the daughters to become glia. Therefore glial cells are the last cell type generated in most areas of the CNS. Other interactions between cell cycle and cell fate need to be explored. Finally, cell fate determinants may interact with the rate of progress though the cell cycle, orientation of cell cleavage and kinetics of cell division and in this way influence the number of cells of particular phenotype(s) being produced at any time in retinal development. Multiple mechanisms likely interact to ensure that the right cell types are formed at the right time and place and in the right quantity to allow the formation of functional retinal circuits.

References

Alexiades, M. R. and Cepko, C. (1996). Quantitative analysis of proliferation and cell cycle length during development of the rat retina. *Dev. Dyn.*, **205**, 293–307.

Austin, C. P., Feldman, D. E., Ida, J. A. J. and Cepko, C. L. (1995). Vertebrate retinal ganglion cells are selected from competent progenitors by the action of Notch. *Development*, **121**, 3637–50.

Bao, Z. Z. and Cepko, C. L. (1997). The expression and function of Notch pathway genes in the developing rat eye. *J. Neurosci.*, **17**, 1425–34.

Barnstable, C. J., Hofstein, R. and Akagawa, K. (1985). A marker of early amacrine cell development in rat retina. *Dev. Brain Res.*, **20**, 286–90.

Bhattacharjee, J. and Sanyal, S. (1975). Developmental origin and early differentiation of retinal Müller cells in mice. *J. Anat.*, **120**, 367–72.

Blanks, J. C. and Bok, D. (1977). An autoradiographic analysis of postnatal cell proliferation in the normal and degenerative mouse retina. *J. Comp. Neurol.*, **174**, 317–28.

Blenkinsopp, W. K. (1968). Duration of availability of tritiated thymidine following intraperitoneal injection. *J. Cell Sci.*, **3**, 89–95.

Bumsted, K., Jasoni, C., Szel, A. and Hendrickson, A. (1997). Spatial and temporal expression of cone opsins during monkey retinal development. *J. Comp. Neurol.*, **378**, 117–34.

Carter-Dawson, L. D. and LaVail, M. M. (1979). Rods and cones in the mouse retina. II. Autoradiographic analysis of cell generation using tritiated thymidine. *J. Comp. Neurol.*, **188**, 263–72.

Carter-Dawson, L., Alvarez, R. A., Fong, S. L. et al. (1986). Rhodopsin, 11-cis vitamin A, and interstitial retinol-binding protein (IRBP) during retinal development in normal and rd mutant mice. *Dev. Biol.*, **116**, 431–8.

Cayouette, M. and Raff, M. (2003). The orientation of cell division influences cell-fate choice in the developing mammalian retina. *Development*, **130**, 2329–39.

Cayouette, M., Whitmore, A. V., Jeffery, G. and Raff, M. (2001). Asymmetric segregation of Numb in retinal development and the influence of the pigmented epithelium. *J. Neurosci.*, **21**, 5643–51.

Chang, W. S and Harris, W. A (1998). Sequential genesis and determination of cone and rod photoreceptors in Xenopus. *J. Neurobiol.*, **35**, 227–44.

Chinen, A., Hamaoka, T., Yamada, Y. and Kawamura, S. (2003). Gene duplication and spectral diversification of cone visual pigments of zebrafish. *Genetics*, **163**, 663–75.

Das, T., Payer, B., Cayouette, M. and Harris, W. A. (2003). In vivo time-lapse imaging of cell divisions during neurogenesis in the developing zebrafish retina. *Neuron*, **37**, 597–609.

De Leeuw, A. M., Gaur, V. P., Saari, J. C. and Milam, A. H. (1990). Immunolocalization of cellular retinol-, retinaldehyde- and retinoic acid-binding proteins in rat retina during pre- and postnatal development. *J. Neurocytol.*, **19**, 253–64.

Denham, S. (1967). A cell proliferation study of the neural retina in the two-day rat. *J. Embryol. Exp. Morphol.*, **18**, 53–66.

Dorsky, R. I., Rapaport, D. H. and Harris, W. A. (1995). Xotch inhibits cell differentiation in the Xenopus retina. *Neuron*, **14**, 487–96.

Dorsky, R. I., Chang, W. S., Rapaport, D. H. and Harris, W. A. (1997). Regulation of neuronal diversity in the Xenopus retina by Delta signaling. *Nature*, **385**, 67–70.

Dräger, U. C. (1985). Birth dates of ganglion cells giving rise to crossed and uncrossed retinal projections in the mouse. *Proc. R. Soc. London B*, **224**, 57–77.

Evans, J. A. and Battelle, B.-A. (1987). Histogenesis of dopamine-containing neurons in the rat retina. *Exp. Eye Res.*, **44**, 407–14.

French, M. B., Koch, U., Shaye, R. E. et al. (2002). Transgenic expression of numb inhibits notch signaling in immature thymocytes but does not alter T cell fate specification. *J. Immunol.*, **168**, 3173–80.

Fujita, S. (1962). Kinetics of cellular proliferation. *Exp. Cell Res.*, **28**, 52–60.

Fujita, S. and Horii, M. (1963). Analysis of cytogenesis in chick retina by tritiated thymidine autoradiography. *Arch. Histol. Jpn.*, **23**, 359–66.

Fukuda, Y., Sawai, H., Watanabe, M., Wakakuwa, K. and Morigiwa, K. (1989). Nasotemporal overlap of crossed and uncrossed retinal ganglion cell projections in the Japanese monkey (*Macaca fuscata*). *J. Neurosci.*, **9**, 2353–73.

Furukawa, T., Mukherjee, S., Bao, Z. Z., Morrow, E. M. and Cepko, C. L. (2000). rax, Hes1, and notch1 promote the formation of Muller glia by postnatal retinal progenitor cells. *Neuron*, **26**, 383–94.

Galli-Resta, L. and Ensini, M. (1996). An intrinsic time limit between genesis and death of individual neurons in the developing retinal ganglion cell layer. *J. Neurosci.*, **16**, 2318–24.

Gardino, P. F., dos Santos, R. M. and Hokoc, J. N. (1993). Histogenesis and topographical distribution of tyrosine hydroxylase immunoreactive amacrine cells in the developing chick retina. *Brain Res. Dev. Brain Res.*, **72**, 226–36.

Gloor, B. P., Rokos, L. and Kaldarar-Pedotti, S. (1985). Cell cycle time and life-span of cells in the mouse eye. *Dev. Ophthalmol.*, **12**, 70–129.

Gonzalez-Fernandez, F. and Healy, J. I. (1990). Early expression of the gene for interphotoreceptor retinol-binding protein during photoreceptor differentiation suggests a critical role for the interphotoreceptor matrix in retinal development. *J. Cell. Biol.*, **111**, 2775–84.

Hamburger, V. and Hamilton, H. C. (1951). A series of normal stages in the development of the chick embryo. *J. Morphol.*, **88**, 49–92.

Harman, A. M. and Beazley, L. D. (1989). Generation of retinal cells in the wallaby, *Setonix brachyurus* (quokka). *Neuroscience*, **28**, 219–32.

Harris, W. A. and Messersmith, S. L. (1992). Two cellular inductions involved in photoreceptor determination in Xenopus retina. *Neuron*, **9**, 357–72.

Hauswirth, W. W., Langerijt, A. V., Timmers, A. M., Adamus, G. and Ulshafer, R. J. (1992). Early expression and localization of rhodopsin and interphotoreceptor retinoid-binding protein (IRBP) in the developing fetal bovine retina. *Exp. Eye Res.*, **54**, 661–70.

Hickey, T. L., Whikehart, D. R., Jackson, C. A., Hitchcock, P. F. and Paduzzi, J. D. (1983). Tritiated thymidine experiments in the cat: a description of techniques and experiments to define the time-course of radioactive thymidine availability. *J. Neurosci. Methods*, **8**, 139–47.

Hollyfield, J. G. (1972). Histogenesis of the retina in the killifish *Fundulus heteroclitus*. *J. Comp. Neurol.*, **144**, 373–80.

Holt, C. E., Bertsch, T. W., Ellis, H. M. and Harris, W. A. (1988). Cellular determination in the Xenopus retina is independent of lineage and birth date. *Neuron*, **1**, 15–26.

Holt, C. E., Garlick, N. and Cornel, E. (1990). Lipofectin of cDNAs in the embryonic vertebrate central nervous system. *Neuron*, **4**, 203–14.

Hu, M. and Easter, S. S. (1999). Retinal neurogenesis: the formation of the initial central patch of post-mitotic cells. *Dev. Biol.*, **207**, 309–21.

Huang, B., Mitchell, C. K. and Redburn-Johnson, D. A. (2000). GABA and GABA(A) receptor antagonists alter developing cone photoreceptor development in neonatal rabbit retina. *Vis. Neurosci.*, **17**, 925–35.

Jacobson, M. (1976). Histogenesis of retina in the clawed frog with implications for the pattern of development of retinotectal connections. *Brain Res.*, **103**, 541–5.

Jacobson, M. (1978). *Developmental Neurobiology*. New York, London: Plenum Press.

Jeffery, G. (1985). The relationship between cell density and the nasotemporal division in the rat retina. *Brain Res.*, **347**, 354–7.

Johnson, P. T., Williams, R. R., Cusato, K. and Reese, B. E. (1999). Rods and cones project to the inner plexiform layer during development. *J. Comp. Neurol.*, **414**, 1–12.

Johnson, P. T., Williams, R. R. and Reese, B. E. (2001). Developmental patterns of protein expression in photoreceptors implicate distinct environmental versus cell-intrinsic mechanisms. *Vis. Neurosci.*, **18**, 157–68.

Kahn, A. J. (1974). An autoradiographic analysis of the time of appearance of neurons in the developing chick neural retina. *Dev. Biol.*, **18**, 163–79.

Kay, J. N., Finger-Baier, K. C., Roeser, T., Staub, W. and Baier, H. (2001). Retinal ganglion cell genesis requires lakritz, a Zebrafish atonal Homolog. *Neuron*, **30**, 725–36.

Kuwabara, T. and Weidman, T. A. (1974). Development of the prenatal rat retina. *Invest. Ophthalmol. Vis. Sci.*, **13**, 725–739.

Kuzmanovic, M., Dudley, V. J. and Sarthy, V. P. (2003). GFAP promoter drives Müller cell-specific expression in transgenic mice. *Invest. Ophthalmol. Vis. Sci.*, **44**, 3606–13.

Larison, K. D. and Bremiller, R. (1990). Early onset of phenotype and cell patterning in the embryonic zebrafish retina. *Development*, **109**, 567–76.

LaVail, M. M., Rapaport, D. H. and Rakic, P. (1991). Cytogenesis in the monkey retina. *J. Comp. Neurol.*, **309**, 86–114.

Lee, M. Y., Shin, S. L., Han, S. H. and Chun, M. H. (1999). The birthdates of GABA-immunoreactive amacrine cells in the rat retina. *Exp. Brain Res.*, **128**, 309–14.

Li, Z., Hu, M., Ochocinska, M. J., Joseph, N. M. and Easter, S. S., Jr. (2000). Modulation of cell proliferation in the embryonic retina of zebrafish (*Danio rerio*). *Dev. Dyn.*, **219**, 391–401.

Livesey, F. J. and Cepko, C. L. (2001). Vertebrate neural cell-fate determination: lessons from the retina. *Nat. Rev. Neurosci.*, **2**, 109–18.

Masland, R. H. (2001). The fundamental plan of the retina. *Nat. Neurosci.*, **4**, 877–86.

Masland, R. H. and Tauchi, M. (1986). The cholinergic amacrine cell. *Trends Neurosci.*, **9**, 218–23.

McConnell, S. K. (1988). Fates of visual cortical neurons in the ferret after isochronic and heterochronic transplantation. *J. Neurosci.*, **8**, 945–74.

McConnell, S. K. and Kaznowski, C. E. (1991). Cell cycle dependence of laminar determination in developing neocortex. *Science*, **254**, 282–5.

Meller, K. and Tetzlaff, W. (1976). Scanning electron microscopic studies on the development of the chick retina. *Cell Tissue Res.*, **170**, 145–59.

Messersmith, E. K. and Redburn, D. A. (1993). The role of GABA during development of the outer retina in the rabbit. *Neurochem. Res.*, **18**, 463–70.

Mitchell, C. K., Rowe-Rendleman, C. L., Ashraf, S. and Redburn, D. A. (1995). Calbindin immunoreactivity of horizontal cells in the developing rabbit retina. *Exp. Eye Res.*, **61**, 691–8.

Mitchell, C. K., Huang, B. and Redburn-Johnson, D. A. (1999). GABA(A) receptor immunoreactivity is transiently expressed in the developing outer retina. *Vis. Neurosci.*, **16**, 1083–8.

Nawrocki, L., BreMiller, R., Streisinger, G. and Kaplan, M. (1985). Larval and adult visual pigments of the zebrafish, *Brachydanio rerio*. *Vis. Res.*, **25**, 1569–76.

Neumann, C. J. and Nuesslein-Volhard, C. (2000). Patterning of the zebrafish retina by a wave of sonic hedgehog activity. *Science*, **289**, 2137–9.

Nieuwkoop, P. D. and Faber, J. (1956). *Normal Table of Xenopus laevis (Daudin)*. Amsterdam: North-Holland Publishing Co.

Noctor, S. C., Flint, A. C., Weissman, T. A., Dammerman, R. S. and Kriegstein, A. R. (2001). Neurons derived from radial glial cells establish radial units in neocortex. *Nature*, **409**, 714–20.

Noctor, S. C., Flint, A. C., Weissman, T. A. *et al.* (2002). Dividing precursor cells of the embryonic cortical ventricular zone have morphological and molecular characteristics of radial glia. *J. Neurosci.*, **22**, 3161–73.

Noctor, S. C., Martinez-Cerdeno, V., Ivic, L. and Kriegstein, A. R. (2004). Cortical neurons arise in symmetric and asymmetric division zones and migrate through specific phases. *Nat. Neurosci.*, **7**, 136–44.

Nowakowski, R. S. and Rakic, P. (1974). Clearance rate of exogenous 3H-thymidine from the plasma of pregnant rhesus monkeys. *Cell Tissue Kinet.*, **7**, 189–94.

Ohnuma, S. and Harris, W. A. (2003). Neurogenesis and the cell cycle. *Neuron*, **40**, 199–208.

Ohnuma, S.-I., Philpott, A., Wang, K., Holt, C. E. and Harris, W. A. (1999). p27Xic1, a Cdk inhibitor, promotes the determination of glial cells in Xenopus retina. *Cell*, **99**, 499–510.

Ohnuma, S., Philpott, A. and Harris, W. A. (2001). Cell cycle and cell fate in the nervous system. *Curr. Opin. Neurobiol.*, **11**, 66–73.

Ohnuma, S., Mann, F., Boy, S., Perron, M. and Harris, W. A. (2002). Lipofection strategy for the study of Xenopus retinal development. *Methods*, **28**, 411–19.

Olney, J. W. (1968). An electron microscopic study of synapse formation, receptor outer segment development, and other aspects of developing mouse retina. *Invest. Ophthalmol.*, **7**, 250–68.

Perry, V. H. and Silveira, L. C. L. (1988). Functional lamination in the ganglion cell layer of the macaque's retina. *Neuroscience*, **25**, 217–23.

Peterson, R. E., Fadool, J. M., McClintock, J. and Linser, P. J. (2001). Müller cell differentiation in the zebrafish neural retina: evidence of distinct early and late stages in cell maturation. *J. Comp. Neurol.*, **429**, 530–40.

Prada, C., Puga, J., Pérez-Méndez, L., López, R. and Ramirez, G. (1991). Spatial and temporal patterns of neurogenesis in the chick retina. *Eur. J. Neurosci.*, **3**, 559–69.

Provis, J. and Watson, C. R. (1981). The distributions of ipsilaterally and contralaterally projecting ganglion cells in the retina of the pigmented rabbit. *Exp. Brain. Res.*, **44**, 82–92.

Rakic, P. (1981). Neuronal-glial interaction during brain development. *Trends Neurosci.*, **4**, 184–7.

Rapaport, D. H. and Dorsky, R. I. (1998). Inductive competence, its significance in retinal cell fate determination and a role for Delta-Notch signaling. *Semin. Cell Dev. Biol.*, **9**, 241–7.

Rapaport, D. H. and Stone, J. (1983). The topography of cytogenesis in the developing retina of the cat. *J. Neurosci.*, **3**, 1824–34.

Rapaport, D. H. and Vietri, A. (1991). Identity of cells produced by two stages of cytogenesis in the postnatal cat retina. *J. Comp. Neurol.*, **312**, 341–52.

Rapaport, D. H., Fletcher, J., LaVail, M. M. and Rakic, P. (1992). Genesis of neurons in the retinal ganglion cell layer of the monkey. *J. Comp. Neurol.*, **322**, 577–88.

Rapaport, D. H., LaVail, M. M. and Rakic, P. (1996). Spatiotemporal gradients of cell genesis in the monkey retina. *Perspect. Dev. Neurobiol.*, **3**, 147–60.

Rapaport, D. H., Wong, L. L., Wood, E. D., Yasumura, D. and LaVail, M. M. (2004). Timing and topography of cell genesis in the rat retina. *J. Comp. Neurol.*, **474**, 304–24.

Redburn, D. A. and Madtes, P., Jr. (1986). Postnatal development of 3H-GABA-accumulating cells in rabbit retina. *J. Comp. Neurol.*, **243**, 41–57.

Reese, B. E. and Colello, R. J. (1992). Neurogenesis in the retinal ganglion cell layer of the rat. *Neuroscience*, **46**, 419–29.

Reese, B. E., Guillery, R. W. and Mallarino, C. (1992). Time of ganglion cell genesis in relation to the chiasmatic pathway choice of retinofugal axons. *J. Comp. Neurol.*, **324**, 336–42.

Reese, B. E., Thompson, W. F. and Peduzzi, J. D. (1994). Birthdates of neurons in the retinal ganglion cell layer of the ferret. *J. Comp. Neurol.*, **341**, 464–75.

Reh, T. A. (1989). The regulation of neuronal production during retinal neurogenesis. In *Development of the Vertebrate Retina*, ed, B. L. Finlay and D. R. Sengelaub. New York: Plenum, pp. 43–67.

Rodieck, R. W. and Marshak, D. W. (1992). Spatial density and distribution of choline acetyltransferase immunoreactive cells in human, macaque, and baboon retinae. *J. Comp. Neurol.*, **321**, 46–64.

Sarthy, P. V., Fu, M. and Huang, J. (1991). Developmental expression of the *glial fibrillary acidic protein (GFAP)* gene in the mouse retina. *Cell. Mol. Neurobiol.*, **11**, 623–37.

Sauer, M. E. and Chittenden, A. C. (1959). Deoxyribonucleic acid content of cell nuclei in the neural tube of the chick embryo: evidence for interkinetic migration of nuclei. *Exp. Cell Res.*, **16**, 1–6.

Sengelaub, D. R., Dolan, R. P. and Finlay, B. L. (1986). Cell generation, death, and retinal growth in the development of the hamster retinal ganglion cell layer. *J. Comp. Neurol.*, **246**, 527–43.

Sharma, S. C. and Ungar, F. (1980). Histogenesis of the goldfish retina. *J. Comp. Neurol.*, **191**, 373–82.

Sidman, R. L. (1961). Histogenesis of the mouse retina studied with thymidine 3-H. In *The Structure of the Eye*, ed. G. K. Smelser. New York: Academic Press, pp. 487–506.

Sidman, R. L. (1970). *Cell Proliferation, Migration, and Interaction in the Developing Mammalian Central System*. New York: The Rockefeller University Press.

Silva, A. O., Ercole, C. E. and McLoon, S. C. (2002). Plane of cell cleavage and numb distribution during cell division relative to cell differentiation in the developing retina. *J. Neurosci.*, **22**, 7518–25.

Stenkamp, D. L., Hisatomi, O., Barthel, L. K., Tokunaga, F. and Raymond, P. A. (1996). Temporal expression of rod and cone opsins in embryonic goldfish retina predicts the spatial organization of the cone mosaic. *Invest. Ophthalmol. Vis. Sci.*, **37**, 363–76.

Stenkamp, D. L., Barthel, L. K. and Raymond, P. A. (1997). Spatiotemporal coordination of rod and cone photoreceptor differentiation in goldfish retina. *J. Comp. Neurol.*, **382**, 272–84.

Stiemke, M. M. and Hollyfield, J. G. (1995). Cell birthdays in *Xenopus laevis* retina. *Differentiation*, **58**, 189–93.

Stone, J. (1966). The naso-temporal division of the cat's retina. *J. Comp. Neurol.*, **126**, 585–600.

Stone, J. (1983). *Parallel Processing in the Visual System: The Classification of Retinal Ganglion Cells and its Impact on the Neurobiology of Vision*. New York, London: Plenum Press.

Stone, J., Leicester, J. and Sherman, S. (1973). The nasotemporal division of the the monkey's retina. *J. Comp. Neurol.*, **150**, 333–48.

Strettoi, E. and Masland, R. H. (1995). The organization of the inner nuclear layer of the rabbit retina. *J. Neurosci.*, **15**, 875–88.

Tibber, M. S., Kralj-Hans, I., Savage, J., Mobbs, P. G. and Jeffery, G. (2004). The orientation and dynamics of cell division within the plane of the developing vertebrate retina. *Eur. J. Neurosci.*, **19**, 497–504.

Turner, D. L. and Cepko, C. L. (1987). A common progenitor for neurons and glia persists in rat retina late in development. *Nature*, **328**, 131–6.

Uga, S. and Smelser, G. K. (1973). Electron microscopic study of the development of retinal Müllerian cells. *Invest. Ophthalmol. Vis. Sci.*, **12**, 295–307.

Vaney, D. I. (1990). The mosaic of amacrine cells in the mammalian retina. In *Progress in Retinal Research*, ed. N. N. Osborne and G. J., Chader, Oxford: Pergamon Press, pp. 50–100.

Vaney, D. I., Peichl, L. and Boycott, B. B. (1981). Matching populations of amacrine cells in the inner nuclear layer and ganglion cell layers of the rabbit retina. *J. Comp. Neurol.*, **199**, 373–91.

Versaux-Botteri, C., Pochet, R. and Nguyen-Legros, J. (1989). Immunohistochemical localization of GABA-containing neurons during postnatal development of the rat retina. *Invest. Ophthalmol. Vis. Sci.*, **30**, 652–9.

Vetter, M. L. and Moore, K. B. (2001). Becoming glial in the neural retina. *Dev. Dyn.*, **221**, 146–53.

Vihtelic, T. S., Doro, C. J. and Hyde, D. R. (1999). Cloning and characterization of six zebrafish photoreceptor opsin cDNAs and immunolocalization of their corresponding proteins. *Vis. Neurosci.*, **16**, 571–85.

Waid, D. K. and McLoon, S. C. (1995). Immediate differentiation of ganglion cells following mitosis in the developing retina. *Neuron*, **14**, 117–24.

Wakamatsu, Y., Maynard, T. M., Jones, S. U. and Weston, J. A. (1999). NUMB localizes in the basal cortex of mitotic avian neuroepithelial cells and modulates neuronal differentiation by binding to NOTCH-1. *Neuron*, **23**, 71–81.

Walsh, C. and Polley, E. H. (1985). The topography of ganglion cell production in the cat's retina. *J. Neurosci.*, **5**, 741–50.

Walsh, C., Polley, E. H., Hickey, T. L. and Guillery, R. W. (1983). Generation of cat retinal ganglion cells in relation to central pathways. *Nature*, **302**, 611–14.

Weidman, T. A. and Kuwabara, T. (1968). Postnatal development of the rat retina. An electron microscopic study. *Arch. Ophthalmol.*, **79**, 470–84.

Wetts, R. and Fraser, S. E. (1988). Multipotent precursors can give rise to all major cell types of the frog retina. *Science*, **239**, 1142–5.

Wikler, K. C., Perez, G. and Finlay, B. L. (1989). Duration of retinogenesis: its relationship to retinal organization in two cricetine rodents. *J. Comp. Neurol.*, **285**, 157–76.

Williams, R. W. and Goldowitz, D. (1992). Structure of clonal and polyclonal cell arrays in chimeric mouse retina. *Proc. Natl. Acad. Sci. U.S.A.*, **89**, 1184–8.

Young, R. W. (1985a). Cell proliferation during postnatal development of the retina in the mouse. *Brain Res.*, **353**, 229–39.

Young, R. W. (1985b). Cell differentiation in the retina of the mouse. *Anat. Rec.*, **212**, 199–205.

Zhang, D. R. and Yeh, H. H. (1990). Histogenesis of corticotropin releasing factor-like immunoreactive amacrine cells in the rat retina. *Brain Res. Dev. Brain Res.*, **53**, 194–9.

Zhong, W., Feder, J. N., Jiang, M. M. Jan, L. Y. and Jan, Y. N. (1996). Asymmetric localization of a mammalian numb homolog during mouse cortical neurogenesis. *Neuron*, **17**, 43–53.

Zimmerman, R. P., Polley, E. H. and Fortney, R. L. (1988). Cell birthdays and rate of differentiation of ganglion cells of the developing cat's retina. *J. Comp. Neurol.*, **274**, 77–90.

4
Cell migration

Leanne Godinho
Washington University School of Medicine, St Louis, USA
(Currently at Harvard University, Cambridge, USA)

Brian Link
Medical College of Wisconsin, Milwaukee, USA

4.1 Introduction

Like most parts of the CNS, retinal cells are generated some distance from where they will ultimately reside. Migration to the correct place at the right time is vital for their ability to make appropriate synaptic connections and function normally. Understanding how each of the seven retinal cell types migrate to their appropriate layer is critical to understanding how this CNS structure becomes organized during development.

The entire retinal neuroepithelium is a proliferative zone early in development. Retinal neuroepithelial cells with cytoplasmic processes that extend from the outer limiting membrane (OLM) to the inner limiting membrane (ILM) engage in interkinetic nuclear migration, a process by which their nuclei migrate within the cytoplasm, undergoing different phases of the cell cycle at different depths within the neuroepithelium (see Chapter 3). Thus, neuroepithelial cells in S-phase have their nuclei positioned near the ILM, and they enter M-phase at the OLM. Consequently, following a final mitotic divison, when cells leave the cell cycle they do so adjacent to the OLM. Newborn postmitotic cells therefore need to migrate varying distances to take up residence in one of the three prospective cellular layers. Cells destined for the ganglion cell layer (GCL), for example, have comparatively longer distances to travel than rod and cone photoreceptors. Birthdating studies in diverse species have shown that the first cohort of cells to become postmitotic are ganglion cells (Prada *et al.*, 1991; Rapaport *et al.*, 1996, 2004; Hu and Easter, 1999) (see Chapter 3). The GCL is the earliest detectable layer in the developing retina. Thus, in addition to being the first generated cell type, ganglion cells are the first cohort of cells to complete their migration. The appearance of the cellular layers proceeds from the ILM to the OLM. Thus, the appearance of the GCL is followed by the inner nuclear layer (INL), and subsequently the outer nuclear layer (ONL). However, cells destined for each of these layers are not strictly generated in a sequence that reflects this organization. Instead, cells destined for different layers are often generated concurrently. For example, following ganglion cell generation, progenitors fated to become amacrine, horizontal and cone photoreceptor cells become postmitotic, destined

Retinal Development, ed. Evelyne Sernagor, Stephen Eglen, Bill Harris and Rachel Wong.
Published by Cambridge University Press. © Cambridge University Press 2006.

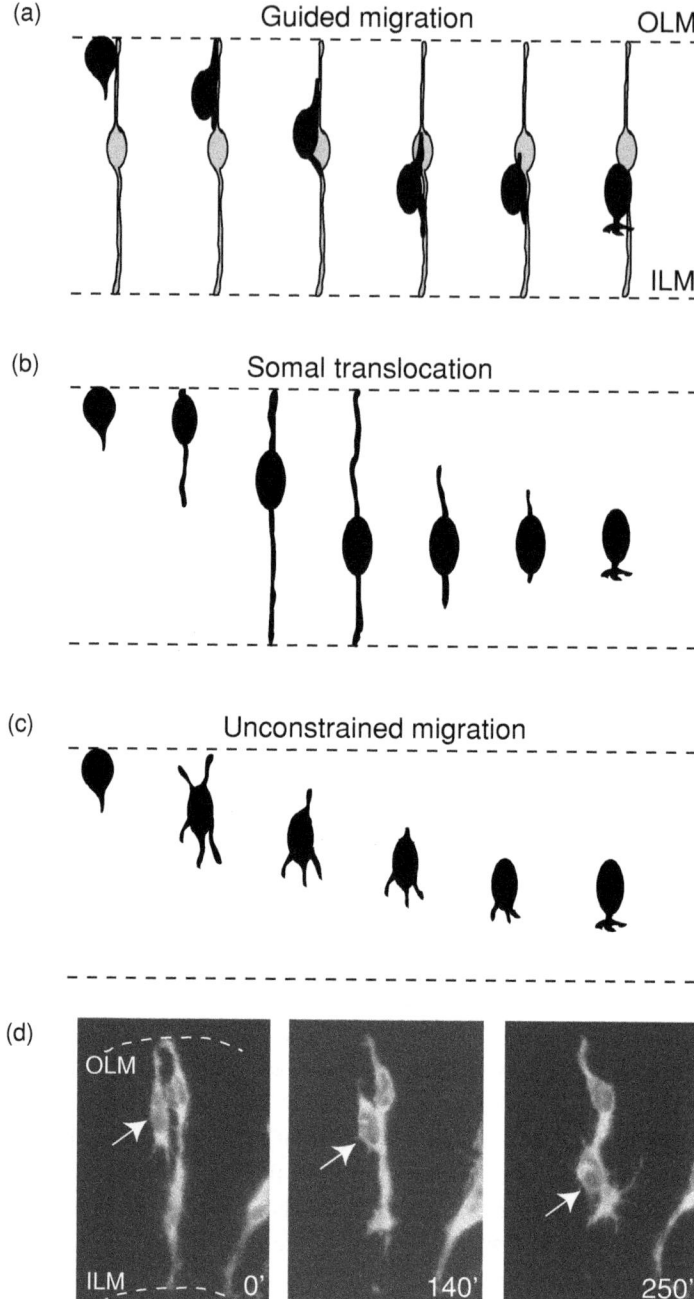

Figure 4.1 Proposed migratory modes for a retinal cell. (a) Guided migration: having left the cell cycle at the OLM a postmitotic retinal cell (black) might use a cellular substrate (grey) as a scaffold to migrate upon until it reaches a depth in the retina appropriate for its phenotype. The identity of the cellular substrate as Müller glial cells remains contentious. Cycling neuroepithelial cells, with cytoplasmic processes spanning the depth of the retina from the OLM to the ILM might be suitable

for the INL and ONL. This is followed by the generation of bipolar cells and Müller glial cells. Rod photoreceptor cells are generated over an extended developmental time, overlapping with the periods of cytogenesis of most other cell types (Sidman, 1961; Carter-Dawson and LaVail, 1979; Young, 1985a,b).

How do retinal cells that are generated concurrently migrate appropriately to different layers? Clues about the cellular mechanisms newborn retinal cells use to migrate to their definitive positions come largely from morphological observations at progressive stages of development. In addition, gene expression and genetic studies are beginning to shed light on the molecular machinery that directs cell migration and positioning in the retina.

4.2 Cellular mechanisms of migration

Drawing from static observations in the retina and from studies of migration in the neocortex, the CNS structure morphologically most analogous to the retina, three cellular mechanisms have been proposed to account for the way in which newborn retinal cells migrate to their final positions: glial-guided migration, somal translocation and unconstrained migration.

4.2.1 Glial-guided migration

Postmitotic retinal neurons have been suggested to migrate to their definitive positions along Müller glial cells. Like radial glial cells that span the thickness of the neocortex and have been shown to mediate neuronal migration, Müller glial cells span the retinal neuroepithelium and have therefore been regarded as good candidates to provide a scaffold for retinal cell migration (Meller and Tetzlaff, 1976; Wolburg et al., 1991) (Figure 4.1a). Radial glial cells are present at the earliest time-points of cortical development; their somata lie in the ventricular zone and their processes span the entire thickness of the developing cerebral wall. Thus, temporally and spatially, radial glial cells are good candidates for a guidance role. In addition, electron microscopy studies suggest a close apposition between postmitotic neurons and radial glial fibres (Rakic, 1972; Gadisseux et al., 1990; Misson et al., 1991) and observations of this apposition both in vivo (Miyata et al., 2001; Noctor

---

Figure 4.1 (cont.) candidates for such a role. (b) Somal translocation: in this mode of migration, a retinal cell might extend a cytoplasmic process basally until it reaches the ILM. It then translocates its cell body within this process until it reaches the appropriate lamina, while still being anchored at the OLM. Upon arriving at its destination, the cell would retract its apically and basally directed processes. (c) Unconstrained migration: with no attachments to the apical or basal surfaces, cells may migrate to their proper position using their neurites to explore the environment along the way. (d) Retinal cell exhibiting unconstrained migration in zebrafish retina. A presumed postmitotic retinal cell (arrow) labelled using an α-tubulin promoter driving green fluorescent protein (GFP) was followed by time-lapse confocal microscopy in vivo. With no cytoplasmic attachments to either the OLM or ILM, the cell moved vitreally, extending highly motile neurites as it did so. The GFP-labelled cell to the right appeared to lose its cytoplasmic attachment to the ILM (140', 250') while maintaining an attachment to the OLM. (L. Godinho and J. S. Mumm, unpublished images.)

et al., 2001) and in vitro (Edmondson and Hatten, 1987; O'Rourke et al., 1992; Anton et al., 1996) lend strong support for radial glial-guided migration. Time-lapse observations of DiI-labelled cortical neurons in slice cultures undergoing glial-guided migration revealed cells with short leading and trailing cytoplasmic processes displaying saltatory movement (Edmondson and Hatten, 1987; Nadarajah et al., 2001).

Birthdating studies suggest that Müller cells are generated late during the period of cell genesis, after many other retinal cells, including amacrine and ganglion cells, have already migrated to their appropriate layer. It therefore seems unlikely that Müller cells provide a guidance role, at least at the earliest time-points in development. Immunoreactivity for vimentin, a glia-specific antigen, and electron-microscopic observations were both used to suggest that radial glial cells are present in the immature retina, earlier than suggested by birthdating studies (Meller and Tetzlaff, 1976; Wolburg et al., 1991). However, vimentin may not be a specific marker for glial cells early in development and electron microscopy may not indisputably identify glial cells (Lemmon and Rieser, 1983; Bennett and DiLullo, 1985). As an alternative to Müller glia, mitotically active neuroepithelial cells, with their cytoplasmic processes contacting both ends of the epithelium, have also been proposed as candidates to act as guide posts for migrating cells (Malicki, 2004) (Figure 4.1a). However, evidence in support of a direct role for both Müller glia or cycling neuroepithelial cells in mediating migration is lacking.

4.2.2 Somal translocation

Translocation of their somata is another means by which retinal cells have been proposed to migrate to the appropriate laminar position (Morest, 1970; Snow and Robson, 1994, 1995). In this scenario, a newly postmitotic cell at the apical surface extends a cytoplasmic process towards the ILM, while maintaining contact with the OLM. The cell body then translocates within this process, until it reaches a position in the depth of the retina appropriate for its phenotype after which it retracts its cytoplasmic extensions from both the apical and basal sides of the neuroepithelium (Figure 4.1b). Evidence for this mode of migration in the retina comes from studies of the morphology of retinal neuroblasts by Golgi impregnations (Morest, 1970; Prada et al., 1981). Somal translocation is further supported by observations of ganglion cell morphology at progressive developmental stages. Taking advantage of the fact that ganglion cells extend an axon soon after becoming postmitotic, several studies used applications of the carbocyanine dye DiI (Snow and Robson, 1994, 1995) or horseradish peroxidase (Dunlop, 1990) onto the optic nerve to retrogradely label these cells. At early developmental ages ganglion cells with a bipolar morphology were found. Cell bodies were found at various depths within the developing neuroepithelium, with an axon extending toward the ILM and a cytoplasmic process attached to the apical surface. Apically directed cytoplasmic processes persisted even when ganglion cell somata reached their prospective final destination in the inner retina. However, these processes were subsequently lost when dendritic growth was initiated. Similar observations were made when ganglion cell-specific antibodies (McLoon and Barnes, 1989) and other immunohistochemical markers (Brittis *et al.*, 1995) were used to label ganglion cells during development.

The first hints of somal translocation in the cortex came from morphological observations during development (Morest, 1970; Brittis *et al.*, 1995). Time-lapse experiments in mouse embryonic brain slices permitted a direct observation of this migratory mode (Nadarajah *et al.*, 2001). Cortical cells undergoing somal translocation have a process extending to the pial surface while still maintaining an attachment in the ventricular zone. As the nucleus moves through the cytoplasm towards the pia, the leading process shortens and the apical cytoplasmic attachment to the ventricular surface is gradually lost. Compared with cells engaged in glial-guided migration, the movement of translocating cells in the cortex was found to be continuous towards the pial surface and at greater average speeds (Nadarajah *et al.*, 2001).

4.2.3 Unconstrained migration

Both translocating cells and cells using substrates as guides to migrate upon are restricted in their migratory path. In contrast to this, some retinal cells are believed to engage in what has been called 'free' migration (Prada *et al.*, 1987). Following exit from the mitotic cycle, these cells are thought to simply travel to their definitive layer without the aid of cytoplasmic anchors at the OLM and ILM (Figure 4.1c). The first suggestions of this unconstrained migration came from electron microscopy studies that described ganglion, amacrine and horizontal cell precursors, each bearing morphological hallmarks of freely migrating cells (Hinds and Hinds, 1978, 1979, 1983). Golgi impregnations of amacrine cell neuroblasts also hinted at free migration (Prada *et al.*, 1987). In this study, two morphologically distinct amacrine cell neuroblasts were described: first, a bipolar-shaped cell type with short processes directed sclerally and vitreally and second, a multipolar cell type with multiple shorter cytoplasmic processes. As development progressed, both cell types were found at locations closer to their prospective destination within the INL (Prada *et al.*, 1987). Recent advances in the ability to transgenically label cells in the zebrafish retina have permitted the direct observation of freely migrating cells in the zebrafish retina in vivo (Figure 4.1d). With a lack of anchorage to either limiting membrane, it is likely that such migrating cells move through the retinal neuroepithelium interacting with other cells as well as the extracellular matrix (ECM). The morphological features of freely migrating cells suggest an ability to explore the environment through their neurites. Such explorations would allow for the detection of migratory cues.

In the cortex, unconstrained migratory cells have been observed in acute slice preparations (Nadarajah *et al.*, 2003; Tabata and Nakajima, 2003). These cells are characterized by abundant, highly motile cytoplasmic processes and consequently were referred to as 'multipolar' (Tabata and Nakajima, 2003) or 'branching' (Nadarajah *et al.*, 2003). Interestingly, such multipolar cells were also found to display an ability to move laterally. These cells were not found closely apposed to radial glial fibres or to tangentially oriented axon bundles, suggesting that they do not use a cellular substrate for their migration (Tabata and Nakajima, 2003).

4.3 Migration trajectories

4.3.1 Radial and tangential trajectories of migration

Two of the proposed migratory mechanisms described above, guided migration and somal translocation, by their very nature constrain the trajectory of migration to a radial one. Support for this strict movement in the radial axis came from studies in which retroviral constructs were used to mark small numbers of progenitor cells early in development (Turner and Cepko, 1987; Turner *et al.*, 1990; Fekete *et al.*, 1994). When the progeny of these labelled progenitors were examined in the mature retina, they were distributed in tightly organized radial columns spanning the thickness of the retina. This distribution pattern was taken to suggest that newborn retinal cells migrate from their birth place in a strict radial axis. This pattern, however, does not appear to be the case for all retinal cells.

Using a similar experimental paradigm of labelling progenitors early and examining their progeny at maturity, but with transgenic techniques that permitted greater numbers of labelled progenitors, a small percentage of retinal cells was found to be capable of tangential movement (Williams and Goldowitz, 1992; Reese *et al.*, 1995, 1999) (Figure 4.2a). These small numbers of cells, which had perhaps gone undetected with other techniques, were found outside of radial columns and were therefore regarded to have dispersed tangentially. Interestingly, rod photoreceptors, bipolar cells and Müller glia were found exclusively in radial columns and therefore said to be radially dispersing. Cone photoreceptors, horizontal cells, amacrine cells and ganglion cells were the cell types found to stray from the boundaries of radial columns and therefore believed to be capable of tangential dispersion (Reese *et al.*, 1995) (Figure 4.2a). This tight link between cell phenotype and mode of dispersion highlights the importance of migration for the proper placement of cells. All the tangentially dispersing cell classes are regularly spaced across the retinal surface. It has been suggested that tangential dispersion might be the way this regularity is achieved during development (Reese *et al.*, 1999) (see Chapter 10).

Developmental studies suggest that the tangentially dispersing cell classes first migrate radially to their appropriate layers before moving tangentially (Galli-Resta *et al.*, 1997; Reese *et al.*, 1999). The distances over which tangential dispersion occurs are relatively small (Reese *et al.*, 1999) and one could speculate that these cells simply translocate their somata to appropriate positions through laterally oriented cytoplasmic processes. Alternatively, the entire cell may move laterally using exploratory neurites to search for cues. Time-lapse imaging of tangentially migrating cells could help distinguish between these two possibilities.

4.3.2 Unidirectional or bidirectional trajectories of migration?

Recent studies combining static observations and time-lapse analyses have begun to reveal unexpected ways in which some retinal cells travel. The expression pattern of an early

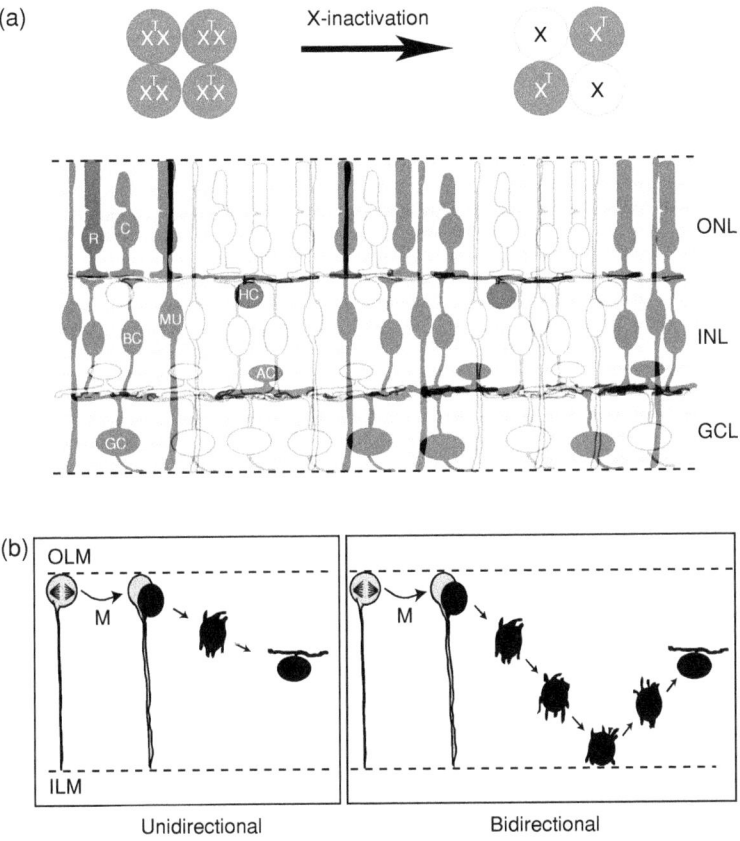

Figure 4.2 Trajectories of migration. (a) Radial and tangential trajectories: schematic representation of a transgenic mouse model used to study cell dispersion patterns in the retina (Reese *et al.*, 1995). A transgenic mouse line was created in which *lacZ* was inserted into one of the X chromosomes (X^T). All cells express the transgene protein product early in development (grey cells). Random inactivation of one of the two X chromosomes, which occurs in all female mammals, results in half of all progenitor cells being transgene-positive. The progeny of transgene-expressing cells maintain *lacZ* expression into maturity. Radial columns of transgene-expressing (grey) and non-expressing (white) retinal cells can be seen in the mature retina, indicative of radial dispersion. Occasional transgene-positive cells in transgene-negative columns and vice versa suggest that some cells are capable of tangential dispersion as well. Rod (R), cone (C), horizontal (HC), bipolar (BC), Müller glia (MU), amacrine (AC), ganglion (GC) cell. (b) Unidirectional trajectory: following mitosis (M) at the apical surface a postmitotic cell (black) is believed to migrate to its definitive location. Bidirectional trajectory: recent observations of horizontal cells in the chick and mouse retina suggest that these cells migrate basally until the ILM before moving apically to their definitive location.

horizontal cell-specific marker, the homeobox-containing transcription factor Lim1, in the mouse (Liu *et al.*, 2000) and chick retina (Edqvist and Hallbook, 2004), was found to be suggestive of a bidirectional mode of movement. Observations of Lim1-expressing cells at progressive time-points suggested that rather than migrating a short distance to their definitive position in the outer INL, newborn horizontal cells first move to the vitreal side of the retina, adjacent to the ILM, before migrating back in the scleral direction to arrive at their definitive locations (Figure 4.2b). Time-lapse recordings aimed at confirming the existence of this bidirectional movement were only able to verify migration in the scleral direction, as these were performed in explants of chick retina, which cannot be maintained for long periods (Edqvist and Hallbook, 2004). The ILM has been implicated as a rich source of cues for migrating cells. The unusual route horizontal cells take permits contact with the ILM and consequently exposure to such cues. It remains to be seen if other retinal cells also employ this bidirectional mode of movement.

4.4 Fundamentals of signalling in neuronal cell migration

Underlying the cellular modes of migration by cells of the developing retina, and migratory cells in general, are extrinsic and intrinsic molecules that guide and facilitate cellular motility. For each of the cell behavioural modes discussed, cellular migration requires communication between the migrating cell, adjacent cells and the ECM. These interactions may be either instructive, including both attractive and repulsive cues, or simply permissive for motility or somal translocation. Signals from the outside environment must then be transferred and integrated within the cell to coordinate cytoskeletal assembly and disassembly. In many instances, adhesion proteins mediate the communication between cells and their environment. Experimental evidence has shown that specific classes of adhesion molecules play pivotal roles in transducing signals across the plasma membrane, which cascade within the cell and lead to rapid changes in the cytoskeleton. In the next section, the types of extracellular cues that influence retinal cell migration will be presented. This will be followed by a discussion of the intracellular molecules that facilitate cytoskeletal changes that ultimately impact migratory decisions and the final laminar position of retinal neurons. As much of the work in the area of cell migration has been accomplished outside of the retina, the extrinsic and intrinsic factors that regulate cell migration in general will be described, while highlighting those molecules and pathways with proven roles in retinal cell migration and positioning.

4.4.1 Extracellular guidance cues

From a general perspective, in order to achieve differential location of newly postmitotic retinal cells there must be either an asymmetric distribution of guidance cues within the developing retinal neuroepithelium or a polarized localization of the proteins within the cell that respond to such cues at the onset of migration. Expression analysis and experimental challenges in several species have shown that both strategies exist for directing migration within the developing retina.

Inner limiting membrane

The earliest migratory cues established within the developing optic cup are associated with the ILM, a basal lamina (Figure 4.3a). The ILM lies at the border between the basal surface

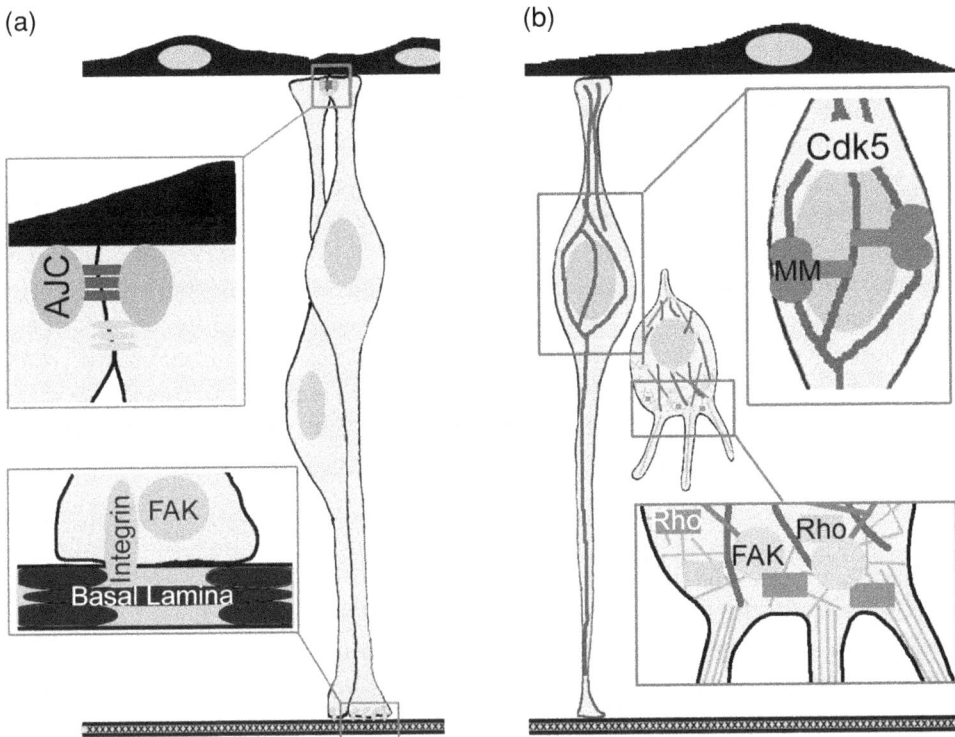

Figure 4.3 Factors that establish and regulate retinoblast migration. (a) Cycling retinal neuroepithelial cells and (b) postmitotic migratory retinal neuroblasts. Retinal neuroblasts are shown in grey, pigment epithelial cells are in black and the ILM (basal lamina) is blue. In (a), upper inset shows components of the apical junctional complex (AJC). For the proliferating neuroblasts, tight junctions are represented in dark blue rectangles and adherens junctions in light blue ovals. For the migratory neuroblasts (b), actin is represented by green lines, microtubules with blue lines and actin:myosin stress fibres by red lines. Lower inset diagrams components of the basal lamina and neuroepithelial endfeet proteins, which interact with this extracellular matrix (integrins and focal adhesion kinase (FAK)). In (b), a post-mitotic retinal neuroblast undergoing somal translocation (*left*) and another undergoing unconstrained migration (*right*) are schematized. Intrinsic regulatory proteins for each type of migratory mode are shown in insets: cyclin-dependent kinase 5 (Cdk5, yellow oval); microtubule motors (MM, red shape); FAK (orange circle); Rho GTPases (Rho, green and blue boxes to represent multiple subtypes). For colour version, see Plate 4.

of the retinal neuroepithelium and the developing vitreous and lens. Like other basal lamina, the ILM is an ECM sheet, approximately 50 nanometres thick (Halfter *et al.*, 2000; Dong *et al.*, 2002). The ILM components are secreted into the vitreous by cells of the ciliary body, lens and optic disc (Halfter *et al.*, 2000). The retinal neuroepithelium, however, is important for formation and maintenance of the ILM. Endfeet of retinal neuroepithelial cells express receptors that bind and immobilize proteins of the ILM. Once nucleating components of the ILM are stabilized, the structure can self-assemble based on mutual affinities between the various proteins (Colognato and Yurchenco, 2000). The importance of the ILM for cell migration has been demonstrated by several experiments including transient enzymatic disruption, which resulted in the mispositioning of neurons throughout the retina and local disruptions to lamination (Halfter, 1998).

The significant role of the ILM on retinal cell migration is also demonstrated by manipulations to subunits of receptors to ILM components. Integrins, a major family of cell surface receptors that mediate cell–cell and cell–matrix interactions, interact with ILM components. Several experiments address a role for integrins in retinal cell migration. For example, when β1-integrin is functionally disabled within the developing eye of several species, retinal cell migration is dramatically altered (Svennevik and Linser, 1993; Li and Sakaguchi, 2004). Additionally, mice with null mutations in α6- or α3-integrin show retinal ectopias and lamination defects consistent with altered cell migration (Georges-Labouesse *et al.*, 1998). Cumulatively, these experiments demonstrate the important role of the ILM as well as receptors for this basement membrane for retinal cell migration.

The apical polarity complex

An apically localized protein complex has also recently been shown to be essential for directing retinal cell migration (Figure 4.3a). This complex, which shows homology to the *Drosophila* epithelial cell polarity complex, is associated with junctional complexes of retinal neuroepithelial cells during proliferative and migratory stages of development. Experiments, primarily in zebrafish, have shown the essential nature of this signalling complex in directing retinal cell migration (Malicki and Driever, 1999; Horne-Badovinac *et al.*, 2001; Peterson *et al.*, 2001; Jensen and Westerfield, 2004). These genetic experiments suggest that the polarity complex controls either the localized secretion of guidance/adhesion molecules, or the complex regulates the signalling/adhesive capabilities of such secreted factors.

Cell adhesion molecules

Several secreted and transmembrane proteins (cell adhesion molecules) localized to retinal neuroepithelial cells have been shown to regulate retinal cell migration. Foremost among these is the classical cadherin 2, CDH2 (previously named neural- or N-cadherin). This is a transmembrane cell surface protein that mediates Ca^{2+}-dependent cell adhesion as well as cell–cell and cell–matrix signalling. Function-deleting manipulations to CDH2 within the retina have demonstrated an essential role for this molecule in retinal cell migration (Matsunaga *et al.*, 1988; Lele *et al.*, 2002; Malicki *et al.*, 2003; Masai *et al.*, 2003). Interestingly, CDH2 does not appear to be localized in an asymmetric fashion within the retinal neuroepithelium when newly postmitotic cells migrate (Matsunaga *et al.*, 1988; Malicki *et al.*, 2003). Because CDH2 is uniformly expressed on all retinal progenitors as they migrate, additional molecules must interact to provide differential adhesion or to modulate the intracellular signalling activities. Additional experiments are needed to identify these factors.

4.4.2 Intracellular mechanisms and modes of retinal cell positioning

While the extrinsic cues of retinal cell migration have begun to be experimentally addressed, the signal transduction pathways and intrinsic cytoskeletal regulation that facilitate migration remain largely uncharacterized for the developing retina. However, a general framework for the intracellular mechanisms involved has emerged based on in vitro studies and in vivo experiments with migratory cells outside of the retina.

Directed cell motility and intracellular signals

A large array of intracellular signalling molecules has been implicated in directed motility of migrating neuroblasts (Figure 4.3b) (reviewed in Meyer and Feldman, 2002; Fukata *et al.*, 2003). These include mitogen-activated protein kinases (MAPK), lipid-activated kinases, phospholipases, serine/threonine kinases, cytoplasmic tyrosine kinases, and a multitude of scaffold proteins. One particular class of proteins, however, appears to be central in coordinating intracellular signalling during directed cell migration in all motile cells: the RhoGTPases. In general RhoGTPases function to promote the assembly of actin:myosin stress fibres, leading to cell contraction and an increase in focal adhesions. Focal adhesions are multi-protein complexes that link the extracellular matrix to the actin cytoskeleton, usually via integrins and focal adhesion kinases. For the RhoGTPases, multiple subtypes exist and detailed characterization of expression, localization and activation specificity is just beginning. As central facilitators of directed cell migration, the RhoGTPases are tightly regulated by three main classes of proteins: guanine nucleotide exchange factors (GEFs), GTPase-activating proteins (GAPs) and guanine nucleotide dissociation inhibitors (GDIs). Proteins in these families are thought to provide cell type specificity for the widely expressed RhoGTPases. In addition, these regulators of RhoGTPases have been shown to provide integrative links between extracellular signalling molecules and the proteins that mediate actin-cytoskeletal changes. Within the developing retina, the role of RhoGTPases and their effectors on cell migration has not been evaluated. However, components of this signalling pathway are expressed in the retina during migration (Malosio *et al.*, 1997; Ruchhoeft *et al.*, 1999; Wong *et al.*, 2000).

Mechanisms of somal translocation

The intracellular mechanisms and regulation of somal translocation, have not been directly investigated in retinal cells. However, other cell types, including other CNS neurons that exhibit nuclear translocation, have been studied to establish models and candidate factors for how this might occur in the retina (Figure 4.3b). In general these include microtubule-based motors and associated proteins such as those of the dynein complex. In addition, nuclear translocation is dependent on signalling molecules, such as the kinases Cdk5 and focal adhesion kinases, which associate with the centrosomes and the microtubule-based nuclear cage, respectively (reviewed in Tsai and Gleeson, 2005).

4.5 Concluding remarks

The study of cell migration in the retina is in its infancy. While the cellular mechanisms proposed for migrating retinal cells are plausible, evidence to date has only been suggestive. Extrapolations made from studies of migration in the cortex are informative. However, migratory modes employed by cortical cells may not be precisely recapitulated in the retina. The retina is a much thinner structure and cells may have simpler requirements to get to their final destination. In the cortex, somal translocation was found to be the predominant migratory mode at early developmental ages when the cortical wall is thin (Nadarajah *et al.* 2001). The width of the retina may suggest that this mode of migration alone suffices.

Morphological descriptions of postmitotic retinal neurons at progressive developmental ages and elegant lineage studies provide a tantalizing view of what might be happening in vivo. However, direct observations of retinal cell migration by time-lapse analysis are necessary to confirm what cellular mechanisms are in fact used by retinal cells. Time-lapse imaging, used extensively in studies of cortical cell migration, has only recently begun to be applied to the study of cell migration in the retina (Edqvist and Hallbook, 2004). Vertebrate model organisms such as the zebrafish (see Chapter 17) offer the ability to conduct time-lapse studies in vivo (Koster and Fraser, 2001; Das *et al.*, 2003; Gilmour *et al.*, 2004; Kay *et al.*, 2004), circumventing problems associated with maintaining explanted retinal tissue. The transparency and rapid development of zebrafish not only allow for the imaging of retinal migration in vivo but also the ability to follow the same cells through their entire course of development, from the time they become postmitotic until their arrival in their definitive layer. Using such imaging approaches, one possible result that might be found is that retinal cells do not exclusively use one mode of migration. For instance, a cell may use somal translocation to get to a depth appropriate for its type, retract its cytoplasmic attachments from apical and basal surfaces, and then more accurately position itself by unconstrained migration. This use of two modes of migration by individual cells has been observed for cortical cells that first use glial-guides and subsequently somal translocation to migrate to the right layer (Nadarajah *et al.* 2001). In addition, a particular mode of travel may be used more predominantly at certain times in development.

Lineage studies have shown that a cell's ultimate phenotype is tightly correlated with whether it chooses a radial or tangential trajectory to get to its destination (Reese *et al.*, 1995, 1999). This opens up the possibility that a particular cell fate may direct that a certain migratory trajectory be taken. Alternatively, a cell may acquire its fate as a result of its migration path, receiving cues from its extracellular environment along the way.

Much progress has been made studying the molecular machinery needed for cell migration in general and this knowledge can be applied to the retina. However, gene expression studies aimed at identifying the molecular players within the retina should not be the sole criterion for their implication in retinal cell migration. One notable example is the importance of the glycoprotein Reelin in cortical cell migration. Reelin is also expressed in the retina. However, in its absence retinal cells appear to migrate appropriately, while migration in the cortex is severely disrupted (Rice and Curran, 2000; Rice et al., 2001). Thus, only through direct investigation can the molecular cues that mediate retinas cell migration become clear.

References

Anton, E. S., Cameron, R. S. and Rakic, P. (1996). Role of neuron-glial junctional domain proteins in the maintenance and termination of neuronal migration across the embryonic cerebral wall. *J. Neurosci.*, **16**, 2283–93.

Bennett, G. S. and DiLullo, C. (1985). Transient expression of a neurofilament protein by replicating neuroepithelial cells of the embryonic chick brain. *Dev. Biol.*, **107**, 107–27.

Brittis, P. A., Meiri, K., Dent, E. and Silver, J. (1995). The earliest patterns of neuronal differentiation and migration in the mammalian central nervous system. *Exp. Neurol.*, **134**, 1–12.

Carter-Dawson, L. D. and LaVail, M. M. (1979). Rods and cones in the mouse retina. II. Autoradiographic analysis of cell generation using tritiated thymidine. *J. Comp. Neurol.*, **188**, 263–72.

Colognato, H. and Yurchenco, P. D. (2000). Form and function: the laminin family of heterotrimers. *Dev. Dyn.*, **218**, 213–34.

Das, T., Payer, B., Cayouette, M. and Harris, W. A. (2003). In vivo time-lapse imaging of cell divisions during neurogenesis in the developing zebrafish retina. *Neuron*, **37**, 597–609.

Dong, S., Landfair, J., Balasubramani, M. *et al.* (2002). Expression of basal lamina protein mRNAs in the early embryonic chick eye. *J. Comp. Neurol.*, **447**, 261–73.

Dunlop, S. A. (1990). Early development of retinal ganglion cell dendrites in the marsupial *Setonix brachyurus*, quokka. *J. Comp. Neurol.*, **293**, 425–47.

Edmondson, J. C. and Hatten, M. E. (1987). Glial-guided granule neuron migration in vitro: a high-resolution time-lapse video microscopic study. *J. Neurosci.*, **7**, 1928–34.

Edqvist, P. H. and Hallbook, F. (2004). New-born horizontal cells migrate bi-directionally across the neuroepithelium during retinal development. *Development*, **131**, 1343–51.

Fekete, D. M., Perez-Miguelsanz, J., Ryder, E. F. and Cepko, C. L. (1994). Clonal analysis in the chicken retina reveals tangential dispersion of clonally related cells. *Dev. Biol.*, **166**, 666–82.

Fukata, M., Nakagawa, M. and Kaibuchi, K. (2003). Roles of Rho-family GTPases in cell polarisation and directional migration. *Curr. Opin. Cell. Biol.*, **15**, 590–7.

Gadisseux, J. F., Kadhim, H. J., van den Bosch de Aguilar, P., Caviness, V. S. and Evrard, P. (1990). Neuron migration within the radial glial fiber system of the developing murine cerebrum: an electron microscopic autoradiographic analysis. *Brain Res. Dev. Brain. Res.*, **52**, 39–56.

Galli-Resta, L., Resta, G., Tan, S. S. and Reese, B. E. (1997). Mosaics of islet-1-expressing amacrine cells assembled by short-range cellular interactions. *J. Neurosci.*, **17**, 7831–8.

Georges-Labouesse, E., Mark, M., Messaddeq, N. and Gansmuller, A. (1998). Essential role of alpha 6 integrins in cortical and retinal lamination. *Curr. Biol.*, **8**, 983–6.

Gilmour, D., Knaut, H., Maischein, H. M. and Nusslein-Volhard, C. (2004). Towing of sensory axons by their migrating target cells *in vivo*. *Nat. Neurosci.*, **7**, 491–2.

Halfter, W. (1998). Disruption of the retinal basal lamina during early embryonic development leads to a retraction of vitreal end feet, an increased number of ganglion cells, and aberrant axonal outgrowth. *J. Comp. Neurol.*, **397**, 89–104.

Halfter, W., Dong, S., Schurer, B. et al. (2000). Composition, synthesis, and assembly of the embryonic chick retinal basal lamina. *Dev. Biol.*, **220**, 111–28.

Hinds, J. W. and Hinds, P. L. (1978). Early development of amacrine cells in the mouse retina: an electron microscopic, serial section analysis. *J. Comp. Neurol.*, **179**, 277–300.

Hinds, J. W. and Hinds, P. L. (1979). Differentiation of photoreceptors and horizontal cells in the embryonic mouse retina: an electron microscopic, serial section analysis. *J. Comp. Neurol.*, **187**, 495–511.

Hinds, J. W. and Hinds, P. L. (1983). Development of retinal amacrine cells in the mouse embryo: evidence for two modes of formation. *J. Comp. Neurol.*, **213**, 1–23.

Horne-Badovinac, S., Lin, D., Waldron, S. et al. (2001). Positional cloning of *heart and soul* reveals multiple roles for PKC in zebrafish organogenesis. *Curr. Biol.*, **11**, 1492–502.

Hu, M. and Easter, S. S. (1999). Retinal neurogenesis: the formation of the initial central patch of post-mitotic cells. *Dev. Biol.*, **207**, 309–21.

Jensen, A. M. and Westerfield, M. (2004). Zebrafish *mosaic eyes* is a novel FERM protein required for retinal lamination and retinal pigmented epithelial tight junction formation. *Curr. Biol.*, **14**, 711–17.

Kay, J. N., Roeser, T., Mumm, J. S. et al. (2004). Transient requirement for ganglion cells during assembly of retinal synaptic layers. *Development*, **131**, 1331–42.

Koster, R. W. and Fraser, S. E. (2001). Direct imaging of *in vivo* neuronal migration in the developing cerebellum. *Curr. Biol.*, **11**, 1858–63.

Lele, Z., Folchert, A., Concha, M. et al. (2002). *parachute*/n-cadherin is required for morphogenesis and maintained integrity of the zebrafish neural tube. *Development*, **129**, 3281–94.

Lemmon, V. and Rieser, G. (1983). The development distribution of vimentin in the chick retina. *Brain Res.*, **313**, 191–7.

Li, M. and Sakaguchi, D. S. (2004). Inhibition of integrin-mediated adhesion and signaling disrupts retinal development. *Dev. Biol.*, **275**, 202–14.

Liu, W., Wang, J. H. and Xiang, M. (2000). Specific expression of the LIM/homeodomain protein Lim-1 in horizontal cells during retinogenesis. *Dev. Dyn.*, **217**, 320–5.

Malicki, J. (2004). Cell fate decisions and patterning in the vertebrate retina: the importance of timing, asymmetry, polarity and waves. *Curr. Opin. Neurobiol.*, **14**, 15–21.

Malicki, J. and Driever, W. (1999). *oko meduzy* mutations affect neuronal patterning in the zebrafish retina and reveal cell–cell interactions of the retinal neuroepithelial sheet. *Development*, **126**, 1235–46.

Malicki, J., Jo, H. and Pujic, Z. (2003). Zebrafish N-cadherin, encoded by the *glass onion* locus, plays an essential role in retinal patterning. *Dev. Biol.*, **259**, 95–108.

Malosio, M. L., Gilardelli, D., Paris, S., Albertinazzi, C. and de Curtis, I. (1997). Differential expression of distinct members of Rho family GTP-binding proteins during neuronal development: identification of Rac1B, a new neural-specific member of the family. *J. Neurosci.*, **17**, 6717–28.

Masai, I., Lele, Z., Yamaguchi, M. et al. (2003). N-cadherin mediates retinal lamination, maintenance of forebrain compartments and patterning of retinal neurites. *Development*, **130**, 2479–94.

Matsunaga, M., Hatta, K. and Takeichi, M. (1988). Role of N-cadherin cell adhesion molecules in the histogenesis of neural retina. *Neuron*, **1**, 289–95.

McLoon, S. C. and Barnes, R. B. (1989). Early differentiation of retinal ganglion cells: an axonal protein expressed by premigratory and migrating retinal ganglion cells. *J. Neurosci.*, **9**, 1424–32.

Meller, K. and Tetzlaff, W. (1976). Scanning electron microscopic studies on the development of the chick retina. *Cell Tissue Res.*, **170**, 145–59.

Meyer, G. and Feldman, E. L. (2002). Signalling mechanisms that regulate actin-based motility processes in the nervous system. *J. Neurochem.*, **83**, 490–503.

Misson, J. P., Austin, C. P., Takahashi, T., Cepko, C. L. and Caviness, V. S., Jr. (1991). The alignment of migrating neural cells in relation to the murine neopallial radial glial fiber system. *Cereb. Cortex*, **1**, 221–9.

Miyata, T., Kawaguchi, A., Okano, H. and Ogawa, M. (2001). Asymmetric inheritance of radial glial fibers by cortical neurons. *Neuron*, **31**, 727–41.

Morest, D. K. (1970). The pattern of neurogenesis in the retina of the rat. *Z. Anat. Entwicklungsgesch*, **131**, 45–67.

Nadarajah, B., Brunstrom, J. E., Grutzendler, J., Wong, R. O. and Pearlman, A. L. (2001). Two modes of radial migration in early development of the cerebral cortex. *Nat. Neurosci.*, **4**, 143–50.

Nadarajah, B., Alifragis, P., Wong, R. O. and Parnavelas, J. G. (2003). Neuronal migration in the developing cerebral cortex: observations based on real-time imaging. *Cereb. Cortex*, **13**, 607–11.

Noctor, S. C., Flint, A. C., Weissman, T. A., Dammerman, R. S. and Kriegstein, A. R. (2001). Neurons derived from radial glial cells establish radial units in neocortex. *Nature*, **409**, 714–20.

O'Rourke, N. A., Dailey, M. E., Smith, S. J. and McConnell, S. K. (1992). Diverse migratory pathways in the developing cerebral cortex. *Science*, **258**, 299–302.

Peterson, R. T., Mably, J. D., Chen, J. N. and Fishman, M. C. (2001). Convergence of distinct pathways to heart patterning revealed by the small molecule concentramide and the mutation heart-and-soul. *Curr. Biol.*, **11**, 1481–91.

Prada, C., Puelles, L. and Genis-Galvez, J. M. (1981). A Golgi study on the early sequence of differentiation of ganglion cells in the chick embryo retina. *Anat. Embryol. (Berl.)*, **161**, 305–17.

Prada, C., Puelles, L., Genis-Galvez, J. M. and Ramirez, G. (1987). Two modes of free migration of amacrine cell neuroblasts in the chick retina. *Anat. Embryol. (Berl.)*, **175**, 281–7.

Prada, C., Puga, J., Perez-Mendez, L., Lopez, R. and Ramirez, G. (1991). Spatial and temporal patterns of neurogenesis in the chick retina. *Eur. J. Neurosci.*, **3**, 559–69.

Rakic, P. (1972). Mode of cell migration to the superficial layers of fetal monkey neocortex. *J. Comp. Neurol.*, **145**, 61–83.

Rapaport, D. H., Rakic, P. and LaVail, M. M. (1996). Spatiotemporal gradients of cell genesis in the primate retina. *Perspect. Dev. Neurobiol.*, **3**, 147–59.

Rapaport, D. H., Wong, L. L., Wood, E. D., Yasumura, D. and LaVail, M. M. (2004). Timing and topography of cell genesis in the rat retina. *J. Comp. Neurol.*, **474**, 304–24.

Reese, B. E., Harvey, A. R. and Tan, S. S. (1995). Radial and tangential dispersion patterns in the mouse retina are cell-class specific. *Proc. Natl. Acad. Sci. U. S. A.*, **92**, 2494–8.

Reese, B. E., Necessary, B. D. and Tam, P. P., Faulkner-Jones, B. and Tan, S. S. (1999). Clonal expansion and cell dispersion in the developing mouse retina. *Eur. J. Neurosci.*, **11**, 2965–78.

Rice, D. S. and Curran, T. (2000). Disabled-1 is expressed in type AII amacrine cells in the mouse retina. *J. Comp. Neurol.*, **424**, 327–38.

Rice, D. S., Nusinowitz, S., Azimi, A. M. *et al.* (2001). The reelin pathway modulates the structure and function of retinal synaptic circuitry. *Neuron*, **31**, 929–41.

Ruchhoeft, M. L., Ohnuma, S., McNeill, L., Holt, C. E. and Harris, W. A. (1999). The neuronal architecture of Xenopus retinal ganglion cells is sculpted by rho-family GTPases *in vivo*. *J. Neurosci.*, **19**, 8454–63.

Sidman R. L. (1961). Histogenesis of mouse retina with thymidine-H3. In *Structure of the Eye*, ed. G. K. Smelser. New York: Academic Press, pp. 487–506.

Snow, R. L. and Robson, J. A. (1994). Ganglion cell neurogenesis, migration and early differentiation in the chick retina. *Neuroscience*, **58**, 399–409.

Snow, R. L. and Robson, J. A. (1995). Migration and differentiation of neurons in the retina and optic tectum of the chick. *Exp. Neurol.*, **134**, 13–24.

Svennevik, E. and Linser, P. J. (1993). The inhibitory effects of integrin antibodies and the RGD tripeptide on early eye development. *Invest. Ophthalmol. Vis. Sci.*, **34**, 1774–84.

Tabata, H. and Nakajima, K. (2003). Multipolar migration: the third mode of radial neuronal migration in the developing cerebral cortex. *J. Neurosci.*, **23**, 9996–10 001.

Tsai, L. H. and Gleeson, J. G. (2005). Nucleokinesis in neuronal migration. *Neuron*, **46**, 383–8.

Turner, D. L. and Cepko, C. L. (1987). A common progenitor for neurons and glia persists in rat retina late in development. *Nature*, **328**, 131–6.

Turner, D. L., Snyder, E. Y. and Cepko, C. L. (1990). Lineage-independent determination of cell type in the embryonic mouse retina. *Neuron*, **4**, 833–45.

Williams, R. W. and Goldowitz, D. (1992). Structure of clonal and polyclonal cell arrays in chimeric mouse retina. *Proc. Natl. Acad. Sci. U. S. A.*, **89**, 1184–8.

Wolburg, H., Willbold, E. and Layer, P. G. (1991). Müller glia end-feet, a basal lamina and the polarity of retinal layers form properly in vitro only in the presence of marginal pigmented epithelium. *Cell Tissue Res.*, **264**, 437–51.

Wong, W. T., Faulkner-Jones, B. E., Sanes, J. R. and Wong, R. O. (2000). Rapid dendritic re-modeling in the developing retina: dependence on neurotransmission and reciprocal regulation by Rac and Rho. *J. Neurosci.*, **20**, 5024–36.

Young, R. W. (1985a). Cell differentiation in the retina of the mouse. *Anat. Rec.*, **212**, 199–205.

Young, R. W. (1985b). Cell proliferation during postnatal development of the retina in the mouse. *Brain Res.*, **353**, 229–39.

5

Cell determination

Michalis Agathocleous and William A. Harris
University of Cambridge, Cambridge, UK

5.1 Introduction

The sheet of retinal neuroepithelial cells resulting from the specification of the eye field is transformed into a layered array of differentiated cells by the simultaneous processes of cell division, apoptosis, differentiation and migration. The production of the six major cell types, with their multiple subtypes, in the correct numbers and at the appropriate time is essential for normal development. The retina has been studied extensively as a model for cell determination in the vertebrate nervous system for a number of reasons. It is easily accessible to genetic and embryological manipulations in vivo because of its position and large size and can also be studied in vitro because cells in retinal explant cultures faithfully follow in vivo differentiation programmes. Numerous genes involved in cell determination do not affect other processes and their disruption does not cause early lethality. The different major cell types can be readily distinguished by their laminar position, their distinct morphologies and by cell-specific markers. The persistence of a proliferating ciliary marginal zone in amphibians, fish and avian species provides a model that recapitulates embryonic proliferation and differentiation and facilitates the examination of gene expression and function (Perron et al., 1998).

Retinal progenitors are multipotent and vary greatly with respect to their clonal compositions, both in terms of the cell types produced and the number of progeny. Several studies in the past 15 years have demonstrated key aspects of cell fate decisions: the different clonal compositions that progenitors give are probably accounted for both by exposure to different environments and by a heterogeneity in the internal programmes in place at the beginning of neurogenesis. The histogenetic order of birth is again due to a temporally changing environment as well as changing internal programmes. The two sides of signalling interact to push a cell towards a particular fate. We will also see how cell fate decisions must be coordinated with proliferation and how numerous molecules influence both the cell cycle and cell fate. Different models have been proposed to integrate a wealth of data into a general scheme for fate determination.

Retinal Development, ed. Evelyne Sernagor, Stephen Eglen, Bill Harris and Rachel Wong.
Published by Cambridge University Press. © Cambridge University Press 2006.

5.2 Histogenesis

Early birthdating studies show that the different cellular types are born in a conserved order (Young, 1985; LaVail *et al.*, 1991). This aspect of retinal development is discussed in detail in Chapter 3. Ganglion cells are generated first, followed by horizontals, cones and amacrines. Rods and bipolars are generated later followed by Müller cells. Short window-labelling experiments, using sequential administration of ^{3}H-thymidine and 5'-Bromo-2'-deoxy-uridine (BrdU) in chicks, demonstrate that at specific short time intervals during this process various cell types may arise simultaneously, showing that there is considerable overlap of the periods in which specific cell types are born (Repka and Adler, 1992; Belecky-Adams *et al.*, 1996). The same appears to be true in most other vertebrate species studied. Lineage analysis demonstrates that retinal progenitors are multipotent. Clones show great variety of cellular compositions, in terms of both number and type (Turner and Cepko, 1987; Holt *et al.*, 1988; Wetts and Fraser, 1988). However, in line with the constraints imposed by the process of histogenesis, later progenitors generally produce clones of smaller size that consist of late cell types.

Two major questions arise from these observations. First, how do the initial seemingly homogeneous multipotent early progenitors diversify to give the great number of different clonal compositions? Second, how do the fates of daughter cells produced by a single progenitor cell change over time? The answer to both these questions appears to lie in the integration of extrinsic signals from a dynamic environment with the intrinsic regulators of maturing progenitors. These processes are all acting in the context of a bidirectional relationship with the cell cycle, so that some progenitors pull out of the cell cycle early and assume early fates, while others pull out later and assume later fates.

5.3 Extrinsic signalling

5.3.1 *Feedback signalling from postmitotic neurons affects progenitors*

In amphibian embryos, early lineage studies showed that even small clones are composed of near random assortments of cell types. The most obvious explanation for these results is that the changing environment extrinsically influences cell fate postmitotically. Other possibilities are that there is an intrinsic stochastic process or that there is a large number of varied fixed lineages. One idea in favour of the extrinsic possibility is that the addition of new differentiated cells would in itself change the local retinal environment, and these postmitotic cells could feed signals back to the dividing progenitors and influence the fate of their daughters. Experimental evidence supports a feedback inhibition mechanism, for example when Negishi *et al.* (1982) destroyed the dopaminergic amacrine cells in developing goldfish, they found increased production of these cells from the proliferating germinal zone after three months. The presence of an amacrine-derived inhibitor for further amacrine production was also suggested by cell-mixing experiments using amacrine-enriched or amacrine-depleted cellular environments (Belliveau and Cepko, 1999). Moreover, Reh and Tully (1986) found that the increase was specific for the amacrine cell subtype that was destroyed in the frog retina. Kainic acid injection in the frog and labelling of subsequently born neurons with ^{3}H-thymidine showed an increase in the newborn cells in each layer proportional to the number of cells initially destroyed in that layer, suggesting cell type-specific feedback regulation operates more generally during differentiation (Reh, 1987). Other evidence for feedback inhibition came from co-cultivation of younger progenitors with older retinas. Such experiments show ganglion cells produce a diffusible factor that inhibits the production of more ganglion cells (Waid and McLoon, 1998).

5.3.2 Diffusible extracellular signals modulate cell fate choices

Other studies demonstrated that there might be a multitude of extrinsic regulators for the production of various cell fates in both early and late retinal environments, and that not all of them necessarily fit the feedback inhibition mechanism. Postmitotic early progenitors in the chick become photoreceptors if dissociated one day after their terminal S-phase, but if left in vivo for three more days they become other types of retinal neurons, suggesting that there may be instructive signals in the changing environment (Adler and Hatlee, 1989). A diffusible factor from cultures of postnatal day (P1) retinal cells in the mouse increases the probability that embryonic day (E)15 progenitors will give rise to rod cells (Watanabe and Raff, 1992). Similarly, when embryonic and postnatal mouse retinal cells are cocultured, signals in the postnatal retinas inhibit amacrine and favour cone production from embryonic progenitors, whereas signals in the embryonic retinas inhibit rod and favour bipolar generation from postnatal progenitors (Belliveau and Cepko, 1999; Belliveau et al., 2000).

What molecular progress has been made on these extrinsic determination factors? Sonic hedgehog (Shh) has been identified as one factor that has the potential to be both a feedback inhibitory signal for the production of ganglion cells (Zhang and Yang, 2001) and a positive factor for the proper differentiation of other retinal cell types (Stenkamp et al., 2002; Shkumatava et al., 2004). Taurine, an unusual amino acid, is an extrinsic factor produced from P0 rat retinal cultures and its addition to retinal explants promotes rod differentiation in a stage-sensitive manner, having its most pronounced effect at E20 and P0 explants (Altshuler et al., 1993). Taurine seems to act via the α2 glycine receptor (GlyRα2) and the γ-aminobutyric acid (GABA)$_A$ receptor, as knock-down of the GlyRα2 in late progenitors in vivo reduces rod production whereas misexpression in early progenitors in explants increases rod production in a taurine-dependent manner (Young and Cepko, 2004). It has also been shown that isolated progenitors differentiate to rods or cones according to the relative amounts of retinoic acid and thyroid hormone (Kelley et al., 1995, 1999), and endogenous retinoic acid inhibition results in a reduction in rod differentiation (Hyatt et al., 1996). A dominant negative form of the fibroblast growth factor (FGF) receptor, which blocks FGF signalling, results in a decrease in photoreceptor and increase in Müller cells when overexpressed in vivo in the Xenopus retina, whereas a block of non-FGF-mediated FGF receptor signalling has the opposite effect (McFarlane et al., 1998). Epidermal growth factor (EGF) receptor is yet another factor that affects fate, and it is thought to do so by affecting the balance between proliferation and differentiation, so that appropriate numbers of each differentiated cell type will be generated. Responsiveness to extrinsic cues also varies with development. Late, but not early, progenitors with high levels of the EGF receptor respond to transforming growth factor-α (TGF-α) by switching from rod to Müller glia production (Lillien and Cepko, 1992; Lillien, 1995; Lillien and Wancio, 1998).

The same factors may have differing effects at different times during the histogenetic process and they may also have different effects in different species. For example, a mutation in the ciliary neurotrophic factor (CNTF) receptor results in increased rod differentiation in mouse retinal explants and CNTF addition may re-specify rods into bipolars in the rat retina (Ezzeddine et al., 1997; although see Neophytou et al., 1997). In the chick retina however, CNTF was found to stimulate photoreceptor rather than bipolar differentiation (Fuhrmann et al., 1995). This may indicate a species-specific difference between the mouse rod-dominated and the chick cone-dominated retina.

5.3.3 *The Notch–Delta pathway regulates the competence to differentiate*

A final example of how extrinsic signals affect cell fate in the retina and contribute to histogenesis is the Notch pathway. Activation of the Notch receptor by the Delta ligand leads to the downregulation of certain proneural basic helix-loop-helix (bHLH) transcription factors. In addition to promoting the differentiation of neurons, these bHLH factors upregulate Delta expression. So cells that express large amounts of Delta inhibit their neighbours, via Notch, from differentiating as neurons, while they themselves have Notch signalling removed as their neighbours express less Delta.

Experiments in the frog retina show that early progenitors with extra Delta activity, and thus less Notch pathway activation, tend to differentiate as the earliest cell fates, ganglion cells and cone photoreceptors. Cells released from Notch pathway activation slightly later, via Delta overexpression, tend to differentiate as rods. On the contrary, cells that continue to receive Notch activation take Müller glial fates and may even remain neuroepithelial (Dorsky et al., 1995, 1997) (Figure 5.1). Similarly, in the chick retina, in vivo transfection of Delta causes the cells that receive the Delta signal to retain a neuroepithelial fate, whilst cells released from Notch pathway activation differentiate as neurons (Henrique et al., 1997) (Figure 5.1). Experimental direct activation of the Notch pathway intrinsically inhibits the early-born retinal ganglion cell production (Austin et al., 1995). These results suggest Delta–Notch signalling ensures that progenitors differentiate in the sequential manner of histogenesis and generate the diversity of both early- and late-born cells. In the absence of this signalling system, all cells could differentiate at the same time in the same environment, and take the same fates!

As illustrated from the examples above, cell fate is influenced by extrinsic signals that may act to promote particular fates, to inhibit specific fates by feedback, to affect the balance between differentiation and proliferation, and to regulate temporally the competence to differentiate at all. These signals may sometimes be instructive or permissive, allowing the development of an immature differentiated cell towards a cell fate to which it is already inclined. An important question, to which we do not yet know the answer, is whether the multitude of the extrinsic signals are sufficient to generate the full diversity of cell fates in

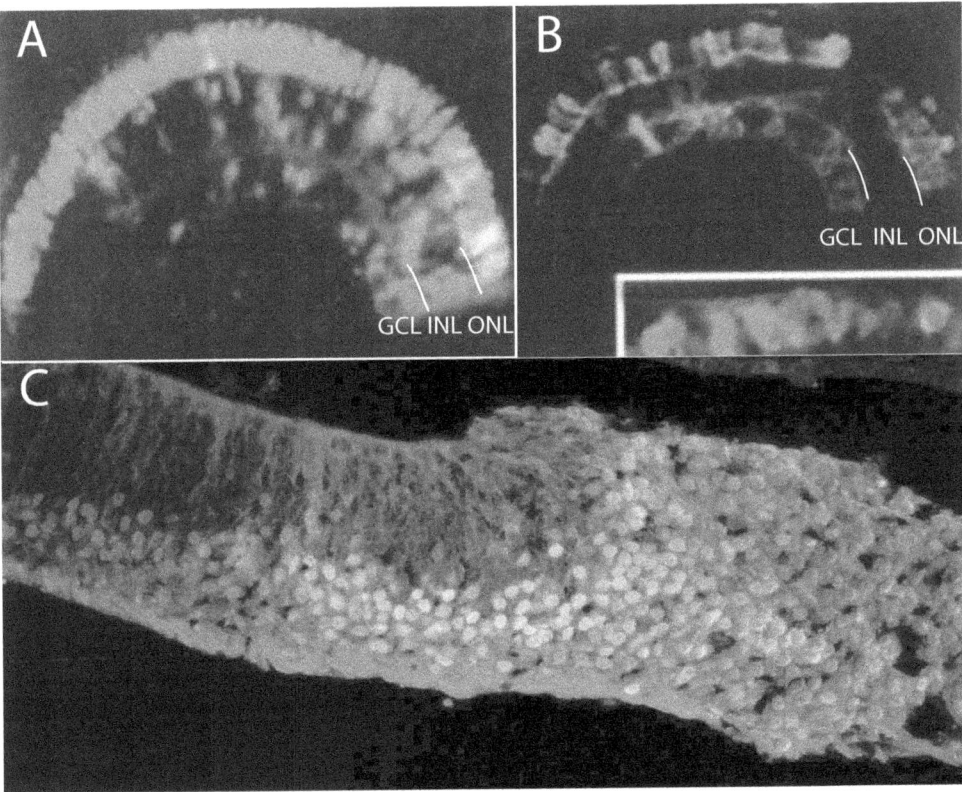

Figure 5.1 Action of the Delta–Notch pathway on cell fate determination. (a) Green fluorescent protein (GFP)-overexpressing cells (green) in the *Xenopus* retina are distributed in all three layers at Stage 41. The red marker in this and the following panel is a cone marker. (b) On the contrary, single cells overexpressing Delta assume the earliest fates available and become ganglion cells or cone photoreceptors. GCL, ganglion cell layer; INL, inner nuclear layer; ONL, outer nuclear layer. (A and B from Dorsky *et al.*, 1997). (c) Retroviral infection in the chick E5 embryo of a truncated Delta form, DeltaSTU, which inhibits the Notch pathway cell-autonomously. Infected cells (green) assume neuronal fates as indicated by islet-1, islet-2 immunoreactivity (red) (Henrique *et al.*, 1997). For colour version, see Plate 5.

the retina and the normal sequence of histogenesis. The fact that at any given time interval several neuronal classes are being born suggests that progenitor-intrinsic regulators may also influence the cell fate decision.

5.4 Intrinsic signalling

5.4.1 bHLH proneural genes direct the generation of multiple cell fates

Ultimately, extrinsic signals need to be translated into an internal code that will guide a cell towards one fate or another. Several transcription factors are involved in generating this diversity of fates. The bHLH proneural genes, mentioned above as targets of Notch signalling, are prominent amongst these. They have the ability to cross-activate each other and thus participate in genetic cascades. They also activate genes specific for particular cell fates, and interact with the cell cycle machinery. Moreover, their expression and activity is modified with developmental time, pointing strongly to a role in cell fate determination. Proneural genes are expressed in the mitotically active ciliary margin zone (CMZ) of the adult retinas of amphibians and fish, but they are also expressed in specific subsets of neurons with a pattern pointing to a potential combinatorial code dictating cell fate (Perron et al., 1998).

A prime example of a bHLH proneural gene is *Ath5*. The *Xenopus* homologue, *Xath5*, promotes retinal ganglion cell (RGC) genesis when overexpressed in vivo and participates in a regulatory network with other proneural genes (Kanekar et al., 1997). It can induce the expression of BarH1, a homeobox transcription factor in RGCs, which in turn can induce Brn3.0, a POU domain transcription factor specific in ganglion cells (Hutcheson and Vetter, 2001; Liu et al., 2001; Poggi et al., 2004). When the *Ath5* gene is non-functional, such as in the zebrafish mutant *lakritz* (Kay et al., 2001) or in *Math5* mutant mice (Brown et al., 2001; Wang et al., 2001), there is a huge depletion of ganglion cells. *Ath5* has an interesting effect on the cell cycle as cells that overexpress *Xath5* not only tend to become ganglion cells but also tend to exit the cell cycle earlier than cells that do not. Indeed, *Xath5*-overexpressing cells tend to leave the cell cycle at the appropriate time for RGC histogenesis (Ohnuma et al., 2002). In mutants that lack *Ath5* function, the cells that were to become RGCs appear to remain in the cell cycle for an extra round of division while on their way to becoming other cell types (Kay et al., 2001). Recent advances in time-lapse microscopy in zebrafish allow the observation of cells that express *Ath5* going through their terminal division and differentiating as ganglion cells (Figure 5.2).

Other bHLH transcription factors have different profiles of activity with respect to cell determination in the retina. NeuroD, for example, promotes amacrine over bipolar cell fate in the rodent retina and favours photoreceptor survival (Morrow et al., 1999). Overexpression of *Xath3* promotes ganglion and photoreceptor over bipolar fate and overexpression of *Neurogenin1* promotes photoreceptor over bipolar fate (Perron et al., 1999). As is the case for *Xath5*, the cells transfected with these proneural genes also tend to exit the cell cycle at the appropriate histogenetic time to give rise to the cells types that these transcription factors intrinsically favour.

All the proneural genes seem to share the ability to promote neuronal over glial cell fate. Thus, cells transfected with any of the proneural bHLH transcription factor genes have a reduced probability of becoming Müller glia (Cai et al., 2000) whereas mice mutant for proneural bHLH factors show increased numbers of Müller glia (Tomita et al., 1996; Akagi et al., 2004). This finding fits well with the discovery that Notch signalling seems to promote Müller glial fate, probably by inhibiting the expression and activity of proneural bHLH genes (Furukawa et al., 2000; Scheer et al., 2001; Ohnuma et al., 2002). The gliogenic activity of Notch appears to be context-dependent, as coexpression of activated Notch and Xath5 favours the RGC fate, probably due to the effect of Notch on the cell cycle, as we shall see later (Ohnuma et al., 2002).

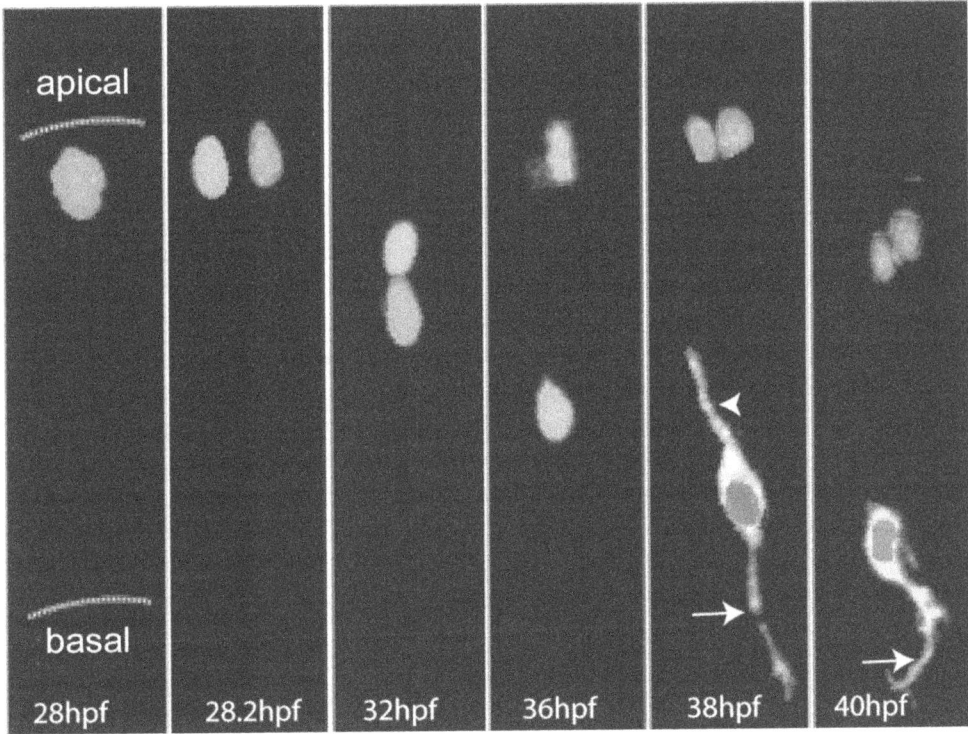

Figure 5.2 Time-lapse analysis of retinal progenitor division. A time-lapse video analysis of a dividing progenitor cell in the zebrafish retina from 28 hours post-fertilization (hpf) to 40 hpf. All cells are labelled with histone-GFP, which marks the nucleus (shown in dark grey). As one of the two daughter cells migrates towards the basal (lens) side, membrane-bound GFP starts to be expressed under the control of the *Ath5* promoter. The *Ath5*-positive cell retracts its apical process (arrowhead) and extends a basal process (arrow) indicating that it is differentiating as a ganglion cell. The second daughter cell initiates another cell cycle and finally undergoes mitosis in the apical surface (courtesy of Lucia Poggi).

5.4.2 Homeobox and other family transcription factors also participate in fate determination

The generation of retinal cell types is also regulated by the action of different homeobox genes. Mice lacking Crx, an Otx family transcription factor, exhibit deficits in photoreceptor outer-segment formation. *Crx* overexpression in P0 progenitors in vivo promotes photoreceptor and inhibits amacrine fate (Furukawa *et al.*, 1997) and *Crx* activates photoreceptor-specific genes (Furukawa *et al.*, 1997; Blackshaw *et al.*, 2001). In vivo overexpression of *Otx5b*, a close homologue of *Crx* in *Xenopus*, promotes photoreceptor cell fate (Viczian *et al.*, 2003). Bipolar cell differentiation requires *Chx10* (Burmeister *et al.*, 1996). *Otx2* overexpression in *Xenopus* suppresses photoreceptor and promotes bipolar fate (Viczian *et al.*, 2003). In mice, however, conditional knockout of *Otx2* prevents photoreceptor differentiation and these cells seem instead either to become amacrine cells or undergo apoptosis, whilst *Otx2* overexpression at P0 activates *Crx* expression and increases photoreceptors (Nishida et al., 2003). The discrepancy in the role of *Otx2* in the mouse and *Xenopus* studies may reflect differences between species in the role of related homeobox genes, or, alternatively, differences in the timing of action of *Otx2* between the two experiments.

Prox1 null mice lack horizontal cells and *Prox1* overexpression promotes horizontal cell fate (Dyer et al., 2003). *Rax*, a gene involved in retinal progenitor proliferation, favours Müller cell production in the mouse retina and, as might be expected from the above, activates Notch (Furukawa et al., 2000). *Pax6*, a gene involved initially in setting up the early eye field, appears to be at the top of a genetic hierarchy activating several bHLH genes which subsequently go on to generate diversity in cell fates. Mice with conditional inactivation of *Pax6* in the retinal progenitors lose expression of many bHLH proneural genes such as *Ngn2*, *Mash1* and *Math5*, but they retain *NeuroD* expression allowing the retinal progenitors to become only amacrine cells (Marquardt et al., 2001). Foxn4, a winged/helix forkhead transcription factor, also works at the early stages of this hierarchy. Mutations in *Foxn4* result in the elimination of horizontal cells and a great reduction of amacrines. *Foxn4* overexpression activates *Math3*, *NeuroD* and *Prox1*, and favours an amacrine fate (Li et al., 2004).

5.4.3 *Multiple interactions between intrinsic factors regulate their function*

Importantly, during retinal development, it seems that several different transcription factors may be expressed in the same progenitor cells, a fact that suggests the possibility of a combinatorial mode of action. There are several findings that support such an idea. Double mutants of *Math3* and *NeuroD* have no amacrine cells in their retinas, with a corresponding increase in ganglion and glial cells, whereas single mutants of either gene exhibit normal amacrine numbers. Overexpression of either *Math3* or *NeuroD* in murine retinal explants results in an increase in rods, however, both produce amacrine cells when coexpressed with either *Pax6* or *Six3* (Inoue et al., 2002). Similarly, *Mash1* or *Math3* overexpression alone increases rods and *Chx10* overexpression alone causes a general increase in inner nuclear cell types, predominantly Müller cells; however, *Mash1/Math3* double mutants exhibit depleted bipolar cells while overexpression of *Chx10* with either *Mash1* or *Math3* promotes bipolar over Müller fate (Hatakeyama et al., 2001). One proposal to account for these results is that these genes do not act on their own to specify bipolar cells, but that *Mash1* and *Math3* may regulate the neuronal versus glial fate while *Chx10* may favour the inner nuclear layer fate. Consequently, their combined action favours bipolar cell production. Another aspect of a combinatorial mode of action is a possible antagonistic relationship of different proneural genes. *Math5* is upregulated in retinas lacking *NeuroD* and *Math3*, resulting in an increase in ganglion cells (Inoue et al., 2002) and for the three proneural genes *Mash1*, *Math3* and *Ngn2* the absence of expression of one gene leads to the upregulation of the other two

(Akagi *et al.*, 2004). A direct repressive activity is indicated by the dominant-negative effect exerted by *Ash1* on the *Ath5* promoter, reducing its response to other activating proneural genes (Matter-Sadzinski *et al.*, 2001). The *Ath5* promoter provides a good example of the relationships at the transcriptional level between different fate-determining factors, as it is regulated by other bHLH genes as well as by *Ath5* itself (Hutcheson *et al.*, 2005; Matter-Sadzinski *et al.*, 2005). Moreover, bHLH and homeobox proteins could cooperate at the molecular level by activating the same gene (e.g. *NeuroD* and *Pitx1* physically interact to direct enhanced synergistic transcription from the pro-opiomelanocortin locus in pituitary cells (Poulin *et al.*, 2000)), functioning in a common cascade (e.g. *Pax6* activating expression of proneural genes in the retina) or directly binding to each other (e.g. *XSix3* binds to *Xath5*, *XNeuroD*, *Xash1* and *Xath3*; Tessmar *et al.*, 2002). It should, however, be noted that a purely combinatorial transcriptional code does not seem to be sufficient to direct cell specification (Wang and Harris, 2005).

In summary, we have mentioned two major sets of transcription factors, bHLH and homeobox, that have specific effects on cell type determination alone or in combination. Their activity is influenced by context, notably by the presence of other transcription factors in the same cells, as well as by extrinsic signals, most clearly seen in the case of Notch signalling. It seems, therefore, that there is some merit in examining the way the intrinsic and extrinsic signalling work together to produce the cell types and numbers appropriate both spatially and temporally.

5.5 The competence model: a balance of intrinsic and extrinsic cues

Despite the abundance of intrinsic and extrinsic factors involved in retinal differentiation, the relationship between the two remains largely unclear and so does their relative contribution to the cell fate decision. Apart from the differences in the early and late environments, it is clear that early and late progenitors are also intrinsically different. Dissociated early progenitors differentiate primarily into RGCs whereas late progenitors differentiate into rods in vitro (Reh and Kljavin, 1989). Early progenitors are not competent to give photoreceptors (Harris and Messersmith, 1992) and the results from the cell-mixing experiments, although showing that extracellular factors can influence fate choice, suggest, however, that progenitors do not produce specific cell types at histogenetically inappropriate times (Belliveau and Cepko, 1999; Belliveau *et al.*, 2000). Moreover, the kinetics of opsin expression differ in cells coming from early and late progenitors and they cannot be changed by exposure to a heterochronic environment either in vivo (Rapaport *et al.*, 2001) or in vitro (Morrow *et al.*, 1998), suggesting an intrinsic programme dictating differentiation irrespectively of external cues. In some cases this restriction may not be absolute. Late rat retinal progenitors were found to increase their production of ganglion cells when in contact with early retinal cells in culture (James *et al.*, 2003), however, this effect appeared to be dependent on the exposure of the progenitors to in vitro cues.

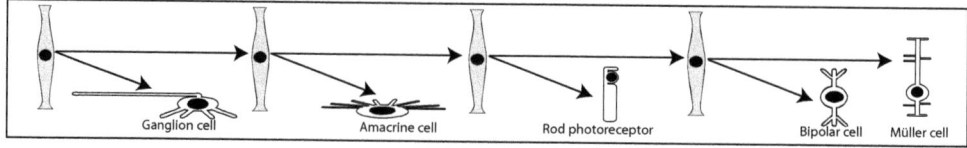

(a) The competence model: a dynamic environment and changing progenitor intrinsic characteristics

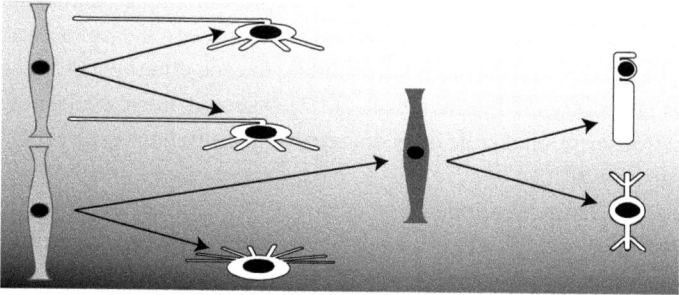

(b) The progenitor mosaic model: intrinsic programmes only in a heterogeneous progenitor population

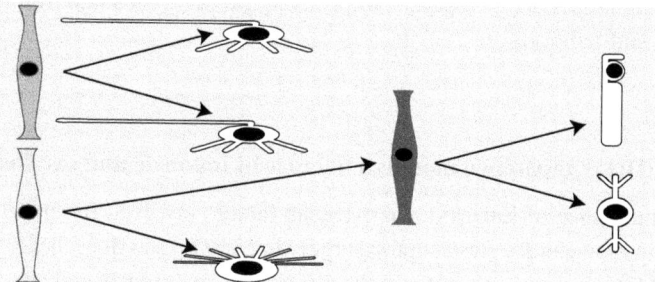

(c) The stochastic model: intrinsic programmes in similar progenitors confer biases for different fates

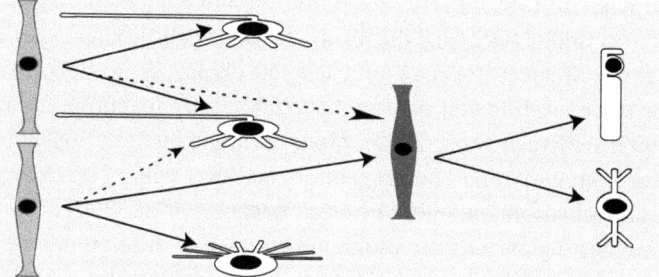

Figure 5.3 Models for retinal cell fate determination. Top box: a diagram showing the birth order of various retinal cells during histogenesis. (a–c) Examples of clonal compositions of two retinal progenitors and three models to explain their pattern. The background shading denotes the environment of the cells, whereas the progenitors' shade reflects their intrinsic characteristics.

The competence model (see Figure 5.3a) put forward to explain the majority of the above results suggests that the intrinsic competence of progenitor cells to respond to cues changes with time such that at any given point they have a limited repertoire of cell fates available for their progeny (Watanabe and Raff, 1990; Livesey and Cepko, 2001; Rapaport et al., 2001). This concept could have many explanations at the molecular level. The receptor composition and downstream events conveying extracellular signals may change with time, as for example the GlyRα2 expression is upregulated in late progenitors enabling rod production (Lillien, 1995; Young and Cepko, 2004). Extracellular signals may regulate the competence to produce different fates by modulating the activity of intrinsic cell fate determinants, as in the case of Notch signalling. The secreted molecule Gdf11 inhibits the production of RGCs by limiting the temporal window of *Math5* expression. *Gdf11* mutant mice exhibit an increase in RGCs, a prolonged *Math5* expression and a delay in the expression of *NeuroD* and *Mash1*, bHLH genes that favour later neuronal fates, without any changes in progenitor proliferation (Kim et al., 2005). Glycogen synthase kinase-3β (GSK 3β) phosphorylates and inactivates *NeuroD* in early progenitors; a GSK 3β-insensitive form of *NeuroD* transfected early on favours ganglion cell production, not amacrine (Moore et al., 2002). Activity of GSK 3β can be modulated by extrinsic cues such as Wnt signalling and tyrosine kinase receptor activation, thus providing a possible link between a changing environment and the role of intrinsic factors. Finally, the cross-regulatory effects of transcription factors mentioned above could alter their relative expression levels with time, changing the favoured cell fates. These different aspects of cellular competence appear to change in parallel with

Figure 5.3 (*cont.*) (a) The competence model: progenitor cells are initially equivalent and competent to produce only early-born cells. The two progenitors produce different early-born neurons by virtue of their exposure to different environments. For example, the top progenitor might receive much less Notch signalling and hence give rise to two ganglion cells, whereas the bottom one gives rise to another progenitor and an amacrine cell. In late histogenesis, the remaining progenitor has different intrinsic characteristics and also receives signals from a changed environment, thus becoming competent to give only late-born cells. (b) The progenitor mosaic model: at the beginning of differentiation the progenitor cells have already diversified, and they step through a series of preprogrammed differentiated cell production. In this case, our two progenitors differ greatly in their intrinsic programmes, for example the top one but not the bottom strongly expresses at the beginning of differentiation transcription factors that favour a RGC fate. The environment regulates cell survival and the maturation of the differentiated phenotype but not the cell fate decisions. (c) Stochastic choice model: intrinsic factors bias each progenitor towards the generation of particular cell types, with different probabilities for each type. Random fluctuations in the activity of different fate-influencing molecules can be amplified so that initially similar progenitors can produce different clones and the overall cell numbers and clonal compositions are determined by the probability for generation of each type. In this case again the environment does not play a significant role in fate determination.

A combination of the above models could explain the available data. For example, a progenitor population with some heterogeneity may initially be dependent on extrinsic cues for the stochastic production of early cell types and diversification of the progenitor pool. With time the now diversified progenitor population becomes more and more dependent on intrinsic cues.

a dynamic environment, as illustrated above, but it is not clear what the relative importance of intrinsic fate determination programmes is compared to external cues.

5.6 Contribution of progenitor heterogeneity to cell fate determination

Recent work puts into doubt the significance of the extrinsic cues, and consequently of the competence model. Cayouette *et al.* (2003) found that dissociated progenitors of mouse E16/17 retinas go on to produce progeny in similar proportions and clonal compositions to the in vivo progenitors. They suggested that extrinsic cues are not critical for the cell fate determination, which is dictated solely by an intrinsic programme, different for each progenitor, that is already in place by at least as early as E16 (see Figure 5.3b). The multiple effects of the extrinsic factors described previously could simply be fine-tuning mechanisms, regulating proliferation and differentiation in space and time (e.g. by local, transient feedback inhibition) and influencing the late stages of differentiation of committed, immature precursors (e.g. leukaemia inhibitory factor release from Müller cells may regulate the timing of rhodopsin expression in immature but committed rod precursors (Neophytou *et al.*, 1997)). However, the number of different cell types, their relative order of birth and the clonal composition would be determined autonomously for each progenitor by an internal motor, driving progenitors in successive rounds of division to produce specific cell types. It should be noted that it is unclear, nevertheless, how this hypothesis can be reconciled with findings of cell fate changes when extracellular signals are inhibited (e.g. Young and Cepko, 2004). One might argue that the extracellular signals are primarily needed as permissive rather than instructive factors for differentiation of certain cell types, and that what changes with time and therefore dictates the cell fate is the intrinsic, autonomous change in the reception of the signal. However, the fact that progenitors were examined only after E16 does nothing to exclude the possibility that extrinsic signals have a role in diversifying the progenitor population during early differentiation.

The above results suggest the presence of a rich mosaic of heterogeneous retinal progenitors at least during later differentiation in the mouse. Several other pieces of evidence suggested a degree of heterogeneity exists from the very beginning of retinal differentiation. Fate-mapping studies have shown correlations between progenitor origin and expression of molecules in subsets of progenitors and the fate of their progeny. Moody and colleagues found that blastomere origin of *Xenopus* retinal neurons biases their cell fate choice. The large majority of neuropeptide Y- or dopamine-positive amacrines originate from specific blastomeres (Huang and Moody, 1995) and, critically, transplantations of individual blastomeres showed that some, but not all, are intrinsically biased even at this early cleavage stage to produce an amacrine subtype (Moody *et al.*, 2000). Particular blastomeres have a bias in producing specific classes of 5-hydroxytryptamine amacrines despite the fact that these cells have no spatial bias in their distribution in the retinal antero posterior and dorsoventral quadrants (Huang and Moody, 1997).

Progenitors carrying certain markers were found to be biased in the cell fates they produce. Alexiades and Cepko (1997) labelled progenitor cells and their progeny permanently by using fluorescent latex microspheres coated to antibodies against VC1.1 and syntaxin-1a, markers of differentiated amacrine and horizontal cells. By co-labelling with ^{3}H-thymidine, they showed that embryonic progenitors expressing VC1.1 and syntaxin were biased towards giving more amacrine and horizontal cells, whereas VC1.1-negative progenitors gave more cones. Similarly, later progenitors labelled thus gave amacrine and rod cells. p27Kip1 and p57Kip2, two cyclin-dependent kinase (cdk) inhibitors, are expressed in complementary domains in the mouse retina and amacrine cells seem to originate from the p57-positive population (Dyer and Cepko, 2000b, 2001).

Further indication for heterogeneity comes from results that suggest some fate-influencing genes are expressed in subsets of progenitors. In a recent genomic study, Livesey *et al.* (2004) identified a small number of genes such as *Otx2* and *SFRP2*, which are expressed in subsets of progenitors; however, the authors note this pattern may reflect cell cycle variation of expression as opposed to heterogeneity. The vast majority of progenitor-specific genes seem to have a ubiquitous expression pattern.

Apart from the predominant multipotent progenitor demonstrated in the cell lineage experiments (Turner and Cepko, 1987; Holt *et al.*, 1988; Wetts and Fraser, 1988) could there be also unipotent progenitors? Studies in the mature teleost retina have shown the presence of a rod-only precursor (Mack and Fernald, 1995) and there has been a report of a mouse E14 progenitor producing a clone of 33 rods (Turner *et al.*, 1990).

If there is heterogeneity among retinal progenitors that contributes to their production of particular daughter cell types, a question arises about whether this heterogeneity is deliberately programmed or comes about as a matter of chance and early environmental exposure (see Figure 5.3). There may, for example, be fluctuations in the levels of *Ath5* and other intrinsic determinants that neighbouring progenitors have in their nucleus and these different levels may influence the probability that the progenitors will produce a ganglion cell. Similarly, progenitor cells may have been exposed to small variations in signalling through the Notch or other pathways also resulting in slight heterogeneities in their potential. In this view, retinal progenitors are not developmentally programmed to be different from each other, but differences arise as a consequence of unprogrammed variations in the way similar cells develop.

5.7 The mechanism of cell cycle progression and cell fate determination

A highly conserved molecular mechanism underlies progression through the cell cycle. Coordination between cell cycle and fate determination is essential, not least because of the birthdate effect on cell fate and the need to generate the appropriate numbers of early- and late-born neurons. Moreover the observation that, in most cases, differentiated cells have permanently exited the cycle and cannot divide suggests that the process of differentiation is somehow linked to the cell cycle (for a possible exception in the retina see Dyer and Cepko, 2000a; Fischer and Reh, 2001).

Classical transplantation studies in the ferret cortex (McConnell and Kaznowski, 1991) suggested that an environmental signal acts before the terminal S-phase of the cell cycle to impart the fate of the postmitotic daughter cell. So one important aspect of this issue is the timing of the cell fate decision during the cell cycle, indeed whether there is such a decision by which a cell irreversibly commits to a fate.

The second important aspect is how the coordination between cycling and differentiation is achieved. Is differentiation dictating cell cycle exit or does the cessation of the cycle allow differentiation towards particular fates to take place? Recent studies in many systems, including the retina, have illustrated a bidirectional relationship between components of the cell cycle machinery and cell fate determinants.

5.7.1 Cell cycle components influence fate determination

Inhibition of cell cycle progression using the DNA synthesis inhibitors aphidicolin and hydroxyurea still resulted in the generation of diverse neuronal cell types in the *Xenopus* CNS, suggesting that cell cycle components are not essential for the generation of particular fates (Harris and Hartenstein, 1991). What roles then have been demonstrated for these components in determination? Ohnuma *et al.* (1999) found that p27Xic1, an inhibitor of the G1S-phase transition, promotes Müller cell fate over bipolars when overexpressed in the *Xenopus* retina and its inhibition reduces Müller cells. This effect is mediated by the cdk/cyclin-binding domain of p27Xic1, however it is worth noting that a mutant form of the mammalian homologue p21Cip1, which exhibits minimal cell cycle-inhibitory activity, has nevertheless the same Müller-promoting effect when overexpressed in *Xenopus*, suggesting the gliogenic effect of p27Xic1 is distinct from its action on the cell cycle. The authors hypothesized that the increasing p27Xic1 levels with development may push later cells to exit the cycle as Müller cells (Figure 5.4). On the other hand, cell cycle inhibition by p27Xic1 when *Xath5* is overexpressed enhances the ganglion cell fate-promoting effects of *Xath5*, suggesting that early cell cycle exit promotes early cell-fates (Ohnuma *et al.*, 2002). In mice, a mammalian homologue p27Kip1 was not found to affect cell fate, as mice lacking p27Kip1 did not have altered proportions of retinal cell types (Dyer and Cepko, 2001). Mice mutant in p57Kip2 however had increased calbindin-positive amacrine cells (Dyer and Cepko, 2000b), reflecting species-specific or cell-specific differences in the role of cell cycle inhibitors.

The tumour suppressor Retinoblastoma (Rb), which regulates cell cycle progression by repressing E2F-mediated-S-phase initiation, also has an effect in retinal differentiation. Mice with a conditional Rb mutation show deficits in rod differentiation, which indicates Rb has an effect on rod phenotype maturation subsequent to the cell fate decision; however, microarray analysis suggests Rb may be upstream of *Nrl*, a gene implicated in cell fate choice (Zhang *et al.*, 2004). Overexpression of *Prox1*, the vertebrate homologue of *prospero*, forces progenitors to exit the cycle and promotes horizontal cell fate, whereas *Prox1* mutant mice do not have horizontal cells (Dyer *et al.*, 2003). *Chx10*, necessary for bipolar development, also regulates progenitor proliferation and its mutation causes the mouse ocular retardation phenotype (Burmeister *et al.*, 1996); however, its effects on cellular proliferation and cell

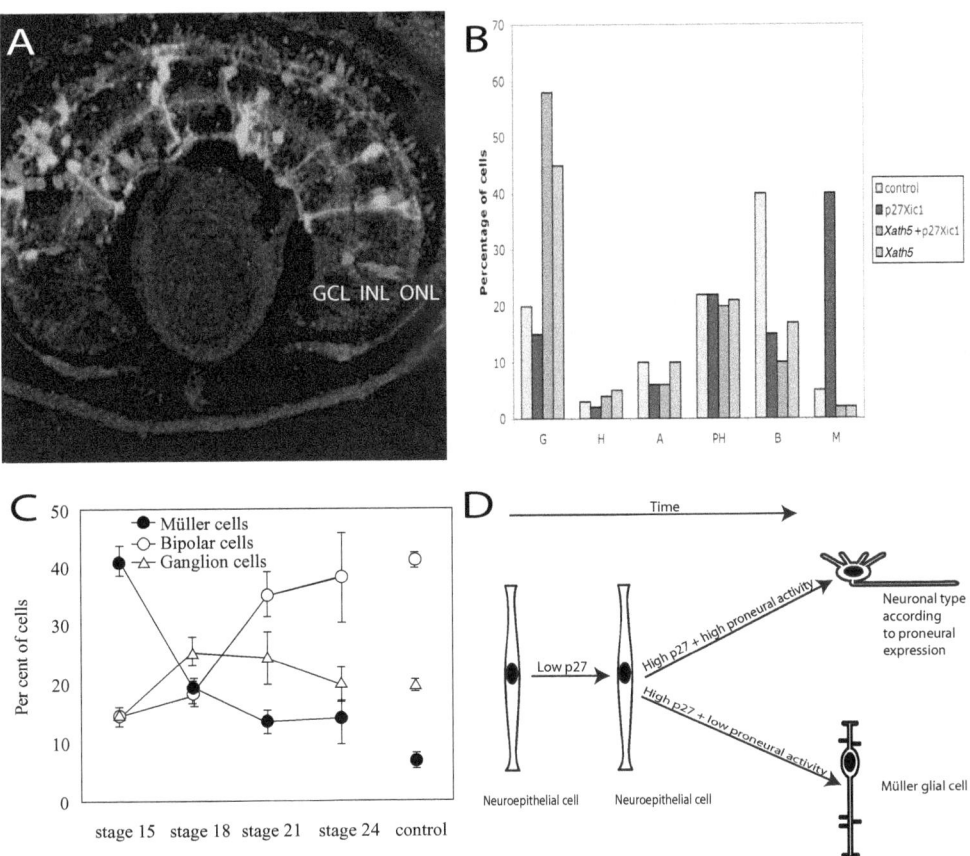

Figure 5.4 Action of p27Xic1, a cell cycle inhibitor, on cell fate determination. (A, B) Lipofection of p27Xic1 in the *Xenopus* retina results in an increase in Müller cells and decrease in bipolar cells. Co-lipofection of p27Xic1 and *Xath5* results instead in an increase in the ganglion cell-promoting effect of *Xath5*. GCL, ganglion cell layer; INL, inner nuclear layer; ONL, outer nuclear layer. (A from Ohnuma *et al.*, 1999; B adapted from Ohnuma *et al.*, 1999, 2002.) (C) Lipofection of p27Xic1 at later points in histogenesis results in fewer Müller cells generated because p27Xicl is now acting in the context of a higher proneural activity (from Ohnuma *et al.*, 1999). (D) A model for the effect of p27 on cell fate: low p27 initially will allow a neuroepithelial cell to stay in the cycle after division. As the p27 levels rise, cells with high proneural activity will exit the cycle and assume the neuronal fates appropriate for their proneural profile and the histogenetic time of birth. High p27 levels will also bias low proneural activity cells to differentiate as Müller cells and exit the cycle.

fate decisions appear to be distinct, as cell number, but not the bipolar fate, can be rescued in the *Chx10/p27Kip1* double mutant (Green *et al.*, 2003). *Six3* promotes amacrine fate when coexpressed with *Math3* and *NeuroD* (Inoue *et al.*, 2002), physically interacts with bHLH proteins (Tessmar *et al.*, 2002) and promotes progenitor proliferation by binding to the replication initiation inhibitor geminin (Del Bene *et al.*, 2004).

5.7.2 Several fate determination factors impact on the cell cycle

Many fate-influencing extrinsic factors are mitogens and they have a dose- and receptor-composition-dependent effect on the coordination of proliferation and differentiation (TGF-α: Lillien and Cepko, 1992; Lillien, 1995; Lillien and Wancio, 1998; TGF-β: Anchan and Reh, 1995; Shh: Jensen and Wallace, 1997; Zhang and Yang, 2001). Notch signalling in the presence of proneural activity induces early cell cycle exit (Ohnuma *et al.*, 2002). Retinal neurons in *Ath5* zebrafish mutants exhibit delayed cell cycle exit (Kay *et al.*, 2001) and there is some evidence from other systems that bHLH genes may activate expression of cell cycle inhibitors (Farah *et al.*, 2000).

Apart from the transcriptional regulation of the two pathways, less studied mechanisms might involve post-translational events. For example, determination factors may be targets of Cdk-phosphorylation or protein degradation in other systems (e.g. Reynaud *et al.*, 1999).

5.7.3 When is cell fate determined?

Studies on the timing of cell fate commitment have given inconsistent results. Adler and Hatlee (1989) found a postmitotic effect of the chick retinal environment biasing cells away from a photoreceptor fate. Ciliary neurotrophic factor was found to be able to divert rat postmitotic cells from a rod to a bipolar fate up until the time of opsin expression by the rod cell (Ezzeddine *et al.*, 1997), although Neophytou *et al.* (1997) reported that this effect was a reversible rod differentiation arrest and not a respecification to the bipolar fate.

On the other hand, cells destined to become RGCs were found to express an RGC-specific marker, RA4, 15 minutes after terminal mitosis, suggesting ganglion cell specification occurred before or during M-phase (Waid and McLoon, 1995). Moreover, the negative feedback mechanism for amacrine production appears to act before the progenitor terminal M-phase (Belliveau and Cepko, 1999). These incongruities may reflect cell-type-specific differences or, at least for some cases, may point to a treatment-related plasticity of the cell fate even after the cell is on its way to a particular fate.

5.8 Comparison with other systems

The cell fate specification in the vertebrate retina presents similarities in the types of molecular mechanisms used in other systems, and seems to employ a variety of strategies each of which, but maybe not all together, are found elsewhere. In the *Drosophila* eye, for example, the Atonal–Notch lateral inhibition pathway specifies the first photoreceptor, R8 (similarly it contributes to RGC specification in the vertebrates); however, the other seven photoreceptors in a cluster are recruited in a stereotypical sequential fashion. The Notch and Receptor Tyrosine Kinase–Ras–Mitogen-Activated Protein Kinase pathways are instrumental in this process and cooperate with or may induce different intracellular factors such as rough and lozenge, which go on to specify the fate (reviewed in Voas and Rebay, 2004). The *Drosophila Prox1* homologue *Prospero* and the *Otx/Crx* homologue *Otd* are also involved in photoreceptor differentiation. In general, parallels pointing to common evolutionary origins have been drawn between first-born retinal neurons in several species. *Ath*, an *atonal/Ath5* homologue, also specifies the first photoreceptors to differentiate in the larval *Platynereis*, a species of Polychaeta, with two simple eye structures each having one photoreceptor and one pigment cell and the vertebrate RGC is similar to the invertebrate rhabdomeric photoreceptor because of similarities in their axonal projections, *atonal/Ath5* specification of their fate, *Brn3* and *BarH* expression and expression of r-opsins (vertebrate melanopsin) (Arendt, 2003).

Development of the *Drosophila* CNS resembles the vertebrate retina in that multipotent progenitors go through a series of divisions giving cell types dependent on their birthdate. Doe and colleagues developed an elegant model involving a fixed, sequential, transient expression of heterochronic genes in each neuroblast dictating the fate of the progeny born at the time of expression. They hypothesized these genes may work via chromatin modifications to translate a temporary neuroblast code to a permanent progeny identity (Isshiki *et al.*, 2001; Pearson and Doe, 2003). Such a mechanism has not yet been found in the vertebrate eye, and the study of epigenetic modifications in cycling progenitors may prove a fruitful area of investigation.

In the vertebrate spinal cord, cell fate is determined by gradients of dorsoventral and rostrocaudal signalling molecules. This mechanism is clearly different from the ones that operate in the retina, however other aspects are more similar, as illustrated for example by the differentiation of multipotent progenitors into both motor neurons and oligodendrocytes. The sequential generation is dictated by the change in expression of transcription factors whose combined action dictates neuronal versus glial fate (Novitch *et al.*, 2001; Zhou *et al.*, 2001). In general, the employment of a transcriptional combinatorial code of homeobox genes and perhaps bHLH proteins to define neuronal classes along the dorsoventral axis is the predominant theme of cell fate specification in the spinal cord and resembles the action of intrinsic factors in the retina. Moreover, there is an influence of the birthdate on cell subtype within the motor neuron domain brought about through intercellular interactions of the daughter cells (reviewed in Jessell, 2000).

5.9 Concluding remarks

Although several mechanisms have been found to operate in directing the multipotent retinal progenitors to produce the different cellular fates, integration of these aspects in a single framework has been lagging. One important area of investigation will therefore be the establishment of connections between the different mechanisms, particularly how signalling pathways downstream of extrinsic cues interact with intrinsic factors and what their relative contribution to the fate decision is. Does a precursor move from an early, largely uncommitted, extrinsic signal-dependent programme to a later autonomous intrinsic programme and if so, when does this happen? Are these programmes strictly dictating the clonal composition of each progenitor, or are we dealing with a more stochastic determination, with the different molecules imparting biases to the progenitors for one fate or the other?

Related to this, it will be important to clarify the extent to which the decision for the cell fate is linked to a gradual differentiation process, and whether the competence of a precursor is restricted in a step-wise fashion so that the determination process initially leads to a generic cell type, such as a ganglion or bipolar cell and subsequently specifies subtypes. For example, Harris and Messersmith (1992) suggested that two extrinsic inductive events in the retina specify first a photoreceptor fate and subsequently a rod or cone fate and, in the same vein, the transcription factor Nrl may be one factor dictating rod versus cone specification, as mice mutant for Nrl lose their rods and display an increase in S-cone-like cells (Mears *et al.*, 2001). On the other hand, the production of certain amacrine subtypes is influenced by blastomere origin, suggesting that subtype specification does not strictly follow a generic type specification (Moody *et al.*, 2000) and, similarly, manipulation of *BarHl2* expression affects the production of glycinergic amacrines at the expense of Müller, bipolar or photoreceptor numbers but has no effect on γ-aminobutyric acid (GABA) ergic amacrines (Mo *et al.*, 2004).

Ultimately, the answer to many of these questions must link the fate-determining factors described in this chapter to the mature cellular phenotype. Some progress has already been made, particularly in studies of rhodopsin expression regulation; however, important matters are still open. Are there 'master' regulatory genes integrating the different signalling pathways to regulate all aspects of cell differentiation or are multiple parallel pathways acting in concert to produce the mature phenotype? What is the connection between the expression of genes specific to one cell type and the repression of genes specific to another cell type or to pluripotency?

Powerful new tools, including reverse genetics in the mouse, forward genetics in zebrafish, genomic analysis and in vivo time-lapse imaging, promise exciting developments in the coming years.

References

Adler, R. and Hatlee, M. (1989). Plasticity and differentiation of embryonic retinal cells after terminal mitosis. *Science*, **243**, 391–3.

Akagi, T., Inoue, T., Miyoshi, G. *et al.* (2004). Requirement of multiple basic helix-loop-helix genes for retinal neuronal subtype specification. *J. Biol. Chem.*, **279**, 28 492–8.

Alexiades, M. R. and Cepko, C. L. (1997). Subsets of retinal progenitors display temporally regulated and distinct biases in the fates of their progeny. *Development*, **124**, 1119–31.

Altshuler, D., Lo Turco, J. J., Rush, J. and Cepko, C. (1993). Taurine promotes the differentiation of a vertebrate retinal cell type in vitro. *Development*, **119**, 1317–28.

Anchan, R. M. and Reh, T. A. (1995). Transforming growth factor-beta-3 is mitogenic for rat retinal progenitor cells in vitro. *J. Neurobiol.*, **28**, 133–45.

Arendt, D. (2003). Evolution of eyes and photoreceptor cell types. *Int. J. Dev. Biol.*, **47**, 563–71.

Austin, C. P., Feldman, D. E., Ida, J. A., Jr and Cepko, C. L. (1995). Vertebrate retinal ganglion cells are selected from competent progenitors by the action of Notch. *Development*, **121**, 3637–50.

Belecky-Adams, T., Cook, B. and Adler, R. (1996). Correlations between terminal mitosis and differentiated fate of retinal precursor cells *in vivo* and in vitro: analysis with the 'window-labelling' technique. *Dev. Biol.*, **178**, 304–15.

Belliveau, M. J. and Cepko, C. L. (1999). Extrinsic and intrinsic factors control the genesis of amacrine and cone cells in the rat retina. *Development*, **126**, 555–66.

Belliveau, M. J., Young, T. L. and Cepko, C. L. (2000). Late retinal progenitor cells show intrinsic limitations in the production of cell types and the kinetics of opsin synthesis. *J. Neurosci.*, **20**, 2247–54.

Blackshaw, S., Fraioli, R. E., Furukawa, T. and Cepko, C. L. (2001). Comprehensive analysis of photoreceptor gene expression and the identification of candidate retinal disease genes. *Cell*, **107**, 579–89.

Brown, N. L., Patel, S., Brzezinski, J. and Glaser, T. (2001). Math5 is required for retinal ganglion cell and optic nerve formation. *Development*, **128**, 2497–508.

Burmeister, M., Novak, J., Liang, M. Y. *et al.* (1996). Ocular retardation mouse caused by Chx10 homeobox null allele: impaired retinal progenitor proliferation and bipolar cell differentiation. *Nat. Genet.*, **12**, 376–84.

Cai, L., Morrow, E. M. and Cepko, C. L. (2000). Mis-expression of basic helix-loop-helix genes in the murine cerebral cortex affects cell fate choices and neuronal survival. *Development*, **127**, 3021–30.

Cayouette, M., Barres, B. A. and Raff, M. (2003). Importance of intrinsic mechanisms in cell fate decisions in the developing rat retina. *Neuron*, **40**, 897–904.

Del Bene, F., Tessmar-Raible, K. and Wittbrodt, J. (2004). Direct interaction of geminin and Six3 in eye development. *Nature*, **427**, 745–9.

Dorsky, R. I., Rapaport, D. H. and Harris, W. A. (1995). Xotch inhibits cell differentiation in the Xenopus retina. *Neuron*, **14**, 487–96.

Dorsky, R. I., Chang, W. S., Rapaport, D. H. and Harris, W. A. (1997). Regulation of neuronal diversity in the Xenopus retina by Delta signaling. *Nature*, **385**, 67–70.

Dyer, M. A. and Cepko, C. L. (2000a). Control of Müller glial cell proliferation and activation following retinal injury. *Nat. Neurosci.*, **3**, 873–80.

Dyer, M. A. and Cepko, C. L. (2000b). p57(Kip2) regulates progenitor cell proliferation and amacrine interneuron development in the mouse retina. *Development*, **127**, 3593–605.

Dyer, M. A. and Cepko, C. L. (2001). p27Kip1 and p57Kip2 regulate proliferation in distinct retinal progenitor cell populations. *J. Neurosci.*, **21**, 4259–71.

Dyer, M. A., Livesey, F. J., Cepko, C. L. and Oliver, G. (2003). Prox1 function controls progenitor cell proliferation and horizontal cell genesis in the mammalian retina. *Nat. Genet.*, **34**, 53–8.

Ezzeddine, Z. D., Yang, X., DeChiara, T., Yancopoulos, G. and Cepko, C. L. (1997). Postmitotic cells fated to become rod photoreceptors can be respecified by CNTF treatment of the retina. *Development*, **124**, 1055–67.

Farah, M. H., Olson, J. M., Sucic, H. B. *et al.* (2000). Generation of neurons by transient expression of neural bHLH proteins in mammalian cells. *Development*, **127**, 693–702.

Fischer, A. J. and Reh, T. A. (2001). Müller glia are a potential source of neural regeneration in the postnatal chicken retina. *Nat. Neurosci.*, **4**, 247–252.

Fuhrmann, S., Kirsch, M. and Hofmann, H. D. (1995). Ciliary neurotrophic factor promotes chick photoreceptor development in vitro. *Development*, **121**, 2695–706.

Furukawa, T., Morrow, E. M. and Cepko, C. L. (1997). *Crx*, a novel otx-like homeobox gene, shows photoreceptor-specific expression and regulates photoreceptor differentiation. *Cell*, **91**, 531–41.

Furukawa, T., Mukherjee, S., Bao, Z. Z., Morrow, E. M. and Cepko, C. L. (2000). rax, Hes1, and notch1 promote the formation of Müller glia by postnatal retinal progenitor cells. *Neuron*, **26**, 383–94.

Green, E. S., Stubbs, J. L. and Levine, E. M. (2003). Genetic rescue of cell number in a mouse model of microphthalmia: interactions between Chx10 and G1-phase cell cycle regulators. *Development*, **130**, 539–52.

Harris, W. A. and Hartenstein, V. (1991). Neuronal determination without cell division in Xenopus embryos. *Neuron*, **6**, 499–515.

Harris, W. A. and Messersmith, S. L. (1992). Two cellular inductions involved in photoreceptor determination in the Xenopus retina. *Neuron*, **9**, 357–72.

Hatakeyama, J., Tomita, K., Inoue, T. and Kageyama, R. (2001). Roles of homeobox and bHLH genes in specification of a retinal cell type. *Development*, **128**, 1313–22.

Henrique, D., Hirsinger, E., Adam, J. *et al.* (1997). Maintenance of neuroepithelial progenitor cells by Delta-Notch signaling in the embryonic chick retina. *Curr. Biol.*, **7**, 661–70.

Holt, C. E., Bertsch, T. W., Ellis, H. M. and Harris, W. A. (1988). Cellular determination in the Xenopus retina is independent of lineage and birth date. *Neuron*, **1**, 15–26.

Huang, S. and Moody, S. A. (1995). Asymmetrical blastomere origin and spatial domains of dopamine and neuropeptide Y amacrine sub-types in Xenopus tadpole retina. *J. Comp. Neurol.*, **360**, 442–53.

Huang, S. and Moody, S. A. (1997). Three types of serotonin-containing amacrine cells in tadpole retina have distinct clonal origins. *J. Comp. Neurol.*, **387**, 42–52.

Hutcheson, D. A. and Vetter, M. L. (2001). The bHLH factors Xath5 and XNeuroD can upregulate the expression of XBrn3d, a POU-homeodomain transcription factor. *Dev. Biol.*, **232**, 327–38.

Hutcheson, D. A., Hanson, M. I., Moore, K. B. *et al.* (2005). bHLH-dependent and -independent modes of *Ath5* gene regulation during retinal development. *Development*, **132**, 829–39.

Hyatt, G. A., Schmitt, E. A., Fadool, J. M. and Dowling, J. E. (1996). Retinoic acid alters photoreceptor development *in vivo*. *Proc. Natl. Acad. Sci. U. S. A.*, **93**, 13 298–303.

Inoue, T., Hojo, M., Bessho, Y. *et al.* (2002). Math3 and NeuroD regulate amacrine cell fate specification in the retina. *Development*, **129**, 831–42.

Isshiki, T., Pearson, B., Holbrook, S. and Doe, C. Q. (2001). Drosophila neuroblasts sequentially express transcription factors which specify the temporal identity of their neuronal progeny. *Cell*, **106**, 511–21.

James, J., Das, A. V., Bhattacharya, S. *et al.* (2003). In vitro generation of early-born neurons from late retinal progenitors. *J. Neurosci.*, **23**, 8193–203.

Jensen, A. M. and Wallace, V. A. (1997). Expression of Sonic hedgehog and its putative role as a precursor cell mitogen in the developing mouse retina. *Development*, **124**, 363–71.

Jessell, T. M. (2000). Neuronal specification in the spinal cord: inductive signals and transcriptional codes. *Nat. Rev. Genet.*, **1**, 20–9.

Kanekar, S., Perron, M., Dorsky, R. *et al.* (1997). *Xath5* participates in a network of bHLH genes in the developing Xenopus retina. *Neuron*, **19**, 981–94.

Kay, J. N., Finger-Baier, K. C., Roeser, T., Staub, W. and Baier, H. (2001). Retinal ganglion cell genesis requires lakritz, a Zebrafish atonal Homolog. *Neuron*, **30**, 725–36.

Kelley, M. W., Turner, J. K. and Reh, T. A. (1995). Ligands of steroid/thyroid receptors induce cone photoreceptors in vertebrate retina. *Development*, **121**, 3777–85.

Kelley, M. W., Williams, R. C., Turner, J. K., Creech-Kraft, J. M., Reh, T. A. (1999). Retinoic acid promotes rod photoreceptor differentiation in rat retina *in vivo*. *NeuroReport*, **10**, 2389–94.

Kim, J., Wu, H. H., Lander, A. D. *et al.* (2005). GDF11 controls the timing of progenitor cell competence in developing retina. *Science*, **308**, 1927–30.

LaVail M. M., Rapaport, D. H. and Rakic, P. (1991). Cytogenesis in the monkey retina. *J. Comp. Neurol.*, **309**, 86–114.

Li, S., Mo, Z., Yang, X. *et al.* (2004). Foxn4 controls the genesis of amacrine and horizontal cells by retinal progenitors. *Neuron*, **43**, 795–807.

Lillien, L. (1995). Changes in retinal cell fate induced by over-expression of EGF receptor. *Nature*, **377**, 158–62.

Lillien, L. and Cepko, C. (1992). Control of proliferation in the retina: temporal changes in responsiveness to FGF and TGF alpha. *Development*, **115**, 253–66.

Lillien, L. and Wancio, D. (1998). Changes in epidermal growth factor receptor expression and competence to generate glia regulate timing and choice of differentiation in the retina. *Mol. Cell. Neurosci.*, **10**, 296–308.

Liu, W., Mo, Z. and Xiang, M. (2001). The *Ath5* proneural genes function upstream of *Brn3* POU domain transcription factor genes to promote retinal ganglion cell development. *Proc. Natl. Acad. Sci. U. S. A.*, **98**, 1649–54.

Livesey, F. J. and Cepko, C. L. (2001). Vertebrate neural cell-fate determination: lessons from the retina. *Nat. Rev. Neurosci.*, **2**, 109–18.

Livesey, F. J., Young, T. L. and Cepko, C. L. (2004). An analysis of the gene expression program of mammalian neural progenitor cells. *Proc. Natl. Acad. Sci. U. S. A.*, **101**, 1374–79.

Mack, A. F. and Fernald, R. D. (1995). New rods move before differentiating in adult teleost retina. *Dev. Biol.*, **170**, 136–41.

Marquardt, T., Ashery-Padan, R., Andrejewski, N. *et al.* (2001). *Pax6* is required for the multipotent state of retinal progenitor cells. *Cell*, **105**, 43–55.

Matter-Sadzinski, L., Matter, J. M., Ong, M. T., Hernandez, J. and Ballivet, M. (2001). Specification of neurotransmitter receptor identity in developing retina: the chick ATH5 promoter integrates the positive and negative effects of several bHLH proteins. *Development*, **128**, 217–31.

Matter-Sadzinski, L., Puzianowska-Kuznicka, M., Hernandez, J., Ballivet, M. and Matter, J. M. (2005). A bHLH transcriptional network regulating the specification of retinal ganglion cells. *Development*, **132**, 3907–21.

McConnell, S. K. and Kaznowski, C. E. (1991). Cell cycle dependence of laminar determination in developing neocortex. *Science*, **254**, 282–5.

McFarlane, S., Zuber, M. E. and Holt, C. E. (1998). A role for the fibroblast growth factor receptor in cell fate decisions in the developing vertebrate retina. *Development*, **125**, 3967–75.

Mears, A. J., Kondo, M., Swain, P. K. *et al.* (2001). Nrl is required for rod photoreceptor development. *Nat. Genet.*, **29**, 447–52.

Mo, Z., Li, S., Yang, X. and Xiang, M. (2004). Role of the *Barhl2* homeobox gene in the specification of glycinergic amacrine cells. *Development*, **131**, 1607–18.

Moody, S. A., Chow, I. and Huang, S. (2000). Intrinsic bias and lineage restriction in the phenotype determination of dopamine and neuropeptide Y amacrine cells. *J. Neurosci.*, **20**, 3244–53.

Moore, K. B., Schneider, M. L. and Vetter, M. L. (2002). Posttranslational mechanisms control the timing of bHLH function and regulate retinal cell fate. *Neuron*, **34**, 183–95.

Morrow, E. M., Belliveau, M. J. and Cepko, C. L. (1998). Two phases of rod photoreceptor differentiation during rat retinal development. *J. Neurosci.*, **18**, 3738–48.

Morrow, E. M., Furukawa, T., Lee, J. E. and Cepko, C. L. (1999). NeuroD regulates multiple functions in the developing neural retina in rodent. *Development*, **126**, 23–36.

Negishi, K., Teranishi, T. and Kato, S. (1982). New dopaminergic and indoleamine-accumulating cells in the growth zone of goldfish retinae after neurotoxic destruction. *Science*, **216**, 747–9.

Neophytou, C., Vernallis, A. B., Smith, A. and Raff, M. C. (1997). Müller-cell-derived leukaemia inhibitory factor arrests rod photoreceptor differentiation at a post-mitotic pre-rod stage of development. *Development*, **124**, 2345–54.

Nishida, A., Furukawa, A., Koike, C. *et al.* (2003). *Otx2* homeobox gene controls retinal photoreceptor cell fate and pineal gland development. *Nat. Neurosci.*, **6**, 1255–63.

Novitch, B. G., Chen, A. I. and Jessell, T. M. (2001). Coordinate regulation of motor neuron subtype identity and pan-neuronal properties by the bHLH repressor Olig2. *Neuron*, **31**, 773–89.

Ohnuma, S., Philpott, A., Wang, K., Holt, C. E. and Harris, W. A. (1999). p27Xic1, a Cdk inhibitor, promotes the determination of glial cells in Xenopus retina. *Cell*, **99**, 499–510.

Ohnuma, S., Hopper, S., Wang, K. C., Philpott, A. and Harris, W. A. (2002). Co-ordinating retinal histogenesis: early cell cycle exit enhances early cell fate determination in the Xenopus retina. *Development*, **129**, 2435–46.

Pearson, B. J. and Doe, C. Q. (2003). Regulation of neuroblast competence in Drosophila. *Nature*, **425**, 624–8.

Perron, M., Kanekar, S., Vetter, M. L. and Harris, W. A. (1998). The genetic sequence of retinal development in the ciliary margin of the Xenopus eye. *Dev. Biol.*, **199**, 185–200.

Perron, M., Opdecamp, K., Butler, K., Harris, W. A. and Bellefroid, E. J. (1999). X-ngnr-1 and Xath3 promote ectopic expression of sensory neuron markers in the neurula ectoderm and have distinct inducing properties in the retina. *Proc. Natl. Acad. Sci. U. S. A.*, **96**, 14 996–5001.

Poggi, L., Vottari, T., Barsacchi, G., Wittbrodt, J. and Vignali, R. (2004). The homeobox gene *Xbh1* cooperates with proneural genes to specify ganglion cell fate within the Xenopus neural retina. *Development*, **131**, 2305–15.

Poulin, G., Lebel, M., Chamberland, M., Paradis, F. W. and Drouin, J. (2000). Specific protein–protein interaction between basic helix-loop-helix transcription factors and homeoproteins of the Pitx family. *Mol. Cell. Biol.*, **20**, 4826–37.

Rapaport, D. H., Patheal, S. L. and Harris, W. A. (2001). Cellular competence plays a role in photoreceptor differentiation in the developing Xenopus retina. *J. Neurobiol.*, **49**, 129–41.

Reh, T. A. (1987). Cell-specific regulation of neuronal production in the larval frog retina. *J. Neurosci.*, **7**, 3317–24.

Reh, T. A. and Kljavin, I. J. (1989). Age of differentiation determines rat retinal germinal cell phenotype: induction of differentiation by dissociation. *J. Neurosci.*, **9**, 4179–89.

Reh, T. A. and Tully, T. (1986). Regulation of tyrosine hydroxylase-containing amacrine cell number in larval frog retina. *Dev. Biol.*, **114**, 463–9.

Repka, A. M. and Adler, R. (1992). Accurate determination of the time of cell birth using a sequential labeling technique with [3H]-thymidine and bromodeoxyuridine ('window labelling'). *J. Histochem. Cytochem.*, **40**, 947–53.

Reynaud, E. G., Pelpel, K., Guillier, M., Leibovitch, M. P. and Leibovitch, S. A. (1999). p57(Kip2) stabilizes the MyoD protein by inhibiting cyclin E-Cdk2 kinase activity in growing myoblasts. *Mol. Cell. Biol.*, **19**, 7621–29.

Scheer, N., Groth, A., Hans, S. and Campos-Ortega, J. A. (2001). An instructive function for Notch in promoting gliogenesis in the zebrafish retina. *Development*, **128**, 1099–107.

Shkumatava, A., Fischer, S., Müller, F., Strahle, U. and Neumann, C. J. (2004). Sonic hedgehog, secreted by amacrine cells, acts as a short-range signal to direct differentiation and lamination in the zebrafish retina. *Development*, **131**, 3849–58.

Stenkamp, D. L., Frey, R. A., Mallory, D. E. and Shupe, E. E. (2002). Embryonic retinal gene expression in sonic-you mutant zebrafish. *Dev. Dyn.*, **225**, 344–50.

Tessmar, K., Loosli, F. and Wittbrodt, J. (2002). A screen for co-factors of *Six3*. *Mech. Dev.*, **117**, 103–13.

Tomita, K., Nakanishi, S., Guillemot, F. and Kageyama, R. (1996). *Mash1* promotes neuronal differentiation in the retina. *Genes Cells*, **1**, 765–74.

Turner, D. L. and Cepko, C. L. (1987). A common progenitor for neurons and glia persists in rat retina late in development. *Nature*, **328**, 131–6.

Turner, D. L., Snyder, E. Y. and Cepko, C. L. (1990). Lineage-independent determination of cell type in the embryonic mouse retina. *Neuron*, **4**, 833–45.

Viczian, A. S., Vignali, R., Zuber, M. E., Barsacchi, G. and Harris, W. A. (2003). *XOtx5b* and *XOtx2* regulate photoreceptor and bipolar fates in the Xenopus retina. *Development*, **130**, 1281–94.

Voas, M. G. and Rebay, I. (2004). Signal integration during development: insights from the Drosophila eye. *Dev. Dyn.*, **229**, 162–75.

Waid, D. K. and McLoon, S. C. (1995). Immediate differentiation of ganglion cells following mitosis in the developing retina. *Neuron*, **14**, 117–24.

Waid, D. K. and McLoon, S. C. (1998). Ganglion cells influence the fate of dividing retinal cells in culture. *Development*, **125**, 1059–66.

Wang, J. C. and Harris, W. A. (2005). The role of combinational coding by homeodomain and bHLH transcription factors in retinal cell fate specification. *Dev. Biol.*, **285**, 101–15.

Wang, S. W., Kim, B. S., Ding, K. *et al.* (2001). Requirement for *math5* in the development of retinal ganglion cells. *Genes Dev.*, **15**, 24–9.

Watanabe, T. and Raff, M. C. (1990). Rod photoreceptor development in vitro: intrinsic properties of proliferating neuroepithelial cells change as development proceeds in the rat retina. *Neuron*, **4**, 461–7.

Watanabe, T. and Raff, M. C. (1992). Diffusible rod-promoting signals in the developing rat retina. *Development*, **114**, 899–906.

Wetts, R. and Fraser, S. E. (1988). Multipotent precursors can give rise to all major cell types of the frog retina. *Science*, **239**, 1142–45.

Young, R. W. (1985). Cell differentiation in the retina of the mouse. *Anat. Rec.*, **212**, 199–205.

Young, T. L. and Cepko, C. L. (2004). A role for ligand-gated ion channels in rod photoreceptor development. *Neuron*, **41**, 867–79.

Zhang, J., Gray, J., Wu, L. *et al.* (2004). Rb regulates proliferation and rod photoreceptor development in the mouse retina. *Nat. Genet.*, **36**, 351–60.

Zhang, X. M. and Yang, X. J. (2001). Regulation of retinal ganglion cell production by Sonic hedgehog. *Development*, **128**, 943–57.

Zhou, Q., Choi, G. and Anderson, D. J. (2001). The bHLH transcription factor Olig2 promotes oligodendrocyte differentiation in collaboration with Nkx2.2. *Neuron*, **31**, 791–807.

6

Neurotransmitters and neurotrophins

Rachael A. Pearson

University College London, London, UK

6.1 Introduction

In addition to intrinsic control mechanisms (see Chapter 5 and Cepko *et al.*, 1996), the production of neurons by progenitor cells and the determination of their fate are regulated via an array of diffusible factors, two families of which are considered in this chapter: neurotransmitters and neurotrophins. Neurotrophins are now known to play an essential role in both the formation and the maintenance of the nervous system throughout development and adult life. There is growing evidence that besides their role as molecules mediating communication between nerve cells in the mature nervous system, a variety of both slow and fast neurotransmitters also play important roles during neuronal development. This chapter reviews recent evidence that demonstrates that a number of non-synaptic neurotransmitter release mechanisms, together with many neurotransmitters and their receptors, are present in the developing retina prior to the onset of synapse formation and that these early neurotransmitters act to modulate a range of events in neural development. Their precise mechanisms of action are still being elucidated but, as described here, the ability to modulate $[Ca^{2+}]_i$ is one feature common to these early neurotransmitter systems, and is thought to underlie a number of their developmental actions. It is becoming clear that both neurotransmitters and neurotrophins play important regulatory roles in the early stages of retinal development, including the modulation of proliferation, differentiation, cell survival and circuit formation.

6.2 Neurotransmitters in early retinal development

The adult retina expresses a number of neurotransmitters. Four are considered in this chapter with respect to their roles in development, acetylcholine (ACh), adenosine triphosphate (ATP), γ-aminobutyric acid (GABA) and glutamate.

6.2.1 Mechanisms of neurotransmitter release in development

The classical mechanism of neurotransmitter release is synaptic vesicular exocytosis, whereby neurotransmitters are stored at the presynaptic terminal of a neuron in intracellular

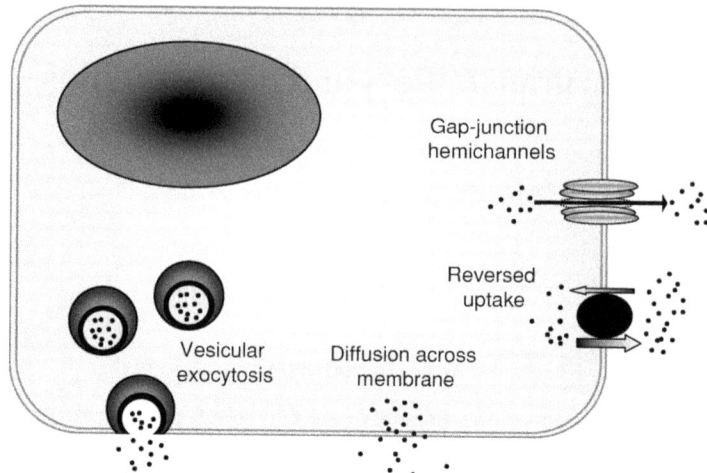

Figure 6.1 Mechanisms of neurotransmitter release during development. A number of potential mechanisms could be involved in the release of neurotransmitters prior to synaptogenesis. These include vesicular release, diffusion or 'leakage' across the cell membrane, carrier proteins and reversed action of uptake transporters and transient opening of gap-junction hemichannels.

vesicles, which transiently fuse with the cell membrane, discharging their contents into the synaptic cleft. Neurotransmitters diffuse across the cleft to the postsynaptic terminal of a target neuron and activate the appropriate receptors. However, at the stage of development considered in this chapter, synaptic connections have yet to form, which raises the question of how neurotransmitters are released. There are, in fact, several mechanisms that could enable neurotransmitter release prior to the maturation of synapses (Figure 6.1).

Elements of the presynaptic exocytotic machinery are expressed very early in development. In the chick, synaptotagmin, syntaxin, synaptic vesicle-2 and dynamin, all components of the vesicular release mechanism, are expressed in immature retinal ganglion cells (RGCs) as early as embryonic day (E)4 (Bergmann et al., 1999, 2000; Grabs et al., 2000). Specific vesicular transporters for GABA (VGAT) and glutamate (VGLUT1 and VGLUT2), which are required to package GABA or glutamate into vesicles ready for release, are also expressed before synapse formation in the rodent retina (Johnson et al., 2003; Sherry et al., 2003). Thus, differentiating cells in the immature retina express many components of the vesicular release mechanism and may be capable of vesicular release prior to synaptogenesis.

Neurotransmitters can also be released by non-vesicular mechanisms. These include 'leakage', or diffusion, of neurotransmitter molecules across the plasma membrane (Katz and Miledi, 1977), transport on carrier proteins and release through membrane channels (Figure 6.1). In the mature nervous system, diffusion across the plasma membrane has little effect on neuronal signalling since this 'leaked' transmitter is largely restricted to the synaptic cleft and rapidly broken down by enzymes or sequestered by uptake mechanisms.

However, during development, when the mechanisms that terminate transmitter action are still maturing, this leakage may represent a significant source of release into the extracellular environment.

Non-vesicular release of GABA and glutamate has been observed in several preparations and is thought to occur via the reversed action of uptake transporters (Attwell et al., 1993; Schwartz, 1987) or exchangers (Warr et al., 1999), which normally act to terminate neurotransmitter action by removing it from the extracellular space and that are expressed during early retinal development (see above). Another non-vesicular GABA-release mechanism has been described for the immature cortex (Demarque et al., 2002). Here, release was unaffected by traditional blockers of vesicular release, such as voltage-gated Na^+ and Ca^+ channel blockers and botulinum toxin. Demarque et al. (2002) also considered reversed-uptake and exchange-mediated release improbable, since inhibition of GAT-1 had no effect on release. However, since other transporters are expressed at these early times, release mediated by exchangers should not be ruled out yet and more work is required to determine the precise mechanisms involved. Recently, Yang and Kunes (2004) demonstrated that, in the developing *Drosophila* retina, ACh is also released via a non-vesicular mechanism and is important in the regulation of photoreceptor (PR) axon guidance, although the details of the release mechanism were not determined.

A novel mechanism of transmitter release, involving gap-junction hemichannels, has been revealed by recent investigations. Gap junctions are large intercellular channels formed by the docking of two connexin hemichannels, one contributed by each cell. The opening of 'undocked' hemichannels permits communication between a cell's cytoplasm and the extracellular space, and the passage of a variety of molecules, including ATP (Cotrina et al., 1998, 2000; Stout et al., 2002; Bennett et al., 2003). We now know that hemichannels open spontaneously under physiological conditions during retinal and cortical development. In the embryonic chick retina, hemichannels on the retinal-facing surface of the retinal pigment epithelium (RPE) open and release ATP into the subretinal space, which subsequently stimulates proliferation of the underlying progenitor cells (Pearson et al., 2005a). Similarly, in the embryonic cortex, radial glial cells release ATP via a hemichannel-dependent mechanism, which acts to promote their proliferation (Weissman et al., 2004).

6.2.2 *Sources of neurotransmitters in development*

The immature retina is responsive to neurotransmitters from very early stages of development. For example, the embryonic chick retina can respond to cholinergic stimulation by E3 (see Section 6.2.5). However, for these responses to be physiologically relevant, neurotransmitters must be present at the same stage. Where might these neurotransmitters come from? One major potential source is the population of immature neurons. Neurogenesis begins as early as E2 in the chick retina and, as described above, neurotransmitters can be released prior to synapse formation. Immunocytochemical and biochemical studies have shown that ACh is the predominant neurotransmitter in a subpopulation of amacrine cells (ACs) called 'starburst' amacrines (Baughman and Bader, 1977; Hayden et al., 1980), which are involved in the processing of direction and movement and born from E3 in the chick. During development, immature horizontal cells (HCs) may also release ACh, as cholinergic markers are transiently expressed in the outer mammalian retina prior to synaptogenesis (Kim et al., 1999, 2000). Other markers of cholinergic neurons, such as acetylcholinesterase, can also be found by E3 in the chick (Layer, 1991).

In the adult retina, GABA is synthesized from glutamate by the enzyme, glutamic acid decarboxylase, and released by ACs and HCs. da Costa Calaza et al. (2000) examined the developmental profile of glutamic acid decarboxylase expression in chick retina. This enzyme is present at low levels between E3 and E6, as revealed by immunohistochemical staining, but increases rapidly after E6, a time that corresponds with the peak of AC and HC birth. Similarly, autoradiographic studies of GABA uptake (the mechanism by which GABA is removed from the extracellular space) has shown that cell bodies and processes in the central regions of the retina label at E6, but no labelling occurs in the ventricular zone (VZ) (Frederick, 1987), and immunolabelling studies have shown that the GAT-1 GABA transporter is present at E8 (Catsicas and Mobbs, 2001). Consistent with these findings, Frederick further demonstrated that HC, AC and RGCs all label positively for GABA uptake by E8, although progenitor cells remain unlabelled (Frederick, 1987). Thus, the expression of GABA transporters and synthesizing enzymes appear to be features of differentiating, but not proliferating, cells.

Purine nucleotides play important roles as neurotransmitters and neuromodulators in the mature CNS. Potential sources of extracellular purines include both neurons and glia. In the adult nervous system, ATP is co-released with other neurotransmitters from adrenergic, cholinergic, glutamatergic and GABAergic neurons (for review, see Burnstock, 2004). Adenosine triphosphate is also released via a synaptic-independent mechanism (see Section 6.2.1) from the embryonic RPE (Pearson et al., 2005a) and Müller glia (Newman, 2001) and, in the embryonic cortex, from radial glia (Weissman et al., 2004).

6.2.3 Modes of action of neurotransmitters

Neurotransmitters exert their actions largely by activating one of two families of membrane-bound receptors, ionotropic and metabotropic receptors. Ionotropic receptors are predominantly involved in fast synaptic transmission and the receptor-binding site is directly coupled to an ion channel. Neurotransmitter binding causes the ion channel to open, allowing the passage of particular ions, such as Na^+, K^+, Ca^{2+} or Cl^-. Depending on which ions pass through the channel, a neurotransmitter may depolarize the cell (e.g. ACh acting on the nicotinic ACh receptor (nAChR)) and hence be excitatory, or hyperpolarize it (e.g. GABA acting on the $GABA_A$ receptor), and exert an inhibitory effect. Metabotropic receptors are G-protein-coupled receptors. Transmitter binding leads to the activation of a G-protein that, depending on the type of G-protein, activates one of several effector systems including the phospholipase C/inositol(1,4,5)-triphosphate (IP_3)/Ca^{2+} system, the adenylate cyclase/cAMP system, or by direct action on ion channels such as K^+ and Ca^{2+} channels.

6.2.4 Neurotransmitters, $[Ca^{2+}]_i$ and development

How do neurotransmitters exert their effects on early developmental events? Neurotransmitters act to alter a number of aspects of cell function including membrane voltage, enzyme activity and $[Ca^{2+}]_i$. The latter is of particular interest since all the early embryonic transmitter systems act to modulate $[Ca^{2+}]_i$. It is also one of the responses most amenable to study and manipulation, and thus much of the research into the role of neurotransmitters in retinal development has focused on these transmitter-evoked $[Ca^{2+}]_i$ changes and their downstream consequences. Intracellular Ca^{2+} transients are frequent in the developing retina and can be patterned temporally, as periodic oscillations, and spatially, as transients occurring in single cells, as synchronized events occurring in small groups of neighbouring cells, or as large-scale propagating waves (see below and Figure 6.4).

Calcium has a key influence on many developmental events in the CNS and is implicated in the regulation of proliferation, migration, differentiation and circuit formation. Intracellular Ca^{2+} transients may be required for progression through several steps in the proliferative cell cycle including the G1/S-phase transition, S-phase itself, entry into mitosis and key points within mitosis including the metaphase–anaphase transition and induction of cytokinesis (see Berridge, 1995; Santella, 1998; Santella et al., 1998; Whitaker and Larman, 2001). Changes in $[Ca^{2+}]_i$ are also required for the movement of progenitor cells between the VZ and vitreal surface in G1 and G2 of the cell cycle (Pearson et al., 2005b).

Intracellular Ca^{2+} levels play a major role in the regulation of neuronal differentiation. Imaging studies have shown that multiple spontaneous $[Ca^{2+}]_i$ transients occur during the maturation of neurons (Gu and Spitzer, 1993, 1995; Gu et al., 1994) and that they control a variety of developmental events via information encoded in their frequency, amplitude and duration. These include the maturation of K^+ currents, neurotransmitter synthesis and receptor expression (Spitzer et al., 1993), and the regulation of the rate of neurite extension, the outgrowth of which is inversely related to the frequency of growth cone-restricted $[Ca^{2+}]_i$ transients. Growth-cone stalling, axon retraction and growth-cone turning have all been found to be associated with subtle changes in the frequency and amplitude of $[Ca^{2+}]_i$ transients (Gu and Spitzer, 1995; Dolmetsch et al., 1998; Zheng et al., 1994, Zheng, 2000) although the details of the molecular mechanisms underlying these processes remain to be determined. Changes in $[Ca^{2+}]_i$ are also required for the translocation of immature neurons, as described in the developing cortex (Komuro and Rakic, 1998). The possibility that $[Ca^{2+}]_i$ transients might provide a route by which neurotransmitters could influence retinal development is considered in greater detail later in this chapter (see p. 111).

6.2.5 Early expression of neurotransmitter receptors in the embryonic retina

ACh receptors in the developing retina

Acetylcholine, the classic fast excitatory neurotransmitter of the peripheral nervous system, acts at (1) nAChRs, which are ligand-gated ionotropic receptors that are selectively activated by nicotine-like ligands, and permeable to Na^+ and K^+, and (2) muscarinic (m) AChRs.

The five subtypes of mAChR belong to the superfamily of metabotropic G-protein-coupled receptors. The M_1, M_3 and M_5 receptors cause the release of Ca^{2+} from IP_3-sensitive intracellular stores, whilst the M_2 and M_4 receptor subtypes reduce cAMP levels via the inhibition of adenylate cyclase.

In the chick, nAChR mRNA is present by E4.5 (Hamassaki-Britto et al., 1994). Its expression appears to be restricted to differentiating, rather than proliferating, cells; imaging studies have shown that nicotinic agonists are unable to evoke responses from progenitor cells (Wong, 1995a; Pearson et al., 2002), the responses instead arising from immature RGCs and ACs in the ganglion cell layer (GCL) and inner nuclear layer (INL) (Wong, 1995a). The expression of nAChRs by these cells is maintained in the adult (Keyser et al., 1988; Wada et al., 1989; Cauley et al., 1990). In the adult retina, mAChRs are expressed in AC, RGC and bipolar cells (BCs) and, in the case of M_3, in the RPE (Fischer et al., 1998). They are also present in both the avian and mammalian retina from early in development (McKinnon and Nathanson, 1995; Wong, 1995a; Pearson et al., 2002), although the receptor subtypes expressed change with time. Immunoprecipitation and immunoblot analyses in the chick show that the M_4 subtype predominates during the first week of development, whilst the expression of M_3 increases moderately, and that of M_2 dramatically, during the second week of development (E7 to E14) (Nadler et al., 1999).

Non-confocal Ca^{2+}-imaging studies demonstrate that stimulation of mAChRs leads to marked increases in $[Ca^{2+}]_i$ as early as E3 in the chick (Figure 6.2a; Yamashita and Fukuda, 1993; Yamashita et al., 1994; R.A.P, unpublished observations), and E20 in the rabbit (Wong, 1995a), retina. Non-confocal imaging does not permit the identification of the individual cells that respond to a given agonist since it also records changes in fluorescence from outside the focal plane. However, confocal-imaging studies have further demonstrated that mAChRs are expressed by cells throughout the immature neural retina, including dividing progenitor cells (Movie 6.1 online; Pearson et al., 2002). The activation of mAChRs leads to the release of Ca^{2+} from intracellular stores, causing oscillatory changes in $[Ca^{2+}]_i$ (Yamashita et al., 1994; Pearson et al., 2002). The proportion of progenitor cells responding to muscarinic agonists is maximal during the peak of neurogenesis (~E6) but declines rapidly thereafter. The response is virtually absent by E8 in the chick (Figure 6.2a) (Yamashita et al., 1994; Pearson et al., 2002; R.A.P., unpublished results) and postnatal day (P)11 in the rabbit (Wong, 1995a), remaining only in differentiating neurons in the inner retina.

ATP receptors in the developing retina

Adenosine triphosphate and its derivatives stimulate P2 receptors, of which there are two classes, P2Y and P2X. The P2Y receptors are G-protein-coupled receptors, predominantly coupled to the IP_3/Ca^{2+} cascade, although some isoforms are linked with cyclic adenosine monophosphate (cAMP) and the stimulation of tyrosine kinases and mitogen-activated protein kinases (MAPKs). The P2X receptors are ligand-gated ion channels, permeable to Na^+, K^+ and Ca^{2+}.

Purines play important roles in development from the moment of fertilization. The ATP receptors are amongst the earliest functionally active membrane receptors, present in the

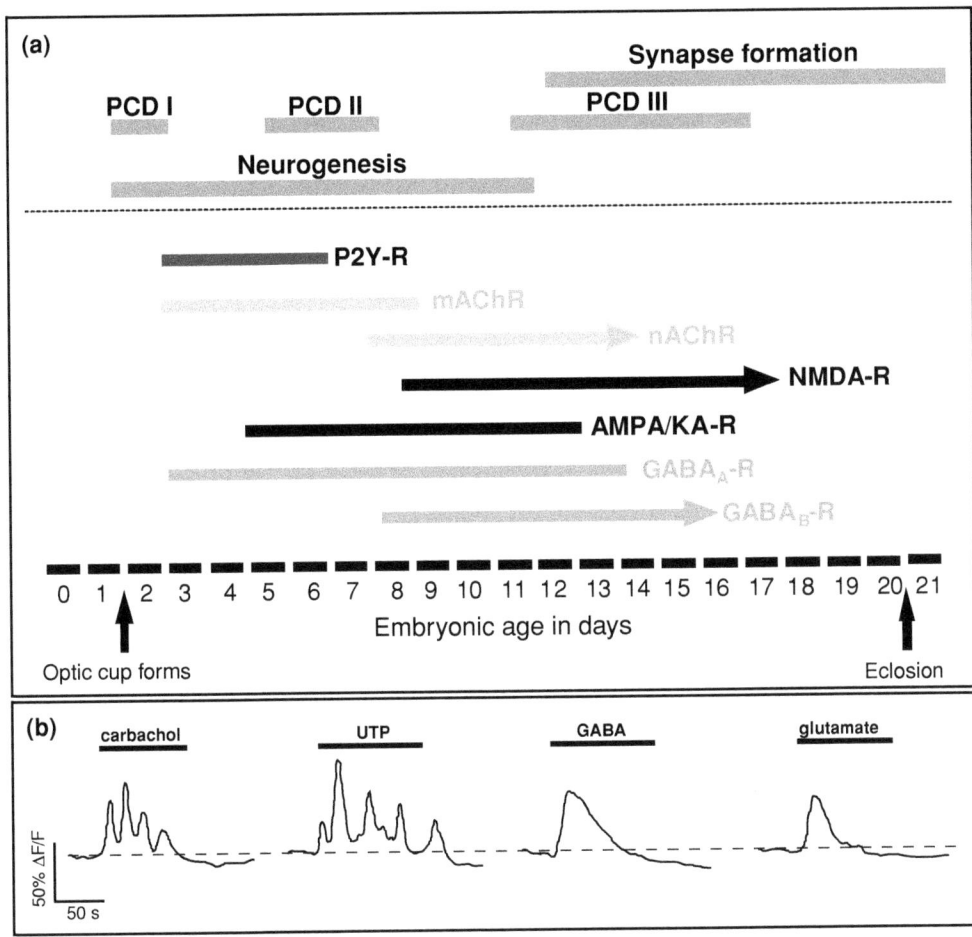

Figure 6.2 Neurotransmitter-evoked changes in $[Ca^{2+}]_i$ in the embryonic retina. (a) The neurotransmitters ACh, ATP, GABA and glutamate evoke increases in $[Ca^{2+}]_i$ at different times during embryonic retinal development. Diagram illustrates when the early immature neural retina is responsive to these neurotransmitters, as determined from Ca^{2+}-imaging studies in the chick retina (see text). It should be noted that ACh and ATP play different and essential roles in the proper functioning of the adult tissue and may also modulate $[Ca^{2+}]_i$ in differentiated cells. NB. Arrowheads indicate the latest time point in development studied in the current literature, for a given neurotransmitter receptor. It is possible that these responses continue later into development but await further investigation. AMPA, α-amino-3-hydroxy-5-methyl-4-isoxazolepropionate; KA, kainate; NMDA, N-methyl-D-aspartate; PCD, programmed cell death; UTP, uridine triphosphate. (b) Example traces of the changes in $[Ca^{2+}]_i$, as measured by a change in fluorescence intensity (ΔF/F), evoked in chick retinal progenitor cells by muscarinic (carbachol), purinergic (UTP), GABAergic and glutamatergic agonists. Adapted from Pearson *et al.* (2002).

embryo at the time of germ layer formation (Laasberg, 1990). However, there is little systematic data on the developmental expression patterns of the different receptor subtypes. Purine 2Y-like receptors are expressed throughout the proliferative period of development in the embryonic chick neural retina (Sakaki et al., 1996; Sugioka et al., 1996; Pearson et al., 2002). In the adult rat retina, glial cells express $P2Y_2$ and $P2Y_4$ receptors, and $P2X_2$, $P2X_3$, $P2X_4$ and $P2X_7$ receptors have been identified in RGCs (Greenwood et al., 1997; Wheeler-Schilling et al., 2001) and in the Müller cell population of the mammalian and human retina (Greenwood et al., 1997; Brändle et al., 1998; Pannicke et al., 2000; Wheeler-Schilling et al., 2001). Purinergic receptors are also present in the embryonic (Pearson et al., 2005a) and adult (Mitchell, 2001) RPE.

In the developing retina, ATP acts largely on progenitor cells, rather than immature neurons, evoking Ca^{2+} mobilization by activating P2Y receptors (Figure 6.2; Pearson et al., 2002; Sugioka and Yamashita, 2003). In the chick, the number of progenitor cells responding to ATP is largest at E3, declines dramatically towards E6 and is minimal from E8 onwards (Figure 6.2a; Sakaki et al., 1996; Sugioka et al., 1996; Pearson et al., 2002). This developmental decrease parallels the time course of proliferation, which is highest in the first few days of development and ceases around E8 in the chick retina (Prada et al., 1991), raising the possibility that purines may be involved in the regulation of proliferation.

GABA receptors in the developing retina

γ-Aminobutyric acid, the major inhibitory neurotransmitter of the CNS, acts on three receptor subclasses, $GABA_A$, $GABA_B$ and $GABA_C$. The $GABA_A$ and $GABA_C$ receptors are ligand-gated ion channels and demonstrate great pharmacological and functional diversity (Rudolph et al., 2001). In the adult CNS, GABA is an inhibitory neurotransmitter; activation of the $GABA_A$ receptor opens an integral Cl^--permeable channel, and causes the cell to hyperpolarize. In early development, $[Cl^-]_i$ is high and $GABA_A$ receptor activation results in Cl^- leaving the cell, causing a depolarization that activates L-type voltage-gated Ca^{2+} channels and Ca^{2+} influx (Segal and Barker 1984; Cherubini et al., 1991). γ-Aminobutyric acid is understood to exert trophic actions via this rise in $[Ca^{2+}]_i$. The high $[Cl^-]_i$ occurs due to a delay in the maturation of the Cl^- extrusion system, relative to Cl^- uptake mechanisms (Nishi et al., 1974; Frambach and Misfeldt, 1983; Zhang et al., 1991; Reichling et al., 1994). The ability of GABA to cause increases in $[Ca^{2+}]_i$ via secondary activation of voltage-gated Ca^{2+} channels decreases as the extrusion mechanism develops. Like the $GABA_A$ receptor, $GABA_C$ receptors are linked to a Cl^- channel whilst $GABA_B$ receptors are linked indirectly to Ca^{2+} and K^+ channels via G-protein activation.

All three GABA receptor subtypes are found in the adult retina. The $GABA_A$ receptors are expressed by subpopulations of BC, AC and RGCs (Greferath et al., 1994; Karne et al., 1997). They are also present during development, prior to synaptogenesis, and their abundance closely parallels neurogenesis (see Barker et al., 1998). The $GABA_B$ receptors are present by E8 in the chick and are located on RGCs and ACs (Catsicas and Mobbs, 2001). The $GABA_C$ receptor subunits are not detected until much later in development. In the mouse, they appear at the end of neurogenesis (postnatal (P)6) and are functional from P9 (Greka et al., 2000). This correlates with the period of BC differentiation as well as eye opening.

The precise timing of when retinal cells become responsive to GABA is unclear. Responses to GABA appear early in retinal development, during the proliferative period, but the evidence for their presence on progenitor cells is equivocal. Non-confocal imaging of whole retina has shown that GABA, acting via GABA$_A$ receptors, can evoke changes in $[Ca^{2+}]_i$ from as early as E3 in the chick retina (Yamashita and Fukuda, 1993), although the response peaks at around E8 (Allcorn et al., 1996; Catsicas and Mobbs, 2001; R.A.P. unpublished observations) before disappearing by E14 (Figure 6.2; Allcorn et al., 1996; Catsicas and Mobbs, 2001), when the Cl^- extrusion mechanisms mature and GABA begins to exert a hyperpolarizing, rather than depolarizing action. From E6, the GABA-evoked Ca^{2+} response is largely restricted to immature neurons in the GCL and INL (Allcorn et al., 1996; Catsicas and Mobbs, 2001), suggesting that GABA sensitivity may be a property of differentiated neurons. However, in studies of the rat neocortex, whole-cell electrophysiological recordings of VZ cells (presumed to be progenitors, although this was not determined) indicated that they too respond to GABAergic stimulation (LoTurco et al., 1995), as do progenitor-like cells of the regenerating newt retina (Ohmasa and Saito, 2004). It is possible that GABA$_A$ receptors are expressed by progenitor cells in both tissues but only shortly prior to terminal division. Consistent with this, more recent studies in the cortex have demonstrated that GABA-evoked changes in membrane potential and $[Ca^{2+}]_i$ are confined to cells either in the final round of division or undergoing differentiation; only a very small fraction of the responses arose from cells identified as neural progenitor cells (Maric et al., 2001). Similarly, combined confocal and immunohistochemical studies of the embryonic retinal VZ suggest that, whilst a subpopulation of VZ cells do respond to GABA, it is comprised largely of postmitotic, rather than progenitor, cells (Pearson et al., 2002).

Glutamate receptors in the developing retina

Glutamate, the predominant fast excitatory neurotransmitter of the CNS, activates metabotropic (mGluRs) and ionotropic (N-methyl-D-aspartate (NMDA) and non-NMDA) receptors. Non-NMDA receptors are further divided into α-amino-3-hydroxy-5-methyl-4-isoxazolepropionate (AMPA)-preferring and kainate-preferring subtypes. The metabotropic mGluR1 and mGluR5 receptors are linked to the IP_3/Ca^{2+} signalling cascade (Abe et al., 1992; Aramori and Nakanishi, 1992), whilst NMDA receptor channels are directly permeable to Na^+, K^+ and Ca^{2+}. The Ca^{2+} permeability of the NMDA receptor underlies its roles in development, cell cytotoxicity and in learning and memory. Magnesium blocks the channel in a voltage-dependent manner and is relieved by depolarization (Bliss and Collingridge, 1993). The AMPA receptors are permeable to Na^+ and K^+ and usually have low permeability to Ca^{2+}, although certain isoforms may be highly permeable to Ca^{2+}; AMPA receptors lacking an edited version of the GluR2 subunit, or made up of GluR1, GluR3 and GluR4 subunits, are Ca^{2+} permeable (Murphy and Miller, 1989; Pruss et al., 1991;

Burnashev *et al.*, 1992). Interestingly, the Ca^{2+}-permeable form of the AMPA receptor is prevalent during early development but decreases with maturation (Pellegrini-Giampietro *et al.*, 1992), suggesting that this isoform may have a specific role(s) in development.

Glutamate is the main neurotransmitter of RGCs, BCs and PRs in the vertebrate retina. The distribution of glutamate receptors has been described for the adult retina of several species, although the precise distribution may vary between species. In the postnatal chick, all subunits of the AMPA/kainate receptors and the NR1 subunit of the NMDA receptor are expressed in the INL and the GCL, predominantly by ACs and RGCs (Santos Bredariol and Hamassaki-Britto, 2001). Expression of the majority of glutamate receptor subtypes is low prior to synaptogenesis, increasing as development progresses (Cristovao *et al.*, 2002a,b). However, Western blot analyses have shown that a number of glutamate subunits are also present much earlier in development (E5) (Santos Bredariol and Hamassaki-Britto, 2001), although their expression is largely restricted to immature neurons.

Interestingly, electrophysiological recordings from retinal cell cultures, and Ca^{2+}-imaging studies in whole-mount retinas, have shown that the Ca^{2+}-permeable forms of the AMPA/kainate receptor are functional from E5 to E6 in the chick retina (Allcorn *et al.*, 1996; Sugioka *et al.*, 1998). The AMPA/kainate receptor-mediated Ca^{2+} rises are maximal at E9 to E10, before declining towards E12 and the onset of synaptogenesis (Figure 6.2; Sugioka *et al.*, 1998). The cells that respond to glutamate at these early stages are largely immature RGCs and ACs, located in the presumptive INL and GCL (Rorig and Grantyn, 1994; Wong, 1995b; Allcorn *et al.*, 1996; Sugioka *et al.*, 1998). Studies of transdifferentiating RPE (which can grow to reform the neural retina) and retinal re-aggregation models have similarly found that a response to glutamate is only prominent once differentiation has commenced (Naruoka *et al.*, 2003).

Glutamate-evoked Ca^{2+} changes have been observed in some cells in the VZ of the E6 chick (Allcorn *et al.*, 1996; Pearson *et al.*, 2002), and E20 rabbit (Wong, 1995b), retina. However, the response to glutamate, like that to GABA, appears to be mainly restricted to a population of non-dividing cells (Pearson *et al.*, 2002). Maric *et al.* (2000) have shown, using whole-cell patch clamping and Ca^{2+} imaging, that glutamate-evoked inward currents and associated increases in $[Ca^{2+}]_i$ are only seen in differentiating cortical neurons. In contrast, ionotropic agonists failed to evoke changes in voltage in progenitor cells. However, a subpopulation of cells that stained positively for both 5'-Bromo-2'-deoxy-uridine (BrdU; a marker of proliferation) and TuJ-1 (an antibody against neuron-specific β-tubulin and a marker of differentiated neurons) did show increases in $[Ca^{2+}]_i$ in response to AMPA, suggesting that functional ionotropic glutamate receptors are expressed at the time of terminal cell division and early differentiation (Maric *et al.*, 2000). A similar situation may pertain in the retina, although further studies will be required to confirm this. In contrast, NMDA receptor-mediated responses to glutamate are absent during early retinal development (Allcorn *et al.*, 1996; Sugioka *et al.*, 1998; Pearson *et al.*, 2002) and appear later, in differentiated cells, near the onset of synaptogenesis (Somahano *et al.*, 1988; Wong, 1995b; Duarte *et al.*, 1996; Catsicas *et al.*, 2001). On the basis of these findings, it seems likely that NMDA and non-NMDA glutamate receptors adopt different roles during retinal development, and that non-NMDA receptors alone are involved in early developmental processes.

6.2.6 The role of neurotransmitters in early retinal development

Regulation of progenitor cell cycle by cholinergic and purinergic neurotransmitters

During early development, progenitor cells divide rapidly to expand the pool of retinal cells. The onset of neurogenesis overlaps with this proliferative period (Prada *et al.*, 1991). That muscarinic and purinergic neurotransmitters are expressed at these times (see Section 6.2.2) raises the possibility that they might modulate proliferation and/or differentiation. Interestingly, the response to purinergic stimulation is maximal during the proliferative period (Sakaki *et al.*, 1996; Sugioka *et al.*, 1996; Pearson *et al.*, 2002), whilst the muscarinic response, although prevalent at early times, is maintained throughout the peak of neurogenesis (Yamashita *et al.*, 1994; Wong, 1995a; Sakaki *et al.*, 1996; Pearson *et al.*, 2002), suggesting that these two neurotransmitters might play different roles in development. Consistent with this, activation of purinoceptors increases both retinal progenitor cell DNA synthesis (a measure of the number of cells re-entering the cell cycle) (Sugioka *et al.*, 1999; Sanches *et al.*, 2002) and speeds their mitosis, effects that lead to enhanced proliferation and bigger eyes (Pearson *et al.*, 2002, 2005a) (Figure 6.3). In contrast to the proliferative actions of ATP, muscarinic stimulation appears to act as a brake on mitosis, almost doubling the time it takes for cells to divide (Pearson *et al.*, 2002). Muscarinic agonists similarly reduce DNA synthesis in embryonic rat retinal cultures (Santos *et al.*, 2003), whilst in the chick, more prolonged exposure to muscarinic agonists or antagonists leads to smaller or larger eyes, respectively, and corresponding changes in the number of proliferating cells (Pearson *et al.*, 2002). However, there are reports that earlier exposure to muscarinic agonists can increase eye size in the chick (Angelini *et al.*, 1998) and affect nuclear status in the sea urchin (Harrison *et al.*, 2002). The reasons for these apparently opposing effects of ACh are not yet clear.

In the embryonic cortex, GABA and glutamate have been shown to influence proliferation by changing the duration of the progenitor cell cycle; both GABA and glutamate increase the size of cortical VZ clones (Haydar *et al.*, 2000). In contrast, LoTurco *et al.* (1995) have shown that GABA and glutamate decrease the number of dissociated, embryonic cortical cells synthesizing DNA. When applied alone, $GABA_A$ and AMPA/kainate receptor antagonists increase proliferation by increasing DNA synthesis (suggesting that GABA or glutamate might act to repress proliferation), but decrease it when applied together. LoTurco and colleagues suggest that glutamate and GABA bring about their actions through depolarization-evoked $[Ca^{2+}]_i$ increases, similar to those evoked by these transmitters in the embryonic retina (Yamashita and Fukuda 1993; Wong, 1995b; Pearson *et al.*, 2002). However, in contrast to findings in the cortex, the Ca^{2+} influxes evoked by GABA and glutamate do not appear to regulate mitosis in the retina (Pearson *et al.*, 2002), a finding perhaps consistent with the idea that GABA and glutamate receptor expression might be

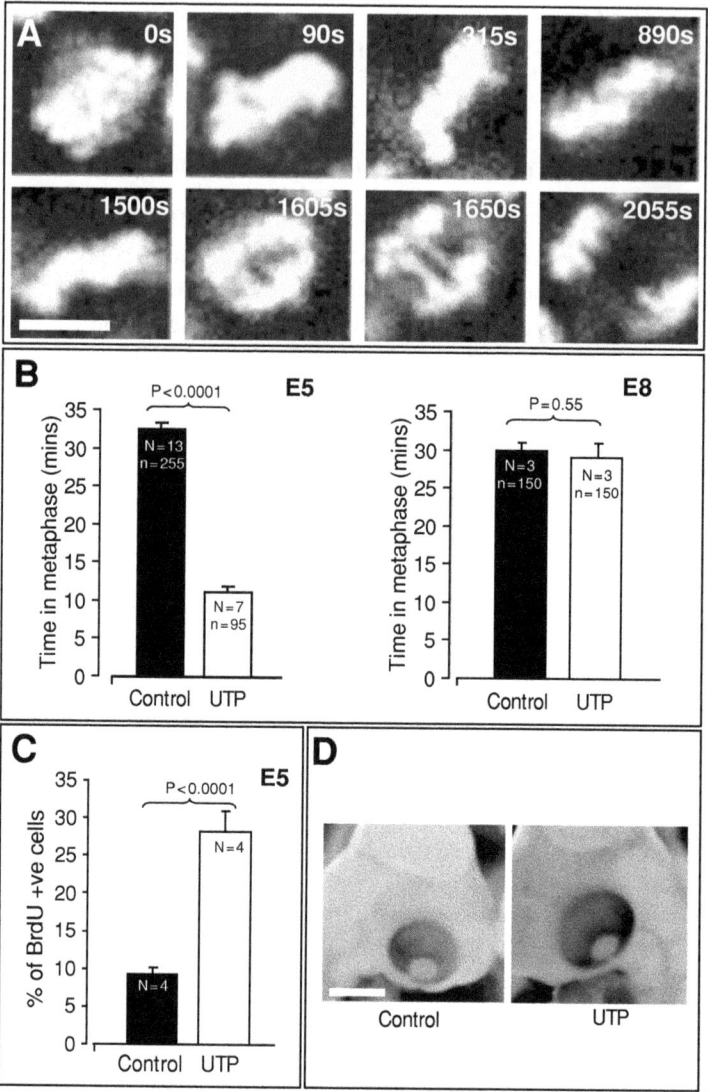

Figure 6.3 Neurotransmitters can modulate proliferation in early retinal development. ATP acts at P2Y receptors to stimulate the proliferation of neural retinal progenitor cells during early development. (A) Example images from a confocal time-lapse series of mitosis in the chick retinal VZ. The chromatin of the mitotic cell progresses from prophase (0 s), through metaphase (90 to 1500 s) and into anaphase (1650 s onwards). Scale bar 5 μm. (B) Mitosis in the embryonic neural retinal VZ is faster in the presence of purinergic agonists. Histogram shows the time spent in mitosis by progenitor cells in either control solution or in the presence of the purinergic P2Y-R agonist, UTP (*left*). The proliferative actions of purinergic agonists decrease with age. Histogram (*right*) shows that the time progenitor cells spend in mitosis is no longer affected by purinergic agonists at E8. N, number of retinas; n, number of cells. (C) Purinergic agonists increase proliferation in the early embryonic neural retina. Histogram shows the % of BrdU-positive cells found following 8 hour exposure to UTP, compared with controls, in the E5 retina. (D) Increased proliferation following exposure to purinergic agonists can lead to increases in eye size, compared with controls. Scale bar 2 mm. Adapted from Pearson *et al.* (2005a).

correlated with differentiation, rather than proliferation (Wong, 1995b; Maric et al., 2000, 2001; Pearson et al., 2002). An interesting possibility is that, in the neocortex, the actions of GABA and glutamate may be to depolarize differentiated neurons and bring about the release of some other factor, such as ACh, which then acts as a brake on the cell cycle. This does not appear to be the case in the retina, however, for both glutamatergic, and GABAergic, stimulation and blockade were without effect on mitosis or eye development (Pearson et al., 2002). Nevertheless, since transmitter systems interact to regulate other aspects of development (Catsicas et al., 1998; Wong et al, 2000), this possibility is worthy of further investigation.

Neurotransmitters and the modulation of Ca^{2+} transients in the proliferating retina

It is possible, as mentioned above, that neurotransmitters exert their effects on the cell cycle via their ability to modulate $[Ca^{2+}]_i$. Both $[Ca^{2+}]_i$ transients in single cells and locally synchronized events between adjacent progenitors in the retina (Pearson et al., 2002, 2004; Sugioka and Yamashita, 2003), and the cortex (Owens and Kriegstein, 1998), occur during the proliferative period of development (Figure 6.4). Spontaneous $[Ca^{2+}]_i$ transients in single cells (Figure 6.4b, Movie 6.2 online) are independent of action potentials, as demonstrated by their insensitivity to the Na^+ channel blocker tetrodotoxin (Owens and Kriegstein, 1998; Pearson et al., 2002), but are modulated by neurotransmitters. In the chick retina, both muscarinic and purinergic receptor antagonists reduce the frequency of these transients (Pearson et al., 2002) suggesting that, in vivo, endogenous purines and ACh normally act to evoke $[Ca^{2+}]_i$ transients in progenitor cells. Further, exogenous ATP and ACh generate conspicuous changes in $[Ca^{2+}]_i$ in retinal progenitor cells and, since their actions are prevented by buffering $[Ca^{2+}]_i$ changes, their actions on cell division (described above) are thought to be mediated via a Ca^{2+}-dependent mechanism (Pearson et al., 2002; Sugioka and Yamashita, 2003). In glial cultures, ATP-induced proliferation is similarly accompanied by $[Ca^{2+}]_i$ oscillations (Morita et al., 2003), the duration of which regulates cell division (Moll et al., 2002).

Thus, it is possible that the $[Ca^{2+}]_i$ signal evoked in progenitors by neurotransmitters, such as ATP, exerts a direct effect on the cell cycle. If this were so, it would provide a mechanism by which molecules such as neurotransmitters could regulate cell division and modulate both cell number and the timing of the production of the different types of CNS cells. For example, progenitor cells initially cycle rapidly, dividing to increase their number. The cell cycle then slows down as neurogenesis proceeds (Caviness et al., 1995). As we have seen, ATP acts to stimulate proliferation in the embryonic retina (Sugioka et al., 1999; Pearson et al., 2002, 2005a), whilst ACh has the opposite effect (Pearson et al., 2002; Santos et al., 2003). Thus, during early retinal development, ATP may act to promote proliferation and expansion of the progenitor pool, whilst later on a decrease or down-regulation of the purinergic signal (see Figure 6.3B) could allow the muscarinic signal to exert a stronger effect, producing a slowing of the cell cycle. Glutamatergic and GABAergic antagonists also caused a reduction in $[Ca^{2+}]_i$ transients but, interestingly, only in a population of

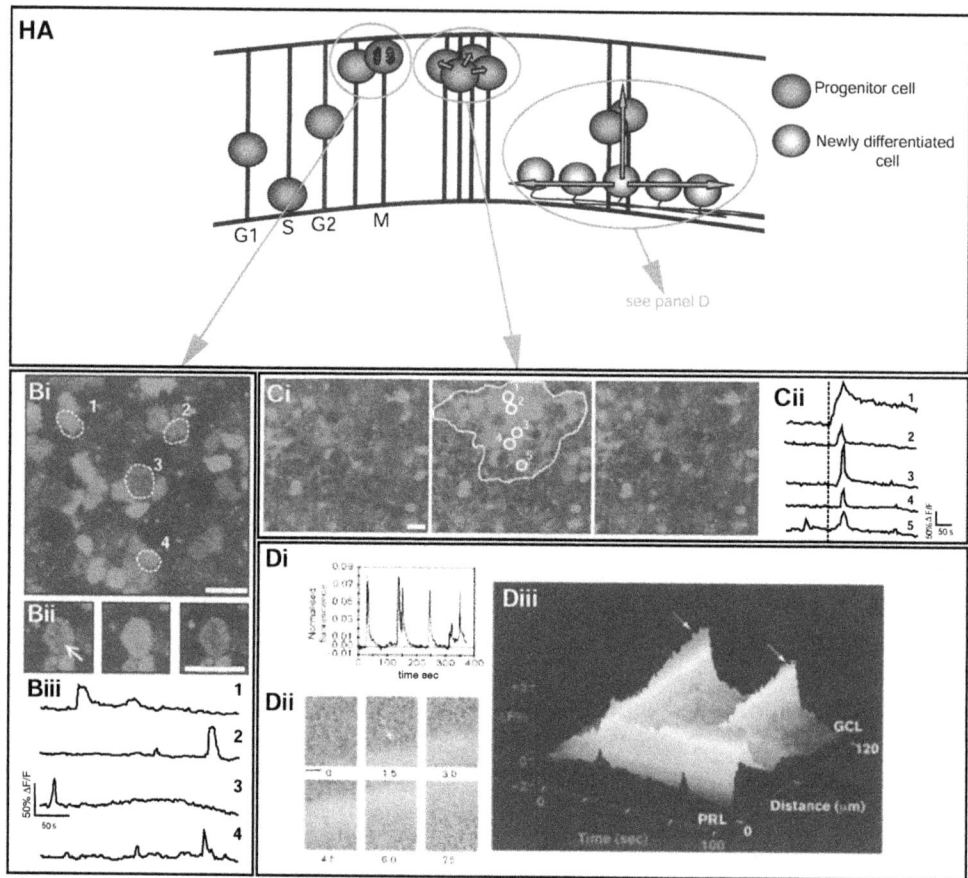

Figure 6.4 Spontaneous $[Ca^{2+}]_i$ transients in the embryonic chick retina. (A) Schematic diagram of the immature retina showing (*left*) the proliferative cell cycle and (*right*) types of Ca^{2+} activity. (B) spontaneous $[Ca^{2+}]_i$ in single cells in the VZ. (i) Confocal section through the VZ of an E5 chick retina labelled with the Ca^{2+} indicator, Fluo-4, (ii) an individual mitotic cell undergoing a single transient, (iii) traces showing transient changes in $[Ca^{2+}]_i$ in interphase (cells 1, 2 and 4) and a mitotic cell (cell 3) (chromatin labelling not shown). Scale bar 10 μm. (C) Coordinated spontaneous $[Ca^{2+}]_i$ in groups of adjacent cells. (i) Confocal section through the VZ of an E5 chick retina labelled with Fluo-4, (ii) traces showing the coordinated rise in $[Ca^{2+}]_i$ in neighbouring cells. Scale bar 10 μm. (D) spontaneous $[Ca^{2+}]_i$ waves propagate both laterally through differentiating neurons and back into the depth of the retina towards the VZ. (i) Spontaneous $[Ca^{2+}]_i$ waves in the prospective GCL of an E8 retina, (ii) propagation of a Ca^{2+} wave across the retina. Fluorescence (non-confocal) images taken through the GCL in an E11 retina loaded with Fura-2. Scale bar 200 μm, (iii) Spontaneous $[Ca^{2+}]_i$ waves propagate throughout the depth of the retina. The three-dimensional pseudo-colour line-image graph shows how $[Ca^{2+}]_i$ changes through the depth of a transverse retinal slice from an E11 embryo as a function of time. Warmer colours reflect higher $[Ca^{2+}]_i$. Two fast $[Ca^{2+}]_i$ transients (white arrows) can be seen spanning the depth of the retina from the GCL to the VZ (labelled here PRL, or prospective PR layer). %ΔF/F corresponds to the percentage change in the average normalized fluorescence over the field of view. Adapted from Pearson *et al.* (2002, 2004) and Catsicas *et al.* (1998). For colour version, see Plate 6.

non-dividing cells (Pearson et al., 2002). This is consistent with the findings from the pharmacological studies described earlier, that these receptors appear to be expressed by differentiating, rather than proliferating cells (Maric et al., 2000, 2001; Pearson et al., 2002).

Transient changes in $[Ca^{2+}]_i$ may also be involved in the regulation of another phase of the cell cycle, a process called interkinetic nuclear migration, where progenitor cell nuclei move between the VZ and the vitreal surface during G1 and G2. Like migrating immature cortical neurons (Komuro and Rakic, 1996), translocation of the nucleus in progenitor cells requires transient changes in $[Ca^{2+}]_i$ (Pearson et al., 2005b). Glutamate and GABA receptors have been shown to regulate the migration of immature neurons in the cortex (Komuro and Rakic, 1993; Rakic and Komuro, 1995). Whilst it seems likely that neurotransmitters could modulate the spatiotemporal features of $[Ca^{2+}]_i$ events in interkinetic nuclear migration, this possibility remains to be investigated.

One signal, but opposing actions?

The Ca^{2+} signals generated by both muscarinic and purinergic stimulation arise from the release of Ca^{2+} from intracellular stores (Sugioka et al., 1996; Pearson et al., 2002). How then might they exert different effects on proliferation? Such differences may be attributable to one or more of several factors; they may be linked to the production of different second messengers or effectors in addition to, or downstream of, Ca^{2+}. Alternatively, different receptors may couple to Ca^{2+} stores in different locations within the same cell (Short et al., 2000). It is also possible that the ACh and ATP released endogenously produce $[Ca^{2+}]_i$ signals of a different magnitude or frequency that code their different effects (Dolmetsch et al., 1997). Support for the first hypothesis comes from studies that demonstrate that MAPK, which is downstream of the IP_3/Ca^{2+} second messenger cascade, is activated in ATP-induced retinal proliferation (Sanches et al., 2002) but is not involved in the muscarinic-mediated inhibition of proliferation (Santos et al., 2003).

Neurotransmitters and the modulation of local Ca^{2+} events in the proliferating retina

In addition to the $[Ca^{2+}]_i$ transients in single cells, small groups of adjacent progenitor cells also undergo synchronized rises in $[Ca^{2+}]_i$ during the proliferative period of development in the retina (Figure 6.4C; Pearson et al., 2004, 2005b) and the cortex (Owens and Kriegstein, 1998). These local $[Ca^{2+}]_i$ waves are likely mediated via a combination of neurotransmitter release and gap-junction coupling between neighbouring cells (Owens and Kriegstein, 1998; Pearson et al., 2002, 2004; Weissman et al., 2004). Recently, Kriegstein and colleagues have shown that waves with similar spatiotemporal characteristics occur in proliferating radial glial cells (which act as progenitor cells during early development). These $[Ca^{2+}]_i$ waves are mediated by the spread of ATP released via gap-junction hemichannels from individual glia and act to modulate radial glial proliferation (Weissman et al., 2004). A similar mechanism may exist in the proliferating retina since synchronized $[Ca^{2+}]_i$ transients also occur between retinal progenitors (Figure 6.4C; Pearson et al., 2002, 2004). The importance of local synchronization of Ca^{2+} changes in cell proliferation has yet to be determined.

Neurotransmitters and the modulation of propagating Ca^{2+} waves in the differentiating retina

The third type of Ca^{2+} activity observed during early retinal development comprises large-scale propagating waves, which spread rapidly across large regions of the immature retina (Shatz, 1996; Wong et al., 1998; Wong, 1999). These waves are first seen during the peak of neurogenesis (as opposed to proliferation) (Figure 6.4D and Movie 6.3 online; Catsicas et al., 1998; Syed et al., 2004a,b) and, until recently, were thought to be restricted to differentiated cells, and primarily involved in the process of synaptogenesis (see Chapter 13 for a comprehensive discussion). However, elegant experiments by Syed et al. (2004a,b) that combined Ca^{2+} imaging and electrophysiology have demonstrated that progenitor cells in the VZ can also take part in these spontaneous, propagating waves, which appear distinct from the relatively small-scale events occurring earlier in development (Pearson et al., 2002, 2004).

These propagating waves first occur at E22 in the rabbit (Syed et al., 2004b). The waves are apparently initiated by cholinergic stimulation of nAChRs on differentiated cells in the inner retina and then propagate laterally, to adjacent neurons and radially, back to the VZ. The precise mechanism of propagation is unclear but involves gap-junction coupling and requires the activation of nAChRs, mAChRs and adenosine receptors. In contrast, neither GABAergic nor glutamatergic transmission are involved in the generation of inner retinal or VZ waves. Similar large-scale propagating Ca^{2+} waves have been reported for the chick retina (Figure 6.4D; Catsicas et al., 1998). These first occur at E8 and also appear to propagate back into the VZ. Similarly, they require gap-junction coupling and ACh, dopamine and glycine, but not GABA or glutamate, modulate the spatial and temporal properties of the waves at this time.

The function of these 'retrograde' waves is unclear. Syed et al. (2004a,b) suggest that they may represent a signal from differentiated neurons to progenitor cells. It is conceivable that these mAChRs-mediated waves might act to coordinate and synchronize cell cycle events (Cai et al., 1997; Owens and Kriegstein, 1998; Pearson et al., 2004, 2005b) or prompt cohorts of progenitor cells to exit the cell cycle and differentiate. The precise role of such signalling between differentiated neurons and progenitor cells presents an intriguing area for future research.

Neurotransmitters and the control of neuronal differentiation

In addition to their recently discovered roles in the regulation of proliferation, neurotransmitters have also been implicated in the regulation of differentiation. One specific example is the role of glutamate in the modulation of neurite outgrowth. The maturation of retinal neurons requires the extension and stabilization of dendrites prior to synaptogenesis. In the embryonic chick retina, glutamate inhibits neurite outgrowth in retinal neurons via Ca^{2+}-permeable AMPA receptors (Catsicas et al., 2001). This inhibition may form part of a 'stop' signal from developing BCs (a major source of glutamate in the adult retina) at a time when BC–RGC synaptic partners first enter into contact with one another at the onset of synaptogenesis, thus preventing neurites from growing past their targets (Catsicas *et al.*, 2001; Liets and Chalupa, 2001). Glutamate receptor activation has been shown to have similar stabilizing actions on dendritic growth in the retinotectal projection and its targets (Wu and Cline, 1998; Rajan *et al.*, 1999), whilst both NMDA and non-NMDA receptors are involved in RGC dendritic remodelling (Wong *et al.*, 2000) (see Chapter 12).

6.3 Neurotrophins in early retinal development

The neurotrophins, a family of growth factors consisting of nerve growth factor (NGF), brain-derived neurotrophic factor (BDNF), neurotrophin-3 (NT-3) and neurotrophin-4/5 (NT-4/5), are critical for the correct specification and survival of a number of classes of neurons in the central and peripheral nervous system (Lewin and Barde, 1996). A large body of experimental work led to the neurotrophic theory (Purves *et al.*, 1988); neurotrophins are expressed by the target areas innervated by axons and only those neurons that successfully compete for the neurotrophin source will survive, whilst the rest die (see Chapter 10; Barde, 1989; Oppenheim *et al.*, 1991). However, neurotrophins, like neurotransmitters, are now understood to have important roles in earlier stages of development since they and their receptors are expressed before neurons have even sprouted axons, let alone contacted their target areas. Two classes of receptor mediate the actions of neurotrophins, the Trk tyrosine kinase receptors (TrkA, TrkB and TrkC) and the neurotrophin receptor p75 (Barbacid, 1994; Lewin and Barde, 1996). All four neurotrophins bind with low affinity to the p75 receptor (Rodríguez-Tébar *et al.*, 1990, 1992) and with high affinity to one of the Trk receptors; NGF binds to TrkA, BDNF and NT-4/5 bind to TrkB and NT-3 binds to TrkC. Whilst all the neurotrophins are involved in cell survival during the period of target innervation (see Chapter 11), BDNF, NGF and NT-3 are each expressed much earlier, in both the avian and mammalian retina, during the peak of neurogenesis.

6.3.1 NT-3 and retinal differentiation

During the proliferative period of development, NT-3 mRNA is localized to the RPE, but not the neural retina. As development progresses the expression of NT-3 in the neural retina increases, and concomitantly decreases in the RPE. Expression of the TrkC NT-3 receptor in the chick retina occurs in two successive waves (Rodríguez-Tébar *et al.*, 1993); the first occurs at E6 to E7 and the second around E12. The latter correlates with the period of synaptogenesis, whilst the first is coincident with the onset of neuronal differentiation (Prada *et al.*, 1991).

A number of in vitro studies have indicated that NT-3, together with its receptor TrkC, plays a role in the generation of neurons. Neurotrophin-3 causes the premature arrest of proliferation and a reduction in the number of neurons in sensory ganglia (Ockel *et al.*, 1996) and initiates the differentiation of neural crest cells (Chalazonitis *et al.*, 1994, 1998; Hapner *et al.*, 1998). Similarly, NT-3, together with another growth factor, insulin-like growth factor-1 (de la Rosa *et al.*, 1994), stimulates the birth of new neurons in dissociated embryonic neural retinal cultures, whilst BDNF and NGF are both unable to mimic this effect (de la Rosa *et al.*, 1994). The actions of NT-3 have been further investigated using a specific antibody to neutralize endogenous NT-3 (Bovolenta *et al.*, 1996). In this study, inhibiting NT-3 activity at E6 led to a marked decrease in retinal neuronal differentiation, the RGC population being most notably affected. Overexpression of a truncated isoform of the NT-3 receptor, TrkC, which inhibits TrkC signalling, results in a reduction of all differentiated retinal cell types (Das *et al.*, 2000). Additionally, this impairment of NT-3 signalling causes a decrease in the clonal expansion of cells derived from a single retinal progenitor cell, suggesting that NT-3 might target retinal precursor cells rather than differentiated cell types. Thus, NT-3 appears to play an important role in both the onset of differentiation and the cessation of proliferation.

6.3.2 BDNF and NGF and early cell death

Programmed cell death (PCD) is a widespread phenomenon and is essential for the normal development of the nervous system (see Chapter 11). The precise developmental roles of this process are not well understood but include the regulation of final cell number and the elimination of aberrant connections during synaptogenesis. Three phases of PCD have been identified (Frade *et al.*, 1999). The first occurs before the onset of neurogenesis, the second coincides with the peak of neurogenesis and the third occurs during the period of synaptogenesis and projection refinement. Neurotrophins are not thought to be expressed at the time of the first wave (Frade *et al.*, 1999), thus only the second wave is considered here. The second wave of PCD primarily affects cells in the inner retina, but not differentiated RGCs. The location of the dying cells and the period of development in which this wave of PCD occurs suggest it most likely affects postmitotic, immature neurons that are migrating towards their final destination (Frade *et al.*, 1997).

Nerve Growth Factor mRNA is found from E4 onwards in the chick neural retina. Although NGF's high-affinity receptor TrkA is not found until after the peak of neurogenesis (Frade *et al.*, 1996), the low-affinity receptor p75 is expressed during this period. Nerve growth factor has no apparent effect on either proliferation or differentiation (de la Rosa *et al.*, 1994). However, selective inhibition of the p75 receptor decreased the levels of retinal cell death, as did reducing endogenous levels of NGF by the use of an anti-NGF antibody (Frade *et al.*, 1996). Thus, retinal-derived NGF acts to kill postmitotic cells during the second wave of PCD. The only source of NGF in the retina are the RGCs. Studies by Gonzalez-Hoyuela *et al.* (2001) have proposed that stratified RGCs use NGF, acting via the pro-apoptotic p75 receptor, to regulate their own numbers by inhibiting the generation of new RGCs and killing incoming migratory postmitotic RGCs. Selective ablation of RGCs in the chick retina resulted in the repopulation of the GCL by new cells and a large decrease in cell death. Conversely, application of NGF reversed these effects by increasing the levels of PCD and preventing the repopulation of RGCs. Layered RGCs may survive the apoptotic effect of NGF by expressing the high-affinity TrkA receptor, which switches the NGF signal from a pro-apoptotic to a neurotrophic one (Gonzalez-Hoyuela *et al.*, 2001). Recent studies in the mouse have suggested that interactions between NGF and the p75 receptor also account for early RGC cell death although other mechanims may also be involved at later stages (Harada et al., 2006).

Brain-derived neurotrophic factor counteracts the killing actions of NGF during the early period of PCD (E4) (Frade et al., 1997). At this time, BDNF, like NT-3, is predominantly expressed in the RPE, whilst its receptor TrkB is expressed in the chick neural retina (Frade et al., 1997). Expression of TrkB also occurs in two waves, again correlating with the two later waves of PCD, at E6 and at E12. Early reports indicated that BDNF provided trophic support to neural crest cells that give rise to the primordial of dorsal root ganglia, and protected them from cell death (Kalcheim et al., 1987). The exogenous application of BDNF to chick embryos led to a significant reduction in the number of cells dying during the second wave of PCD and concomitantly increased the number of RGCs by a similar proportion (Frade et al., 1997). Thus, a balance appears to exist between the trophic actions of BDNF and the apoptotic properties of NGF, which serves to control the number of newborn neurons that reach their final destination in the retina.

6.4 Concluding remarks

The role of neurotransmitter and neurotrophin signalling in the establishment and refinement of neural circuits is a dominant theme in developmental neurobiology. There is now growing evidence that even before synapses form these signalling systems are involved in the regulation of processes including cell proliferation, differentiation and survival. The growth factors BDNF and NGF play important roles in the regulation of cell number in the early period of cell death, whilst NT-3 is involved in neuronal differentiation. Neurotransmitters including ATP, ACh, GABA and glutamate are released, via a variety of vesicular and non-vesicular mechanisms, into the early developing retinal environment where they act to modulate a range of developmental effects. The purinergic and muscarinic neurotransmitter systems, for example, are present during the early proliferative period of retinal development, and may act to modulate proliferation and the expansion of the progenitor pool. The precise roles of GABA and glutamate in early retinal development have yet to be fully elucidated but may include the regulation of neuronal migration and differentiation, and neurite outgrowth and extension. Understanding how these neurotransmitters are linked to specific downstream consequences remains an issue for future research. One potential mechanism by which neurotransmitters could modulate developmental processes is through the second messenger, Ca^{2+}. Intracellular Ca^{2+} has a key influence on developmental events in the CNS and has been implicated in the regulation of differentiation, migration, cell fate and circuit formation. The ability to modulate $[Ca^{2+}]_i$ is a feature common to all the early neurotransmitters systems. In the developing retina, neurotransmitters can evoke increases in $[Ca^{2+}]_i$ and modulate the frequency and spatial characteristics of spontaneous $[Ca^{2+}]_i$ events including single cell transients, locally synchronized transients and propagating waves. The exact role of these early neurotransmitter signals, and how $[Ca^{2+}]_i$ transients are 'decoded' to influence retinal development, remain to be defined.

Acknowledgements

I am very grateful to Professor Peter Mobbs for his helpful comments on this chapter.

References

Abe, T., Sugihara, H., Nawa, H. et al. (1992). Molecular characterization of a novel metabotropic glutamate receptor mGluR5 coupled to inositol phosphate/Ca^{2+} signal transduction. *J. Biol. Chem.*, **267**, 13361–8.

Allcorn, S., Catsicas, M. and Mobbs, P. (1996). Developmental expression and self-regulation of Ca^{2+} entry via AMPA/KA receptors in the embryonic chick retina. *Eur. J. Neurosci.*, **8**, 2499–510.

Angelini, C., Costa, M., Morescalchi, F. et al. (1998). Muscarinic drugs affect cholinesterase activity and development of eye structures during early chick development. *Eur. J. Histochem.*, **42**(4), 309–20.

Aramori, I. and Nakanishi, S. (1992). Signal transduction and pharmacological characteristics of a metabotropic glutamate receptor, mGluR1, in transfected CHO cells. *Neuron*, **8**, 757–65.

Attwell, D., Barbour, B. and Szatkowski, M. (1993). Nonvesicular release of neurotransmitter. *Neuron*, **11**, 401–7.

Barbacid, M. (1994). The Trk family of neurotrophin receptors. *J. Neurobiol.*, **25**, 1386–403.

Barde, Y. A. (1989). Trophic factors and neuronal survival. *Neuron*, **2**, 1525–34.

Barker, J. L., Behar, T., Li, Y. X. et al. (1998). GABAergic cells and signals in CNS development. *Perspect. Dev. Neurobiol.*, **5**, 305–22.

Baughman, R. W. and Bader, C. R. (1977). Biochemical characterization and cellular localization of the cholinergic system in the chicken retina. *Brain Res.*, **138**, 469–85.

Bennett, M. V., Contreras, J. E., Bukauskas, F. F. and Saez, J. C. (2003). New roles for astrocytes: gap junction hemichannels have something to communicate. *Trends Neurosci.*, **26**, 610–17.

Bergmann, M., Grabs, D. and Rager, G. (1999). Developmental expression of dynamin in the chick retinotectal system. *J. Histochem. Cytochem.*, **47**, 1297–306.

Bergmann, M., Grabs, D. and Rager, G. (2000). Expression of pre-synaptic proteins is closely correlated with the chronotopic pattern of axons in the retinotectal system of the chick. *J. Comp. Neurol.*, **418**, 361–372.

Berridge, M. J. (1995). Calcium signaling and cell proliferation. *BioEssays*, **17**, 491–500.

Bliss, T. V. and Collingridge, G. L. (1993). A synaptic model of memory: long-term potentiation in the hippocampus. *Nature*, **361**, 31–9.

Bovolenta, P., Frade, J. M., Marti, E. et al. (1996). Neurotrophin-3 antibodies disrupt the normal development of the chick retina. *J. Neurosci.*, **16**, 4402–10.

Brändle, U., Guenther, E., Irrle, C. and Wheeler-Schilling, T. H. (1998). Gene expression of the P2x receptors in the rat retina. *Brain Res. Mol. Brain Res.*, **59**(2), 269–72.

Burnashev, N., Monyer, H., Seeburg, P. H. and Sakmann, B. (1992). Divalent ion permeability of AMPA receptor channels is dominated by the edited form of a single subunit. *Neuron*, **8**, 189–98.

Burnstock, G. (2004). Cotransmisson. *Curr. Opin. Pharmacol.*, **4**(1), 47–52.

Cai, L., Hayes, N. L. and Nowakowski, R. S. (1997). Synchrony of clonal cell proliferation and contiguity of clonally related cells: production of mosaicism in the ventricular zone of developing mouse neocortex. *J. Neurosci.*, **17**, 2088–100.

Catsicas, M. and Mobbs, P. (2001). GABAb receptors regulate chick retinal calcium waves. *J. Neurosci.*, **21**, 897–910.

Catsicas, M., Bonness, V., Becker, D. and Mobbs, P. (1998). Spontaneous Ca^{2+} transients and their transmission in the developing chick retina. *Curr. Biol.*, **8**, 283–6.

Catsicas, M., Allcorn, S. and Mobbs, P. (2001). Early activation of Ca^{2+}-permeable AMPA receptors reduces neurite outgrowth in embryonic chick retinal neurons. *J. Neurobiol.*, **49**, 200–11.

Cauley, K., Agranoff, B. W. and Goldman, D. (1990). Multiple nicotinic acetylcholine receptor genes are expressed in goldfish retina and tectum. *J. Neurosci.*, **10**, 670–83.

Caviness, V. S., Jr. Takahashi, T. and Nowakowski, R. S. (1995). Numbers, time and neocortical neuronogenesis: a general developmental and evolutionary model. *Trends Neurosci.*, **18**, 379–383.

Cepko, C. L., Austin, C. P., Yang, X., Alexiades, M. and Ezzeddine, D. (1996). Cell fate determination in the vertebrate retina. *Proc. Natl. Acad. Sci. U. S. A.*, **93**, 589–95.

Chalazonitis, A., Rothman, T. P., Chen, J. *et al.* (1994). Neurotrophin-3 induces neural crest-derived cells from fetal rat gut to develop in vitro as neurons or glia. *J. Neurosci.*, **14** (11 Pt 1), 6571–84.

Chalazonitis, A., Rothman, T. P., Chen, J. and Gershon, M. D. (1998). Age-dependent differences in the effects of GDNF and NT-3 on the development of neurons and glia from neural crest-derived precursors immunoselected from the fetal rat gut: expression of GFRalpha-1 in vitro and in vivo. *Dev. Biol.*, **204**(2), 385–406.

Cherubini, E., Gaiarsa, J. L. and Ben Ari, Y. (1991). GABA: an excitatory transmitter in early postnatal life. *Trends Neurosci.*, **14**, 515–19.

Cotrina, M. L., Kang, J., Lin, J. H. *et al.* (1998). Astrocytic gap junctions remain open during ischemic conditions. *J. Neurosci.*, **18**(7), 2520–37.

Cotrina, M. L., Lin, J. H., Lopez-Garcia, J. C., Naus, C. C. and Nedergaard, M. (2000). ATP-mediated glia signaling. *J. Neurosci.*, **20**(8), 2835–44.

Cristovao, A. J., Oliveira, C. R. and Carvalho, C. M. (2002a). Expression of AMPA/kainate receptors during development of chick embryo retina cells: in vitro versus in vivo studies. *Int. J. Dev. Neurosci.*, **20**(1), 1–9.

Cristovao, A. J., Oliveira, C. R. and Carvalho, C. M. (2002b). Expression of functional N-methyl-D-aspartate receptors during development of chick embryo retina cells: in vitro versus in vivo studies. *Brain Res. Mol. Brain Res.*, **99**(2), 125–33.

da Costa Calaza, K., Hokoc, J. N. and Gardino, P. F. (2000). Neurogenesis of GABAergic cells in the click retina. *Int. J. Dev. Neurosci.*, **18**(8), 721–6.

Das, I., Sparrow, J. R., Lin, M. I. *et al.* (2000). Trk C signaling is required for retinal progenitor cell proliferation. *J. Neurosci.*, **20**(8), 2887–95. Erratum in *J. Neurosci.* 2000 July 15, **201**(14), 5574.

de la Rosa, E. J., Bondy, C. A., Hernandez-Sanchez, C. *et al.* (1994). Insulin and insulin-like growth factor system components gene expression in the chicken retina from early neurogenesis until late development and their effect on neuroepithelial cells. *Eur. J. Neurosci.*, **6**(12), 1801–10.

Demarque, M., Represa, A., Becq, H. *et al.* (2002). Paracrine intercellular communication by a Ca^{2+}- and SNARE-independent release of GABA and glutamate prior to synapse formation. *Neuron*, **36**, 1051–61.

Dolmetsch, R. E., Lewis, R. S., Goodnow, C. C. and Healy, J. I. (1997). Differential activation of transcription factors induced by Ca^{2+} response amplitude and duration. *Nature*, **386**(6627), 855–8. Eratum in *Nature* 1997 July 17, **388**(6639), 308.

Dolmetsch, R. E., Xu, K. and Lewis, R. S. (1998). Calcium oscillations increase the efficiency and specificity of gene expression. *Nature*, **392**, 933–6.

Duarte, C. B., Santos, P. F., Sanchez-Prieto, J. and Carvalho, A. P. (1996). On-line detection of glutamate release from cultured chick retinospheroids. *Vis. Res.*, **36**, 1867–72.

Fischer, A. J., McKinnon, L. A., Nathanson, N. M. and Stell, W. K. (1998). Identification and localization of muscarinic acetylcholine receptors in the ocular tissues of the chick. *J. Comp. Neurol.*, **392**, 273–84.

Frade, J. M., Rodriguez-Tebar, A. and Barde, Y. A. (1996). Induction of cell death by endogenous nerve growth factor through its p75 receptor. *Nature*, **383**, 166–8.

Frade, J. M., Bovolenta, P., Martinez-Morales, J. R. *et al.* (1997). Control of early cell death by BDNF in the chick retina. *Development*, **124**, 3313–20.

Frade, J. M., Bovolenta, P. and Rodriguez-Tebar, A. (1999). Neurotrophins and other growth factors in the generation of retinal neurons. *Microsc. Res. Tech.*, **45**, 243–51.

Frambach, D. A. and Misfeldt, D. S. (1983). Furosemide-sensitive Cl transport in embryonic chicken retinal pigment epithelium. *Am. J. Physiol.*, **244**, F679–85.

Frederick, J. M. (1987). The emergence of GABA-accumulating neurons during retinal histogenesis in the embryonic chick. *Exp. Eye Res.*, **45**, 933–45.

Gonzalez-Hoyuela, M., Barbas, J. A. and Rodriguez-Tebar, A. (2001). The autoregulation of retinal ganglion cell number. *Development*, **128**(1), 117–24.

Grabs, D., Bergmann, M. and Rager, G. (2000). Developmental expression of amphiphysin in the retinotectal system of the chick: from mRNA to protein. *Eur. J. Neurosci.*, **12**, 1545–53.

Greenwood, D., Yao, W. P. and Housley, G. D. (1997). Expression of the P2X2 receptor subunit of the ATP-gated ion channel in the retina. *NeuroReport*, **8**, 1083–8.

Greferath, U., Grunert, U., Muller, F. and Wassle, H. (1994). Localization of GABA$_A$ receptors in the rabbit retina. *Cell Tissue Res.*, **276**(2), 295–307.

Greka, A., Lipton, S. A. and Zhang, D. (2000). Expression of GABA$_C$ receptor rho1 and rho2 subunits during development of the mouse retina. *Eur. J. Neurosci.*, **12**(10), 3575–82.

Gu, X. and Spitzer, N. C. (1993). Low-threshold Ca^{2+} current and its role in spontaneous elevations of intracellular Ca^{2+} in developing Xenopus neurons. *J. Neurosci.*, **13**, 4936–48.

Gu, X. and Spitzer, N. C. (1995). Distinct aspects of neuronal differentiation encoded by frequency of spontaneous Ca^{2+} transients. *Nature*, **375**, 784–7.

Gu, X., Olson, E. C. and Spitzer, N. C. (1994). Spontaneous neuronal calcium spikes and waves during early differentiation. *J. Neurosci.*, **14**, 6325–35.

Hamassaki-Britto, D. E., Gardino, P. F., Hokoc, J. N. *et al.* (1994). Differential development of alpha-bungarotoxin-sensitive and alpha-bungarotoxin-insensitive nicotinic acetylcholine receptors in the chick retina. *J. Comp. Neurol.*, **347**, 161–70.

Hapner, S. J., Boeshore, K. L., Large, T. H. and Lefcort, F. (1998). Neural differentiation promoted by truncated trkC receptors in collaboration with p75(NTR). *Dev. Biol.*, **201**, 90–100.

Harada, C., Harada, T., Nakamura, K. *et al.* (2006). Effect of p75 (NTR) on the regulation of naturally occurring cell death and retinal ganglion cell number in the mouse eye. *Dev. Biol.*, **290**(1), 57–65.

Harrison, P. K., Falugi, C., Angelini, C. and Whitaker, M. J. (2002). Muscarinic signaling affects intracellular calcium concentration during the first cell cycle of sea urchin embryos. *Cell Calcium*, **31**(6), 289–97.

Haydar, T. F., Wang, F., Schwartz, M. L. and Rakic, P. (2000). Differential modulation of proliferation in the neocortical ventricular and subventricular zones. *J. Neurosci.*, **20**, 5764–74.

Hayden, S. A., Mills, J. W. and Masland, R. M. (1980). Acetylcholine synthesis by displaced amacrine cells. *Science*, **210**(4468), 435–7.

Johnson, J., Tian, N., Caywood, M. S. *et al.* (2003). Vesicular neurotransmitter transporter expression in developing postnatal rodent retina: GABA and glycine precede glutamate. *J. Neurosci.*, **23**, 518–29.

Kalcheim, C., Barde, Y. A., Thoenen, H. and Le Douarin, N. M. (1987). In vivo effect of brain-derived neurotrophic factor on the survival of developing dorsal root ganglion cells. *EMBO J.* **6**(10), 2871–3.

Karne, A., Oakley, D. M. and Wong, G. K. and Wong, R. O. (1997). Immunocytochemical localization of GABA, $GABA_A$ receptors, and synapse-associated proteins in the developing and adult ferret retina. *Vis. Neurosci.*, **14**(6), 1097–108.

Katz, B. and Miledi, R. (1977). Transmitter leakage from motor nerve endings. *Proc. R. Soc. London B Biol. Sci.*, **196**(1122), 59–72.

Keyser, K. T., Hughes, T. E., Whiting, P. J., Lindstrom, J. M. and Karten, H. J. (1988). Cholinoceptive neurons in the retina of the chick: an immunohistochemical study of the nicotinic acetylcholine receptors. *Vis. Neurosci.*, **1**(4), 349–66.

Kim, I. B., Park, D. K., Oh, S. J. and Chun, M. H. (1999). Horizontal cells of the rat retina show choline acetyltransferase- and vesicular acetylcholine transporter-like immunoreactivities during early postnatal developmental stages. *Neurosci. Lett.*, **253**(2), 83–6.

Kim, I. B., Lee, E. J., Kim, M. K., Park, D. K. and Chun, M. H. (2000). Choline acetyltransferase-immunoreactive neurons in the developing rat retina. *J. Comp. Neurol.*, **427**(4), 604–16.

Komuro, H. Rakic, P. (1993). Modulation of neuronal migration by NMDA receptors. *Science*, **260**(5104), 95–7.

Komuro, H. and Rakic, P. (1996). Intracellular Ca^{2+} fluctuations modulate the rate of neuronal migration. *Neuron*, **17**(2), 275–85.

Komuro, H. and Rakic, P. (1998). Orchestration of neuronal migration by activity of ion channels, neurotransmitter receptors, and intracellular Ca^{2+} fluctuations. *J. Neurobiol.*, **37**(1), 110–30. Review.

Laasberg, T. (1990). Ca^{2+}-mobilizing receptors of gastrulating chick embryo. *Comp. Biochem. Physiol. C*, **97**(1), 9–12.

Layer, P. G. (1991). Cholinesterases during development of the avian nervous system. *Cell. Mol. Neurobiol.*, **11**(1), 7–33. Review.

Lewin, G. R. and Barde, Y. A. (1996). Physiology of the neurotrophins. *Annu. Rev. Neurosci.*, **19**, 289–317.

Liets, L. C. and Chalupa, L. M. (2001). Glutamate-mediated responses in developing retinal ganglion cells. *Prog. Brain Res.*, **134**, 1–16.

Lo Turco, J. J., Owens, D. F., Heath, M. J., Davis, M. B. and Kriegstein, A. R. (1995). GABA and glutamate depolarise cortical progenitor cells and inhibit DNA synthesis. *Neuron*, **15**(6), 1287–98.

Maric, D., Liu, Q. Y., Grant, G. M. *et al.* (2000). Functional ionotropic glutamate receptors emerge during terminal cell division and early neuronal differentiation of rat neuroepithelial cells. *J. Neurosci. Res.*, **61**(6), 652–62.

Maric, D., Liu, Q., Maric, I. *et al.* (2001). GABA expression dominates neuronal lineage progression in the embryonic rat neocortex and facilitates neurite outgrowth via $GABA_A$ autoreceptor/Cl^- channels. *J. Neurosci.*, **21**(7), 2343–60.

McKinnon, L. A. and Nathanson, N. M. (1995). Tissue-specific regulation of muscarinic acetylcholine receptor expression during embryonic development. *J. Biol. Chem.*, **270**(35), 20 636–42.

Mitchell, C. H. (2001). Release of ATP by a human retinal pigment epithelial cell line: potential for autocrine stimulation through subretinal space. *J. Physiol.*, **534**, 193–202.

Moll, V., Weick, M., Milenkovic, I. *et al.* (2002). P2Y receptor-mediated stimulation of Müller glial DNA synthesis. *Invest. Ophthalmol. Vis. Sci.*, **43**(3), 766–73.

Morita, M., Higuchi, C., Moto, T. *et al.* (2003). Dual regulation of calcium oscillation in astrocytes by growth factors and pro-inflammatory cytokines via the mitogen-activated protein kinase cascade. *J. Neurosci.*, **23**(34), 10 944–52.

Murphy, S. N. and Miller, R. J. (1989). Two distinct quisqualate receptors regulate Ca^{2+} homeostasis in hippocampal neurons in vitro. *Mol. Pharmacol.*, **35**(5), 671–80.

Nadler, L. S., Rosoff, M. L., Hamilton, S. E. *et al.* (1999). Molecular analysis of the regulation of muscarinic receptor expression and function. *Life Sci.*, **64**(6–7), 375–9. Review.

Naruoka, H., Kojima, R., Ohasa, M., Layer, P. G. and Saito, T. (2003). Transient muscarinic calcium mobilisation in transdifferentiating as in reaggregating embryonic chick retinae. *Brain Res. Dev. Brain Res.*, **134**(2), 233–44.

Newman, E. A. (2001). Propagation of intercellular calcium waves in retinal astrocytes and Müller cells. *J. Neurosci.*, **21**(7), 2215–23.

Nishi, S., Minota, S. and Karczmar, A. G. (1974). Primary afferent neurons: the ionic mechanism for GABA-mediated depolarisation. *Neuropharmacology*, **13**, 215–19.

Ockel, M., Lewin, G. R. and Barde, Y. A. (1996). In vivo effects of neurotrophin-3 during sensory neurogenesis. *Development*, **122**(1), 301–7.

Ohmasa, M. and Saito, T. (2004). $GABA_A$-receptor-mediated increase in intracellular Ca^{2+} concentration in the regenerating retina of adult newt. *Neurosci. Res.*, **49**, 219–27.

Oppenheim, R. W., Prevette, D., Yin, Q. W., Collins, F. and MacDonald, J. (1991). Control of embryonic motoneuron survival *in vivo* by ciliary neurotrophic factor. *Science*, **251**(5001), 1616–18.

Owens, D. F. and Kriegstein, A. R. (1998). Patterns of intracellular calcium fluctuation in precursor cells of the neocortical ventricular zone. *J. Neurosci.*, **18**(14), 5374–88.

Pannicke, T., Fischer, W., Biedermann, B. *et al.* (2000). P2X7 receptors in Müller glial cells from the human retina. *J. Neurosci.*, **20**, 5965–72.

Pearson, R., Catsicas, M., Becker, D. and Mobbs, P. (2002). Purinergic and muscarinic modulation of the cell cycle and calcium signaling in the chick retinal ventricular zone. *J. Neurosci.*, **22**, 7569–79.

Pearson, R. A., Catsicas, M., Becker, D. L. *et al.* (2004). Ca^{2+} signaling and gap junction coupling within and between pigment epithelium and neural retina in the developing chick. *Eur. J. Neurosci.*, **19**, 2435–45.

Pearson, R. A., Dale, N., Llaudet, E. and Mobbs, P. (2005a). ATP released via gap junction hemichannels from the pigment epithelium regulates neural retinal progenitor proliferation. *Neuron*, **46**, 731–44.

Pearson, R. A., Luneborg, N. L., Becker, D. and Mobbs, P. (2005b). Gap junctions modulate interkinetic nuclear migration in retinal progenitor cells. *J. Neurosci.*, **25**(46), 10 803–14.

Pellegrini-Giampietro, D. E., Bennett, M. V. and Zukin, R. S. (1992). Are Ca^{2+}-permeable kainate/AMPA receptors more abundant in immature brain? *Neurosci. Lett.*, **144**(1–2), 65–9.

Prada, C., Puga, J., Perez-Mendez, L., Lopez, R. and Ramirez, G. (1991). Spatial and temporal patterns of neurogenesis in the chick retina. *Eur. J. Neurosci.*, **3**, 559–69.

Pruss, R. M., Akeson, R. L., Racke, M. M. and Wilburn, J. L. (1991). Agonist-activated cobalt uptake identifies divalent cation-permeable kainate receptors on neurons and glial cells. *Neuron*, **7**(3), 509–18.

Purves, D., Snider, W. D. and Voyvodic, J. T. (1988). Trophic regulation of nerve cell morphology and innervation in the autonomic nervous system. *Nature*, **336**(6195), 123–8. Review.

Rajan, I., Witte, S. and Cline, H. T. (1999). NMDA receptor activity stabilizes pre-synaptic retinotectal axons and post-synaptic optic tectal cell dendrites *in vivo*. *J. Neurobiol.*, **38**(3), 357–68.

Rakic, P. and Komuro, H. (1995). The role of receptor/channel activity in neuronal cell migration. *J. Neurobiol.*, **26**(3), 299–315.

Reichling, D. B., Kyrozis, A., Wang, J. and MacDermott, A. B. (1994). Mechanisms of GABA and glycine depolarisation-induced calcium transients in rat dorsal horn neurons. *J. Physiol.*, **476**, 411–21.

Rodríguez-Tébar, A., Dechant, G. and Barde, Y. A. (1990). Binding of brain-derived neurotrophic factor to the nerve growth factor receptor. *Neuron*, **4**(4), 487–92.

Rodríguez-Tébar, A., Dechant, G., Gotz, R. and Barde, Y. A. (1992). Binding of neurotrophin-3 to its neuronal receptors and interactions with nerve growth factor and brain-derived neurotrophic factor. *EMBO J.* **11**(3), 917–22.

Rodríguez-Tébar, A., de la Rosa, E. J. and Arribas, A. (1993). Neurotrophin-3 receptors in the developing chicken retina. *Eur. J. Biochem.*, **211**(3), 789–94.

Rorig, B. and Grantyn, R. (1994). Ligand- and voltage-gated ion channels are expressed by embryonic mouse retinal neurones. *NeuroReport*, **5**(10), 1197–200.

Rudolph, U., Crestani, F. and Mohler, H. (2001). GABA(A) receptor sub-types: dissecting their pharmacological functions. *Trends Pharmacol. Sci.*, **22**(4), 188–94. Review.

Sakaki, Y., Fukuda, Y. and Yamashita, M. (1996). Muscarinic and purinergic Ca^{2+} mobilisations in the neural retina of early embryonic chick. *Int. J. Dev. Neurosci.*, **14**, 691–9.

Sanches, G., de Alencar, L. S. and Ventura, A. L. (2002). ATP induces proliferation of retinal cells in culture via activation of PKC and extracellular signal-regulated kinase cascade. *Int. J. Dev. Neurosci.*, **20**, 21–7.

Santella, L. (1998). The role of calcium in the cell cycle: facts and hypotheses. *Biochem. Biophys. Res. Commun.*, **244**(2), 317–24. Review.

Santella, L., Kyozuka, K., De Riso, L. and Carafoli, E. (1998). Calcium, protease action, and the regulation of the cell cycle. *Cell Calcium*, **23**(2–3), 123–30. Review.

Santos, A. A., Medina, S. V., Sholl-Franco, A. and de Araujo, E. G. (2003). PMA decreases the proliferation of retinal cells in vitro: the involvement of acetylcholine and BDNF. *Neurochem. Int.*, **42**, 73–80.

Santos Bredariol, A. and Hamassaki-Britto, D. (2001). Ionotropic glutamate receptors during the development of the chick retina. *J. Comp. Neurol.*, **441**, 58–70.

Schwartz, E. A. (1987). Depolarisation without calcium can release gamma-aminobutyric acid from a retinal neuron. *Science*, **238**(4825), 350–5.

Segal, M. and Barker, J. L. (1984). Rat hippocampal neurons in culture: properties of GABA-activated Cl^- ion conductance. *J. Neurophysiol.*, **51**(3), 500–15.

Shatz, C. J. (1996). Emergence of order in visual system development. *J. Physiol. Paris*, **90**(3–4), 141–50. Review.

Sherry, D. M., Wang, M. M., Bates, J. and Frishman, L. J. (2003). Expression of vesicular glutamate transporter 1 in the mouse retina reveals temporal ordering in development of rod vs. cone and ON vs. OFF circuits. *J. Comp. Neurol.*, **465**(4), 480–98.

Short, A. D., Winston, G. P. and Taylor, C. W. (2000). Different receptors use inositol trisphosphate to mobilize Ca(2+) from different intracellular pools. *Biochem. J.*, **351**(Pt 3), 683–6.

Somahano, F., Roberts, P. J. and Lopez-Colome, A. M. (1988). Maturational changes in retinal excitatory amino acid receptors. *Dev. Brain Res.*, **42**, 59–67.

Spitzer, N. C., Debaca, R. C., Allen, K. A. and Holliday, J. (1993). Calcium dependence of differentiation of GABA immunoreactivity in spinal neurons. *J. Comp. Neurol.*, **337**(1), 168–75.

Stout, C. E., Costantin, J. L., Naus, C. C., and Charles, A. C. (2002). Intercellular calcium signaling in astrocytes via ATP release through connexin hemichannels. *J. Biol. Chem.*, **277**(12), 10 482–8.

Sugioka, M. and Yamashita, M. (2003). Calcium signaling to nucleus via store-operated system during cell cycle in retinal neuroepithelium. *Neurosci. Res.*, **45**, 447–58.

Sugioka, M., Fukuda, Y. and Yamashita, M. (1996). Ca^{2+} responses to ATP via purinoceptors in the early embryonic chick retina. *J. Physiol.*, **493**(3), 855–63.

Sugioka, M., Fukuda, Y. and Yamashita, M. (1998). Development of glutamate-induced intracellular Ca^{2+} rise in the embryonic chick retina. *J. Neurobiol.*, **34**(2), 113–25.

Sugioka, M., Zhou, W. L., Hofmann, H. D. and Yamashita, M. (1999). Involvement of P2 purinoceptors in the regulation of DNA synthesis in the neural retina of chick embryo. *Int. J. Dev. Neurosci.*, **17**, 135–44.

Syed, M. M., Lee, S., He, S. and Zhou, Z. J. (2004a). Spontaneous waves in the ventricular zone of developing mammalian retina. *J. Neurophysiol.*, **91**(5), 1999–2009.

Syed, M. M., Lee, S., Zheng, J. and Zhou, Z. J. (2004b). Stage-dependent dynamics and modulation of spontaneous waves in the developing rabbit retina. *J. Physiol.*, **560**(2), 533–49.

Wada, E., Wada, K., Boulter, J. *et al.* (1989). Distribution of alpha 2, alpha 3, alpha 4, and beta 2 neuronal nicotinic receptor subunit mRNAs in the central nervous system: a hybridization histochemical study in the rat. *J. Comp. Neurol.*, **284**(2), 314–35.

Warr, O., Takahashi, M. and Attwell, D. (1999). Modulation of extracellular glutamate concentration in rat brain slices by cystine-glutamate exchange. *J. Physiol.*, **514**(3), 783–93.

Weissman, T. A., Riquelme, P. A., Ivic, L., Flint, A. C., and Kriegstein, A. R. (2004). Calcium waves propagate through radial glial cells and modulate proliferation in the developing neocortex. *Neuron*, **43**(5), 647–61.

Wheeler-Schilling, T. H., Marquordt, K., Kohler, K., Guenther, E. and Jabs, R. (2001). Identification of purinergic receptors in retinal ganglion cells. *Brain Res. Mol. Brain Res.*, **92**(1–2), 177–80.

Whitaker, M. and Larman, M. G. (2001). Calcium and mitosis. *Semin. Cell Dev. Biol.*, **12**(1), 53–8.

Wong, R. O. (1995a). Cholinergic regulation of $[Ca^{2+}]i$ during cell division and differentiation in the mammalian retina. *J. Neurosci.*, **15**(4), 2696–706.

Wong, R. O. (1995b). Effects of glutamate and its analogs on intracellular calcium levels in the developing retina. *Vis. Neurosci.*, **12**(5), 907–17.

Wong, R. O. (1999). Retinal waves: stirring up a storm. *Neuron*, **24**(3), 493–5.

Wong, W. T., Sanes, J. R. and Wong, R. O. (1998). Developmentally regulated spontaneous activity in the embryonic chick retina. *J. Neurosci.*, **18**(21), 8839–52.

Wong, W. T., Myhr, K. L., Miller, E. D. and Wong, R. O. (2000). Developmental changes in the neurotransmitter regulation of correlated spontaneous retinal activity. *J. Neurosci.*, **20**(1), 351–60.

Wu, G. Y. and Cline, H. T. (1998). Stabilisation of dendritic arbor structure *in vivo* by CaMKII. *Science*, **279**(5348), 222–6.

Yamashita, M. and Fukuda, Y. (1993). Calcium channels and GABA receptors in the early embryonic chick retina. *J. Neurobiol.*, **24**(12), 1600–14.

Yamashita, M., Yoshimoto, Y. and Fukuda, Y. (1994). Muscarinic acetylcholine responses in the early embryonic chick retina. *J. Neurobiol.*, **25**(9), 1144–53.

Yang, H. and Kunes, S. (2004). Nonvesicular release of acetylcholine is required for axon targeting in the Drosophila visual system. *Proc. Natl. Acad. Sci. U. S. A.*, **101**(42), 15 213–8.

Zhang, L., Spigelman, I. and Carlen, P. L. (1991). Development of GABA-mediated, chloride-dependent inhibition in CA1 pyramidal neurones of immature rat hippocampal slices. *J. Physiol.*, **444**, 25–49.

Zheng, J. Q. (2000). Turning of nerve growth cones induced by localized increases in intracellular calcium ions. *Nature*, **403**(6765), 89–93.

Zheng, J. Q., Felder, M., Connor, J. A. and Poo, M. M. (1994). Turning of nerve growth cones induced by neurotransmitters. *Nature*, **368**(6467), 140–4.

7

Comparison of development of the primate *fovea centralis* with peripheral retina

Anita Hendrickson
University of Washington, Seattle, USA

Jan Provis
The Australian National University, Canberra, Australia

7.1 Introduction

The macula lutea ('yellow spot'), located towards the posterior pole of the human retina was identified grossly in the late eighteenth century, and the *fovea centralis* – located approximately at the centre of the macula – was first described by Soemmerring in 1795. It was H. Müller who produced the first histological description of the human fovea (see Polyak, 1941), along with identification of the retinal layers and a correct analysis of their general place in the retinal circuitry. Knowledge of its anatomical organization was greatly expanded by the Golgi impregnation work of Cajal (1893) and Polyak (1941), which indicated that foveal cones give rise to highly specialized circuits.

Some of the earliest studies of developing primate retina were carried out by Chievitz (1888), Magitot (1910) and Bach and Seefelder (1911, 1912, 1914), from whose work some illustrations are reproduced later in this chapter. These early studies were greatly expanded on by Ida Mann in the first part of the twentieth century and published in monograph form in 1928 (see Mann, 1964, 2nd edition). Little further work was published on foveal development until Hendrickson and Kupfer's study (1976) showing photoreceptor displacement towards the developing fovea. This most recent period of analysis of primate (including human) retinal development (1976 to present) has taken place in the context of an expanding body of knowledge, gleaned largely from investigation of retinal development in non-primate species – much of which is covered in other chapters of this volume – including gene regulation of eye and retinal formation, retinal cell generation, development of target visual nuclei, guidance mechanisms, the importance of neuronal acitivity and the significance of apoptosis. The overwhelming conclusion is that the central region of the primate retina is highly specialized and developmental mechanisms at the fovea are unique amongst mammalian retinas.

Retinal Development, ed. Evelyne Sernagor, Stephen Eglen, Bill Harris and Rachel Wong.
Published by Cambridge University Press. © Cambridge University Press 2006.

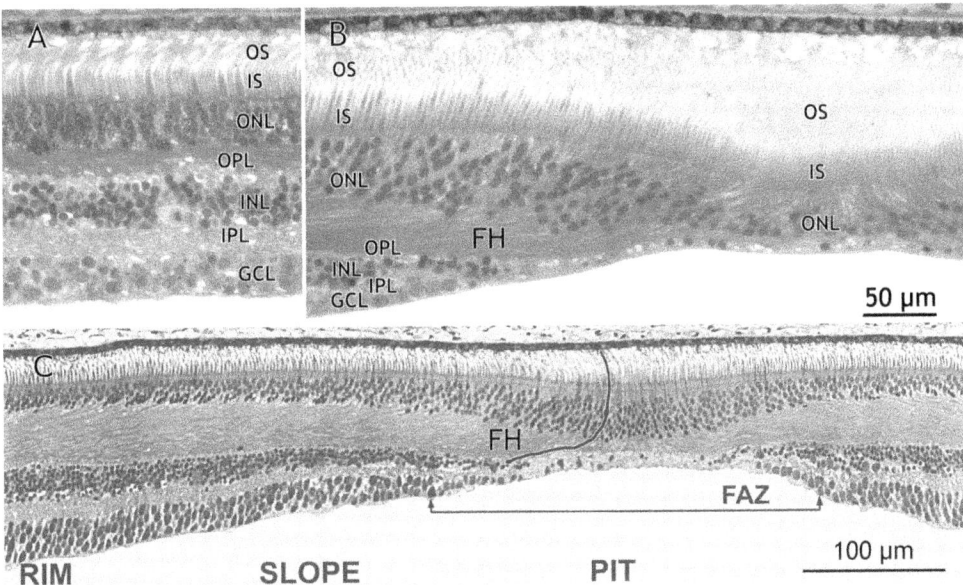

Figure 7.1 Sections through 72-year-old (A,B) and 13-year-old human (C) retinas showing features of the fovea and the foveal subregions. (A) Peripheral retina showing the relative thickness of the layers. (B) The fovea, illustrating the absence of the inner retinal layers in the central fovea, such that the cell bodies of the central cones in the outer nuclear layer (ONL) are close to the inner retinal surface. The axons of foveal cones (Henle fibres, FH) extend away from the fovea, forming a distinct band on the outer aspect of the outer plexiform layer (OPL). Note the very high density of cone inner (IS) and outer (OS) segments at the contralateral fovea. (C) The 'subregions' of the fovea referred to in this text are shown. The full extent of one of the foveal cones is outlined for clarity. By definition the *fovea centralis* extends from rim to rim; however, only the central fovea, where cones reach peak density, is avascular (the foveal avascular zone, FAZ). GCL, ganglion cell layer, INL, inner nuclear layer; IPL, inner plexiform layer.

7.2 Primate retina is characterized by a specialized central retinal region while the peripheral retina resembles the retina of other mammals

The anatomical organization of the peripheral primate retina, whether it is human, Old World *Macaca* monkey or New World monkeys, closely resembles that of any other mammal. The peripheral retina has three neuronal and two synaptic layers (Figure 7.1A). The outer nuclear layer (ONL) contains both rod and cone photoreceptors with rods outnumbering cones 15 to 20:1 (reviewed in Curcio and Hendrickson, 1991). All cone subtypes are present in peripheral retina. In trichromats like humans and macaques there are three cone types – long (L), or red-sensitive ('red'); medium (M), or 'green'; and short (S), or blue-wavelength selective, while dichromatic New World monkeys have either an M- or L-type cone plus S-cones (Nathans *et al.*, 1992; Jacobs and Deegan, 1999). Photoreceptors synapse at the outer plexiform layer (OPL) onto the bipolar (BC) and horizontal (HC) neurons in the inner nuclear layer (INL). In turn the BC and amacrine (AC) neurons in the INL synapse onto the ganglion cells (GCs) via the inner plexiform layer (IPL). All regions of primate retina contain a unique glial cell, the Müller cell, whose cell body is in the INL with processes forming the outer and inner retinal borders. The regional feature that distinguishes the primate retina from other mammals is a central retinal specialization on the temporal side of the optic disc (OD). This area is called the *macula lutea* in the clinical literature, characterized by the presence of the yellow pigment that gives the region its name. The central 1 mm of this area is the *fovea centralis* (Figure 7.1B). The fovea is a depression, or pit, in the inner retinal layers, associated with a very high concentration of cones in the outer layer (Figure 7.1C). Blood vessels and inner retinal layers are absent from the floor of the depression, also known as the foveola. Because the Latin word for a small pit is 'fovea', in modern literature the term is now used to describe this central, highly specialized region.

The foveae of all primates that have been described are remarkably similar (Polyak, 1941; Rohen and Castenholtz, 1967; Hendrickson, 1992; Provis *et al.*, 1998; Franco *et al.*, 2000). One exception is the nocturnal owl monkey *Aotes*, which has a very high rod density but a low central cone density and no clear pit (Rohen and Castenholtz, 1967; Webb and Kaas, 1976; Wikler *et al.*, 1990). Although its exact status within the primates is debated, the nocturnal tarsier has, by contrast, a high rod density but also a cone-rich fovea that resembles those in monkeys (Rohen, 1966; Rohen and Castenholtz, 1967; Hendrickson *et al.*, 2000). The development of the fovea occurs through complex and somewhat unexpected processes, discussed below. In brief, the ganglion cell layer (GCL), IPL and INL are moved peripherally to form the pit during fetal and early neonatal development (Figure 7.1C). This lateral displacement causes a thickening of inner layers around the pit, forming the foveal rim (Figure 7.1C), while the portion between the pit bottom and the peak of the rim is termed the slope (Figure 7.1C). Over the centre of the pit is a very thick ONL comprising tightly packed long thin cones (Figure 7.1B,C). This central ONL is different from peripheral retina in that rods are absent from the central 300 μm. S-cones are absent or sparse in the central 100 μm, and all central photoreceptors have long axons, or fibres of Henle, that run laterally towards their synaptic pedicles, which are moved onto the rim during pit formation (Packer *et al.*, 1989; Curcio *et al.*, 1990, 1991; Curcio and Hendrickson, 1991; Bumsted and Hendrickson, 1999; Martin and Grunert, 1999; Roorda and Williams, 1999; Martin *et al.*, 2000). Another difference, not obvious morphologically, is that the neuronal composition of the foveal slope favours neurons of the midget pathway, consisting of 1 cone → 2 BCs → 2 midget GCs forming a circuit that subserves high-acuity and colour vision (Polyak, 1941; Wässle *et al.*, 1994; Dacey, 1996, 1999). Finally, blood vessels are absent from the central pit – the foveal avascular zone (FAZ) (Fig. 7.1C) but form a plexus on the slope to encircle the pit (Snodderly *et al.*, 1992; Provis *et al.*, 2000; Provis, 2001).

7.3 Developmental events in primate retina begin in the fovea

7.3.1 *The general pattern of cell generation and cell death*

In all mammals, cell differentiation commences in the central retina and proceeds in a wave-like manner towards the periphery (Rapaport and Stone, 1984; Robinson, 1991). For all differentiation processes that have been studied in primates, events begin in the incipient

Table 7.1. *Comparison of the developmental timetables of macaque and human retina*

Developmental event	Macaque		Human	
	Fd	%CP	Fwk	%CP
'Cold-spot' evident	55	44	1–12	39
S-opsin expressed	65–70	52–56	12	43
L-/M-opsin expressed	65–70	52–56	15	53
Rhodopsin expressed	65–70	52–56	15	53
Retinal vessels at disc	70	56	14–15	50–53
Peak GC axon number	80	64	17	60
FAZ defined	100–105	85	~25	89
Pit starts	105–110	85–88	25–26	89–93
Eyelids open	124	100	28	100
Birth	***172***	***138***	***40***	***143***
Excavation complete	170	137	43	153

CP: caecal period is the time between conception and eye opening.
Fd: No of fetal days postconception.
Fwk: No of fetal weeks postconception.
FAZ: foveal avascular zone.
Cold-spot: the term used to refer to the appearance of a region in the ventricular layer where cell division is absent, in primates marking differentiation of the incipient foveal region.

fovea and end weeks to months later at the retinal edge (Mann, 1964; Hendrickson and Kupfer, 1976; Provis *et al.*, 1985a; Provis, 1987; Okada *et al.*, 1994; Dorn *et al.*, 1995; Bumsted *et al.*, 1997; Georges *et al.*, 1999). In this chapter we will refer to human age before birth in fetal weeks (Fwk) while *Macaca* ages will be given in fetal days (Fd). If the species is not stated, it can be identified by the age nomenclature. In Table 7.1 we compare some milestones in human and *Macaca* retinal development, using both chronological age and percentage 'caecal period' (the time between conception and eye opening) as indicators of developmental 'stage'.

In the human fovea, cell division is complete by Fwk 12 (Provis *et al.*, 1985a), and in macaque fovea tritiated-thymidine (^{3}H-TdR) uptake is absent by Fd 55 (LaVail *et al.*, 1991). These are the dates by which all the retinal layers are distinct at the incipient fovea, which can therefore be identified morphologically (Hendrickson, 1992). Cell proliferation is virtually complete throughout the entire human retina by Fwk 30 (Provis *et al.*, 1985a), although ^{3}H-TdR uptake shows that some neurons are born in the far periphery even after birth in macaques (LaVail *et al.*, 1991). The chronological sequence of cell generation in primates is similar to that described in other mammals. Tritiated-thymidine labelling indicates cell differentiation occurs essentially in two phases; an initial phase in which GCs, and cones are generated, and a later phase in which rods and BCs are generated. There is an intermediate period overlapping the two phases in which ACs are born (see Chapter 3). Foveal GCs, HCs and cones, therefore, are the first cells in the primate retina to differentiate and peripheral rods are amongst the last. Müller glial cells appear, on the basis of ^{3}H-TdR incorporation, to be the last retinal cell type generated across the retina (LaVail et al., 1991), although recent findings indicate that Müller cells express the neural progenitor cell marker nestin throughout development and, like the radial glia of cerebral cortex (Noctor et al., 2001), are the end-stage neural progenitor cells of the retina (Fischer et al., 2001; Walcott and Provis, 2003).

There is no clear indication of when the first GCs are generated in humans, although interpolation from macaque ^{3}H-TdR data (Chapter 3) suggests an approximate birthdate of Fwk 7. The first GC axons in the OD also are reported at Fwk 7 (Mann, 1964). Histologically the GCL is distinct at the periphery by Fwk 16, and quantitative analysis suggests that few GC axons are added to the optic nerve after Fwk 17 (Provis et al., 1985a). Triitated-thymidine labelling in macaque GCs is completed by Fd 90 (LaVail et al., 1991) so in both primates GC generation is finished around midgestation. Overlapping with the phase of GC generation is a phase of apoptotic cell death, also initiated in the central GCL early in development, which spreads across the retina to reach the human periphery at around Fwk 30 (Provis, 1987; Georges et al., 1999). Because GCs are added to the retina simultaneous with the period of cell death, the numbers of GCs generated and the numbers lost are difficult to calculate. However, based on optic axon counts, 70% of the axons present in the human optic nerve at Fwk 17 are lost by Fwk 30, when axon numbers stabilize at adult values of 1.1 to 1.3 million (Provis et al., 1985b). Similar studies in macaque optic nerve find a 56% loss before birth (Rakic and Riley, 1983). Analyses of the distribution of pyknotic figures (Provis, 1987) and TUNEL-labelled profiles in the GCL (Georges et al., 1999) show that after Fwk 15 rates of apoptosis are higher in peripheral parts of the retina compared to central. Because foveal pit formation does not begin until around Fwk 25 in humans (Hendrickson and Yuodelis, 1984; Provis et al., 1998), this lack of central cell death suggests that apoptosis is not a key event in formation of the fovea. Furthermore, counts of TUNEL-positive profiles in the central GCL reveal very low levels of apoptosis around the time of formation of the foveal depression (Georges et al., 1999). Studies of cell death in the INL and ONL indicate higher levels of cell death in the BC population, compared with GCs, and very low levels in developing photoreceptors (Georges et al., 1999).

A distinctive feature of primate retina is that certain cell types are never detected in the immature fovea. Analyses of the differentiation of photoreceptor populations in both human and macaque retina show that rods are virtually absent from the centre of the incipient fovea throughout development (Hendrickson and Kupfer, 1976; Diaz-Araya and Provis, 1992). As described in Chapter 3, ^{3}H-TdR labelling, in combination with morphological criteria, indicates that in macaques cones are first generated at Fd 36 to 38, while the earliest rods are generated at Fd 45 (LaVail et al., 1991). When the incipient fovea can be recognized morphologically in macaques at Fd 55, the ONL is a region 500 μm in diameter, containing only cones and surrounded by scattered rods (Hendrickson, 1992). Similarly, rod opsin and the rod-specific nuclear factors Nrl and Nr2e3 are not found within the foveal ONL at any age, but appear on the foveal edge at Fwk 12 (Mears et al., 2001; Bumsted-O'Brien et al., 2003). This absence of rods suggests that rod BCs and AII ACs, essential components of the rod inner retina circuit, also are not generated in the developing foveal region, although this has not been specifically investigated. A similar situation is found for S-cones, which are absent from the fetal and adult human foveal ONL (Curcio and Hendrickson, 1991; Bumsted and Hendrickson, 1999). In the very young fetal human retina, however, both scattered S-cones and rods are present within the fovea, but these are lost with increasing age (Bumsted and Hendrickson, 1999). Programmed cell death of these inappropriate cells may be involved in sculpting the adult foveal photoreceptor mosaic.

7.3.2 The general pattern of synapse formation

In Figure 7.2 we compare retinal development at the *Macaca* fovea, temporal to the OD, and in nasal retina midway between OD and edge, between Fd 64 and 11years. Morphologically, the IPL can be detected at Fd 46 to 48 in the region that will form the fovea and in the human as early as Fwk 8 (Mann, 1964; Hendrickson, 1992). The IPL develops rapidly across the retina, reaching the OD at Fd 68 (or Fwk 12 to 13), and is near the retinal edge at Fd 95 (Fwk 16 to 18). The OPL appears in the foveal region at Fd 50 to 55, or Fwk 10 to 11, but spreads across the retina more slowly than the IPL. Although some cones can be identified peripherally, a distinct OPL is not seen near the OD until Fd 95 (Fwk 16) (Figure 7.2b), and is present in the mid-periphery at Fd 115 (Fwk 24) and at the retinal edge at Fd 150 in the monkey (Figure 7.2c) and shortly before birth in the human. The early differentiation of GCs, HCs and cones might predict that the first synapses should form in the OPL between cones and HCs. However, immunolabelling of synaptic proteins (Okada *et al.*, 1994; Georges *et al.*, 1999) and electron microscopy (EM) morphology (Nishimura and Rakic, 1985, 1987; Linberg and Fisher, 1990; van Driel *et al.*, 1990; Crooks *et al.*, 1995) show that the first synapses formed are in the primate IPL (Figure 7.3a,b). Electron microscopy studies differ as to whether the first synapses are AC conventional synapses (Nishimura and Rakic, 1987), BC ribbon synapses (Crooks *et al.*, 1995) or both (Linberg and Fisher, 1990; van Driel *et al.*, 1990). In part this may be due to slight differences between the youngest ages and the eccentricity of the retina examined; that is, rod-dominated peripheral or cone-dominated fovea. In the foveal IPL of macaque retina, BC ribbon synapses appear first in the outer half where 'OFF' inputs dominate, and slightly later in the inner or 'ON', while AC synapses are found throughout the early fetal IPL (Crooks *et al.*, 1995). In the macaque fovea the OFF-BC synapses are seen as early as Fd 55 and ON-BC synapses are present two to three days later (Figure 7.3b). Ganglion cells retrogradely filled with horseradish peroxidase in Fd 60 monkey have dendrites stratified into OFF and ON sublamina (Kirby and Steineke, 1991) and EM morphology is consistent with many of the early BC contacts terminating on GC dendrites (Crooks *et al.*, 1995). Immunolabelling for synaptic vesicle protein 2 (SV2) (Okada *et al.*, 1994) detects the first IPL labelling as a band consistent with the OFF-BC terminals contacting the OFF-midget and parasol GC, followed one to two days later by an inner band at the level of the ON-midget and parasol GC (Koontz and Hendrickson, 1993). Amacrine conventional synapses are extremely rare at Fd 55 in the foveal IPL, but more than double to 15/100 μm^2 by Fd 90. At the same age, BC ribbon synapses are at adult distribution and density (5.5/100 μm^2) while

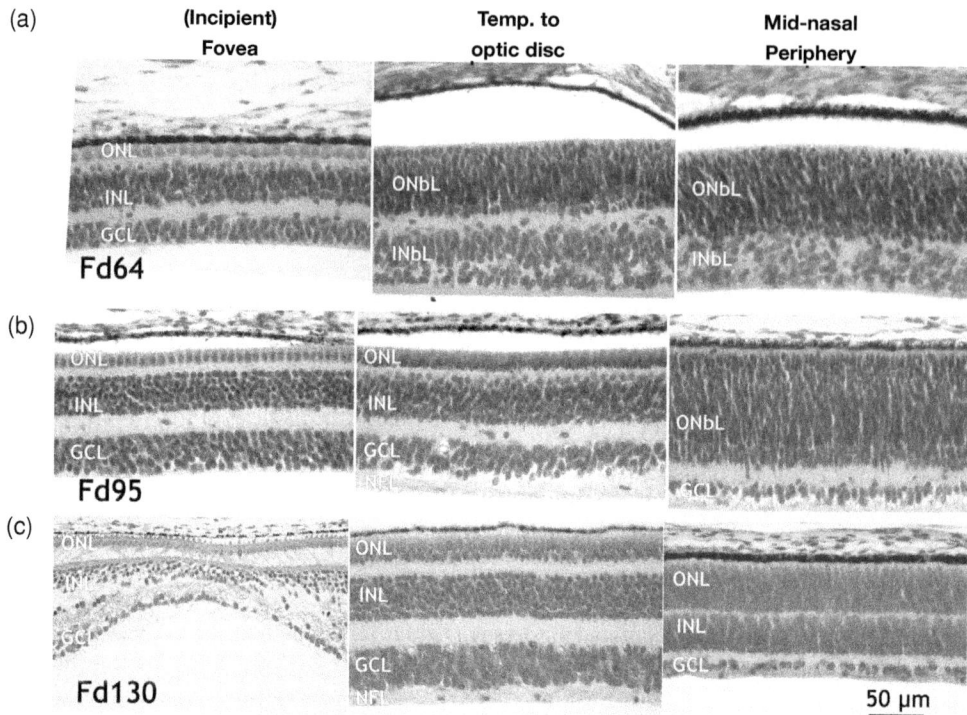

Figure 7.2 Stages in development of the macaque retina. The incipient fovea in central retina (left column) is compared with retina adjacent to the optic disc (OD; middle) and mid-peripheral nasal retina (right) at six different ages. All images at the same magnification. Temp, temporal. (a) Fd 64. The incipient foveal region is the only part of the retina with five distinct layers. Note that the ONL comprises a single layer of cones. At the OD the IPL and GCL are distinct layers but the outer retina is still a single thick outer neuroblastic layer. The immature peripheral retina has an inner and outer neuroblastic layer (InbL and ONbL, respectively). (b) Fd 95. The foveal has changed little, although the retina is slightly thicker. At the OD the retina now has five layers, with rods and cones present in the ONL. The peripheral retina has a distinct IPL but the outer retina is still immature. (c) Fd 130. An early stage of formation of the foveal pit, showing thinning of the inner layers of the retina. The cones are still in a single layer over the developing pit. More rods have been added to the ONL near the OD. The peripheral retina has five distinct layers. (d) Fd 164. Cones are starting to accumulate in several layers thick in the ONL and have developed fibres of Henle (FH). A layer of cone pedicles is still present within the developing pit (arrow). Fibres of Henle are present across the central retina to the OD. Inner and outer segments are present on rods and cones across the retina. Note the thin peripheral retina. (e) Postnatal (P) 12 weeks. The fovea is now quite mature with inner layers absent from the pit centre, which is lined with a thick layer of cone cell bodies. Cone pedicles are displaced onto the slope (arrow). The OD retina has changed little but the peripheral retina is much thinner than before birth. (f) Adult. the centre of the fovea is formed by a very thick layer of cone cell bodies and FH, although scattered neurons remain in the pit centre. The layer of FH is thicker at the OD than in the infant retina, while the peripheral retina is less than half the thickness at the OD. Note the progressive loss of GC in peripheral retina after Fd 130, due to a combination of cell death and retinal stretch with eye growth. Arrow points to the layer of cone pedicles.

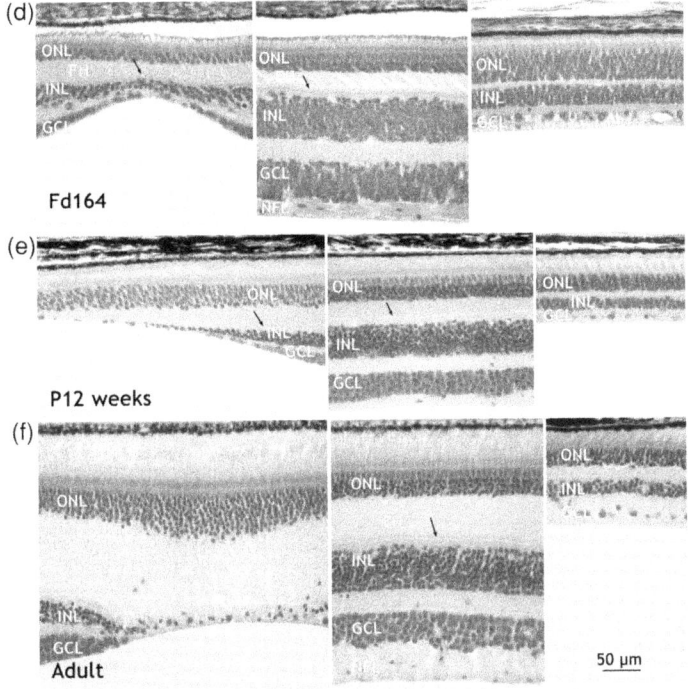

Figure 7.2 (cont.)

AC synaptic density doubles again by Fd 130 when it is almost 30/100 μm², near adult levels (Figure 7.3a). By contrast, in rod-dominated peripheral retina near the OD (Nishimura and Rakic, 1985) there are no IPL synapses until Fd 78 and these are mainly AC (Figure 7.3). Amacrine synaptic density rises rapidly to reach near adult density (16.5/100 μm²) at Fd 114. Bipolar ribbon synapses are not present in significant numbers until Fd 99 but they reach adult density at Fd 150. Immunolabelling shows that the IPL at the *Macaca* retinal edge contains synaptic markers at Fd 90 to 103, but quantitative synaptic density studies are not available for this region.

A striking correlation has been found between the wave of IPL synaptic formation and a closely following wave of BC death (Figure 7.3c) that is two to eight times that of GCs at the same eccentricity (Georges *et al.*, 1999). Why BCs die in such numbers is not clear, but one factor could be that human fetal BCs lack the anti-apoptotic factor Bcl-2 (Georges *et al.* 2006). Bipolar cells also may be at high risk because they have to make a correct synaptic match both for their dendrites in the OPL and their axons in the IPL.

The pattern of synapse formation illustrates the basic rule for primate retina, with synaptic formation beginning in the fovea and spreading into peripheral retina over time. It also shows that IPL synapses form before those in the OPL, so that the retina initially forms its synapses from inner to outer layers. This sequence indicates that IPL synaptic development at any retinal point is delayed until cell generation is nearly over, since BCs are among the last

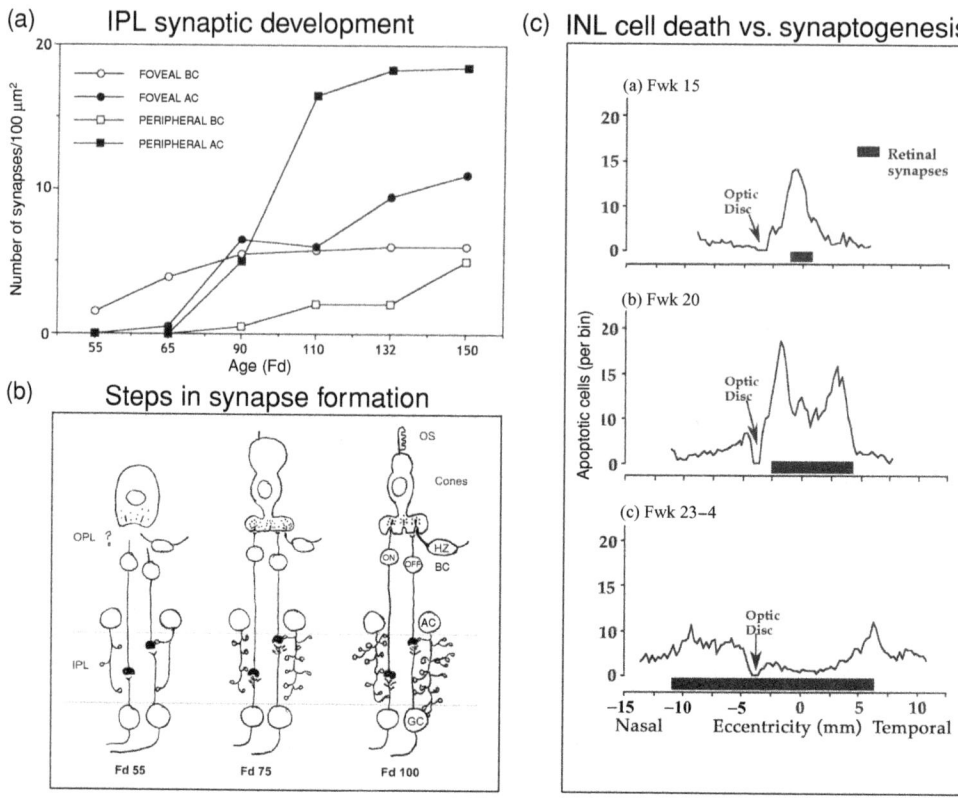

Figure 7.3 Features of synpatic development from macaque and human retina. (a) Synaptic development in the macaque IPL comparing the counts of Crooks *et al.* (1995) and Nishimura and Rakic (1985, 1987). Note that synapses appear in central retina before peripheral. Bipolar (BC) synaptic terminals dominate the early central IPL, and reach adult density by midgestation while amacrine (AC) terminals appear later but increase in density throughout fetal development. Amacrine terminals are dominant in the young peripheral retina but BCs are near adult by Fd 150. (b) A depiction of synaptic development in macaque central retina. Note that BC terminals appear early and mature rapidly while ACs continue to develop for a long period. Cone synapses lag the IPL slightly but appear mature by Fd 100. OS, optic stalk. (c) Counts of apoptotic cells in the INL of human retina, from retinal sections on the horizontal meridian, at three ages indicate that the highest levels of cell death are detected on the margins of the newly forming plexiform layers (horizontal black band). At Fwk 15 peak death in the INL is topographically associated with the newly forming plexiform layers. As the plexiform layers spread toward the periphery, peak cell death in the INL also moves peripheralward.

neurons to be generated (LaVail *et al.*, 1991). At least in the cone-dominated fovea, it also shows that outer (OFF) regions of the IPL develop slightly earlier than inner (ON) regions, that the earliest synapses are made by BCs and that adult BC density is reached long before adult AC density. This pattern is summarized in Figure 7.3b.

In the OPL synaptic development begins very early in the foveal cones. Synaptic ribbons and clustered vesicles labelled for SV2 are present in the cone base of monkey retina at Fd 55 to 60 (Okada *et al.*, 1994), and in humans at Fwk 11 (Linberg and Fisher, 1990).

Numerous processes cluster below the cones in the narrow OPL, but it is difficult to identify their origin at this age. Synaptic vesicle protein labelling in cones spreads across the retina, reaching the optic disc at Fd 85 to 90 and the retinal edge by Fd 125. Thus, cone synaptic formation is initiated even before a distinct OPL can be identified. Although primate rod synaptogenesis has not been reported in detail, rod synapses appear to form later than those of cones at every point on the retina (Hollenberg and Spira, 1973; Okada et al., 1994). The sequence of OPL synaptic types has not been determined for primates, but in other mammals HC processes seem to make the first contact with ribbon synapses (reviewed in Hendrickson (1996)). In monkeys there is a relatively long delay between cone and HC genesis at Fd 36 and synaptic formation at Fd 60. This suggests that synaptic formation is delayed until BC dendrites are present and that, even if HC processes form the earliest contacts, they are not sufficient to initiate synaptogenesis.

7.3.3 The general pattern of expression of phototransduction proteins

Although *Macaca* monkey and human cone and rod photoreceptors seem to be identical in morphology, electrophysiology and biochemistry (Nathans, 1989; Jacobs, 1996; Jacobs and Deegan, 1999), their pattern of opsin expression is not. Riboprobes and antisera have been produced that are specific for S- vs. L-/M-opsin, but none are available that can differentiate L- from M-opsin; therefore the term L-/M- will be used to refer to the red and green opsins (see Bumsted et al., 1997, for discussion). In humans (Xiao and Hendrickson, 2000) the first opsin to appear is in S-cones and can be detected around the fovea at Fwk 12 (Figure 7.4a). In both cones and rods, when opsin is expressed the photoreceptor has formed a short outer segment (Dorn et al., 1995). L-/M-opsin is not detected until Fwk 15 in the foveal centre. S-opsin expression spreads rapidly across the retina so that S-cones are present at all retinal edges by Fwk 22, whereas L-/M-opsin expression is only at the eccentricity of the OD at Fwk 20 and does not reach the retinal edge until near birth. Rod opsin expression in humans follows that of L-/M-opsin, appearing on the foveal edge at Fwk 15 and reaching the retinal edge before birth. By contrast, in *Macaca* monkeys (Bumsted et al., 1997) both S- and L-/M-opsin appear within the foveal region at Fd 65 to 70 (Figure 7.4b). S-opsin expression then spreads rapidly across the retina, reaching the retinal edge by Fd 135 while L-/M-opsin is not present at the edge until Fd 150. Rod opsin also appears around the monkey fovea at Fd 65 to 70 and spreads across the retina to reach its edge by Fd 135 (Dorn et al., 1995).

Comparison of Figures 7.4a and 7.4b suggests a similar pattern for L-/M-opsin in both humans and *Macaca* monkeys, but widely different expression patterns for S- and rod opsin in the two species. Another major difference between species is that there is a transient expression of S-opsin at the onset of L-/M-opsin in some human cones producing transient 'S- + L-/M-' cones that have not been detected in macaques (Cornish et al., 2004b). S-opsin disappears with age in almost all L-/M-cones but this is not associated with cell death; in fact, cell death among the cone population at any age is very low (Georges et al., 1999; Cornish et al., 2004b). Transient expression of S-opsin may in some way direct the future choice of L- or M-opsin. Because a few cones expressing both S- and L-/M-opsins remain throughout all adult human retinas examined, coexpression is compatible with continued

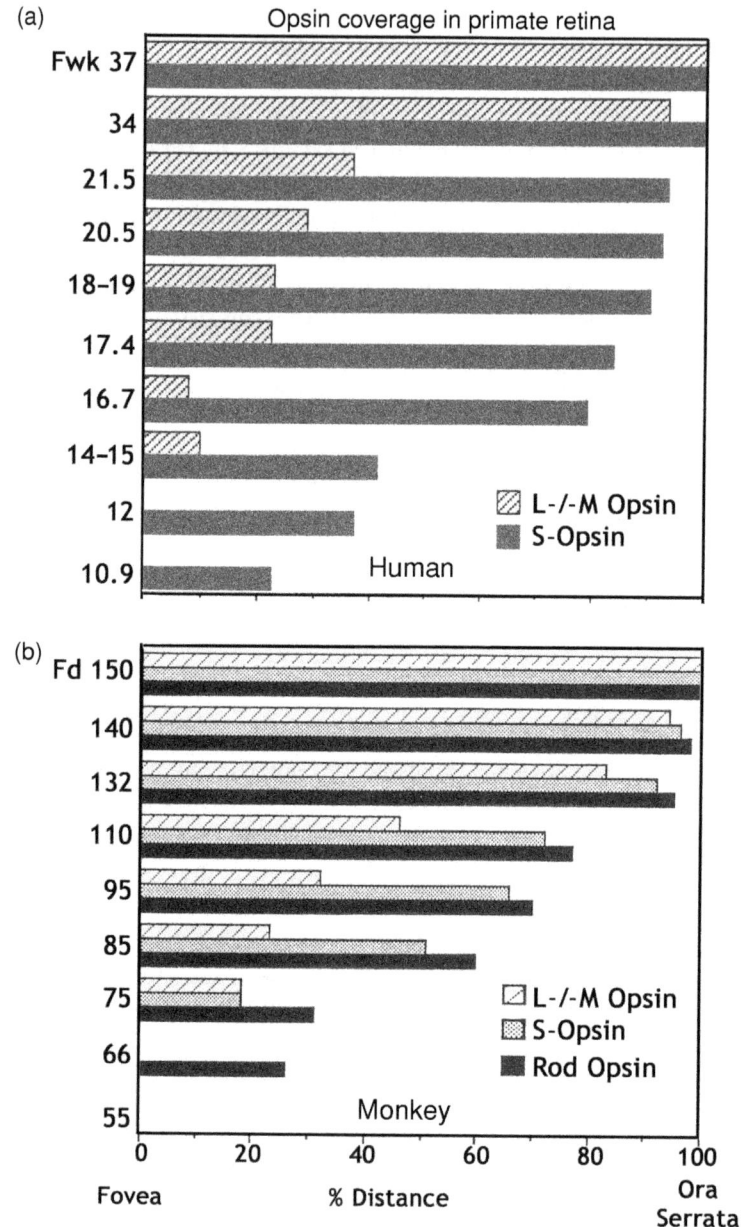

Figure 7.4 A comparison of opsin expression in rod and cone photoreceptors in human (a) and macaque monkey (b) fetal retina. For all photoreceptors opsin first appears in or near the fovea (left side of the graph). Note that short-wavelength-sensitive (S) or 'blue'-cone opsin is the first to appear around the human foveal region at Fwk 10.9 while rod opsin is the first to appear in monkey central retina at Fd 66. Long- and medium-wavelength-sensitive cone opsin (L/M) consistently is detected later (more centrally) than S-opsin at all ages in both primates. All opsins are present at the edge of the retina (right side of the graph) by birth.

survival and presumed function, similar to other mammalian retinas where this occurs (see Cornish et al., 2004b). In both primates, outer-segment length increases for many months after birth. For instance, mid-peripheral rod outer-segment length increases 50% between Fd 115 and birth, and a further 50% between birth and adulthood (Dorn et al., 1995). Foveal cone outer-segment length in humans increases from 3 µm at birth to 50 to 60 µm in the adult (Yuodelis and Hendrickson, 1986).

Expression of proteins involved in the phototransduction cascade has only been followed in *Macaca* monkey cones (Sears et al., 2000). Structural proteins of the outer segment like peripherin appear simultaneously with opsin and morphological optic stalk (OS) in both S- and L-/M-cones. Phototransduction proteins such as cone-specific alpha-transducin, phosphodiesterase and rhodopsin kinase appear shortly after L-/M-opsin expression. Curiously, although S-cones express opsin earlier than L-/M-cones, expression of phototransduction proteins is delayed in S-cones until surrounding L-/M-cones are immunoreactive for both opsin and these proteins.

7.4 The fovea

7.4.1 Formation of the foveal depression

Pre-pit stage

The incipient fovea can be identified temporal to the OD by Fwk 11 in humans and Fd 50 to 55 in macaques, but up to midgestation this region remains relatively unchanged morphologically (Figures 7.3 and 7.5). The incipient fovea is characterized by a single layer of cones in the ONL (Figure 7.5A–D) and has been termed the *rod-free zone* (Hendrickson and Kupfer 1976) because rods are virtually absent from its centre. It also has been called the *pure-cone area* for similar reasons (Springer and Hendrickson, 2004). To emphasize its continuity with the cone mosaic across the retina, it has also been termed the *foveal cone mosaic* (Provis et al., 1998; Cornish et al., 2004a). Here we use the term *foveal cone mosaic* because the fetal ONL, even within the incipient fovea, includes scattered rods. The remaining layers are a thin OPL and thick INL, IPL and GCL. In fact, the early foveal GCL is the thickest in the retina and often bulges into the vitreous so that it can be detected histologically. Another characteristic is that the eye is oval shaped at these stages with a distinct bulge temporal to the OD (not shown), which includes the incipient fovea. In this region the retinal pigment epithelium (RPE) is more pigmented than in the periphery and the retina seems to be tightly attached to the outer eye layers.

A recent study comparing expression patterns as detected by RT-PCR and immunocytochemistry (Bumsted-O'Brien et al., 2003) finds that immunolabelling is more sensitive for detecting small numbers of cells expressing markers early in development of the fovea. By both methods photoreceptor proteins such as phosphodiesterase, CRX, interphotoreceptor retinoid-binding protein, Nrl and S-opsin are expressed at Fwk 9 to 12. All first appear in foveal cones or rods at the foveal edge. L-/M-opsin and rod opsin were detected at Fwk 15 to 16, consistent with immunolabelling (Xiao and Hendrickson 2000). Immunolabelling

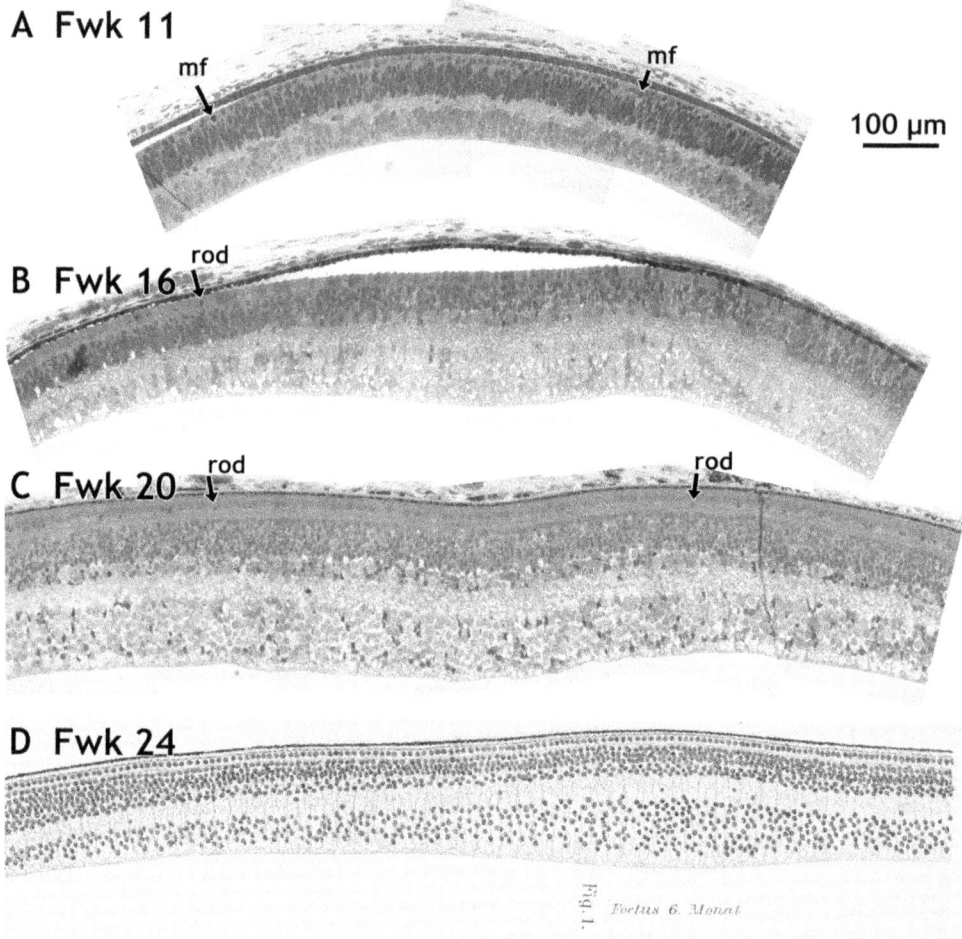

Figure 7.5 Different stages in the development of central retina of the human. A–C, G are photomontages of fetal retinas collected by the authors. D–F, H are drawings from the work of Bach and Seefelder (1911, 1912, 1914) scanned at high resolution by the authors. We found the drawings at all ages to be remarkably accurate, compared with photomontages. (A) At Fwk 11 the differentiated region is less than 1 mm in diameter, and the photoreceptor mosaic comprises cones and virtually no rods. The two mitotic figures indicated (mf) define the full diameter of the differentiated region. Note the relatively even thickness of the retina at this age. (b) At Fwk 16 the differentiated region is some millimetres in diameter, and rods can be easily identified surrounding the foveal cone mosaic (FCM). Note that the retina is somewhat thicker over the FCM (to the right of the rod) than in the adjacent region. (C) The central region at Fwk 20 shows a markedly thickened retina and a FCM (between rods) in which cones are more densely packed than at earlier stages, and is reduced in diameter. (D) The drawing of the central region at Fwk 24 clearly illustrates the thickening of the retina, particularly the GCL, at the FCM.

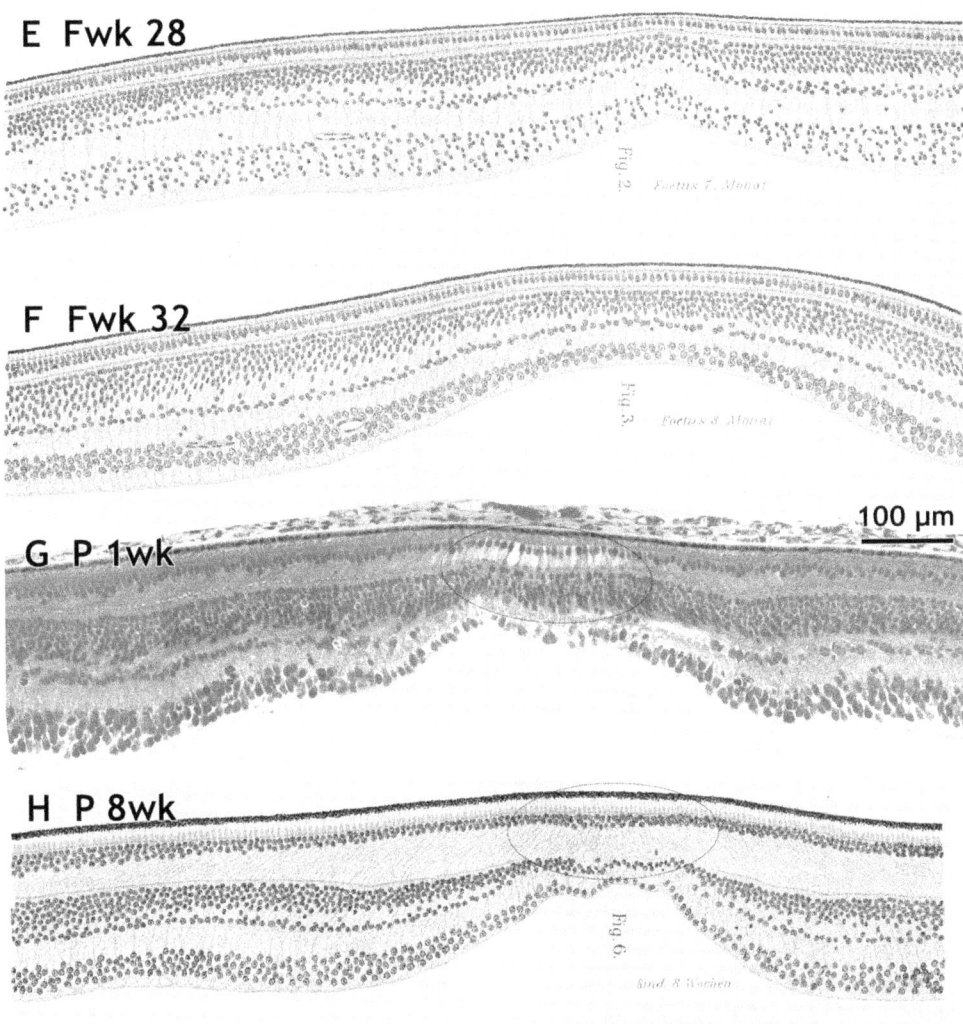

Figure 7.5 (*cont.*) (E,F) These drawings show the early stages of pit formation. The earliest indication is a narrow indentation of the retinal surface that initially affects only the GCL (not shown) but which soon presses into the INL (E). The depression then becomes broader, and its rim is marked by the central-most edge of the developing perifoveal capillary bed, deep in the GCL (F). (G,H) Note the remarkable similarities in the morphology of the foveal depression in these two images. The main changes between P1 wk (G) and P8 wk (H) can be detected within the circled region. In the ONL at P1 wk (G) there is a single layer of cone cell bodies lining the floor of the foveal depression, which do not have distinct Henle fibres; the adjacent INL is several cells deep. At P8 wk (H) there are two tiers of cone cell bodies and a thick layer of Henle fibres intervening between these and a markedly thinned INL. Cones continue to pack into the photoreceptor mosaic of the central fovea over the next few years, before adult densities are reached.

All photomontages are at the same magnification; drawings were scaled to approximate the same degree of magnification.

of synaptic proteins also finds that foveal cones express synaptophysin, SV2, and glutamate vesicular transporter as well as markers for synaptic ribbons by Fwk 11. Electron microscopy studies find SV2 labelling and synaptic ribbons and vesicles in monkey cones by Fd 60 (Okada *et al.*, 1994) while neither S- or L-/M-opsin is detected until Fd 70 (Bumsted *et al.*, 1997). Thus, one distinctive pattern of putative L-/M-cones is that they form synapses very early, and at least a month before they express opsin. S-cones in humans seem to form synapses shortly before they express opsin.

Inner neurons also mature rapidly. As described above (p. 131), IPL synaptogenesis in the fovea is well advanced by Fd 85 in monkeys. Immunolabelling finds that characteristic markers for BCs such as Goα for ON-BCs, recoverin for OFF-BCs and protein kinase α for rod-BCs as well as glutamate receptors 2/3 and 5 are all present in the Fwk 12 to 14 INL, OPL and IPL (Hendrickson, unpublished data). Amacrine cell markers γ-aminobutyric acid (GABA) and glycine also are present in the IPL although adult banding patterns are not yet clear. Thus, by midgestation, at Fwk 20 in humans and Fd 85 in monkeys, although the incipient fovea has not changed much morphologically, its neurons are forming their characteristic synaptic circuits and have expressed phototransduction proteins.

Pit stage

The superior and inferior vascular arcades (Provis *et al.*, 2000; Provis, 2001) grow from the OD to vascularize the central retina. These arcades grow around the fovea, never entering the central region. Rather, they form instead the FAZ (Gariano *et al.*, 1994; Provis *et al.*, 2000). The FAZ remains stable at approximately the diameter of the future pit rim, about 500 μm. The first sign of pit formation is a slight thinning of the GCL at Fwk 25 to 26 in humans (Figure 7.5E–H) and Fd 105 to 110 in monkeys (Figure 7.2C–F). This thinning occurs within the FAZ and is approximately centred on the foveal cone mosaic. The pit forms by a sequential indentation of GCL, IPL, INL and finally the OPL.

Two important points should be emphasized about pit formation. The first is that inner retinal neurons and Müller glia are displaced toward the periphery (centrifugally); they do not die by apoptotic cell death (Georges *et al.*, 1999). The second is that this displacement takes place after synapses have been formed in both the IPL and OPL so that the mechanisms causing this displacement must be able to retain these synaptic connections.

The foveal pit is still relatively immature at birth in humans (Figure 7.5G) with thin IPL and GCL layers and a thick INL. The ONL is still a single layer of elongated cones that tilt toward the centre of the pit, but are not yet tightly packed. On the other hand, the neonatal macaque fovea is relatively much more mature. By Fd 130 (Figure 7.2C) the fovea resembles the neonatal human fovea and by birth (at Fd 174) there is a well-formed pit with a single layer of cells across the bottom (cf. Figure 7.2d). Neonatal monkey cones also have begun to show evidence of packing with a significant elongation of cone cell bodies and cone nuclei packed one to two deep especially on the rim.

7.4.2 Cone packing

Quantitative studies show that both rods and cones are displaced toward the foveal centre (centripetally) during the pre-fovea stage (Diaz-Araya and Provis, 1992). When they first differentiate, cones in the foveal cone mosaic have a packing density of approximately 12 000/mm^2 (Figure 7.6a). By Fwk 30 cone density at the fovea is around 30 000/mm^2, with no addition of new cells (Yuodelis and Hendrickson, 1986; LaVail *et al.*, 1991). This increase is accompanied by a decrease in size of the foveal cone mosaic diameter prenatally – suggesting that cones and rods are displaced towards the pit centre (Hendrickson and Kupfer, 1976; Youdelis and Hendrickson, 1986; Diaz-Araya and Provis, 1992). Analysis of the S-cone population near the OD of human retina shows that photoreceptors even at this relatively peripheral location are displaced towards the foveal cone mosaic between Fwk 17 and 6 weeks postnatal (Cornish *et al.*, 2004a). This emphasizes the protracted period and the large retinal area involved in cone displacement.

A more rapid phase of cone packing begins around birth. Based on thickness of the ONL, this occurs in humans and macaques, but detailed quantitative data comes from study of the macaque retina (Packer *et al*, 1990; Springer & Hendrickson, 2005). Cone density rises from 65 000/mm^2 at 3 weeks to 140 000/mm^2 by one year and will eventually reach 200 000/mm^2 by 1.5 years of age which is equal to a 6 year human. Humans have an even longer developmental time period, with neonatal densities around 36 000/mm^2 and the lowest range of adult density reached at 45 months postnatal (Yuodelis and Hendrickson, 1986), although precisely when final adult density is reached is not certain.

This marked increase in cone density is reflected in a striking change in cone morphology (Hendrickson & Kupfer, 1976; Yuodelis and Hendrickson, 1986; Diaz-Araya and Provis, 1992), shown diagrammatically in Figure 7.6b. Cones in the foveal cone mosaic at Fwk 22 are relatively simple cuboidal cells, 8 to 10 μm in diameter. By birth both an inner and outer segment are present as well as a distinct synaptic pedicle; the inner segment is 5 to 7.5 μm wide and 7.5 to 10 μm long, with a short outer segment 3 to 4 μm in length. By 12 to 15 months the elongated inner segment is half the neonatal width and almost 25 μm long. Adult cones are about 2 μm in diameter with an inner segment 30 to 35 μm long and an outer segment 50 to 65 μm long.

Another striking change in cone morphology involves the change in cone angle and the appearance of an axon or Henle fibre (Figures 7.6b, 7.7). Initially, the synaptic pedicle is located immediately adjacent to the nucleus, and there is only a short axon or Henle fibre (Hendrickson and Yuodelis, 1984). Next, the long axis of the cone tilts, so that the elongating inner segment points toward the centre of the pit and the Henle fibre extends laterally away from the centre of the pit. This change in morphology occurs as the INL neurons are displaced laterally (Figure 7.7). In monkeys this is in the late prenatal stage while in humans it occurs before and just after birth. Elongation of the axon appears to be a morphological adaptation of cones that enables retention of synaptic connections, made in the 'pre-pit' stage, with HCs and BCs. As BCs and HCs move out of the deepening pit,

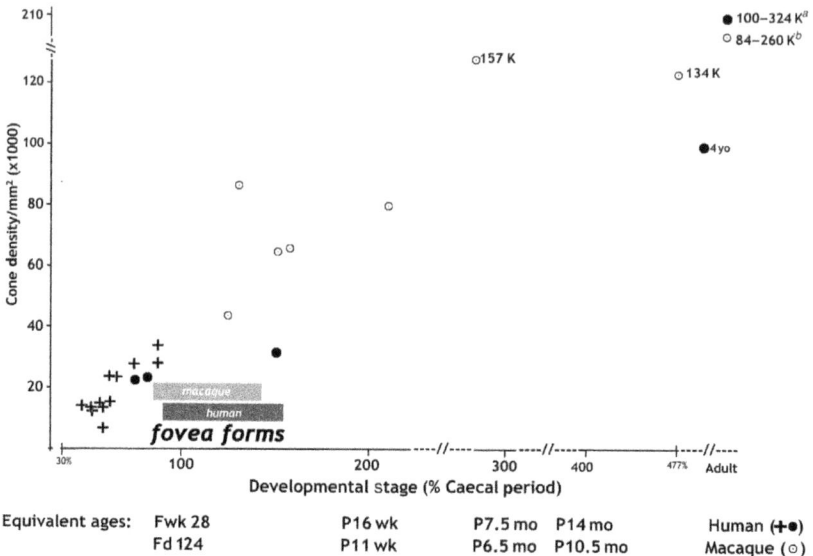

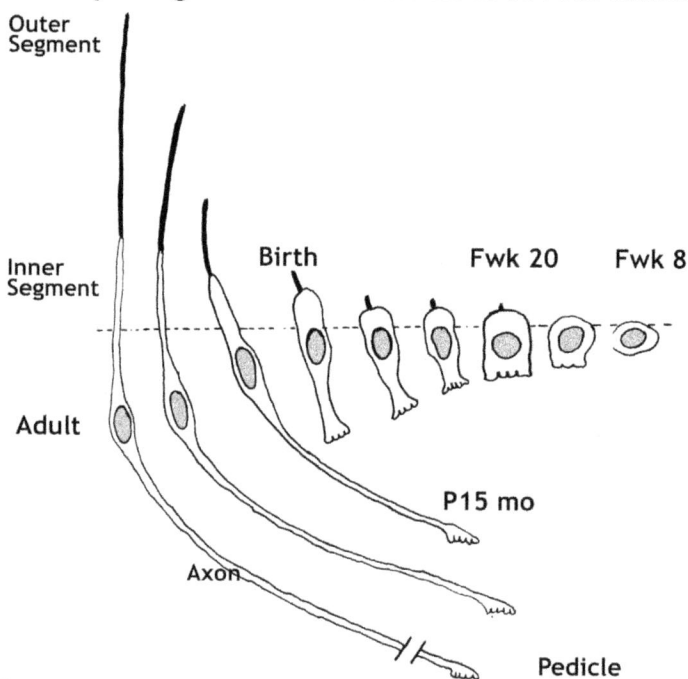

Figure 7.6 Cones pack into the central fovea over a protracted period. This packing is associated with morphological adaptation of individual cones. (a) Cone density at the central fovea in human and macaques is plotted as a function of time, measured as a percentage of the caecal period – the time between conception and eye opening. The period during which the foveal pit forms is also included on the same axis. The graph shows that cone packing takes place over a protracted period, starting before the fovea begins to form and continuing for a considerable time afterwards. [a]Curio et al. (1990); [b]Perry and Cowey (1985), Packer et al. (1989), Wikler et al. (1990). (b) Morphological differentiation of cones takes place over the same time-frame, with much of the specialization occuring postnatally in humans.

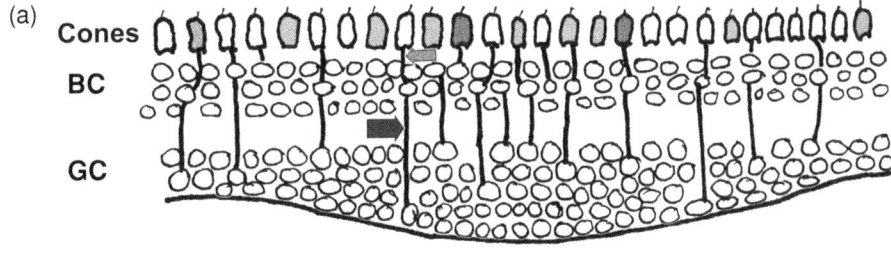
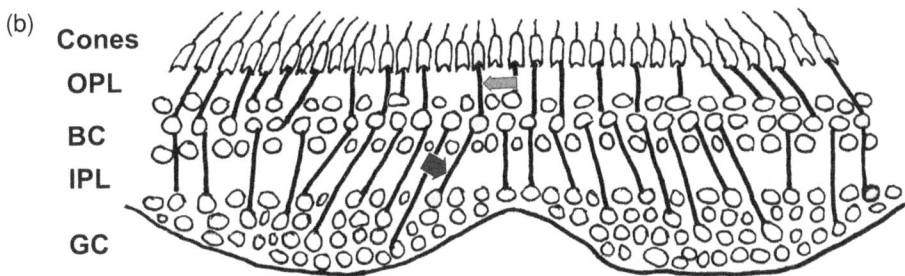
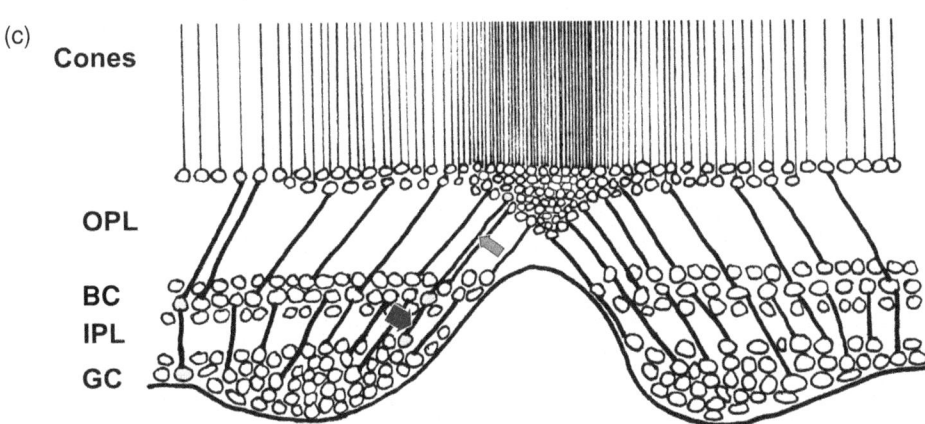

Figure 7.7 A diagrammatic summary of foveal development. (a) In the first stage (Fd 55 to 100 or Fwk 11 to 24) the fovea has five distinct layers. The ONL is a single layer of cones, which have a short Henle axon extending vertically to contact underlying BCs. Bipolar axons are also vertical in the IPL. Arrows indicate the presumed direction of cell displacement. (b) In the second stage (Fd 100 to P1 months or Fwk 24 to P4 to 6 months) the pit invaginates the inner retinal layers, displacing the neurons peripherally. This is reflected by a peripheral slant to the BC axons as GCs are displaced laterally. Note that this slant appears on cone FH on the foveal slope, suggesting this is where cone packing toward the pit centre is initiated. (c) During the final stage (P1 to 12 months or P6 months to 4 years) the pit becomes deeper, all inner retinal neurons are displaced peripherally and cones form a thick layer over the pit so that cone density rises 10×. Cone pedicles are displaced peripherally to retain their synaptic contact with the more peripheral BCs and HCs. For colour version, see Plate 7.

Henle fibres on the edge of the deepening pit elongate further, and begin to develop on cones in the central pit. A similar process can be detected on rods near the foveal cone mosaic, and later on all photoreceptors as far peripheral as the OD, further supporting the view that much of the central photoreceptor mosaic is involved in elevating foveal cone density (Cornish *et al.*, 2004a; Springer and Hendrickson, 2005). A direct measurement of the rod-free zone diameter finds that it is more than 1 mm wide at the beginning of cone packing but only 300 μm wide (or less) in the adult monkey (Hendrickson and Kupfer, 1976). In humans the rod-free region is 1.6 mm at Fwk 22, 1.1 mm at birth, dropping to 300 μm by 4 years of age, while the total number of cones, in the foveal cone mosaic remains constant (Yuodelis and Hendrickson, 1986). The sequential reduction in diameter of the foveal cone mosaic, in conjunction with the narrowing and elongation of individual cones, indicates that photoreceptors are displaced towards, and cones packed into, the foveal cone mosaic during prenatal and postnatal development of central retina.

7.5 Retinal growth patterns

The area of the retina increases dramatically during fetal life and more slowly after birth. In humans retinal area increases from 35 mm^2 at Fwk 11 to 300 mm^2 at Fwk 26 to 907 mm^2 at 6 years (Robb, 1982; Provis, 1985; Provis *et al.*, 1985a). In *Macaca* monkeys retinal area increased from 42 mm^2 at Fd 60 to 400 mm^2 at Fd 150 to 519 mm^2 at 5 months with an additional slow increase to 728 mm^2 by adulthood, which is 5+ years (Robinson and Hendrickson, 1995). Because most cell generation in primate retinas is finished by birth, the postnatal increase in retinal area is due mainly to overall eye-growth-induced retinal stretch (reviewed in Packer *et al.*, 1990; Robinson 1991).

In macaque retina there are two phases of retinal elongation measured in sections cut along the horizontal meridian. The first period extends up to Fd 100 and this is followed by a period of little elongation (Springer and Hendrickson, 2004). The second phase is between P7 and P100 days and is mainly due to stretch exerted on the retina by eye growth. The foveal pit forms when the retina is not elongating, between Fd 100 and P7, prompting the suggestion that intraocular pressure, rather than stretch, is the prime driver for formation of the foveal pit. Furthermore, much cone packing occurs during the second phase of elongation, implicating stretch as a mechanism that supports cone packing (Springer and Hendrickson, 2005). This second phase of stretch has a major effect on the peripheral retina. Peripheral retina thins proportional to its distance from the optic nerve (Springer and Hendrickson 2004) (Figure 7.2d–f). Near the OD the drop in cone density during this phase is minimal, but there is an almost threefold decrease in peripheral cone density (Packer *et al.*, 1990). Similarly, in human peripheral retina GC density drops from 20 000/mm^2 at Fwk 14 to 2500/mm^2 at Fwk 37, while in central retina GC density just outside the fovea remains unchanged (Provis *et al.*, 1985a; Provis, 1987). Retinal stretch plus postnatal growth in the pars plana (the epithelium between the neural retina and ciliary body), thus, adjusts the retina to the enlarging globe (Fischer *et al.*, 2001; Springer and Hendrickson, 2004).

The capacity of these tissues to stretch can be exceeded in abnormal eye elongation, as occurs in myopia. People with a high degree (>6 diopters) of myopia are more likely to suffer peripheral retinal tears or detachments, due to the greatly stretched and thinned peripheral retina. The greatly enlarged eye even affects the retina on the nasal side of their OD, which is dragged peripherally so that only the sclera remains in place.

7.6 Concluding remarks

The macula region of the primate retina is unique among mammals. The two fundamental differences are the presence of a small avascular region containing only cone photoreceptors in excess of 200 000/mm^2 and an inner retinal pit. These differences are established during a protracted period of pre- and postnatal development, and almost certainly involve unique molecules that establish a pure-cone region, inhibit vascular growth and emphasize the creation of midget system BCs and GCs. While we have a clear understanding of the anatomical changes that take place during formation of the fovea, as yet there is no clear understanding of the mechanisms that guide these changes. Researchers have proposed a number of mechanisms that can be broadly classified as either 'molecular' or 'mechanical' in nature (Springer and Hendrickson, 2004; Cornish *et al.*, 2005), but a detailed discussion of them is beyond the scope of this chapter, especially in that few have been subjected to experimental testing. A major goal for the near future is to develop comprehensive theoretical models of how molecular and mechanical forces, alone and in combination, might act to promote the unique vascular and neuronal specializations of the macula region.

References

Bach, L. and Seefelder, R. (1911, 1912, 1914). *Entwicklungsgeschichte des Menschlichen Auges*, Parts 1–3. Leipzig: W. Engelmann.

Bumsted, K. and Hendrickson, A. E. (1999). Distribution and development of short-wavelength cones differ between Macaca monkey and human fovea. *J. Comp. Neurol.*, **403**, 502–16.

Bumsted, K., Jasoni, C., Szél, A. and Hendrickson, A. E. (1997). Spatial and temporal expression of cone opsins during monkey retinal development. *J. Comp. Neurol.*, **378**, 117–34.

Bumsted-O'Brien, K. M., Schulte, D. and Hendrickson, A. E. (2003). Expression of photoreceptor-associated molecules during human fetal eye development. *Mol. Vis.*, **9**, 401–9.

Cajal, S. R. (1893). *The Structure of the Retina: La Retine des Vertebres*. Springfield IL: Thomas Springfield.

Chievitz, J. H. (1888). Entwicklund der fovea centralis retinae. *Anat. Anzeig. Jena Bd*, **III**, 5579.

Cornish, E. E., Hendrickson, A. E. and Provis, J. M. (2004a). Distribution of short wavelength sensitive cones in human fetal and postnatal retina: early development of spatial order and density profiles. *Vis. Res.*, **44**, 2019–26.

Cornish, E. E., Xiao, M., Yang, Z., Provis, J. M. and Hendrickson, A. E. (2004b). The role of opsin expression and apoptosis in determination of cone types in human retina. *Exp. Eye Res.*, **78**, 1143–54.

Cornish, E. E., Madigan, M. C., Natoli, R. C. *et al.* (2005). Gradients of cone differentiation and FGF expression during development of the foveal depression in macaque retina. *Vis. Neurosci.*, **32**, 447–59.

Crooks, J., Okada, M. and Hendrickson, A. E. (1995). Quantitative analysis of synaptogenesis in the inner plexiform layer of macaque monkey fovea. *J. Comp. Neurol.*, **360**, 349–62.

Curcio, C. A. and Hendrickson, A. E. (1991). Organization and development of the primate photoreceptor mosaic. *Prog. Retin. Res. Eye*, **10**, 90–120.

Curcio, C. A., Sloan, K. R., Kalina, R. E. and Hendrickson, A. E. (1990). Human photoreceptor topography. *J. Comp. Neurol.*, **292**, 497–523.

Curcio, C. A., Allen, K. A., Sloan, K. R. *et al.* (1991). Distribution and morphology of human cone photoreceptors stained with anti-blue opsin. *J. Comp. Neurol.*, **312**, 610–24.

Dacey, D. M. (1996). Circuitry for color coding in the primate's retina. *Proc. Natl. Acad Sci. U. S. A.*, **93**, 582–8.

Dacey, D. M. (1999). Primate retina: cell types, circuits and color opponency. *Prog. Retin. Eye Res.*, **18**, 737–63. Erratum in *Prog. Retin. Eye Res.*, 2000 Sep; **19(5)**:following 646.

Diaz-Araya, C. M. and Provis, J. M. (1992). Evidence of photoreceptor migration during early foveal development: a quantitative analysis of human fetal retinae. *Vis. Neurosci.*, **8**, 505–14.

Dorn, E. M., Hendrickson, L. and Hendrickson, A. E. (1995). The appearance of rod opsin during monkey retinal development. *Invest. Ophthalmol. Vis. Sci.*, **36**, 2634–51.

Fischer, A. J., Hendrickson, A. E. and Reh, T. E. (2001). Immunocytochemical characterization of cysts in the peripheral retina and pars plana of the adult primate. *Invest. Ophthalmol. Vis. Sci.*, **42**, 3256–63.

Franco, E. C., Finlay, B. L., Silveira, L. C., Yamada, E. S. and Crowley, J. C. (2000). Conservation of absolute foveal area in New World monkeys. A constraint on eye size and conformation. *Brain Behav. Evol.*, **56**, 276–86.

Gariano, R. F., Iruela-Arispe, M. L. and Hendrickson, A. E. (1994). Vascular development in primate retina: comparison of laminar plexus formation in monkey and human. *Invest. Ophthalmol. Vis. Sci.*, **35**, 3442–55.

Georges, P., Madigan, M. C. and Provis, J. M. (1999). Apoptosis during development of the human retina: relationship to foveal development and retinal synaptogenesis. *J. Comp. Neurol.*, **413**, 198–208.

Georges, P., Cornish, E. E., Provis, J. M. and Medigan, M. C. (2006). Müller cell expression of glutamate cycle related proteins and anti-apoptotic proteins in early human retinal development. *Br. J. Ophthalmol.*, **90**, 223–8.

Hendrickson, A. E. (1992). A morphological comparison of foveal development in man and monkey. *Eye* **6**, 136–44.

Hendrickson, A. E. (1996). Synaptic development in macaque monkey retina and its implications for other developmental sequences. *Perspect. Dev. Neurobiol.*, **3**, 195–201.

Hendrickson, A. E. and Kupfer, C. (1976). The histogenesis of the fovea in the macaque monkey. *Invest. Ophthalmol.*, **15**, 746–56.

Hendrickson, A. E. and Yuodelis, C. (1984). The morphological development of the human fovea. *Ophthalmology*, **91**, 603–12.

Hendrickson, A. E., Djajadi, H. R., Nakamura, L., Possin, D. E. and Sajuthi, D. (2000). Nocturnal tarsier retina has both short and long/medium-wavelength cones in an unusual topography. *J. Comp. Neurol.*, **424**, 718–30.

Hollenberg, M. J. and Spira, A. W. (1973). Human retinal development: ultrastructure of the outer retina. *Am. J. Anat.*, **137**, 357–85.

Jacobs, G. H. (1996). Primate photopigments and primate color vision. *Proc. Natl. Acad. Sci. U. S. A.*, **93**, 577–81.

Jacobs, G. H. and Deegan, J. F., 2nd. (1999). Uniformity of color vision in Old World monkeys. *Proc. R. Soc. London B Biol. Sci.*, **266**, 2023–28.

Kirby, M. A. and Steineke, T. C. (1991). Early dendritic outgrowth of primate retinal ganglion cells. *Vis. Neurosci.*, **7**, 513–30.

Koontz, M. A. and Hendrickson, A. E. (1993). Comparison of immunolocalization patterns for the synaptic vesicle proteins p65 and synapsin I in macaque monkey retina. *Synapse*, **14**, 268–82.

LaVail, M. M., Rapaport, D. H. and Rakic, P. (1991). Cytogenesis in the monkey retina. *J. Comp. Neurol.*, **309**, 86–114.

Linberg, K. A. and Fisher, S. K. (1990). A burst of differentiation in the outer posterior retina of the eleven-week human fetus: an ultrastructural study. *Visual Neurosci.*, **5**, 43–60.

Magitot, M. A. (1910). Etude sur le développement de la rétine humaine. *Annales d'Occulistique*, **143**, 241–82.

Mann, I. (1964). *The Development of the Human Eye, 2nd edn.* New York: Grune and Stratton.

Martin, P. R. and Grunert, U. (1999). Analysis of the short wavelength-sensitive ('blue') cone mosaic in the primate retina: comparison of New World and Old World monkeys. *J. Comp. Neurol.*, **406**, 1–14.

Martin, P. R., Grunert, U., Chan, T. L. and Bumsted, K. (2000). Spatial order in short-wavelength-sensitive cone photoreceptors: a comparative study of the primate retina. *J. Opt. Soc. Am. A Opt. Image Sci. Vis.*, **17**, 557–67.

Mears, A. J., Kondo, M., Swain, P. K. *et al.* (2001). Nrl is required for rod photoreceptor development. *Nat. Genet.*, **29**, 447–52.

Nathans, J. (1989). The genes for color vision. *Sci. Am.*, **260**, 42–9.

Nathans, J., Merbs, S. L., Sung, C. H., Weitz, C. J. and Wang, Y. (1992). Molecular genetics of human visual pigments. *Annu. Rev. Genet.*, **26**, 403–24.

Nishimura, Y. and Rakic, P. (1985). Development of the rhesus monkey retina. I. Emergence of the inner plexiform layer and its synapses. *J. Comp. Neurol.*, **241**, 420–34.

Nishimura, Y. and Rakic, P. (1987). Development of the rhesus monkey retina. II. A three-dimensional analysis of the sequences of synaptic combinations in the inner plexiform layer. *J. Comp. Neurol.*, **262**, 290–313.

Noctor, S. C., Flint, A. C., Weissman, T. A., Dammerman, R. S. and Kriegstein, A. R. (2001). Neurons derived from radial glial cells establish radial units in neocortex. *Nature*, **409**, 714–20.

Okada, M., Erickson, A. and Hendrickson, A. E. (1994). Light and electron microscopic analysis of synaptic development in Macaca monkey retina as detected by immunocytochemical labeling for the synaptic vesicle protein, SV2. *J. Comp. Neurol.*, **339**, 535–58.

Packer, O., Hendrickson, A. E. and Curcio, C. A. (1989). Photoreceptor topography of the retina in the adult pigtail macaque (*Macaca nemestrina*). *J. Comp. Neurol.*, **288**, 165–83.

Packer, O., Hendrickson, A. E. and Curcio, C. A. (1990). Development redistribution of photoreceptors across the *Macaca nemestrina* (pigtail macaque) retina. *J. Comp. Neurol.*, **298**, 472–93.

Perry, V. H. and Cowey, A. (1985). The ganglion cell and cone distributions in the monkey's retina: implications for central magnification factors. *Vis. Res.*, **25**, 1795–1810.

Polyak, S. L. (1941). *The Retina*. Chicago: University of Chicago Press.

Provis, J. M. (1985). Retinal development in humans: the roles of differential growth rates, cell migration and naturally occurring cell death. *Aust. J. Opththalmol.*, **13**, 125–33.

Provis, J. M. (1987). Patterns of cell death in the ganglion cell layer of the human fetal retina. *J. Comp. Neurol.*, **259**, 237–46.

Provis, J. M. (2001). Development of the primate retinal vasculature. *Prog. Retin. Eye Res.*, **20**, 799–821.

Provis, J. M., van Driel, D., Billson, F. A. B. and Russell, P. (1985a). Development of the human retina: patterns of cell distribution and redistribution in the ganglion cell layer. *J. Comp. Neurol.*, **233**, 429–51.

Provis, J. M., van Driel, D., Billson, F. A. B. and Russell, P. (1985b). Human fetal optic nerve: over-production and elimination of retinal axons during development. *J. Comp. Neurol.*, **238**, 92–100.

Provis, J. M., Diaz, C. M. and Dreher, B. (1998). Ontogeny of the primate fovea: a central issue in retinal development. *Prog. Neurobiol.*, **54**, 549–80.

Provis, J. M., Sandercoe, T. and Hendrickson, A. E. (2000). Astrocytes and blood vessels define the foveal rim during primate retinal development. *Invest. Ophthalmol. Vis. Sci.*, **41**, 2827–36.

Rakic, P. and Riley, K. P. (1983). Overproduction and elimination of retinal axons in the fetal rhesus monkey. *Science*, **219**, 1441–4.

Rapaport, D. H. and Stone, J. (1984). The area centralis of the retina in the cat and other mammals: focal point for function and development of the visual system. *Neuroscience*, **11**, 289–301.

Robb, R. (1982). Increase in retinal surface area during infancy and childhood. *J. Pediatr. Ophthalmol. Strabismus*, **19**, 16–20.

Robinson, S. R. (1991). Development of the mammalian retina. *Neuroanatomy of the Visual Pathways and their Development*, Vol. 3, ed. B. Dreher, and S. R. Robinson. London: Macmillan, pp. 69–128.

Robinson, S. R. and Hendrickson, A. E. (1995). Shifting relationships between photoreceptors and pigment epithelial cells in monkey retina: implications for the development of retinal topography. *Vis. Neurosci.*, **12**, 767–78.

Rohen, J. W. (1966). Zur Histologie des Tarsiusauges. *Graefs' Arch. Klin. Exp. Ophthalmol.*, **169**, 299–317.

Rohen, J. W. and Castenholtz, A. (1967). Über die Zentralisation der Retina bei Primaten. *Folia Primatologica*, **5**, 92–147.

Roorda, A. and Williams, D. R. (1999). The arrangement of the three cone classes in the living human eye. *Nature*, **397**, 520–2.

Sears, S., Erickson, A. and Hendrickson, A. E. (2000). The spatial and temporal expression of outer segment proteins during development of Macaca monkey cones. *Invest. Ophthalmol. Vis. Sci.*, **41**, 971–9.

Snodderly, D. M., Weinhaus, R. S. and Choi, J. C. (1992). Neural-vascular relationships in the central retina of macaque monkeys (*Macaca fascicularis*). *J. Neurosci.*, **12**, 1169–93.
Springer, A. and Hendrickson, A. E. (2004). Development of the primate area of high acuity. 2. Quantitative morphological changes associated with retinal and pars plana growth. *Vis. Neurosci.*, **21**, 775–90.
Springer, A. and Hendrickson, A. E. (2005). Development of the primate area of high acuity. 3. Temporal relationships between pit formation, retinal elongation and cone packing. *Vis. Neurosci.*, **22**, 171–86.
van Driel, D., Provis, J. M. and Billson, F. A. (1990). Early differentiation of ganglion, amacrine, bipolar and Müller cells in the developing fovea of human retina. *J. Comp. Neurol.*, **291**, 203–19.
Walcott, J. C. and Provis, J. M. (2003). Müller cells express the neuronal progenitor cell marker nestin in both differentiated and undifferentiated human foetal retina. *Clin. Exp. Ophthalmol.*, **31**, 246–9.
Wässle, H., Grünert, U., Martin, P. and Boycott, B. (1994). Color coding in the primate retina: predictions and constraints from anatomy. *Structural and Functional Organization of the Neocortex*, ed. B. Albowitz, K. Albus, U. Kuhnt, H.-C. Nothdurft and P. Wahle, Berlin, Heidelberg: Springer-Verlag, pp. 94–104.
Webb, S. V. and Kaas, J. H. (1976). The sizes and distribution of ganglion cells in the retina of the owl monkey, *Aotes trivirgatus*. *Vis. Res.*, **16**, 1247–54.
Wikler, K. C., Williams, R. W. and Rakic, P. (1990). Photoreceptor mosaic: number and distribution of rods and cones in the rhesus monkey retina. *J. Comp. Neurol.*, **297**, 499–508.
Xiao, M. and Hendrickson, A. E. (2000). Spatial and temporal expression of short, long/medium, or both opsins in human fetal cones. *J. Comp. Neurol.*, **425**, 545–59.
Yuodelis, C. and Hendrickson, A. E. (1986). A qualitative and quantitative analysis of the human fovea during development. *Vis. Res.*, **26**, 847–55.

8

Optic nerve formation

David W. Sretavan

University of California, San Francisco, USA

8.1 Introduction

The optic nerve is the anatomical pathway through which visual information received in the retina is conveyed along the axons of retinal ganglion cells (RGCs) to central visual targets for processing. In terms of its cellular organization, the optic nerve is relatively simple compared with other white matter tracts in the CNS. Unlike most CNS axon pathways, which typically contain ascending and descending axons from multiple neuronal populations, axons within the optic nerve all originate from RGCs in the eye, and all project in the same direction away from the retina towards the brain. There are no neurons in the optic nerve, and all resident cell nuclei belong to optic nerve glial cells. Given these organizational features, the developing optic nerve is an attractive experimental system and, not surprisingly, has been widely used in studies of axon guidance, glial differentiation, glial migration and myelination. Similarly, the adult optic nerve has also served extremely well as a model for studies of axonal transport and axon regeneration. This chapter describes the developmental mechanisms governing major aspects of optic nerve formation such as the determination of optic stalk cell fate, axon guidance and glia migration. The aim is to highlight our current understanding of these developmental processes, which at a basic level are fundamental to development of all regions of the nervous system.

8.2 Phases of optic nerve development

In considering optic nerve development, it is useful to conceptually divide the process into three phases. The first phase is the determination of optic cup and optic stalk territories, which begins early in neural development following the bilateral evaginations from the anterior region of the neural tube. The second phase is the growth of RGC axons out of the retina towards the brain, a process that transforms the optic stalk into an optic nerve. This phase begins as soon as the very first RGCs undergo differentiation and, in the human, involves the regulation of the pathfinding behaviour of up to several million axons. The last phase is optic nerve glial cell migration, differentiation and myelination. These glia-related

Retinal Development, ed. Evelyne Sernagor, Stephen Eglen, Bill Harris and Rachel Wong.
Published by Cambridge University Press. © Cambridge University Press 2006.

events illustrate the close interactions that must take place between axons and glia before the optic nerve is fully capable of taking on its adult function. These three developmental phases occur sequentially and together cover a protracted period during development. In rodents, for example, the optic stalks are formed before the mid-period of embryonic development, but RGC axon myelination in the optic nerve does not begin until postnatal day (P)7 (Skoff *et al.*, 1976, 1980; Colello *et al.*, 1995). In certain vertebrates such as fish, RGCs are in fact continuously generated even in adult animals, and RGC axons are constantly added to the optic nerve. In these species, optic nerve development can therefore be said to persist throughout life.

8.3 Specifying the optic stalk

The developmental programmes governing optic stalk formation are intimately tied to those that govern the development of the optic cup. Following closure of the neural tube, the prospective eye tissue at the junction of the telencephalon and the diencephalon evaginates and extends laterally. The distal region of this extending tissue rounds up to form the eyecup via a series of morphogenic movements culminating in the fusion of the ventrally located optic fissure. The more proximally located cells in this evaginated tissue eventually become the optic stalk.

8.4 Inductive role of *Sonic hedgehog*

The acquisition of optic stalk cell fate depends on the morphogenic protein Sonic hedgehog (Shh). *Shh* governs optic stalk specification and is found in the embryonic brain tissues close to the ventral diencephalic midline (Macdonald *et al.*, 1995; Hallonet *et al.*, 1999; Take-uchi *et al.*, 2003). In mouse, loss of function mutations in *Shh* result in a lack of optic stalks (Chiang *et al.*, 1996). Cellular responsiveness to the inductive activity of *Shh* is mediated by the transmembrane protein Smoothened, which is essential for *Shh* signal transduction (Chen *et al.*, 2001; Ingham and McMahon, 2001; Varga *et al.*, 2001). However, in the absence of *Shh* activity, the distally located cells of the evaginated tissues still acquire an optic cup cell fate. Although without the proper development of both optic stalks, a single eye field develops resulting in cyclopia.

8.5 Genetic programmes triggered by *Shh*

Since the eyecups are still present in the absence of optic stalks, specification of these two tissues might first appear to be separately regulated. Studies have demonstrated however that the optic cup and optic stalk generation are in fact intimately linked and *Shh* triggers a genetic programme beginning in axial brain structures that then spreads distally to the stalk and into the optic cup. This developmental programme is characterized by the sequential activation and repression of transcription factors in both the optic cup and stalk regions that belong to different families of homeodomain proteins. Amongst the first homeodomain

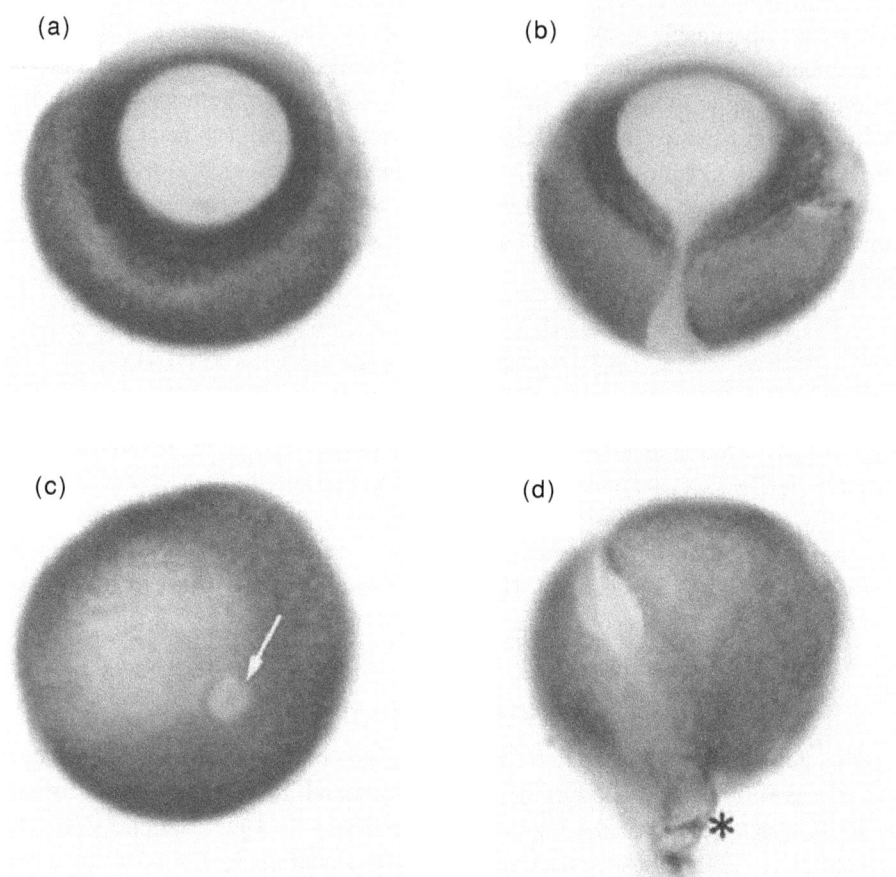

Figure 8.1 *Pax2* is required for the normal closure of the optic fissure and optic nerve development. (a,c) Front and back views respectively of the eye from a wild-type embryonic day (E)17 mouse embryo. (b,d) Front and back views respectively of the eye from a homozygous *Pax2* null E17 mouse embryo. Note the opened optic fissure in the *Pax2* null embryo in (b). The normally well-defined boundary between the pigmented retina and the optic nerve (c, arrow) is lost in the *Pax2* null embryo (d, asterisk). (Modified from Torres et al., 1996.)

proteins to be implicated in optic stalk development was *Pax2* (Torres et al., 1996). *Pax2* is normally expressed in the developing optic stalk and optic nerve head region and is excluded from the optic cup, which expresses *Pax6*. Targeted deletion of the *Pax2* gene results in a lack of proper optic stalk development and the expansion of the *Pax6* expression zone. *Pax2* expression in the optic stalk is normally controlled and activated by *Shh* (Figure 8.1). Thus the disruption of optic stalk cell fate in *Shh* loss of function mutants can be explained in part by *Shh* regulation of *Pax2*.

A second gene that also plays a significant role in optic stalk development is *Vax1*, a member of the *Emx/Not* gene family (Hallonet et al., 1998, 1999; Bertuzzi et al., 1999;

Take-uchi *et al.*, 2003). Loss of *Vax1* function, similar to the loss of *Pax2* function, leads to abnormal optic stalk development characterized by a failure of optic fissure closure and the presence of a coloboma (see below). In *Vax1* loss of function, *Pax2* expression is still maintained along the length of the mutant optic stalk. In *Pax2* loss of function, *Vax1* is likewise still present in the abnormally formed stalk. Thus *Vax1* and *Pax2* appear to be independently regulated. However, normal function of both *Pax2* and *Vax1* are essential for optic stalk development, and the activity of either gene alone is not sufficient.

8.6 Specification of optic stalk and cup tissues

The morphogenic activity of *Shh* and the actions of its downstream effectors, such as *Vax1*, are critical not only for optic stalk specification but also for the delineation of the tissue regions that subsequently acquire optic cup cell fate. During normal development, eyecup and retinal cell fate is characterized by the expression of *Pax6* and *Rx*, two genes that are known to govern eye formation in vertebrates (Hogan *et al.*, 1986; Walther and Gruss, 1991; Grindley *et al.*, 1995; Furukawa *et al.*, 1997; Mathers *et al.*, 1997). *Pax6* and *Rx* are excluded from the optic stalk, and stalk cells are instead characterized by *Pax2* and *Vax1* expression. The complementary patterns of expression of *Pax6/Rx* in the retina and *Pax2/Vax1* in the optic stalk are dependent on a proper level of *Shh* activity and repression by *Vax1*. In *Xenopus*, overexpression of *Shh* causes *Vax1* to be abnormally expressed in the optic cup tissues. This abnormal *Vax1* expression results in the reduction of *Pax6* and the loss of pigmented tissues and neural retina. On the other hand, the loss of *Vax1* from the optic stalk results in the expansion of *Pax6* and retinal tissue into the mutant optic stalk (Bertuzzi *et al.*, 1999; Hallonet *et al.*, 1999; Take-uchi *et al.*, 2003). These results indicate that the normal restriction of *Pax6* to optic cup/retinal tissues is dependent on a normal pattern of *Vax1* activation in the developing optic stalk, which is in turn dependent on *Shh*.

8.7 Malformations of the optic fissure

Developmental abnormalities of optic cup formation can result in an incomplete closure of the optic fissure and the presence of a coloboma. Coloboma (Greek for mutilated) refers to an abnormal opened morphology of the optic nerve head (Brodsky, 1994) and can be diagnosed on routine eye examination. Colobomas are often simply thought of as defects involving the optic nerve head region. In reality, colobomas can range in severity from mild cases involving just the retinal pigment epithelium to severe cases involving eye structures such as the iris. Colobomas are present in a significant fraction of childhood visual disorders. Malformations of the optic nerve head region can affect RGC axon pathfinding out of the eye and are often associated with decreased visual function. A reduction of *Pax2* function in humans results in colobomas, consistent with the role of *Pax2* in optic stalk specification (Sanyanusin *et al.*, 1995; Schimmenti *et al.*, 1997). An understanding of the genes involved in optic stalk and cup development should lead to more comprehensive descriptions of cell biological and molecular mechanisms that mediate cell movement involved in normal fissure closure.

8.8 Axon guidance: from retina into the optic nerve

The transformation of the optic stalk into the optic nerve begins with the entry of the first RGC axons. However, optic nerve development should properly be viewed as starting earlier within the retina itself in the sense that normal RGC axon pathfinding within eye is a prerequisite for the presence of axons in the optic nerve. Developing RGC axons in the retina typically show a stereotypic trajectory that extends straight from their cell bodies towards the optic nerve head region where they exit into the optic stalk/nerve. The direct path RGC axons take to reach the optic nerve head suggests the presence of very robust molecular mechanisms that guide developing axons. Retinal ganglion cell axon development within the retina does not seem to involve an initial process of random axon outgrowth in all directions, followed by the selective maintenance of RGC axons that succeeded in exiting the retina.

In humans, intraretinal pathfinding mechanisms ultimately result in the presence of roughly one million axons in each adult optic nerve. During development, roughly twice this number of RGCs is born and sends their axons into the optic nerve. Through a process of developmental programmed cell death, about half of these RGCs and their axons are subsequently eliminated. In rodents, it is estimated that depending on the specific strain, roughly 100 000 to 150 000 RGC axons find their way into the optic stalk from the retina during embryonic life (Strom and Williams, 1998).

Studies examining the mechanisms of RGC growth from the retina into the optic nerve have focused on deciphering the molecular mechanisms that govern RGC axon behaviours at four locations. (1) How RGCs newly born in peripheral retina orient their axons appropriately towards central retina. (2) How RGC axons maintain their straight trajectories without meandering during their course in mid-retina. (3) Mechanisms that ensure RGC axons accurately target the optic nerve head and enter the nerve. (4) Axon guidance within the optic nerve itself.

8.9 Initial orientation of axons

Studies in which small populations of embryonic RGCs and their axons have been labelled have consistently shown that these developing axons are oriented in straight trajectories towards the optic nerve head. In rodents, studies have pointed to a role for chondroitin-sulphate proteoglycans (CSPG) in establishing this initial orientation of developing axons. Proteoglycans are found quite abundantly in the developing retina and typically are large extracellular molecules consisting of a core protein decorated by negatively charged glycosaminoglycan side chains (Inatani and Tanihara, 2002). There are numerous families of proteoglycans distinguished by the nature of the core protein and side chain modifications.

Certain proteoglycans are known to bind growth factors or function as space occupying molecules.

Chondroitin-sulphate proteoglycans have been reported to be expressed in the peripheral regions of the retina as a ring immediately bordering the newly generated RGCs (Brittis and Silver, 1995; Chung et al., 2000). They have also been shown to inhibit the growth of axons in vitro (Snow et al., 1991; Snow and Letourneau, 1992), and enzymatic digestion of CSPG in eyecups in vitro resulted in aberrant orientation of RGC axons (Brittis et al., 1992). These findings have led to the proposal that inhibition of RGC axon growth by a peripheral ring of CSPG is one mechanism by which RGC axons are initially directed towards central retina.

8.10 Axon collaterals within the retina

Studies of developing RGCs that have been injected intracellularly with fluorescent dyes have reported the presence of axon collaterals (Ramoa et al., 1988). Although in general, these axon collaterals also have straight trajectories towards the optic nerve head, some collaterals grow in different directions away from this exit point. The fact that all of the RGCs that have these 'aberrant' axon collaterals have already extended a main axon all the way into CNS targets suggests that such collaterals develop only after the main axon has correctly found its way into the optic nerve.

8.11 Growth towards central retina

A striking phenomenon that is readily apparent following labelling of embryonic RGCs in the retina is the gathering of RGC axons into small axon bundles (fasciculation) as they traverse the mid region of the retina. This tight axon fasciculation is not so surprising since developing RGC axons express on their membrane surface a number of immunoglobulin superfamily cell adhesion molecules (IgCAMs) that in vitro support axon growth and fasciculation. Some examples include L1, transient axonal glycoprotein-1 (TAG-1), Axonin 1, neural cell adhesion molecule (NCAM) and DM-GRASP (Walsh and Doherty, 1997; Rougon and Hobert, 2003). Experimental support for a role of IgCAMs in governing RGC axon trajectories comes from examination of the role of Neurolin function in zebrafish (homologue of mammalian DM-GRASP). Antibody functional perturbation of Neurolin in vivo in the developing retina (Ott et al., 1998) elicits axon defasciculation and results in abnormal axon trajectories such as U-turns away from the optic nerve head (Figure 8.2).

The function of IgCAMs in axon guidance is not limited to the mediation of fasciculation. L1 has been shown to modulate growth cone responsiveness to Semaphorins, in that growth cones inhibited by Semaphorins appear to be attracted towards these guidance molecules in the presence of L1 (Castellani et al., 2000). Similarly, RGC axons that are inhibited by EphB molecules in the presence of laminin are not responsive to EphB in the presence of L1 (Suh et al., 2004). Thus, IgCAMs expressed by RGC axons confer these axons with

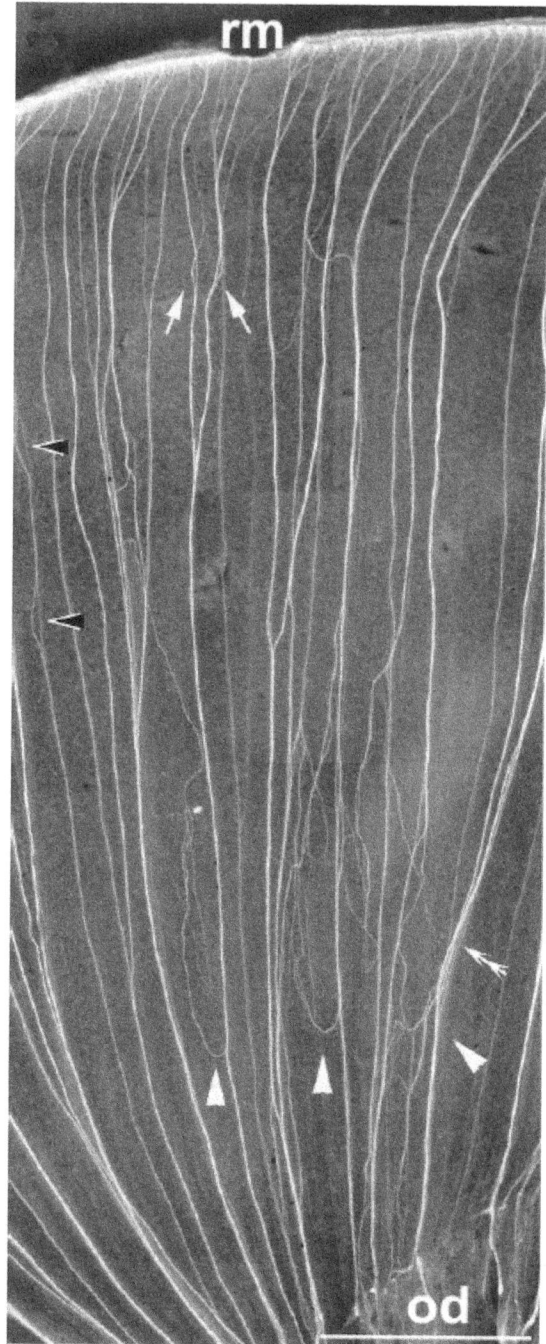

Figure 8.2 Retinal ganglion cell axon pathfinding errors in the retina of goldfish after intraocular injections of anti-neurolin Fab fragments. The pathfinding errors include defasciculation and looping (white arrowheads), failure to fasciculate upon meeting (white arrows), failure to maintain fasciculation (black arrowheads), formation of loose bundles (double arrowheads). Scale 500 μm. od, optic disc; rm, retinal peripheral margin. (Modified from Ott *et al.*, 1998.)

additional capabilities beyond mere axon fasciculation, including an ability to modulate their responsiveness to specific axon guidance molecules.

In addition to IgCAMs, studies of chick retinal development have provided evidence that RGC axons grow towards central retina by using cells that express the axon guidance molecule Slit1 as 'stepping stones' (Jin *et al.*, 2003). Slit proteins are a family of guidance molecules that play a role in multiple developmental processes including axon pathfinding (Ringstedt *et al.*, 2000; Plump *et al.*, 2002) and cell migration (Brose and Tessier-Lavigne, 2000; Nguyen-Ba-Charvet and Chedotal, 2002). Although Slit proteins are thought to act as inhibitory guidance molecules for RGC axons at the anterior optic chiasm (Plump *et al.*, 2002), RGC axons within the retina were found to preferentially grow on Slit-expressing cells. Disruption of Slit1 expression by overexpression of *Irx4*, a homeobox gene that negatively regulates Slit1 expression, leads to perturbations of RGC axon guidance in the retina.

Studies in chick have also suggested that developing retinal tissues possess an intrinsic property that permits axon growth preferentially in the direction from peripheral to central retina (Halfter, 1996). These experiments involved the transplantation of a piece of embryonic retina into a host retina followed by observations on the ability of host-derived RGC axons to extend into the grafted tissues. Results from experiments in which retinal grafts were rotated in their peripheral–central orientation showed that developing retinal tissues allowed axon growth in the peripheral to central direction but not in the opposite direction. The cellular or molecular basis for this polarity is not known.

8.12 Targeting the optic nerve head

After their journey from more peripheral regions of the retina, the accurate targeting of the optic nerve head requires the activity of Eph and ephrin molecules. Ephs and ephrin molecules have been implicated in a number of important developmental events including cell migration, blood vessel formation and axon pathfinding (Adams, 2002; Drescher, 2002; Kullander and Klein, 2002). Eph proteins are receptor tyrosine kinases and are grouped into the EphA and the EphB classes. EphA proteins interact with their ligands, the A ephrins which are glycosylphosphatidylinositol-linked molecules. EphB proteins interact with B ephrins, which are transmembrane molecules. The interaction of EphB with B ephrins is of note since their binding can trigger both forward signalling into the cell expressing EphB, and also reverse signalling in the cell expressing B ephrin (Holland *et al.*, 1996, 1998).

The importance of Eph and ephrin proteins in RGC axon pathfinding was first recognized in the context of retinotopic map formation in the superior colliculus. In this CNS target, a gradient of ephrin A expression exists that is highest at the posterior pole and decreases anteriorly. In the retina, complementary gradients of the corresponding EphA receptor proteins exist such that temporal RGC axons express the highest levels of the EphA receptor while more nasally located RGCs express less of this protein. Since A ephrins trigger an inhibitory response in EphA-expressing RGC axons, the axons from temporal retina are prevented from forming a terminal arborization in posterior superior colliculus and in this

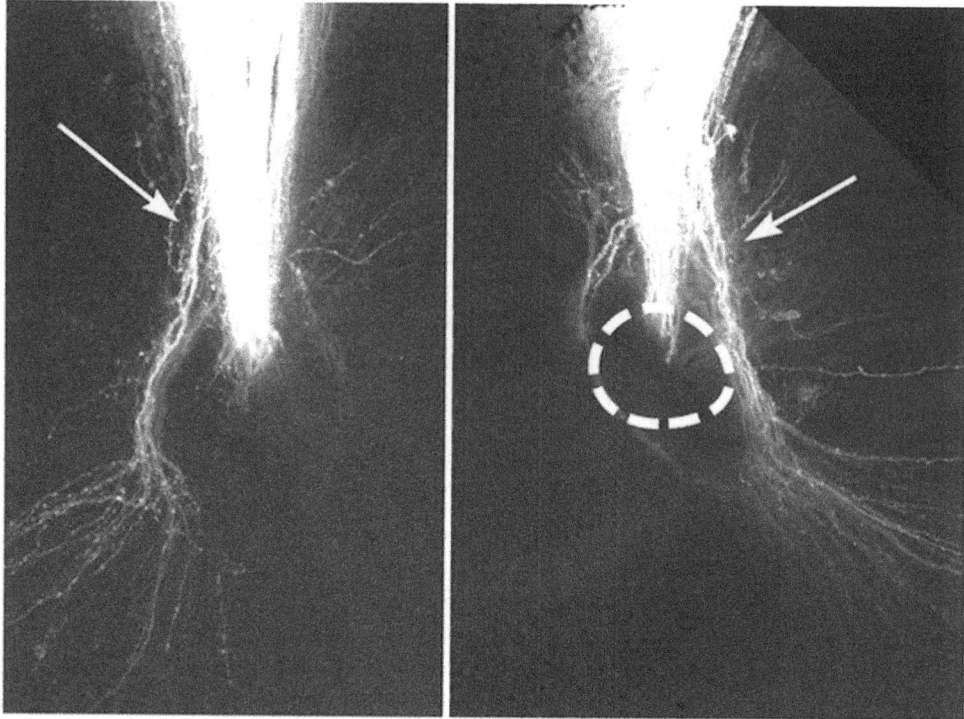

Figure 8.3 Retinal ganglion cell axons pathfinding errors in EphB2 and EphB3 double homozygous null mouse embryos. Arrows point to axon fascicles from dorsal retina that split off from the main axon bundle entering the optic nerve head (dotted circle). Axons that fail to accurately target the nerve head continue to grow inappropriately for substantial distances into ventral retina. Dorsal retina is towards the top. (Modified from Birgbauer *et al.* 2000.)

manner the polarity of the retinotopic map in this CNS target is set up. Along the medial–lateral axis of the superior colliculus, RGC axons from dorsal retina appear to utilize both EphB/B ephrin forward and reverse signalling to organize this second axis of the retinotopic map (Hindges *et al.*, 2002; Mann *et al.*, 2002).

The developing retina also has graded expression patterns of both EphB and B ephrin molecules. In general, EphB proteins are found in a high-ventral to low-dorsal gradient, while B ephrins are found in a complementary high-dorsal to low-ventral gradient. Mouse embryos with targeted deletions of a single EphB molecule do not show any apparent retinal developmental phenotype. However, embryos lacking two different EphB molecules (EphB2 and EphB3) demonstrate a specific phenotype in which RGC axons originating from the dorsal part of the retina fail to find their way accurately to the optic nerve head (Birgbauer *et al.*, 2000) (Figure 8.3). In double mutant animals, RGC axons from ventral retina are not affected. Dorsal RGC axons, but not ventral RGC axons, have been shown to be inhibited in their growth in vitro by the extracellular portion of EphB proteins (Birgbauer *et al.*, 2001). These observations have led to the proposal that normal pathfinding by dorsal retinal

RGC axons as they approach the optic nerve depends on EphB/B ephrin reverse signalling. In this model, RGC axons from dorsal retina normally encounter an increasing gradient of inhibitory EphB proteins as they grow towards the optic nerve head. This increasingly inhibitory environment helps maintain the tight fasciculation of RGC axons and allow them to accurately target this retina exit point.

Mechanisms responsible for the accurate targeting of the nerve head region by RGC axons from the ventral, nasal and temporal retina to the optic nerve head remain unknown. Given that the EphA and ephrin A molecules are also distributed in various gradients along the nasal-temporal retinal axis, RGC axons from these retinal regions may utilize these members of the EphA/ephrin family for pathfinding in the vicinity of the optic nerve head.

8.13 Axon 'avoidance' of the fovea

An important pathfinding task that must be accomplished by developing primate RGC axons is the avoidance of the foveal region, an anatomical specialization found in the primate retina (See Chapter 7 on foveal development). The high visual acuity in these species is due to the presence of a very high density of RGCs and photoreceptors in the fovea where RGCs and photoreceptors are thought to be connected in an almost one-to-one fashion, and the distortion of light impinging on photoreceptors is minimized by the lateral displacement of retinal neurons overlying the photoreceptors. In addition, retinal blood vessels, as well as RGC axons heading centrally, curve around to avoid the fovea on their way to the optic nerve head. These RGC axons with curved trajectories are collectively known as the arcuate fibres. In fetal monkeys, prior to the formation of a histologically defined fovea, RGC axons from temporal retina already grow in a curved fashion on their way to the optic nerve head, avoiding the incipient foveal region (Steineke and Kirby, 1993). This suggests the existence of specific axon guidance molecules within the developing foveal region. The fact that both developing blood vessels and RGC axons avoid the incipient fovea raise the possibility that molecules such as Eph and ephrins, which affect both vessel formation (Adams, 2002) and retinal axon guidance (Kullander and Klein, 2002), may be involved.

8.14 Entry into the optic stalk

Although it would seem that after RGC axons arrive at the optic nerve head, it would be rather straightforward for these axons to then enter the optic nerve, studies show that axon progression into the optic nerve specifically requires the axon guidance molecule Netrin-1 (Deiner *et al.*, 1997). Netrin mRNA and protein are expressed by *Pax2* neuroepithelial cells located at the optic nerve head (Figure 8.4). Exiting RGC axons navigate in between the reticular network formed by the processes of these cells that are decorated with netrin. In vitro, Netrin-1 supports the outgrowth of RGC axons and does so via the Netrin receptor protein Deleted in Colon Cancer (DCC). Mice lacking either Netrin-1 or DCC exhibit a similar RGC axon pathfinding defect characterized by the veering away of RGC axons from the optic nerve head region. Some of these aberrant axons ultimately project inappropriately

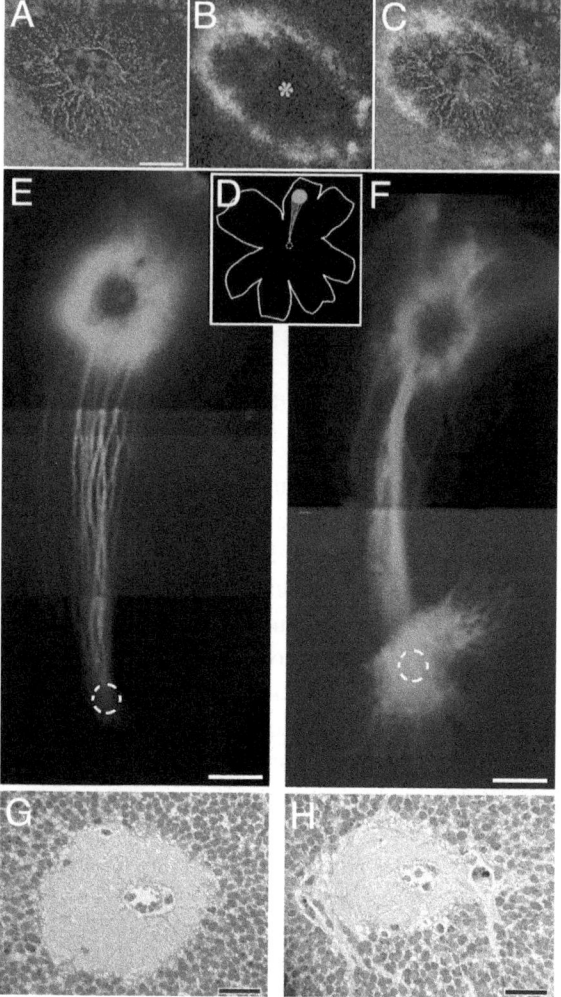

Figure 8.4 Distribution of guidance molecules and axon pathfinding defects at the optic nerve head. (A) Netrin protein visualized by antibody staining (red) is localized to the processes of the neuroepithelial cells at the developing optic nerve head. (B) Semaphorin 5A protein visualized by antibody staining (green) is confined to a peripheral rim surrounding the exiting RGC axons. Asterisk indicates centre of optic nerve head. (C) Overlay of Netrin and Semaphorin 5A protein distributions at the optic nerve head (note that tissue sections in A and B are from different experiments and overlay is for illustrative purposes only). Scale in A, B = 25 μm. (D) Schematic diagram of retina wholemount showing relationship between labelling site and the optic nerve head. (E) In embryonic day (E)15.5 wild-type mouse embryos, RGC axons exhibit straight trajectories to the region of the optic nerve head (dotted circle) where they exit into the optic nerve. (F) In Netrin-1 mutant embryos, RGC axons still find their way to the optic nerve head but fail to exit into the optic nerve. Instead they remain within the retina to grow aberrantly into inappropriate sites. Scale E, F = 100 μm. (G) Histological section through the optic nerve head in a wild-type mouse embryo. (H) Histological section at a similar location through the optic nerve head in a Netrin-1 mutant. Note the reduction in size of the optic nerve and the numerous fascicles of RGC axons that grow aberrantly in the immediate region. Scale G, H = 25 μm. (Modified from Deiner *et al.* (1997) and Oster and Sretavan (2003).) For colour version, see Plate 8.

into the opposite side of the retina and the subretinal space. The inability of RGC axons to enter the optic nerve results in optic nerve hypoplasia (reduction in optic nerve size). The phenotype however can range from no apparent reduction in nerve size to a severe reduction to about 20% of normal. Of note, some RGC axons still manage to find their way into the optic nerve of Netrin-1 mutants. This observation argues for a population of RGC axons that operates independently of Netrin-1 and suggests the presence of other guidance mechanisms that govern RGC axon navigation from the retina into the optic nerve.

Given the importance of Netrin in RGC axon guidance and optic nerve development, the mechanisms regulating Netrin expression are of interest. Netrin expression appears in fact to be triggered by signals derived from the RGC axons (Dakubo *et al.*, 2003). Soon after differentiation, RGCs express *Shh* and neuroepithelial cells at the optic disc region express the hedgehog target gene *Gli*. The selective elimination of *Shh* in RGCs during mouse development results in an inability of optic disc neuroepithelial cells to differentiate and produce Netrin. These animals exhibit an optic nerve hypoplasia resembling the hypoplasia observed in Netrin mutant animals. These results suggest a model in which Shh produced in RGCs is transported along axons to influence development of the optic nerve head region. Exposure to RGC axon-derived Shh triggers cells at the optic nerve head to express the axon guidance molecules necessary for successful axon exit from the retina.

8.15 Growth within the optic nerve

Studies of the optic nerve head have shown that in addition to Netrin-1, a member of the Semaphorin family of axon guidance molecules, Sema5A, is also specifically expressed in this region. Semaphorins form a large family of guidance molecules that generally inhibit axon growth and are found as either secreted or transmembrane proteins (Adams *et al.*, 1996; Raper, 2000). At the developing optic nerve head, transmembrane Sema5A is present as a ring surrounding the inner core formed by the exiting RGC axons and the reticular processes of neuroepithelial cells that express Netrin protein. This localization of Sema5A to an external sheath is maintained within the developing optic nerve. Retinal ganglion cell axon growth is inhibited by Sema5A protein in vitro, and function blocking experiments result in RGC axons separating from the main optic nerve bundle, suggesting that Sema5A normally serves as an inhibitory sheath surrounding and maintaining the integrity of the optic nerve.

A number of other guidance molecules are also present in the optic nerve. This includes Slit1 and 2, secreted molecules that play a role in guidance of axonal and dendritic processes as well as cell migration (Brose and Tessier-Lavigne, 2000; Nguyen-Ba-Charvet and Chedotal, 2002). Slit proteins have been reported to be involved in RGC axon guidance in the anterior part of the optic chiasm (Plump *et al.*, 2002). The role of Slit1 and 2 in developmental events occurring within the optic nerve is unknown although the absence of Slit1, 2 does not cause apparent gross defects in optic nerve development.

8.16 Re-positioning of axons

In mammals, once RGC axons enter the ON, they do not appear to maintain a tight retinotopic organization and become progressively dispersed from their neighbours (Horton *et al.*, 1979; Williams and Rakic, 1985; Chan and Guillery, 1994; Fitzgibbon and Reese, 1996). Through much of their course, RGC axons travel deep to the surface of the optic nerve, but emerge to grow just beneath the surface of the brain at the pre-chiasmatic region. This change in axon position occurs at a site where the glial composition changes from fascicular glia to radially aligned glial cells (Guillery and Walsh, 1987; Colello and Guillery, 1992). The guidance molecules mediating this shift in axon location with respect to the surface of the brain is not known. However, this re-positioning sets the stage for axon interactions with the molecular environment of the chiasm area where axons choose to cross or not cross the midline (Mason and Sretavan, 1997; Williams *et al.*, 2003), and for the substantial reordering of RGC axons into multiple retinotopic maps within the optic tract (Reese, 1996).

8.17 Transient optic nerve axons

Although studies of axon development in the optic nerve have almost entirely focused on axons of RGCs, it should be noted that a transient population of axons exists in the optic nerve prior to the growth of any RGC axons from the retina into the optic stalk. These axons originate from neuronal cell bodies in the diencephalon and grow from the brain towards the retina where they have been reported in the vicinity of the optic nerve head (Reese and Geller, 1995). Given that no similar population of axons is found in the adult optic nerve, these early optic nerve axons are thought to represent a transient axon population of unknown function. One possible role is to serve as a pre-existing track for RGC axons to follow out of the eye.

8.18 Optic nerve malformations

Developmental abnormalities affecting the optic nerve frequently manifest as optic nerve hypoplasia. According to statistics maintained by the Blind Babies Foundation, an institution serving the needs of visually impaired infants and their families, optic nerve developmental abnormalities represent a major cause of childhood visual disorders. Optic nerve hypoplasias are typically sporadic in nature with no clear patterns of inheritance (Taylor, 2005). The causes are generally unknown but have been hypothesized to be due to defects in RGC differentiation or retraction of axons from CNS target tissues resulting in axon and RGC loss. In mice, the loss of the RGC axon guidance molecule Netrin-1 or its receptor DCC also results in optic nerve hypoplasia (Deiner *et al.*, 1997). Optic nerve hypoplasia can also affect specific retinal regions giving rise to entities such as superior segmental optic nerve hypoplasia, which is linked to maternal diabetes (Petersen and Walton, 1977; Kim *et al.*, 1989). One form of optic nerve hypoplasia that occurs together with corpus callosum defects and pituitary disorders is called Septo-Optic Dysplasia (SOD) or De Morsier's syndrome. This has been linked to mutations in the gene encoding HesX1, a transcription factor expressed in the developing brain and the eye (Dattani *et al.*, 2000).

8.19 Optic nerve myelination

Myelination of RGC axons is an essential phase of optic nerve development and prepares the nerve for its duties in the adult visual system. As with other CNS axon tracts, the myelin-forming cell of the optic nerve is the oligodendrocyte. Although myelination by oligodendrocytes begins in earnest only after nearly all RGC axons have grown through the optic nerve, the developmental events that set the stage for myelination such as the migration of oligodendrocyte precursor cells (OPC) actually occur at the same time as axon extension within the optic nerve.

8.20 Origin of optic nerve oligodendrocytes

The ability to migrate over long distances in the nervous system during development is not restricted to neurons or the tips of their growing processes. Long-distance glial cell migration has been documented in several regions of the CNS (Kakita and Goldman, 1999; Tsai *et al.*, 2002). A well-studied example of long-distance glial movement takes place in the optic nerve. The oligodendrocytes that myelinate retinal axons are not intrinsically derived from the optic stalk tissues but instead migrate into the developing optic nerve from embryonic ventral diencephalon tissues. The fact that the optic nerve lacks neuronal cell bodies has made it a useful experimental system for the analysis of glial migration and differentiation.

The diencephalic origin for optic nerve OPCs was first proposed based on experimental data showing that the ventral midline region of the embryonic diencephalon contained detectable OPCs several days ahead of various parts of the optic nerve (Small *et al.*, 1987). Subsequent studies using the fluorescent dye DiI to label cells in the floor of the third ventricle (Ono *et al.*, 1997), or by using focal thymidine labelling of glial precursors in the ventral diencephalon (Sugimoto *et al.*, 2001), and by observation of LacZ-labelled OPCs in transgenic mice (Spassky *et al.*, 2002) (Figure 8.5), have all demonstrated glial cell migration from the embryonic ventral midline brain regions to populate the optic nerve.

8.21 Molecular cues for glial migration

A number of studies have shown that OPCs respond to cell adhesion molecules such as cadherin (Payne and Lemmon, 1993), NCAM (Wang *et al.*, 1994), tenascin (Garcion *et al.*, 2001), and can migrate in the presence of cell-derived extracellular matrix by activation of integrin receptors (Milner *et al.*, 1996; Garcion *et al.*, 2001). In addition, OPCs also show an ability to migrate in response to growth factors such as fibroblast growth factor-2 and platelet-derived growth factor-2 (Milner *et al.*, 1997). Recent work has shown that OPCs may in fact migrate into the optic nerve using the same guidance molecules that RGC axons

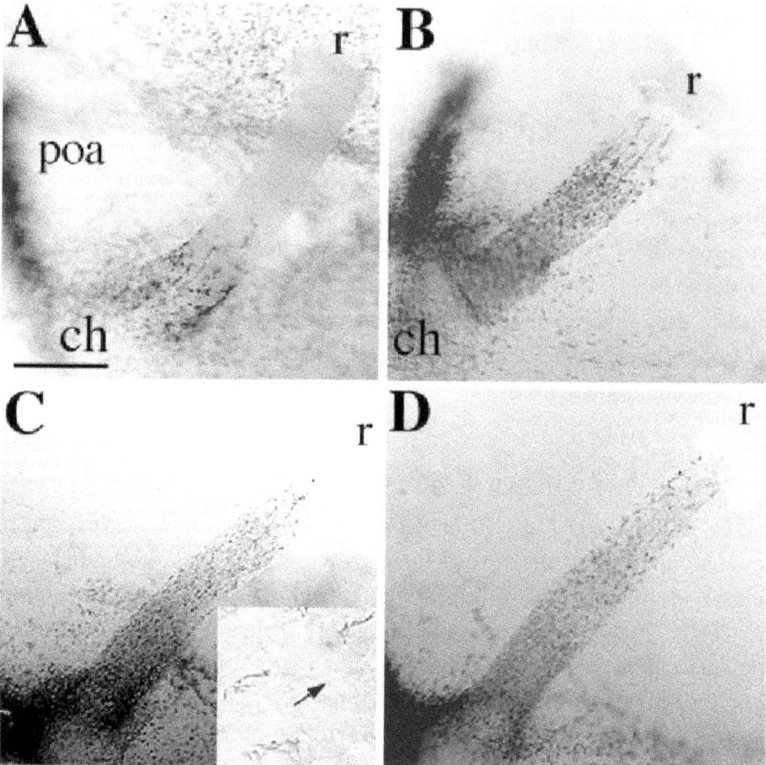

Figure 8.5 Migration of OPCs from the region of the optic chiasm into the optic nerve in mouse embryos. The OPCs were visualized as LacZ-positive cells from a transgenic mouse line in which LacZ expression is under the control of regulatory sequences of the myelin protein plp. (A) E14.5 (B) E 15.5 (C) E16.5 (D) E17.5. Gestation in mouse is 20 days. The inset in C shows the long cytoplasmic extensions of the LacZ-positive cell profiles parallel to the chiasm–retinal axis. poa, pre-optic area; ch, optic chiasm; r, retina. Scale A, B = 110 μm; C, D = 85 μm; C inset = 25 μm. (Modified from Spassky et al., 2002.)

use for pathfinding in the optic nerve and in the ventral diencephalon region (Sugimoto et al., 2001; Spassky et al., 2002). In the case of the OPCs, however, these signals result in cell migration away from the brain towards the retina, whereas RGC axons travel in the opposite direction towards their CNS targets.

Evidence suggests that OPC migration into the optic nerve involves at least two classes of guidance molecules, the Netrins and Semaphorins. Oligodendrocyte precursor cells express Neuropilin 1 and 2, transmembrane proteins involved in mediating cell responsiveness to various Semaphorins (Raper, 2000; Bagri and Tessier-Lavigne, 2002). In addition, OPCs also express DCC and Unc5H1, proteins that serve as Netrin receptors (Keino-Masu et al., 1996; Hong et al., 1999). In vitro, OPCs exposed to sources and gradients of Netrin or Sema3, show a repellant behaviour and move away from these guidance molecules (Sugimoto et al., 2001; Spassky et al., 2002). It is proposed that a source of these guidance molecules is present at the ventral midline regions of the diencephalon and that a gradient of these guidance molecules exists maximally at the brain and drops off in the direction towards the retina. Such a graded distribution of a chemorepellant is thought to force OPCs to migrate away from the ventral diencephalon into the optic nerve towards the retina. Of note, similar repellant mechanisms for glial migration involving Netrin have been proposed in other regions of the CNS (Tsai *et al.*, 2003).

The idea that OPC migration into the optic nerve shares a common molecular basis as RGC axon pathfinding is attractive in that it conserves highly successful mechanisms through which cells sense and respond to their environment and thus limit the number of different molecular system that must be used. However, in vivo evidence for a shared molecular basis with axon pathfinding is more difficult to obtain. Standard targeted disruption of genes encoding guidance molecules or their receptors will likely affect RGC axon guidance. Since RGC axons themselves may act as a source of additional guidance molecules for OPC migration or provide a physical substrate for cell migration, abnormalities of OPC movement in such experiments will require careful interpretation. The use of techniques that allow specific spatiotemporal alteration of gene function in OPCs may be one revealing approach.

8.22 Stop signals at the optic nerve head

A well-known feature of the optic nerve myelination is that RGC axons are not myelinated until some distance proximal, away from the optic nerve head and RGC axons are not myelinated in the retina. (Rabbits are an exception to this rule.) This absence of axon myelination in the retina is thought to be critical for maintaining optical clarity necessary for light transduction. The region of the optic nerve lacking myelination corresponds to the area of the lamina cribosa, a region immediately adjacent to the optic nerve head with a specialized cellular and structural composition (Ffrench-Constant *et al.*, 1988; Perry and Lund, 1990; Ye and Hernandez, 1995). The role of the lamina cribosa in the adult optic nerve is unknown but may involve the provision of mechanical support for this unmyelinated region of the optic nerve. Its importance is highlighted by the fact that alterations in the cellular and molecular composition of the lamina cribosa region is found in glaucoma (Hernandez, 2000), and is thought to contribute to the retrograde RGC cell death characteristic of this disease.

It is possible that structural and molecular elements in the developing lamina cribosa serve a role restricting OPC migration and invasion of the retina (Ffrench-Constant *et al.*, 1988). The nature of such signals is unknown at present. One possible candidate is Netrin, which is highly expressed at the optic nerve head and is responsible for RGC axon exit from the retina into the optic nerve (Deiner *et al.*, 1997). As discussed above, Netrin has also been found to inhibit OPC migration (Sugimoto *et al.*, 2001) and may thus be in a correct location to prevent OPC movement into the retina. In the developing spinal cord, CXCL1 and 2, members of the chemokine family has been shown to act as a stop signal for OPC migration (Tsai *et al.*, 2002).

8.23 Concluding remarks

Optic nerve development represents a microcosm of many of the important events occurring throughout the developing nervous system such as axon extension, glial development, axon–glial interactions and myelination. The specific absence of neurons in the optic nerve provides an added degree of convenience in experimental design and interpretation. Although the topic of this chapter is optic nerve development, one should also not lose sight of the fact that the adult optic nerve is a crucial CNS axon tract whose normal function is essential for vision and, thus, for much of behaviour. Major eye diseases such as glaucoma and various other traumatic insults that compromise optic nerve function represent significant causes of visual impairment. For example, glaucoma is a leading cause of blindness affecting an estimated 70 million individuals worldwide (Quigley, 1996; Fraser and Wormald, 2004). Although the precise disease mechanism is yet to be completely defined, it is widely recognized as an optic neuropathy in which cellular and molecular changes within the optic nerve cause the retrograde apoptotic cell death of retinal ganglion cells (Hernandez, 2000; Fraser and Wormald, 2004). While progress against this disease will surely require efforts on multiple fronts, a better understanding of the mechanisms that work together to construct an optic nerve during embryonic development may inform on aspects of basic biology that can be manipulated as potential therapeutic strategies.

References

Adams, R. H. (2002). Vascular patterning by Eph receptor tyrosine kinases and ephrins. *Semin. Cell Dev. Biol.*, **13**, 55–60.

Adams, R. H., Betz, H. and Puschel, A. W. (1996). A novel class of murine semaphorins with homology to thrombospondin is differentially expressed during early embryogenesis. *Mech. Dev.*, **57**, 33–45.

Bagri, A. and Tessier-Lavigne, M. (2002). Neuropilins as Semaphorin receptors: *in vivo* functions in neuronal cell migration and axon guidance. *Adv. Exp. Med. Biol.*, **515**, 13–31.

Bertuzzi, S., Hindges, R., Mui, S. H., O'Leary, D. D. and Lemke, G. (1999). The homeodomain protein vax1 is required for axon guidance and major tract formation in the developing forebrain. *Genes Dev.*, **13**, 3092–105.

Birgbauer, E., Cowan, C. A., Sretavan, D. W. and Henkemeyer, M. (2000). Kinase independent function of EphB receptors in retinal axon pathfinding to the optic disc from dorsal but not ventral retina. *Development*, **127**, 1231–41.

Birgbauer, E., Oster, S. F., Severin, C. G. and Sretavan, D. W. (2001). Retinal axon growth cones respond to EphB extracellular domains as inhibitory axon guidance cues. *Development*, **128**, 3041–8.

Brittis, P. A. and Silver, J. (1995). Multiple factors govern intra-retinal axon guidance: a time-lapse study. *Mol. Cell. Neurosci.*, **6**, 413–32.

Brittis, P. A., Canning, D. R. and Silver, J. (1992). Chondroitin sulfate as a regulator of neuronal patterning in the retina. *Science*, **255**, 733–6.

Brodsky, M. C. (1994). Congenital optic disk anomalies. *Surv. Ophthalmol.*, **39**, 89–112.

Brose, K. and Tessier-Lavigne, M. (2000). Slit proteins: key regulators of axon guidance, axonal branching, and cell migration. *Curr. Opin. Neurobiol.*, **10**, 95–102.

Castellani, V., Chedotal, A., Schachner, M., Faivre-Sarrailh, C. and Rougon, G. (2000). Analysis of the L1-deficient mouse phenotype reveals cross-talk between Sema3A and L1 signaling pathways in axonal guidance. *Neuron*, **27**, 237–49.

Chan, S. O. and Guillery, R. W. (1994). Changes in fiber order in the optic nerve and tract of rat embryos. *J. Comp. Neurol.*, **344**, 20–32.

Chen, W., Burgess, S. and Hopkins, N. (2001). Analysis of the zebrafish smoothened mutant reveals conserved and divergent functions of hedgehog activity. *Development*, **128**, 2385–96.

Chiang, C., Litingtung, Y., Lee, E. *et al.* (1996). Cyclopia and defective axial patterning in mice lacking Sonic hedgehog gene function. *Nature*, **383**, 407–13.

Chung, K. Y., Shum, D. K. and Chan, S. O. (2000). Expression of chondroitin sulfate proteoglycans in the chiasm of mouse embryos. *J. Comp. Neurol.*, **417**, 153–63.

Colello, R. J. and Guillery, R. W. (1992). Observations on the early development of the optic nerve and tract of the mouse. *J. Comp. Neurol.*, **317**, 357–78.

Colello, R. J., Devey, L. R., Imperato, E. and Pott, U. (1995). The chronology of oligodendrocyte differentiation in the rat optic nerve: evidence for a signaling step initiating myelination in the CNS. *J. Neurosci.*, **15**, 7665–72.

Dakubo, G. D., Wang, Y. P., Mazerolle, C. *et al.* (2003). Retinal ganglion cell-derived sonic hedgehog signaling is required for optic disc and stalk neuroepithelial cell development. *Development*, **130**, 2967–80.

Dattani, M. L., Martinez-Barbera, J., Thomas, P. Q. *et al.* (2000). Molecular genetics of septo-optic dysplasia. *Horm. Res.*, **53 Suppl 1**, 26–33.

Deiner, M. S., Kennedy, T. E., Fazeli, A. *et al.* (1997). Netrin-1 and DCC mediate axon guidance locally at the optic disc: loss of function leads to optic nerve hypoplasia. *Neuron*, **19**, 575–89.

Drescher, U. (2002). Eph family functions from an evolutionary perspective. *Curr. Opin. Genet. Dev.*, **12**, 397–402.

Ffrench-Constant, C., Miller, R. H., Burne, J. F. and Raff, M. C. (1988). Evidence that migratory oligodendrocyte-type-2 astrocyte (O-2A) progenitor cells are kept out of the rat retina by a barrier at the eye-end of the optic nerve. *J. Neurocytol.*, **17**, 13–25.

Fitzgibbon, T. and Reese, B. E. (1996). Organization of retinal ganglion cell axons in the optic fiber layer and nerve of fetal ferrets. *Vis. Neurosci.*, **13**, 847–61.

Fraser, S. and Wormald, R. (2004). Epidemiology of Glaucoma. In *Ophthalmology*, 2nd edn, ed. M. Yanoff and J. S. Duker. St. Louis: Mosby, pp. 1413–1417.

Furukawa, T., Kozak, C. A. and Cepko, C. L. (1997). *rax*, a novel paired-type homeobox gene, shows expression in the anterior neural fold and developing retina. *Proc. Natl. Acad. Sci. U. S. A.*, **94**, 3088–93.

Garcion, E., Faissner, A. and Ffrench-Constant, C. (2001). Knock-out mice reveal a contribution of the extracellular matrix molecule tenascin-C to neural precursor proliferation and migration. *Development*, **128**, 2485–96.

Grindley, J. C., Davidson, D. R. and Hill, R. E. (1995). The role of *Pax-6* in eye and nasal development. *Development*, **121**, 1433–42.

Guillery, R. W. and Walsh, C. (1987). Changing glial organization relates to changing fiber order in the developing optic nerve of ferrets. *J. Comp. Neurol.*, **265**, 203–17.

Halfter, W. (1996). Intraretinal grafting reveals growth requirements and guidance cues for optic axons in the developing avian retina. *Dev. Biol.*, **177**, 160–77.

Hallonet, M., Hollemann, T., Wehr, R. *et al.* (1998). *Vax1* is a novel homeobox-containing gene expressed in the developing anterior ventral forebrain. *Development*, **125**, 2599–610.

Hallonet, M., Hollemann, T. Pieler, T. and Gruss, P. (1999). *Vax1*, a novel homeobox-containing gene, directs development of the basal forebrain and visual system. *Genes Dev.*, **13**, 3106–14.

Hernandez, M. R. (2000). The optic nerve head in glaucoma: role of astrocytes in tissue remodeling. *Prog. Retin. Eye Res.*, **19**, 297–321.

Hindges, R., McLaughlin, T., Genoud, N., Henkemeyer, M. and O'Leary, D. D. (2002). EphB forward signaling controls directional branch extension and arborization required for dorsal-ventral retinotopic mapping. *Neuron*, **35**, 475–87.

Hogan, B. L., Horsburgh, G., Cohen, J. *et al.* (1986). Small eyes (Sey): a homozygous lethal mutation on chromosome 2 which affects the differentiation of both lens and nasal placodes in the mouse. *J. Embryol. Exp. Morphol.*, **97**, 95–110.

Holland, S. J., Gale, N. W., Mbamalu, G. *et al.* (1996). Bi-directional signaling through the EPH-family receptor Nuk and its transmembrane ligands. *Nature*, **383**, 722–5.

Holland, S. J., Peles, E., Pawson, T. and Schlessinger, J. (1998). Cell-contact-dependent signaling in axon growth and guidance: Eph receptor tyrosine kinases and receptor protein tyrosine phosphatase beta. *Curr. Opin. Neurobiol.*, **8**, 117–27.

Hong, K., Hinck, L., Nishiyama, M. *et al.* (1999). A ligand-gated association between cytoplasmic domains of UNC5 and DCC family receptors converts netrin-induced growth cone attraction to repulsion. *Cell*, **97**, 927–41.

Horton, J. C., Greenwood, M. M. and Hubel, D. H. (1979). Non-retinotopic arrangement of fibers in cat optic nerve. *Nature*, **282**, 720–2.

Inatani, M. and Tanihara, H. (2002). Proteoglycans in retina. *Prog. Retin. Eye Res.*, **21**, 429–47.

Ingham, P. W. and McMahon, A. P. (2001). Hedgehog signaling in animal development: paradigms and principles. *Genes Dev.*, **15**, 3059–87.

Jin, Z., Zhang, J., Klar, A. *et al.* (2003). Irx4-mediated regulation of Slit1 expression contributes to the definition of early axonal paths inside the retina. *Development*, **130**, 1037–48.

Kakita, A. and Goldman, J. E. (1999). Patterns and dynamics of SVZ cell migration in the postnatal forebrain: monitoring living progenitors in slice preparations. *Neuron*, **23**, 461–72.

Keino-Masu, K., Masu, M., Hinck, L. *et al.* (1996). Deleted in Colorectal Cancer (DCC) encodes a netrin receptor. *Cell*, **87**, 175–85.

Kim, R. Y., Hoyt, W. F., Lessell, S. and Narahara, M. H. (1989). Superior segmental optic hypoplasia. A sign of maternal diabetes. *Arch. Ophthalmol.*, **107**, 1312–15.

Kullander, K. and Klein, R. (2002). Mechanisms and functions of Eph and ephrin signaling. *Nat. Rev. Mol. Cell. Biol.*, **3**, 475–86.

Macdonald, R., Barth, K. A., Xu, Q. *et al.* (1995). Midline signaling is required for *Pax* gene regulation and patterning of the eyes. *Development*, **121**, 3267–78.

Mann, F., Ray, S., Harris, W. and Holt, C. (2002). Topographic mapping in dorsoventral axis of the Xenopus retinotectal system depends on signaling through ephrin-B ligands. *Neuron*, **35**, 461–73.

Mason, C. A. and Sretavan, D. W. (1997). Glia, neurons, and axon pathfinding during optic chiasm development. *Curr. Opin. Neurobiol.*, **7**, 647–53.

Mathers, P. H., Grinberg, A., Mahon, K. A. and Jamrich, M. (1997). The *Rx* homeobox gene is essential for vertebrate eye development. *Nature*, **387**, 603–7.

Milner, R., Edwards, G., Streuli, C. and Ffrench-Constant, C. (1996). A role in migration for the alpha V beta 1 integrin expressed on oligodendrocyte precursors. *J. Neurosci.*, **16**, 7240–52.

Milner, R., Anderson, H. J., Rippon, R. F. *et al.* (1997). Contrasting effects of mitogenic growth factors on oligodendrocyte precursor cell migration. *Glia*, **19**, 85–90.

Nguyen-Ba-Charvet, K. T. and Chedotal, A. (2002). Role of Slit proteins in the vertebrate brain. *J. Physiol. Paris*, **96**, 91–8.

Ono, K., Yasui, Y., Rutishauser, U. and Miller, R. H. (1997). Focal ventricular origin and migration of oligodendrocyte precursors into the chick optic nerve. *Neuron*, **19**, 283–92.

Oster, S. F. and Sretauan, D. (2003). Connecting the eye to the brain. The molecular basis of ganglion cell axon guidance. *Br. J. Ophthalmol.*, **87**, 639–45.

Ott, H., Bastmeyer, M. and Stuermer, C. A. (1998). Neurolin, the goldfish homolog of DM-GRASP, is involved in retinal axon pathfinding to the optic disk. *J. Neurosci.*, **18**, 3363–72.

Payne, H. R. and Lemmon, V. (1993). Glial cells of the O-2A lineage bind preferentially to N-cadherin and develop distinct morphologies. *Dev. Biol.*, **159**, 595–607.

Perry, V. H. and Lund, R. D. (1990). Evidence that the lamina cribrosa prevents intra-retinal myelination of retinal ganglion cell axons. *J. Neurocytol.*, **19**, 265–72.

Petersen, R. A. and Walton, D. S. (1977). Optic nerve hypoplasia with good visual acuity and visual field defects: a study of children of diabetic mothers. *Arch. Ophthalmol.*, **95**, 254–8.

Plump, A. S., Erskine, L., Sabatier, C. *et al.* (2002). Slit1 and Slit2 cooperate to prevent premature midline crossing of retinal axons in the mouse visual system. *Neuron*, **33**, 219–32.

Quigley, H. A. (1996). Number of people with glaucoma worldwide. *Br. J. Ophthalmol.*, **80**, 389–93.

Ramoa, A. S., Campbell, G. and Shatz, C. J. (1988). Dendritic growth and remodeling of cat retinal ganglion cells during fetal and postnatal development. *J. Neurosci.*, **8**, 4239–61.

Raper, J. A. (2000). Semaphorins and their receptors in vertebrates and invertebrates. *Curr. Opin. Neurobiol.*, **10**, 88–94.

Reese, B. E. (1996). The chronotopic re-ordering of optic axons. *Perspect. Dev. Neurobiol.*, **3**, 233–42.

Reese, B. E. and Geller, S. F. (1995). Precocious invasion of the optic stalk by transient retinopetal axons. *J. Comp. Neurol.*, **353**, 572–84.

Ringstedt, T., Braisted, J. E., Brose, K. *et al.* (2000). Slit inhibition of retinal axon growth and its role in retinal axon pathfinding and innervation patterns in the diencephalon. *J. Neurosci.*, **20**, 4983–91.

Rougon, G. and Hobert, O. (2003). New insights into the diversity and function of neuronal immunoglobulin superfamily molecules. *Annu. Rev. Neurosci.*, **26**, 207–38.

Sanyanusin, P., Schimmenti, L. A., McNoe, L. A. *et al.* (1995). Mutation of the *PAX2* gene in a family with optic nerve colobomas, renal anomalies and vesicoureteral reflux. *Nat. Genet.*, **9**, 358–64.

Schimmenti, L. A., Cunliffe, H. E., McNoe, L. A. *et al.* (1997). Further delineation of renal-coloboma syndrome in patients with extreme variability of phenotype and identical *PAX2* mutations. *Am. J. Hum. Genet.*, **60**, 869–78.

Skoff, R. P., Price, D. L. and Stocks, A. (1976). Electron microscopic autoradiographic studies of gliogenesis in rat optic nerve. II. Time of origin. *J. Comp. Neurol.*, **169**, 313–34.

Skoff, R. P., Toland, D. and Nast, E. (1980). Pattern of myelination and distribution of neuroglial cells along the developing optic system of the rat and rabbit. *J. Comp. Neurol.*, **191**, 237–53.

Small, R. K., Riddle, P. and Noble, M. (1987). Evidence for migration of oligodendrocyte – type-2 astrocyte progenitor cells into the developing rat optic nerve. *Nature*, **328**, 155–7.

Snow, D. M. and Letourneau, P. C. (1992). Neurite outgrowth on a step gradient of chondroitin sulfate proteoglycan (CS-PG). *J. Neurobiol.*, **23**, 322–36.

Snow, D. M., Watanabe, M., Letourneau, P. C. and Silver, J. (1991). A chondroitin sulfate proteoglycan may influence the direction of retinal ganglion cell outgrowth. *Development*, **113**, 1473–85.

Spassky, N., de Castro, F., Le Bras, B. *et al.* (2002). Directional guidance of oligodendroglial migration by class 3 semaphorins and netrin-1. *J. Neurosci.*, **22**, 5992–6004.

Steineke, T. C. and Kirby, M. A. (1993). Early axon outgrowth of retinal ganglion cells in the fetal rhesus macaque. *Brain Res. Dev. Brain Res.*, **74**, 151–62.

Strom, R. C. and Williams, R. W. (1998). Cell production and cell death in the generation of variation in neuron number. *J. Neurosci.*, **18**, 9948–53.

Sugimoto, Y., Taniguchi, M., Yagi, T. *et al.* (2001). Guidance of glial precursor cell migration by secreted cues in the developing optic nerve. *Development*, **128**, 3321–30.

Suh, L. H., Oster, S. F., Soehrman, S. S., Grenningloh, G. and Sretavan, D. W. (2004). L1/Laminin modulation of growth cone response to EphB triggers growth pauses and regulates the microtubule destabilising protein SCG10. *J. Neurosci.*, **24**, 1976–86.

Take-uchi, M., Clarke, J. D. and Wilson, S. W. (2003). Hedgehog signaling maintains the optic stalk-retinal interface through the regulation of *Vax* gene activity. *Development*, **130**, 955–68.

Taylor, D. (2005). Optic nerve axons: life and death before birth. *Eye*, **19**, 499–527.

Torres, M., Gomez-Pardo, E. and Gruss, P. (1996). *Pax2* contributes to inner ear patterning and optic nerve trajectory. *Development*, **122**, 3381–91.

Tsai, H. H., Frost, E., To, V. *et al.* (2002). The chemokine receptor CXCR2 controls positioning of oligodendrocyte precursors in developing spinal cord by arresting their migration. *Cell*, **110**, 373–83.

Tsai, H. H., Tessier-Lavigne, M. and Miller, R. H. (2003). Netrin 1 mediates spinal cord oligodendrocyte precursor dispersal. *Development*, **130**, 2095–105.

Varga, Z. M., Amores, A., Lewis, K. E. *et al.* (2001). Zebrafish smoothened functions in ventral neural tube specification and axon tract formation. *Development*, **128**, 3497–509.

Walsh, F. S. and Doherty, P. (1997). Neural cell adhesion molecules of the immunoglobulin superfamily: role in axon growth and guidance. *Annu. Rev. Cell. Dev. Biol.*, **13**, 425–56.

Walther, C. and Gruss, P. (1991). *Pax-6*, a murine paired box gene, is expressed in the developing CNS. *Development*, **113**, 1435–49.

Wang, C., Rougon, G. and Kiss, J. Z. (1994). Requirement of polysialic acid for the migration of the O-2A glial progenitor cell from neurohypophyseal explants. *J. Neurosci.*, **14**, 4446–57.

Williams, R. W. and Rakic, P. (1985). Dispersion of growing axons within the optic nerve of the embryonic monkey. *Proc. Natl. Acad. Sci. U. S. A.*, **82**, 3906–10.

Williams, S. E., Mann, F., Erskine, L. *et al.* (2003). Ephrin-B2 and EphB1 mediate retinal axon divergence at the optic chiasm. *Neuron*, **39**, 919–35.

Ye, H. and Hernandez, M. R. (1995). Heterogeneity of astrocytes in human optic nerve head. *J. Comp. Neurol.*, **362**, 441–52.

9

Glial cells in the developing retina

Kathleen Zahs and Manuel Esguerra

University of Minnesota, Minneapolis, USA

9.1 Introduction

Müller cells, the principal glia of vertebrate retinas, are radial glia that span the entire depth of the retina. The distal processes of Müller cells form the external limiting membrane of the retina, while their 'endfeet' form the inner limiting membrane. Müller cell processes surround neuronal cell bodies in the nuclear layers and contact synapses in the plexiform layers (Newman and Reichenbach, 1996). Müller cells play a major role in regulating extracellular K^+ and pH (Newman *et al.*, 1984; Karwoski *et al.*, 1989; Kusaka and Puro, 1997), in neurotransmitter uptake (Pow, 2001) and in glutamine synthesis (Riepe, 1977, 1978; Germer *et al.*, 1997a; Prada *et al.*, 1998), functions performed by astrocytes in other regions of the central nervous system. Müller cells also have some similarities to oligodendrocytes; although they do not form myelin, Müller cell processes wrap the axons of retinal ganglion cells (Holländer *et al.*, 1991; Stone *et al.*, 1995). In addition, intercellular Ca^{2+} waves have been observed among Müller cells (Newman and Zahs, 1997). These waves are increases in glial cytosolic Ca^{2+} that propagate away from the site of initial activation. The arrival of Ca^{2+} waves in retinal glia is correlated with modulation of the light-evoked activity of neighbouring retinal ganglion cells (Newman and Zahs, 1998). Modulation of retinal ganglion cell activity has been shown to be mediated by a variety of factors released by Müller cells, including purine nucleotides (Newman, 2003) and D-serine, a co-agonist at the *N*-methyl-D-aspartate (NMDA) type of glutamate receptor (Stevens *et al.*, 2003). In addition to sending signals to their retinal neighbours, Müller cells can also respond to signals via receptors for a number of transmitters and modulators, including glutamate (Keirstead and Miller, 1997), adenosine triphosphate (ATP) (Reifel Saltzberg *et al.*, 2003), acetylcholine (Wakakura *et al.*, 1998), cytokines (Peterson *et al.*, 2000; Valter *et al.*, 2003) and dopamine (Biedermann *et al.*, 1995).

Vascularized mammalian retinas contain astrocytes in addition to Müller cells. The astrocytes are confined to the vitreal surface, where their processes contact the surface blood vessels (Wolter, 1957; Schnitzer and Karschin, 1986; Robinson and Dreher, 1989, 1997; Holländer *et al.*, 1991; Triviño *et al.*, 1992). Astrocytes have a role in the formation of

Retinal Development, ed. Evelyne Sernagor, Stephen Eglen, Bill Harris and Rachel Wong.
Published by Cambridge University Press. © Cambridge University Press 2006.

blood vessels in the developing retina (Provis *et al.*, 2000), although their role in the adult retina is less certain. Müller cell processes also contact the vessels at the vitreal surface, as well as the capillaries that descend into the inner nuclear layer (Holländer *et al.*, 1991; Distler and Dreher, 1996), and it is not clear how the functions of Müller cells and astrocytes differ. Understanding the role of glia in the development of blood vessels may also provide insights into the mechanisms underlying the abnormal growth of blood vessels in several diseases of the retina, including retinopathy of prematurity and diabetic retinopathy.

In the adult retina, glial cells can best be understood in terms of their morphological and functional relationships with neurons and the vasculature. In this chapter, we have largely taken this approach in reviewing the literature on retinal glial cell development.

9.2 Retinal histogenesis

The formation of the retina proceeds through several steps, including the differentiation of progenitors into glia and several classes of neurons, cell migration and formation of the retinal layers and synaptogenesis. Müller glial cells may influence each of these steps in retinal development.

9.2.1 Origin of retinal glia

Müller cells are generated from progenitor cells within the retina (Turner, 1987), while astrocytes immigrate into the retina from the optic nerve (Watanabe and Raff, 1988). Müller cells leave the cell cycle relatively late in retinal development (Dyer and Cepko, 2000; Vetter and Moore, 2001; Rapaport *et al.*, 2004), after the generation of most classes of retinal neurons. There is no consensus regarding the definition of a 'mature' Müller cell. Müller cells are capable of re-entering the cell cycle during 'reactive gliosis' in response to retinal injury or disease (Dyer and Cepko, 2000; Bringmann and Reichenbach, 2001). Müller cells may even serve as neurogenic progenitor cells in the adult retina (Reh and Fischer, 2001; Fischer and Reh, 2003), a topic addressed in Chapter 15.

9.2.2 Radial glia and retinal histogenesis

In other regions of the central nervous system, radial glia serve as a scaffold for neuronal migration (Hatten, 1990). There is no direct evidence that Müller cells serve this function in the retina, although parallels have been drawn with events in cortex and cerebellum. (For the purpose of this discussion, the term 'Müller cell' will refer to a radial glial cell of indeterminate maturity.) Clonal analyses and observations of static relationships between columns of neurons and single Müller cells have led to speculation that Müller cells are a scaffold for migration of later-generated neurons (Brittis and Silver, 1995; Reichenbach and Robinson, 1995; Sharma and Johnson, 2000).

Damage to Müller cells in the developing retina in vivo disrupts subsequent cellular organization. Administration of the glial poison DL-α-aminoadipic acid to the neonatal mouse retina leads to local interruptions of the adherent junctions between Müller cells and photoreceptors, accompanied by aberrant migration of photoreceptors into the subretinal space (Rich *et al.*, 1995). In a transgenic mouse line, overexpression of human Bcl-2 protein (a regulator of apoptosis) in Müller cells causes early postnatal Müller cell death, followed by disorganization of the outer nuclear layer, photoreceptor apoptosis, then retinal degeneration. Interestingly, not all Müller cells overexpress Bcl-2 in this transgenic line, and the retina appears normal in regions where Müller cells are preserved (Dubois-Dauphin *et al.*, 2000).

Studies of retinas in culture provide further evidence for a role for Müller cells in the proper formation of retinal layers. In vitro, retinal lamination occurs under culture conditions that promote correctly oriented Müller cells and the formation of inner and outer limiting membranes (Willbold *et al.*, 2000), while retinal lamination is lost when Müller cells are damaged by exposure to glial poisons (Germer *et al.*, 1997b).

Müller cells may also limit the lateral migration of neurons in the developing retina. When embryonic chick and quail retinas are dissociated into cell suspensions and allowed to re-aggregate, the cells form distinct columns composed of groups of chick or quail neurons and Müller cells. When Müller cells are damaged with glial poisons, neurons migrate out of their columns (Willbold *et al.*, 1995).

9.2.3 Effects of glia on neuronal process outgrowth

Glia may also influence the polarity and/or outgrowth of processes from retinal neurons. Substrates derived from the membranes of Müller cell endfeet are permissive for axon outgrowth from cultured retinal ganglion cells, while substrates derived from the membranes of Müller cell somata inhibit axon outgrowth (Bauch *et al.*, 1998; Stier and Schlosshauer, 1998, 1999) but permit the formation of dendrites (Bauch *et al.*, 1998). These results suggest that Müller cells have distinct domains with regard to cell surface molecules that mediate cell–cell interactions.

9.2.4 Molecules that mediate glial–neuronal interactions during retinal histogenesis

The pathways through which Müller cells influence retinal histogenesis are not yet known. However, Müller cells in the developing retina express cell-recognition and adhesion molecules that mediate glial–neuronal interactions in other parts of the developing nervous system, including $\beta 1$ integrins (Cann *et al.*, 1996; Li and Sakaguchi, 2002); glycosaminoglycans and their binding partners (Threlkeld *et al.*, 1989; Chaitin *et al.*, 1996; Normand *et al.*, 1998; Aricescu *et al.*, 2002); and L1/neural-glial cell adhesion molecule (NgCAM), 5A11, F11, embryonic avian polypeptide and claustrin (reviewed in Sharma and Johnson, 2000).

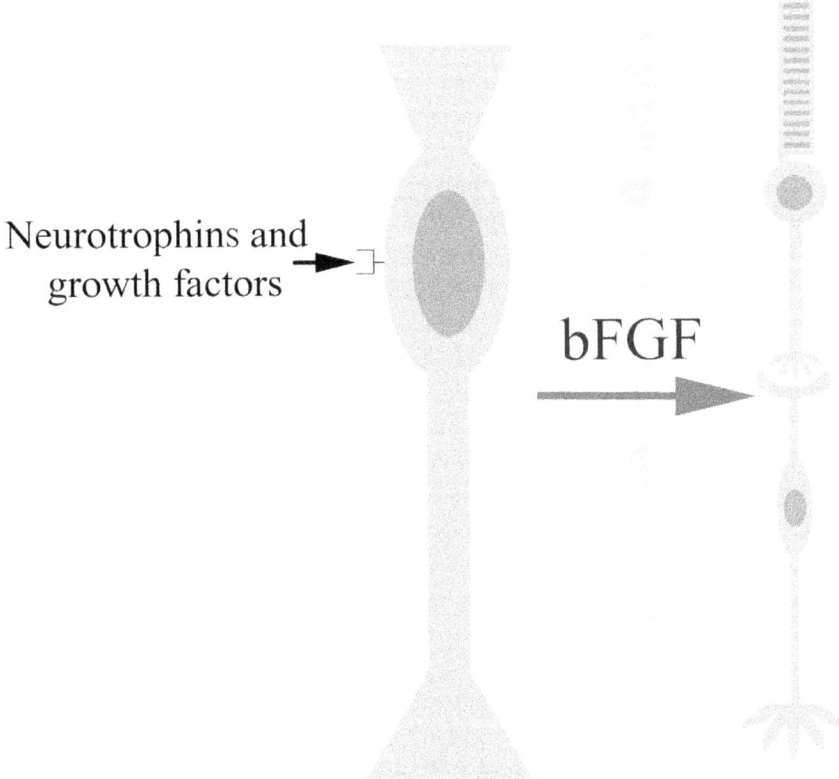

Figure 9.1 Hypothetical rescue of neurons by Müller cells in the degenerating retina. In this example, activation of Trk or growth factor receptors on a Müller cell leads to release of neuroprotective basic fibroblast growth factor (bFGF) (Wexler et al., 1998; Harada et al., 2003).

Müller cells also produce trophic factors that are known to promote neuronal survival and influence differentiation (Figure 9.1). Cultured Müller cells synthesize the neurotrophins nerve growth factor (NGF), brain-derived neurotrophic factor (BDNF,) neurotrophin (NT)-3 and NT-4 (Oku et al., 2002; Taylor et al., 2003), but evidence for glial synthesis of these molecules in the developing retina in vivo is lacking. Müller cells in vivo have been shown to produce ciliary neurotrophic factor (CNTF) in developing (Kirsch et al., 1997) as well as damaged retinas (Walsh et al., 2001). Ciliary neurotrophic factor has cell- and species-specific effects in the developing retina (Kirsch et al., 1997; Jo et al., 1999), where this cytokine appears to act directly on neurons and their precursors (Rhee and Yang, 2003).

Müller cells may also support rod function through the trafficking of docosahexaenoic acid (DHA), the major fatty acid in mammalian rod outer segments (Giusto et al., 2000). Dietary deficiency in this essential fatty acid leads to visual impairments (reviewed in Jeffrey et al., 2001), while dietary supplementation protects against photoreceptor apoptosis in rodent models of retinal degeneration (Moriguchi et al., 2004) and may be beneficial for retinal function in patients with X-linked retinitis pigmentosa (Hoffman et al., 2004). In vitro, DHA protects against photoreceptor apoptosis (Rotstein et al., 2003) and may promote the exit from the cell cycle of cultured neuroblasts (Fernanda Insua et al., 2003). The mechanisms through which DHA exerts its neuroprotective effects are not yet known. In cocultures of rat photoreceptors and Müller cells, DHA appears to be taken up by the Müller cells, incorporated into phospholipids and channelled to photoreceptors (Politi et al., 2001). In the frog retina, retinal pigment epithelium (RPE) cells take up DHA from the circulation and release it into extracellular space adjacent to photoreceptors (Bazan et al., 1992). However, it is not yet known whether the DHA first passes through Müller cells before subsequent uptake into photoreceptors in vivo.

Müller cells are also targets of growth factors and cytokines present in the developing and injured retina. Müller cells possess the high-affinity neurotrophin receptors TrkB and TrkC (Rohrer et al., 1999; Harada et al., 2000), the low-affinity neurotrophin receptor p75 (Ding et al., 2001) and elements of the signalling pathways activated by these receptors (Wahlin et al., 2000). Glial cells commonly react to central nervous system injury or disease by up-regulating their expression of the intermediate-filament protein glial fibrillary acidic protein (GFAP). Brain-derived neurotrophic factor attenuates this response in Müller cells in vivo (Lewis et al., 1999) and in organotypic culture (Pinzon-Duarte et al., 2004). Pigment epithelial-derived growth factor is also protective towards Müller cells in organotypic culture (Jablonski et al., 2001).

Trophic factors produced by Müller cells may prove to be critical for the rescue of neurons in diseased or damaged retinas. In particular, basic fibroblast growth factor (bFGF) is neuroprotective in vitro and in several models of retinal degeneration in vivo, and other trophic factors may exert their neuroprotective effects in retina by stimulating Müller cells to produce bFGF (Zack, 2000). In vitro, BDNF stimulates the production of bFGF by Müller cells, which, in turn, promotes the survival of rod bipolar cells (Wexler et al., 1998). In vivo, constant exposure of rodent retinas to bright light leads to the death of photoreceptors. Injection into the vitreous of either BDNF (LaVail et al., 1992; Ikeda et al., 2003) or NT-3 (Harada et al., 2000) protects rat retinas from such light-induced photoreceptor loss. Neurotrophin-3 activates TrkC receptors on Müller cell processes adjacent to the photoreceptors, stimulating the Müller cells to produce bFGF, which is neuroprotective towards photoreceptors (Harada et al., 2000). Similar to the neurotrophins, glial-derived neurotrophic factor (GDNF) may exert its neuroprotective effects by stimulating Müller cells to produce bFGF (Frasson et al., 1999; McGee Sanftner et al., 2001; Harada et al., 2003; Jomary et al., 2004).

Ciliary neurotrophic factor is also neuroprotective in light-damaged retinas (LaVail et al., 1992). Several lines of evidence suggest that Müller cells mediate this effect. Components of the CNTF receptor complex are found in Müller cells (Valter et al., 2003), and Müller cells respond to intraocular injections of CNTF with activation of Janus-activated kinase–signal transducer and activator of transcription (JAK–STAT) and mitogen-activated protein kinase (MAPK) signalling pathways, extracellular signal-regulated kinase (ERK) phosphorylation and transcription of the immediate early gene c-fos (Peterson et al., 2000; Wahlin et al.,

2000). Interestingly, these pathways in retinal glia are also activated by exposure to non-damaging bright light and penetrating retinal injury, stimuli that protect photoreceptors from light-induced damage (Peterson *et al.*, 2000).

Müller cells can also produce factors toxic to neurons, and these factors may contribute to the progression of certain retinal degenerative diseases. Evidence suggests that the inherited retinal dystrophy in Royal College of Surgeons rats results from abnormal reactivity of Müller cells to stimuli that lead to production of factors toxic to photoreceptors (de Kozak *et al.*, 1997).

9.3 Glutamatergic neurotransmission

Glial cells are important regulators of extracellular glutamate, the major excitatory neurotransmitter in the central nervous system. In the retina, glutamate is released at the photoreceptor-to-bipolar cell and bipolar cell-to-retinal ganglion cell synapses. Rapid uptake of synaptically released glutamate is necessary for terminating synaptic responses (Higgs and Lukasiewicz, 1999). In the adult retina, Müller cells take up synaptically released glutamate and convert it to glutamine, which is then released back to the neurons for recycling into glutamate (Pow and Robinson, 1994). High levels of extracellular glutamate are present early in retinal development, when glutamate might act globally as a regulatory signal rather than as a localized neurotransmitter (Haberecht *et al.*, 1997) (see Chapter 6). Extracellular glutamate levels decline dramatically (<1 μM) during the peak of synaptogenesis (Haberecht and Redburn, 1996; Haberecht *et al.*, 1997), after Müller cells have acquired glutamate transport functions.

9.3.1 Expression of glutamate transporters by Müller cells

Glutamate aspartate transporter (GLAST also known as EAAT-1, excitatory amino acid transporter-1) is the major glutamate transporter in retinal glia (Izumi *et al.*, 2002). In the rat retina, GLAST is present on the day of birth and its amounts increase rapidly at the beginning of the second postnatal week. This is just after the time of eye opening, when Müller cell processes have surrounded the developing synapses in the inner plexiform layer (Pow and Barnett, 1999). Glutamatergic terminals in the inner plexiform layer become capable of vesicular release at this time (see Chapter 13), inviting speculation that synaptically released glutamate regulates GLAST expression by glia. Indeed, exogenous glutamate increases GLAST expression in bullfrog Müller cells (Xu *et al.*, 2004) and cultured rat Müller cells (Taylor *et al.*, 2003).

9.3.2 Development of glutamine synthetase expression in Müller cells

The expression of glutamine synthetase (GS), which catalyzes the conversion of glutamate to glutamine, is a characteristic feature of Müller cells in the adult vertebrate retina

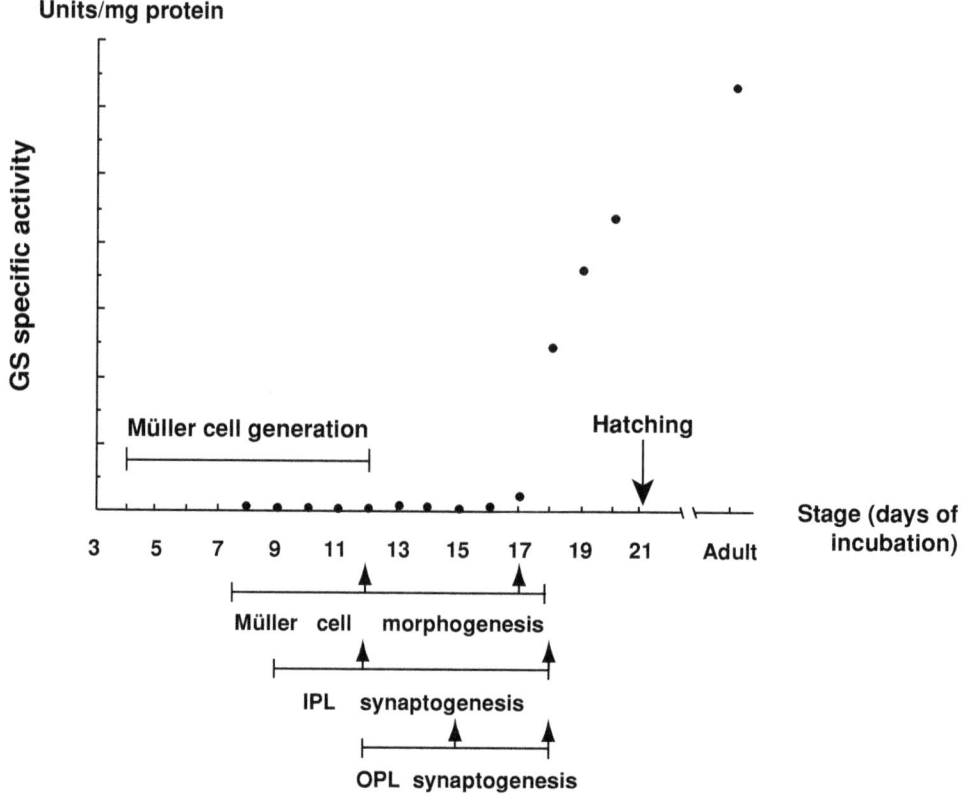

Figure 9.2 Glutamine synthetase (GS) activity in retinal glia increases following the peak of synaptogenesis in outer plexiform layer (OPL) of developing chick retina. IPL, inner plexiform layer. (Prada et al., 1998. Used with permission of the publisher.)

(Figure 9.2). In rodent and rabbit retinas, GS levels rapidly increase during the early postnatal period, when synaptogenesis takes place (Chader, 1971; LaVail and Reif-Lehrer, 1971; Riepe, 1978; Germer et al., 1997a). In the chick retina, GS activity increases 40-fold from embryonic day (E)16 to E20, following the peak of synaptogenesis in the outer plexiform layer (Prada et al., 1998).

Indirect evidence suggests that interactions with neurons are required for the induction of GS expression in Müller cells. In the retina of a cichlid fish, which grows continuously, the amount of GS per Müller cell increases in parallel with the increase in number of photoreceptors (Mack et al., 1998). Müller cells in vitro maintain their expression of GS only if they maintain contact with neurons (Morris and Moscona, 1970; Linser and Moscona, 1979; Moscona and Linser, 1983; Vardimon et al., 1988; Prada et al., 1998). In explant cultures of neonatal rabbit retinas, GS levels are increased by exposure to conditioned medium from cultured RPE cells, to glutamate or to ammonia (Germer et al., 1997a).

9.3.3 Other modulators of glutamatergic transmission in retina

Kynurenic acid is an endogenous antagonist of NMDA and non-NMDA glutamate receptors (Perkins and Stone, 1982; Scharfman *et al.*, 1999). Kynurenic acid and its synthetic enzymes, kynurenine aminotransferases (KAT) I and II, are present in Müller cells in the rat retina (Rejdak *et al.*, 2001, 2004). While KAT-immunoreactivity is restricted to the Müller cell endfeet in the adult retina, KAT I-immunoreactivity is present in the inner plexiform layer during the first two postnatal weeks (Rejdak *et al.*, 2004), when synaptogenesis occurs in the rat retina (Grunder *et al.*, 2000). Chick Müller cells also contain KAT I, and enzyme activity peaks at the time of synaptogenesis in the OPL (Rejdak *et al.*, 2003).

Among the several types of glutamate receptors found in the retina is the NMDA receptor, which is found on retinal ganglion cells and amacrine cells (Lukasiewicz and McReynolds, 1985; Zhou and Dacheux, 2004). Glycine acts as an obligatory co-agonist at NMDA receptors; therefore modulation of glycine levels is a potential mechanism for glial regulation of NMDA receptor functions in development. In Müller cells cultured from embryonic chick retina, ATP-stimulated Ca^{2+} waves increase glycine uptake (Gadea *et al.*, 2002). D-Serine, which also binds to the 'glycine site' of NMDA receptors, is synthesized by retinal glia and modulates NMDA receptor-mediated synaptic transmission in larval-phase amphibian and adult rodent retinas (Stevens *et al.*, 2003). The developmental regulation of these systems has yet to be studied in the retina.

9.4 K^+ spatial buffering

Homeostatic regulation of extracellular K^+ levels is essential to maintain neuronal and glial membrane potential, and uptake of K^+ released by active neurons is a major function of glial cells. Buffering of K^+ in the retina occurs by the mechanism of 'K^+ siphoning' (Newman *et al.*, 1984; Brew *et al.*, 1986), where K^+ released in the plexiform layers enters Müller cells and is then ejected into the vitreous or perivascular spaces, which serve as K^+ 'sinks'. To accomplish K^+ buffering, glial cells express different types of K^+ channels that vary in their biophysical properties and subcellular distribution. In the adult mouse retina, Kir4.1 channels are located on the Müller cell endfeet and perivascular processes, where K^+ leaves the cell, while Kir2.1 channels are located on Müller cell processes that contact neurons, where K^+ enters the cell (Kofuji *et al.*, 2002).

9.4.1 Developmental changes in the expression of Müller cell K^+ channels

Immature Müller cells have a more depolarized membrane potential than do mature Müller cells because they lack inwardly rectifying K^+ channels (Bringmann *et al.*, 1999a, 2000a; Felmy *et al.*, 2001; Pannicke *et al.*, 2002; Schopf *et al.*, 2004). The hyperpolarized membrane potentials characteristic of mature glial cells appear in parallel with the developmental increase in inwardly rectifying K^+ channels (Felmy *et al.*, 2001; Schopf *et al.*, 2004). The increase in inwardly rectifying K^+ channels in rabbit Müller cells is coincident with the

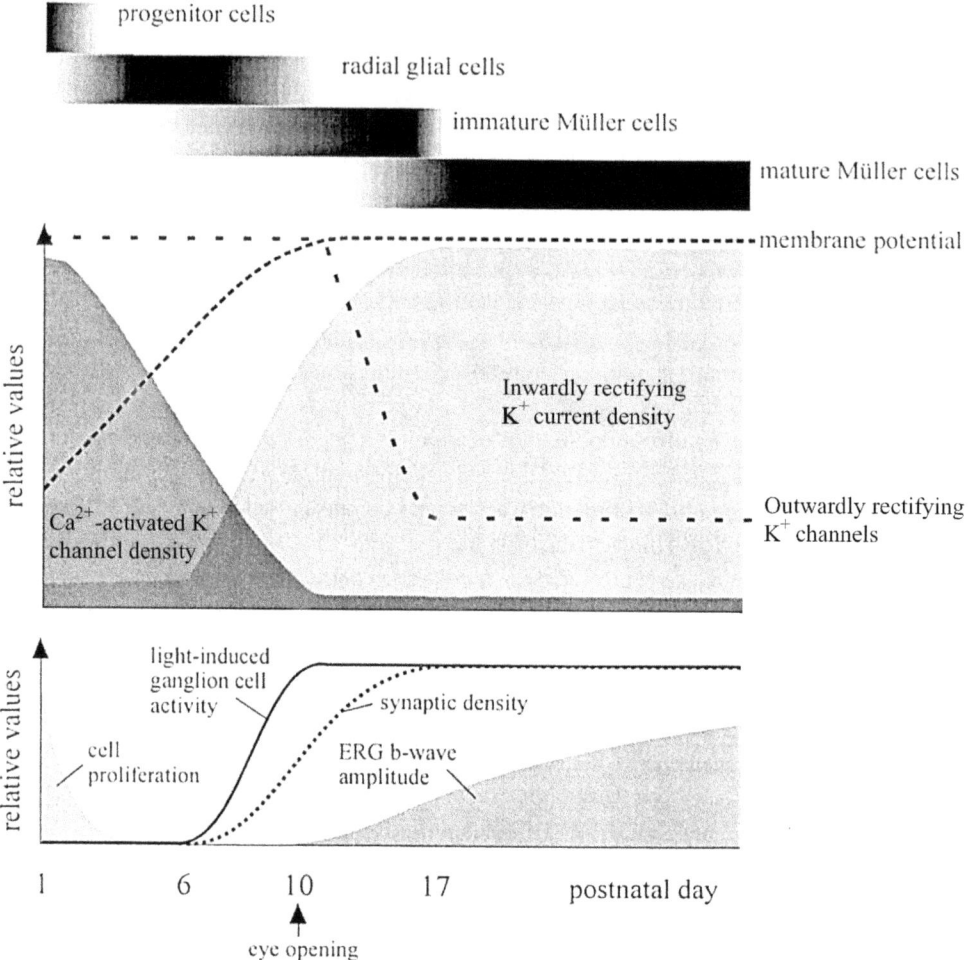

Figure 9.3 The expression of different types of K^+ channels in Müller cells changes during retinal development. Before the appearance of light responses in ganglion cells, Müller cells predominantly express outwardly rectifying and Ca^{2+}-activated K^+ channels. These are largely replaced by inwardly rectifying K^+ channels as light responsiveness and neuronal activity increase. (Adapted from Bringmann et al., 2000b. Used with permission of the publisher.)

onset of light-induced retinal ganglion cell activity (Bringmann et al., 2000b; Figure 9.3). Although there is as yet no direct evidence that neuronal activity up-regulates the expression of inwardly rectifying K^+ channels in retinal glia, neurons have been shown to up-regulate the expression of these channels in glia from other regions of the central nervous system (Barres et al., 1990).

Efficient K^+ buffering probably requires the presence of glial Kir channels (Bringmann et al., 2000a). In the immature retina, before the increased expression of glial Kir channels, retinal ganglion cell activity might result in relatively large accumulations of extracellular

K$^+$, as has been shown in the developing optic nerve (Connors *et al.*, 1982). Since elevated levels of extracellular K$^+$ promote the synchronization of neuronal activity, inefficient K$^+$ buffering may contribute to the propagation of spontaneous waves of neuronal activity in the developing retina (Burgi and Grzywacz, 1994a,b; see also Chapter 13).

The relatively depolarized membrane potential of immature Müller cells allows the opening of low-voltage-activated Ca^{2+} channels, which may have a role in the proliferation and/or differentiation of radial glial/Müller cells (Bringmann *et al.*, 2000b). The appearance of inwardly rectifying K$^+$ channels in maturing Müller cells, and the resulting hyperpolarization of the glial membrane potential, prevents low-voltage-activated Ca^{2+} channels from opening (Bringmann *et al.*, 2000b).

9.4.2 K$^+$ channels in retinal pathology

In a variety of human retinal diseases, including retinal detachment, secondary glaucoma, melanoma and perforating injury, Müller cells lose inwardly rectifying K$^+$ channels. A dramatic down-regulation of inwardly rectifying K$^+$ channels occurs in pathological states where Müller cells proliferate, compared with states where Müller cells do not proliferate (Bringmann *et al.*, 2000a). This loss of inwardly rectifying K$^+$ channels is accompanied by Müller cell depolarization and increased membrane resistance. In Müller cells from patients with proliferative vitreoretinopathy, the decrease in inwardly rectifying K$^+$ channels has been shown to lead to an increase in large-conductance Ca^{2+}-activated K$^+$ currents (Bringmann *et al.*, 1999b). It has been hypothesized that activation of voltage-gated Ca^{2+} channels, after down-regulation of inwardly rectifying K$^+$ channels, contributes to Müller cell proliferation (Bringmann *et al.*, 2000a).

9.5 Neurotransmitter receptors on developing Müller cells

Differentiated Müller glia express receptors for a variety of retinal transmitters and modulators, but few studies have directly investigated the distribution or functions of these receptors during development. Gene expression analysis has been used to identify mRNA transcripts in single adult Müller cells (Wahlin *et al.*, 2004), and will potentially prove a powerful tool in identifying receptors expressed by developing retinal glia.

9.5.1 Glutamate

Differentiated Müller cells of several species express metabotropic glutamate receptors (Keirstead and Miller, 1997; Lopez *et al.*, 1998). Of the ionotropic glutamate receptors, only NMDA receptors have been reported in Müller cells. In cultured human Müller cells, NMDA receptor activation induces proliferation (Uchihori and Puro, 1993), probably via NMDA-R1 subunits (Puro *et al.*, 1996). High extracellular glutamate ($\geq$100 µM), which appears to be present early in retinal development, up-regulates BDNF, NGF, NT-3, NT-4,

GDNF and GLAST expression in cultured rat Müller cells (Taylor et al., 2003). Whether such regulation occurs in vivo is still an open question.

9.5.2 Adenosine triphosphate

Purinergic receptors, particularly P2Y receptors, have been identified in Müller cells in many species, including rabbit (Liu and Wakakura, 1998), rat (Jabs et al., 2000; Li et al., 2001), human (Bringmann et al., 2001), larval tiger salamander (Reifel Saltzberg et al., 2003) and chicken (Sanches et al., 2002; see also Supplementary movie 9.1 online). Müller cells from human patients with proliferative vitreoretinopathy express an increased density of P2X7-like and P2Y ATP receptors, which may allow increased Ca^{2+} entry, accelerated DNA synthesis and Müller cell proliferation (Bringmann et al., 2001, 2002). Adenosine triphosphate acting through P2Y1 receptors also increases ^{3}H-thymidine uptake in chicken retina at E6 to E8, when the Müller cell proliferation rate is at its highest (Sanches et al., 2002).

9.5.3 Catecholamines

D2 dopamine receptors have been found on Müller cells by electrophysiological (Biedermann et al., 1995) and ligand-binding methods (Muresan and Besharse, 1993). The $\alpha 2$ receptor agonists xylazine and clonidine induce bFGF expression in rat retina, apparently by stimulating ERK phosphorylation in Müller cells; however, it is not known whether these agonists act directly on Müller cells (Peng et al., 1998).

9.5.4 γ-Aminobutyric acid and glycine

γ-Aminobutyric acid A ($GABA_A$) receptors are found in Müller cells from skate (Malchow et al., 1989), salamander (Zhang et al., 2003), baboon (Reichelt et al., 1997) and human (Biedermann et al., 2004). The latter study suggests that the relatively large glial depolarization evoked by $GABA_A$ receptor activation might be sufficient to open Ca^{2+} channels in human Müller cells. γ-Aminobutyric acid B receptors have been found in bullfrog Müller cells (Zhang and Yang, 1999). Müller cells also possess glycine receptors (Cunningham and Miller, 1980), possibly localized to inner plexiform layer-spanning regions (Du et al., 2002).

9.6 Glia in retinal vascular development

The pattern of blood vessels serving the retina varies between species. In all vertebrates, the outer retina is nourished by the vessels of the choroid plexus, located behind the pigment epithelium. In very thin retinas (<140 μm), such as those of amphibians and fish, oxygen and nutrients can diffuse through the entire depth of the retina, and the choroidal circulation is the only blood supply to the retina. In thicker retinas, there is an additional blood supply for the inner retina, composed of a layer of vessels near the vitreal surface and a layer of vessels

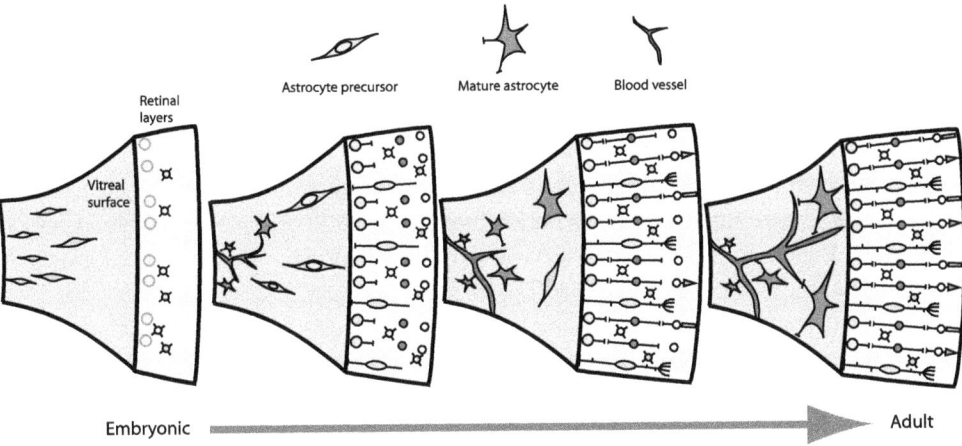

Figure 9.4 Retinal astrocytes and blood vessels influence each other during development. Spindle-shaped astrocyte precursors stimulate the formation of blood vessels on the retinal surface. As blood vessels develop, these precursors mature into adult-appearing astrocytes, which, in turn, induce formation of the blood–retinal barrier.

bracketing and traversing the inner nuclear layer (Michaelson, 1954). As in other regions of the central nervous system, the processes of glial cells surround the retinal vasculature. Both astrocyte and Müller cell processes wrap the surface vasculature, while Müller cell processes wrap the vessels of the inner nuclear layer (Wolter, 1957; Holländer et al., 1991; Distler and Dreher, 1996).

The retinal vasculature and its investment of astrocytes develop in parallel (Figure 9.4). The vascular network arises from elaborate interactions among distinct populations of glial, neuronal and endothelial precursors (Sandercoe et al., 1999). Early in development, a ring of spindle-shaped cells precedes the leading edge of the growing vascular network (Fruttiger, 2002). The nature of these pioneers has been controversial, but examination of cell-specific markers indicates that they are astrocyte precursors. While the developing vascular network labels for VEGF (vascular endothelial growth factor) receptor mRNA, the spindle-cells express the astrocyte proteins GFAP, vimentin and PAX2, and contain mRNA for astrocytic PDGF receptor-α (Fruttiger, 2002). Glial fibrillary acid protein-positive precursors appear to migrate from the optic nerve head, and lead the developing vasculature by 100 to 200 microns (Watanabe and Raff, 1988). However, this apparent migration may actually represent a wave of astrocyte maturation, as GFAP-negative astrocyte precursors cover the retina before the vasculature develops and neighbouring endothelial cells appear to induce precursors to mature into GFAP-positive astrocytes, perhaps by secreting leukaemia inhibitory factor (Mi et al., 2001; Provis, 2001).

In response to increasing neuronal activity, retinal glia generate signals that regulate the differentiation and proliferation of the vasculature. A prominent hypothesis is that, in mammals, neuronal activity induces local hypoxia in non-vascularized areas, which stimulates astrocytes to produce VEGF. Indeed, VEGF mRNA in astrocytes increases dramatically with hypoxia (Provis, 2001). Since maturation of the vasculature reverses this process by bringing oxygen to the tissue, VEGF production is self-limiting, and astrocytic guidance of blood vessels appears to proceed as a delimited wave. Deep-layer hypoxia may also stimulate VEGF production in Müller cells, inducing growth of the deep vasculature (Provis *et al.*, 1997). Glia may also guide vascular development by adding proteins to the extracellular matrix. For example, astrocytes may express fibronectin to guide retinal blood vessels (Risau and Lemmon, 1988), while astrocyte-derived cadherin limits blood vessel growth to areas containing glia (Dorrell *et al.*, 2002).

Astrocytes may induce barrier properties in the retinal vasculature. Cultured astrocytes injected into the rat eye form aggregates that become vascularized and possess an intact blood–tissue barrier. In contrast, injected meningeal cells also aggregate and become vascularized, but do not possess a blood–tissue barrier (Janzer and Raff, 1987). Glial-conditioned medium induces barrier properties in cultured endothelial cells (Raub *et al.*, 1992; Gardner *et al.*, 1997). In rat, loss of astrocytes after hypoxia is correlated with a loss of barrier properties (Provis, 2001).

9.7 Concluding remarks

As this brief review of the literature has shown, interactions among glia, neurons and the vasculature, which are well documented in the adult retina, also play a significant role during retinal development. The story is incomplete, however, and many questions remain to be addressed: do radial glia in the retina guide neuronal migration? What is the definition of a mature Müller cell? How does neuronal activity influence glial maturation, including expression and clustering of K^+ channels, expression of transporters and expression of receptors? Do retinal glia have roles in synaptogenesis? Do glial responses to retinal injury recapitulate events that occur during normal development (e.g. glial proliferation, glial trophic support of retinal neurons)? How do glial functions and properties change during aging?

References

Aricescu, A. R., McKinnell, I. W., Halfter, W. and Stoker, A. W. (2002). Heparan sulfate proteoglycans are ligands for receptor protein tyrosine phosphatase omega. *Mol. Cell. Biol.*, **22**, 1881–92.

Barres, B. A., Koroshetz, W. J., Chun, L. L. Y. and Corey, D. P. (1990). Ion channel expression by white matter glia: the type-1 astrocyte. *Neuron*, **5**, 527–44.

Bauch, H., Stier, H. and Schlosshauer, B. (1998). Axonal versus dendritic outgrowth is differentially affected by radial glia in discrete layers of the retina. *J. Neurosci.*, **18**, 1774–85.

Bazan, N. G., Gordon, W. C. and Rodriguez, de Turco E. B. (1992). Docosahexaenoic acid and metabolism in photoreceptors: retinal conservation by an efficient retinal pigment epithelial cell-mediated recycling process. *Adv. Exp. Med. Biol.*, **318**, 295–306.

Biedermann, B., Frohlich, E., Grosche, J., Wagner, H. J. and Reichenbach, A. (1995). Mammalian Müller (glial) cells express functional D2 dopamine receptors. *NeuroReport*, **6**, 609–12.

Biedermann, B., Bringmann, A. and Franze, K. *et al.* (2004). GABA(A) receptors in Müller glial cells of the human retina. *Glia*, **46**, 302–10.

Brew, H., Gray, P. T. A., Mobbs, P. and Attwell, D. (1986). Endfeet of retinal glial cells have higher densities of ion channels that mediate K^+ buffering. *Nature*, **324**, 466–8.

Bringmann, A. and Reichenbach, A. (2001). Role of Müller cells in retinal degenerations. *Front. Biosci.*, **6**, E72–92.

Bringmann, A., Biedermann, B. and Reichenbach, A. (1999a). Expression of potassium channels during postnatal differentiation of rabbit Müller glial cells. *Eur. J. Neurosci.*, **11**, 2883–96.

Bringmann, A., Francke, M. and Pannicke, T. *et al.* (1999b). Human Müller glial cells: altered potassium channel activity in proliferative vitreoretinopathy. *Invest. Ophthalmol. Vis. Sci.*, **40**, 3316–23.

Bringmann, A., Francke, M. and Pannicke, T. *et al.* (2000a). Role of glial $K^{(+)}$ channels in ontogeny and gliosis: a hypothesis based upon studies on Müller cells. *Glia*, **29**, 35–44.

Bringmann, A., Schopf, S. and Reichenbach, A. (2000b). Developmental regulation of calcium channel-mediated currents in retinal glial (Müller) cells. *J. Neurophysiol*, **84**, 2975–83.

Bringmann, A., Pannicke, T., Moll, V. (2001). Upregulation of P2X(7) receptor currents in Müller glial cells during proliferative vitreoretinopathy. *Invest. Ophthalmol. Vis. Sci.*, **42**, 860–7.

Bringmann, A., Pannicke, T., Weick, M. *et al.* (2002). Activation of P2Y receptors stimulates potassium and cation currents in acutely isolated human Müller (glial) cells. *Glia*, **37**, 139–52.

Brittis, P. A. and Silver, J. (1995). Multiple factors govern intra-retinal axon guidance: a time-lapse study. *Mol. Cell. Neurosci.*, **6**, 413–32.

Burgi, P.-Y. and Grzywacz, N. M. (1994a). Model based on extracellular potassium for spontaneous synchronous activity in developing retinae. *Neural Comput.*, **6**, 983–1004.

Burgi, P.-Y. and Grzywacz, N. M. (1994b). Model for the pharmacological basis of spontaneous synchronous activity in developing retinae. *J. Neurosci.*, **14**, 7426–39.

Cann, G. M., Bradshaw, A. D., Gervin, D. B., Hunter, A. W. and Clegg, D. O. (1996). Widespread expression of beta1 integrins in the developing chick retina: evidence for a role in migration of retinal ganglion cells. *Dev. Biol.*, **180**, 82–96.

Chader, G. J. (1971). Hormonal effects on the neural retina. I. Glutamine synthetase development in the retina and liver of the normal and triiodothyronine-treated rat. *Arch. Biochem. Biophys.*, **144**, 657–62.

Chaitin, M. H., Ankrum, M. T. and Wortham, H. S. (1996). Distribution of CD44 in the retina during development and the rds degeneration. *Brain Res. Mol. Brain Res.*, **94**, 92–8.

Connors, B. W., Ransom, B. R., Kunis, D. M. and Gutnick, M. J. (1982). Activity-dependent K^+ accumulation in the developing rat optic nerve. *Science*, **216**, 1341–3.

Cunningham, R. and Miller, R. F. (1980). Electrophysiological analysis of taurine and glycine action on neurons of the midpuppy retina. I. Intracellular recording. *Brain Res.*, **197**, 123–38.

de Kozak, Y., Cotinet, A., Goureau, O., Hicks, D. and Thillaye-Goldenberg, B. (1997). Tumor necrosis factor and nitric oxide production by resident retinal glial cells from rats presenting hereditary retinal degeneration. *Ocul. Immunol. Inflamm.*, **5**, 85–94.

Ding, J., Hu, B., Tang, L. S. and Yip, H. K. (2001). Study of the role of the low-affinity neurotrophin receptor p75 in naturally occurring cell death during development of the rat retina. *Dev. Neurosci.*, **23**, 390–8.

Distler, C. and Dreher, Z. (1996). Glia cells of the monkey retina-II. Müller cells. *Vis. Res.*, **36**, 2381–94.

Dorrell, M. I., Aguilar, E. and Friedlander, M. (2002). Retinal vascular development is mediated by endothelial filopodia, a preexisting astrocytic template and specific R-cadherin adhesion. *Invest. Ophthalmol. Vis. Sci.*, **43**, 3500–10.

Du, J. L., Xu, L. Y. and Yang, X. L. (2002). Glycine receptors and transporters on bullfrog retinal Müller cells. *NeuroReport*, **13**, 1653–6.

Dubois-Dauphin, M., Poitry-Yamate, C., De Bilbao, F. *et al.* (2000). Early postnatal Müller cell death leads to retinal but not optic nerve degeneration in NSE-HU-BCL-2 transgenic mice. *Neuroscience*, **95**, 9–21.

Dyer, M. A. and Cepko, C. L. (2000). Control of Müller glial cell proliferation and activation following retinal injury. *Nat. Neurosci*, **3**, 873–80.

Felmy, F., Pannicke, T., Richt, J. A., Reichenbach, A. and Guenther, E. (2001). Electrophysiological properties of rat retinal Müller (glial) cells in postnatally developing and in pathologically altered retinae. *Glia*, **34**, 190–9.

Fernanda Insua, M., Garelli, A., Rotstein, N. P. *et al.* (2003). Cell cycle regulation in retinal progenitors by glia-derived neurotrophic factor and docosahexaenoic acid. *Invest. Opthalmol. Vis. Sci.*, **44**, 2235–44.

Fischer, A. J. and Reh, T. A. (2003). Potential of Müller glia to become neurogenic retinal progenitor cells. *Glia*, **43**, 70–6.

Frasson, M., Picaud, S., Leveillard, T. *et al.* (1999). Glial cell line-derived neurotrophic factor induces histologic and functional protection of rod photoreceptors in the rd/rd mouse. *Invest. Ophthalmol. Vis. Sci.*, **40**, 2724–34.

Fruttiger, M. (2002). Development of the mouse retinal vasculature: angiogenesis versus vasculogenesis. *Invest. Ophthalmol. Vis. Sci.*, **43**, 522–7.

Gadea, A., Lopez, E., Hernandez-Cruz, A. and Lopez-Colome, A. M. (2002). Role of Ca^{2+} and calmodulin-dependent enzymes in the regulation of glycine transport in Müller glia. *J. Neurochem.*, **80**, 634–45.

Gardner, T. W., Lieth, E., Khin, S. A. *et al.* (1997). Astrocytes increase barrier properties and ZO-1 expression in retinal vascular endothelial cells. *Invest. Ophthalmol. Vis. Sci.*, **38**, 2423–27.

Germer, A., Jahnke, C., Mack, A., Enzmann, V. and Reichenbach, A. (1997a). Modification of glutamine synthetase expression by mammalian Müller (glial) cells in retinal organ cultures. *NeuroReport*, **8**, 3067–72.

Germer, A., Kuhnel, K., Grosche, J. *et al.* (1997b). Development of the neonatal rabbit retina in organ culture. *Anat. Embryol. (Berl.)*, **196**, 67–79.

Giusto, N. M., Pasquare, S. J., Salvador, G. A. *et al.* (2000). Lipid metabolism in vertebrate retinal rod outer segments. *Prog. Lipid Res.*, **39**, 315–91.

Grunder, T., Kohler, K. and Guenther, E. (2000). Distribution and developmental regulation of AMPA receptor subunit proteins in rat retina. *Invest. Ophthalmol. Vis. Sci.*, **41**, 3600–6.

Haberecht, M. F. and Redburn, D. A. (1996). High levels of extracellular glutamate are present in retina during neonatal development. *Neurochem. Res.*, **21**, 285–91.

Haberecht, M. F., Mitchell, C. K., Lo, G. J. and Redburn, D. A. (1997). N-methyl-D-aspartate-mediated glutamate toxicity in the developing rabbit retina. *J. Neurosci. Res.*, **47**, 416–26.

Harada, C., Harada, T., Quah, H. M. *et al.* (2003). Potential role of glial cell line-derived neurotrophic factor receptors in Müller glial cells during light-induced retinal degeneration. *Neuroscience*, **122**, 229–35.

Harada, T., Harada, C., Nakayama, N. *et al.* (2000). Modification of glial-neuronal cell interactions prevents photoreceptor apoptosis during light-induced retinal degeneration. *Neuron*, **26**, 533–41.

Hatten, M. E. (1990). Riding the glial monorail: a common mechanism for glial-guided neuronal migration in different regions of the developing mammalian brain. *Trends Neurosci.*, **13**, 179–84.

Higgs, M. H. and Lukasiewicz, P. D. (1999). Glutamate uptake limits synaptic excitation of retinal ganglion cells. *J. Neurosci.*, **19**, 3691–700.

Hoffman, D. R., Locke, K. G., Wheaton, D. H. *et al.* (2004). A randomized, placebo-controlled clinical trial of docosahexaenoic acid supplementation for X-linked retinitis pigmentosa. *Am. J. Ophthalmol.*, **137**, 704–18.

Holländer, H., Makarov, F., Dreher, Z. *et al.* (1991). Structure of the macroglia of the retina: sharing and division of labour between astrocytes and Müller cells. *J. Comp. Neurol.*, **313**, 587–603.

Ikeda, K., Tanihara, H., Tatsuno, T., Noguchi, H. and Nakayama, C. (2003). Brain-derived neurotrophic factor shows a protective effect and improves recovery of the ERG b-wave response in light-damage. *J. Neurochem.*, **87**, 290–6.

Izumi, Y., Shimamoto, K., Benz, A. M. *et al.* (2002). Glutamate transporters and retinal excitotoxicity. *Glia*, **39**, 58–68.

Jablonski, M. M., Tombran-Tink, J., Mrazek, D. A. and Iannaccone, A. (2001). Pigment epithelium-derived factor supports normal Müller cell development and glutamine synthetase expression after removal of the retinal pigment epithelium. *Glia*, **35**, 14–25.

Jabs, R., Guenther, E., Marquordt, K. and Wheeler-Schilling, T. H. (2000). Evidence for P2X(3), P2X(4), P2X(5) but not for P2X(7) containing purinergic receptors in Müller cells of the rat retina. *Brain Res. Mol. Brain Res.*, **76**, 205–10.

Janzer, R. C. and Raff, M. C. (1987). Astrocytes induce blood-brain barrier properties in endothelial cells. *Nature*, **325**, 253–7.

Jeffrey, B. G., Weisinger, H. S., Neuringer, M. and Mitchell, D. C. (2001). The role of docosahexaenoic acid in retinal function. *Lipids*, **36**, 859–71.

Jo, S. A., Wang, E. and Benowitz, L. I. (1999). Ciliary neurotrophic factor is an axogenesis factor for retinal ganglion cells. *Neuroscience*, **89**, 579–91.

Jomary, C., Darrow, R. M., Wong, P., Organisciak, D. T. and Jones, S. E. (2004). Expression of neurturin, glial cell line-derived neurotrophic factor, and their receptor components in light-induced retinal degeneration. *Invest. Ophthalmol. Vis. Sci.*, **45**, 1240–6.

Karwoski, C. J., Lu, H. and Newman, E. A. (1989). Spatial buffering of light-evoked potassium increases by retinal Müller (glial) cells. *Science*, **244**, 578–80.

Keirstead, S. A. and Miller, R. F. (1997). Metabotropic glutamate receptor agonists evoke calcium waves in isolated Müller cells. *Glia*, **21**, 194–203.

Kirsch, M., Lee, M., Meyer, V., Wiese, A. and Hofmann, H. (1997). Evidence for multiple, local functions of ciliary neurotrophic factor (CNTF) in retinal development: expression of CNTF and its receptor and in vitro effects on target cells. *J. Neurochem.*, **68**, 979–90.

Kofuji, P., Biedermann, B., Siddharthan, V. *et al.* (2002). Kir potassium channel subunit expression in retinal glial cells: implications for spatial potassium buffering. *Glia*, **39**, 292–303.

Kusaka, S. and Puro, D. G. (1997). Intracellular ATP activates inwardly rectifying K^+ channels in human and monkey retinal Müller (glial) cells. *J. Physiol.*, **500**, 593–604.

LaVail, M. M. and Reif-Lehrer, L. (1971). Glutamine synthetase in the normal and dystrophic mouse retina. *J. Cell Biol.*, **51**, 348–54.

LaVail, M. M., Unoki, K., Yasumura, D. *et al.* (1992). Multiple growth factors, cytokines, and neurotrophins rescue photoreceptors from the damaging effects of constant light. *Proc. Natl. Acad. Sci. U. S. A.*, **89**, 11249–53.

Lewis, G. P., Linberg, K. A., Geller, S. F., Guerin, C. J. and Fisher, S. K. (1999). Effects of the neurotrophin brain-derived neurotrophic factor in an experimental model of retinal detachment. *Invest. Ophthalmol. Vis. Sci.*, **40**, 1530–44.

Li, M. and Sakaguchi, D. S. (2002). Expression patterns of focal adhesion associated proteins in the developing retina. *Dev. Dyn.*, **225**, 544–53.

Li, Y., Holtzclaw, L. A. and Russell, J. T. (2001). Müller cell Ca^{2+} waves evoked by purinergic receptor agonists in slices of rat retina. *J. Neurophysiol.*, **85**, 986–94.

Linser, P. and Moscona, A. A. (1979). Induction of glutamine synthetase in embryonic neural retina: localization in Müller fibers and dependence on cell interactions. *Proc. Natl. Acad. Sci. U. S. A.*, **76**, 6476–80.

Liu, Y. and Wakakura, M. (1998). P1-/P2-purinergic receptors on cultured rabbit retinal Müller cells. *Jpn. J. Ophthalmol.*, **42**, 33–40.

Lopez, T., Lopez-Colome, A. M. and Ortega, A. (1998). Changes in GluR4 expression induced by metabotropic receptor activation in radial glia cultures. *Brain Res. Mol. Brain Res.*, **58**, 40–6.

Lukasiewicz, P. D. and McReynolds, J. S. (1985). Synaptic transmission at N-methyl-D-aspartate receptors in the proximal retina of the mudpuppy. *J. Physiol.*, **367**, 99–115.

Mack, A. F., Germer, A., Janke, C. and Reichenbach, A. (1998). Müller (glial) cells in the teleost retina: consequences of continuous growth. *Glia*, **22**, 306–13.

Malchow, R. P., Qian, H. H. and Ripps, H. (1989). gamma-Aminobutyric acid (GABA)-induced currents of skate Müller (glial) cells are mediated by neuronal-like $GABA_A$ receptors. *Proc. Natl. Acad. Sci. U. S. A.*, **86**, 4326–30.

McGee Sanftner, L. H., Abel, H., Hauswirth, W. W. and Flannery, J. G. (2001). Glial cell line derived neurotrophic factor delays photoreceptor degeneration in a transgenic rat model of retinitis pigmentosa. *Mol. Ther.*, **4**, 622–9.

Mi, H., Haeberle, H. and Barres, B. A. (2001). Induction of astrocyte differentiation by endothelial cells. *J. Neurosci.*, **21**, 1538–47.

Michaelson, I. (1954). *Retinal Circulation in Man and Animals*. Springfield, IL: Charles C. Thomas.

Moriguchi, K., Yoshizawa, K., Shikata, N. *et al.* (2004). Suppression of N-methyl-N-nitrosourea-induced photoreceptor apoptosis in rats by docosahexaenoic acid. *Ophthalmic Res.*, **36**, 98–105.

Morris, J. E. and Moscona, A. A. (1970). Induction of glutamine synthetase in embryonic retina: its dependence on cell interactions. *Science*, **167**, 1736–8.

Moscona, A. A. and Linser, P. (1983). Developmental and experimental changes in retinal glia cells: cell interactions and control of phenotype expression and stability. *Curr. Top. Dev. Biol.*, **18**, 155–88.

Muresan, Z. and Besharse, J. C. (1993). D2-like dopamine receptors in amphibian retina: localization with fluorescent ligands. *J. Comp. Neurol.*, **331**, 149–60.

Newman, E. A. (2003). Glial cell inhibition of neurons by release of ATP. *J. Neurosci.*, **23**, 1659–66.

Newman, E. A. and Reichenbach, A. (1996). The Müller cell: a functional element of the retina. *Trends Neurosci.*, **19**, 307–12.

Newman, E. A. and Zahs, K. R. (1997). Calcium waves in retinal glial cells. *Science*, **275**, 844–7.

Newman, E. A. and Zahs, K. R. (1998). Modulation of neuronal activity by glial cells in the retina. *J. Neurosci.*, **18**, 4022–8.

Newman, E. A., Frambach, D. A. and Odette, L. L. (1984). Control of extracellular potassium levels by retinal glial cell K^+ siphoning. *Science*, **225**, 1174–5.

Normand, G., Hicks, D. and Dreyfus, H. (1998). Neurotrophic growth factors stimulate glycosaminoglycan synthesis in identified retinal cell populations in vitro. *Glycobiology*, **8**, 1227–35.

Oku, H., Ikeda, T., Honma, Y. et al. (2002). Gene expression of neurotrophins and their high-affinity Trk receptors in cultured human Müller cells. *Ophthalmic Res.*, **34**, 38–42.

Pannicke, T., Bringmann, A. and Reichenbach, A. (2002). Electrophysiological characterization of retinal Müller glial cells from mouse during postnatal development: comparison with rabbit cells. *Glia*, **38**, 268–72.

Peng, M., Li, Y., Luo, Z. et al. (1998). Alpha2-adrenergic agonists selectively activate extracellular signal-regulated kinases in Müller cells in vivo. *Invest. Ophthalmol. Vis. Sci.*, **39**, 1721–6.

Perkins, M. N. and Stone, T. W. (1982). An iontophoretic investigation of the actions of convulsant kynurenines and their interaction with the endogenous excitant quinolinic acid. *Brain Res.*, **247**, 184–7.

Peterson, W. M., Wang, Q., Tzekova, R. and Wiegand, S. J. (2000). Ciliary neurotrophic factor and stress stimuli activate the Jak-STAT pathway in retinal neurons and glia. *J. Neurosci.*, **20**, 4081–90.

Pinzon-Duarte, G., Arango-Gonzalez, B., Guenther, E. and Kohler, K. (2004). Effects of brain-derived neurotrophic factor on cell survival, differentiation and patterning of neuronal connections and Müller glia cells in the developing retina. *Eur. J. Neurosci.*, **19**, 1475–84.

Politi, L., Rotstein, N. and Carri, N. (2001). Effects of docosahexaenoic acid on retinal development: cellular and molecular aspects. *Lipids*, **36**, 927–35.

Pow, D. V. (2001). Amino acids and their transporters in the retina. *Neurochem. Int.*, **38**, 463–84.

Pow, D. V. and Barnett, N. L. (1999). Changing patterns of spatial buffering of glutamate in developing rat retinae are mediated by the Müller cell glutamate transporter GLAST. *Cell Tissue Res.*, **297**, 57–66.

Pow, D. V. and Robinson, S. R. (1994). Glutamate in some retinal neurons is derived solely from glia. *Neuroscience*, **60**, 355–66.

Prada, F. A., Quesada, A., Dorado, M. E., Chmielewski, C. and Prada, C. (1998). Glutamine synthetase (GS) activity and spatial and temporal patterns of GS expression in the developing chick retina; relationship with synaptogenesis in the outer plexiform layer. *Glia*, **22**, 221–36.

Provis, J. M. (2001). Development of the primate retinal vasculature. *Prog. Retin. Eye Res.*, **20**, 799–821.

Provis, J. M., Leech, J., Diaz, C. M. *et al.* (1997). Development of the human retinal vasculature: cellular relations and VEGF expression. *Exp. Eye Res.*, **65**, 555–68.

Provis, J. M., Sandercoe, T. and Hendrickson, A. E. (2000). Astrocytes and blood vessels define the foveal rim during primate retinal development. *Invest. Ophthalmol. Vis. Sci.*, **41**, 2827–36.

Puro, D. G., Yuan, J. P. and Sucher, N. J. (1996). Activation of NMDA receptor-channels in human retinal Müller glial cells inhibits inward-rectifying potassium currents. *Vis. Neurosci.*, **13**, 319–26.

Rapaport, D. H., Wong, L. L., Wood, E. D., Yasumura, D. and LaVail, M. M. (2004). Timing and topography of cell genesis in the rat retina. *J. Comp. Neurol.*, **474**, 304–4.

Raub, T. J., Kuentzel, S. L. and Sawada, G. A. (1992). Permeability of bovine brain microvessel endothelial cells in vitro: barrier tightening by a factor released from astroglioma cells. *Exp. Cell Res.*, **199**, 330–40.

Reh, T. A. and Fischer, A. J. (2001). Stem cells in the vertebrate retina. *Brain Behav. Evol.*, **58**, 296–305.

Reichelt W., Hernandez, M., Damian, R. T., Kisaalita, W. S., and Jordan, B. L. (1997). Voltage- and GABA-evoked currents from Müller glial cells of the baboon retina. *NeuroReport*, **8**, 541–4.

Reichenbach, A. and Robinson, S. R. (1995). Phylogenetic constraints on retinal organization and development. *Prog. Retin. Eye Res.*, **15**, 139–71.

Reifel Saltzberg J. M., Garvey, K. A. and Keirstead, S. A. (2003). Pharmacological characterization of P2Y receptor sub-types on isolated tiger salamander Müller cells. *Glia*, **42**, 149–59.

Rejdak, R., Zarnowski, T., Turski, W. A. *et al.* (2001). Presence of kynurenic acid and kynurenine aminotransferases in the inner retina. *NeuroReport*, **12**, 3675–8.

Rejdak, R., Zielinska, E., Shenk, Y. *et al.* (2003). Ontogenic changes of kynurenine aminotransferase I activity and its expression in the chicken retina. *Vis. Res.*, **43**, 1513–7.

Rejdak, R., Shenk, Y., Schuettauf, F. *et al.* (2004). Expression of kynurenine aminotransferases in the rat retina during development. *Vis. Res.*, **44**, 1–7.

Rhee, K. D. and Yang, X. J. (2003). Expression of cytokine signal transduction components in the postnatal mouse retina. *Mol. Vis.*, **9**, 157–22.

Rich, K. A., Figueroa, S. L., Zhan, Y. and Blanks, J. C. (1995). Effects of Müller cell disruption on mouse photoreceptor cell development. *Exp. Eye Res.*, **61**, 235–48.

Riepe, R. E. (1977). Müller cell localisation of glutamine synthetase in rat retina. *Nature*, **268**, 654–5.

Riepe, R. E. (1978). Glutamine synthetase in the developing rat retina: an immunohistochemical study. *Exp. Eye Res.*, **27**, 435–44.

Risau, W. and Lemmon, V. (1988). Changes in the vascular extracellular matrix during embryonic vasculogenesis and angiogenesis. *Dev. Biol.*, **125**, 441–50.

Robinson, S. R. and Dreher, Z. (1989). Evidence for three morphological classes of astrocyte in the adult rabbit retina: functional and developmental implications. *Neurosci. Lett.*, **106**, 261–8.

Rohrer, B., Korenbrot, J. I., LaVail, M. M., Reichardt, L. F. and Xu, B. (1999). Role of neurotrophin receptor TrkB in the maturation of rod photoreceptors and establishment of synaptic transmission to the inner retina. *J. Neurosci.*, **19**, 8919–30.

Rotstein, N. P., Politi, L. E., German, O. L. and Girotti, R. (2003). Protective effect of docosahexaenoic acid on oxidative stress-induced apoptosis of retina photoreceptors. *Invest. Ophthalmol. Vis. Sci.*, **44**, 2252–9.

Sanches, G., de Alencar, L. S. and Ventura, A. L. (2002). ATP induces proliferation of retinal cells in culture via activation of PKC and extracellular signal-regulated kinase cascade. *Int. J. Dev. Neurosci.*, **20**, 21–7.

Sandercoe, T. M., Madigan, M. C., Billson, F. A., Penfold, P. L. and Provis, J. M. (1999). Astrocyte proliferation during development of the human retinal vasculature. *Exp. Eye Res.*, **69**, 511–23.

Scharfman, H. E., Hodgkins, P. S., Lee, S. C. and Schwarcz, R. (1999). Quantitative differences in the effects of *de novo* produced and exogenous kynurenic acid in rat brain slices. *Neurosci. Lett.*, **274**, 111–4.

Schnitzer, J. and Karschin, A. (1986). The shape and distribution of astrocytes in the retina of the adult rabbit. *Cell Tissue Res.*, **246**, 91–102.

Schopf, S., Ruge, H., Bringmann, A., Reichenbach, A. and Skatchkov, S. N. (2004). Switch of K^+ buffering conditions in rabbit retinal Müller glial cells during postnatal development. *Neurosci. Lett.*, **365**, 167–70.

Sharma, R. K. and Johnson, D. A. (2000). Molecular signals for development of neuronal circuitry in the retina. *Neurochem. Res.*, **25**, 1257–63.

Stevens, E. R., Esguerra, M., Kim, P. M., *et al.* (2003). D-serine and serine racemase are present in the vertebrate retina and contribute to the physiological activation of NMDA receptors. *Proc. Natl. Acad. Sci. U. S. A.*, **100**, 6789–94.

Stier, H. and Schlosshauer, B. (1998). Different cell surface areas of polarized radial glia having opposite effects on axonal outgrowth. *Eur. J. Neurosci.*, **10**, 1000–10.

Stier, H. and Schlosshauer, B. (1999). Cross-species collapse activity of polarized radial glia on retinal ganglion cell axons. *Glia*, **25**, 143–53.

Stone, J., Makarov, F. and Holländer, H. (1995). The glial ensheathment of the soma and axon hillock of retinal ganglion cells. *Vis. Neurosci.*, **12**, 273–9.

Taylor, S., Srinivasan, B., Wordinger, R. J. and Roque, R. S. (2003). Glutamate stimulates neurotrophin expression in cultured Müller cells. *Brain Res. Mol. Brain Res.*, **111**, 189–97.

Threlkeld, A., Adler, R. and Hewitt, A. T. (1989). Proteoglycan biosynthesis by chick embryo retina glial-like cells. *Dev. Biol.*, **132**, 559–68.

Triviño, A., Ramirez, J. M., Ramirez, A. I. and Salazar, J. J. (1992). Retinal perivascular astroglia: an immunoperoxidase study. *Vis. Res.*, **32**, 1601–7.

Triviño, A., Ramírez, J. M., Ramírez, A. I., Salazar, J. J. and García-Sánchez, J. (1997). Comparative study of astrocytes in human and rabbit retinae. *Vis. Res.*, **37**, 1707–11.

Turner, D. L. (1987). A common progenitor for neurons and glia persists in rat retina late in development. *Nature*, **328**, 131–6.

Uchihori, Y. and Puro, D. G. (1993). Glutamate as a neuron-to-glial signal for mitogenesis: role of glial N-methyl-D-aspartate receptors. *Brain Res.*, **613**, 212–20.

Valter, K., Bisti, S. and Stone, J. (2003). Location of CNTFRalpha on outer segments: evidence of the site of action of CNTF in rat retina. *Brain Res.*, **985**, 169–75.

Vardimon, L., Fox, L. L., Degenstein, L. and Moscona, A. A. (1988). Cell contacts are required for induction by cortisol of *glutamine synthetase* gene transcription in the retina. *Proc. Natl. Acad. Sci. U. S. A.*, **85**, 5981–5.

Vetter, M. L. and Moore, K. B. (2001). Becoming glial in the neural retina. *Dev. Dyn.*, **221**, 146–53.

Wahlin, K. J., Campochiaro, P. A., Zack, D. J. Adler, R. (2000). Neurotrophic factors cause activation of intracellular signaling pathways in Müller cells and other cells of the inner retina, but not photoreceptors. *Invest. Ophthalmol. Vis. Sci.*, **41**, 927–36.

Wahlin, K. J., Lim, L., Grice, E. A. *et al.* (2004). A method for analysis of gene expression in isolated mouse photoreceptor and Müller cells. *Mol. Vis.*, **10**, 366–75.

Wakakura, M., Utsunomiya-Kawasaki, I. and Ishikawa, S. (1998). Rapid increase in cytosolic calcium ion concentration mediated by acetylcholine receptors in cultured retinal neurons and Müller cells. *Graefe's Arch. Clin. Exp. Ophthalmol.*, **236**, 934–9.

Walsh, N., Valter, K. and Stone, J. (2001). Cellular and sub-cellular patterns of expression of bFGF and CNTF in the normal and light stressed adult rat retina. *Exp. Eye Res.*, **72**, 495–501.

Watanabe, T. and Raff, M. C. (1988). Retinal astrocytes are immigrants from the optic nerve. *Nature*, **332**, 834–7.

Wexler, E. M., Berkovich, O. and Nawy, S. (1998). Role of the low-affinity NGF receptor (p75) in survival of retinal bipolar cells. *Vis. Neurosci.*, **15**, 211–8.

Willbold, E., Reinicke, M., Lance-Jones, C. *et al.* (1995). Müller glia stabilizes cell columns during retinal development: lateral cell migration but not neuropil growth is inhibited in mixed chick-quail retinospheroids. *Eur. J. Neurosci.*, **7**, 2277–84.

Willbold, E., Rothermel, A., Tomlinson, S. and Layer, P. G. (2000). Müller glia cells reorganize reaggregating chicken retinal cells into correctly laminated in vitro retinae. *Glia*, **29**, 45–57.

Wolter, J. R. (1957). Perivascular glia of the blood vessels of the human retina. *Am. J. Ophthalmol.*, **44**, 766–73.

Xu, L. Y., Zhao, J. W. and Yang, X. L. (2004). GLAST expression on bullfrog Müller cells is regulated by dark/light. *NeuroReport*, **15**, 2451–4.

Zack, D. J. (2000). Neurotrophic rescue of photoreceptors: are Müller cells the mediators of survival? *Neuron*, **26**, 285–6.

Zhang, J. and Yang, X. L. (1999). GABA(B) receptors in Müller cells of the bullfrog retina. *NeuroReport*, **10**, 1833–6.

Zhang, J., De Blas, A. L., Miralles, C. P. and Yang, C. Y. (2003). Localization of GABAA receptor subunits alpha 1, alpha 3, beta 1, beta 2/3, gamma 1, and gamma 2 in the salamander retina. *J. Comp. Neurol.*, **459**, 440–53.

Zhou, C. and Dacheux, R. F. (2004). AII amacrine cells in the rabbit retina possess AMPA-, NMDA-, GABA-, and glycine-activated currents. *Vis. Neurosci.*, **21**, 181–8.

10

Retinal mosaics

Stephen J. Eglen
University of Cambridge, Cambridge, UK

Lucia Galli-Resta
Istituto di Neuroscienze CNR, Pisa, Italy

10.1 Introduction

One of the most striking aspects of the architecture of the retina is its highly organized structure. Retinal neurons are positioned in three different layers, at different depths. Usually, all cells of a particular type are found in just one of those layers. When the spatial distribution of one type of cells within a layer can be observed, the cell bodies are arranged in a semi-regular pattern, rather than distributed randomly across the surface (Figure 10.1). These patterns are often termed 'retinal mosaics', due to the way that the cell bodies and dendrites of a type of neuron tend to tile the retina.

This regular arrangement of cells is thought to ensure that the visual field is evenly sampled, avoiding any perceptual blind spots in the visual field. The retina is assembled as an array of functional units, each detecting, processing and conveying to the brain information about a limited portion of the visual scene. The presence of regular arrays of neurons of the same type has long been considered a consequence of this functional design. However, recent studies have shown that retinal mosaics form early in development, before all the elements of the functional units have been born. This chapter reviews our present knowledge of the various mechanisms by which retinal mosaics emerge during development, and summarizes the mathematical techniques used to analyse mosaics.

10.2 Mechanisms of development

To date, the general lack of markers to selectively label neurons of a specific type early enough in development has meant that often the investigators have to infer what happened in development by examining older tissue. However, markers are now available for specific cell types, such as horizontal and cholinergic amacrine cells, that can label cells early enough in development; these cell types have therefore been most studied. Ongoing advances in time-lapse imaging techniques should also mean that in the coming years we should be able to discover much more about the formation of mosaics. In this section, the main mechanisms underlying formation of different retinal mosaics are considered.

Retinal Development, ed. Evelyne Sernagor, Stephen Eglen, Bill Harris and Rachel Wong.
Published by Cambridge University Press. © Cambridge University Press 2006.

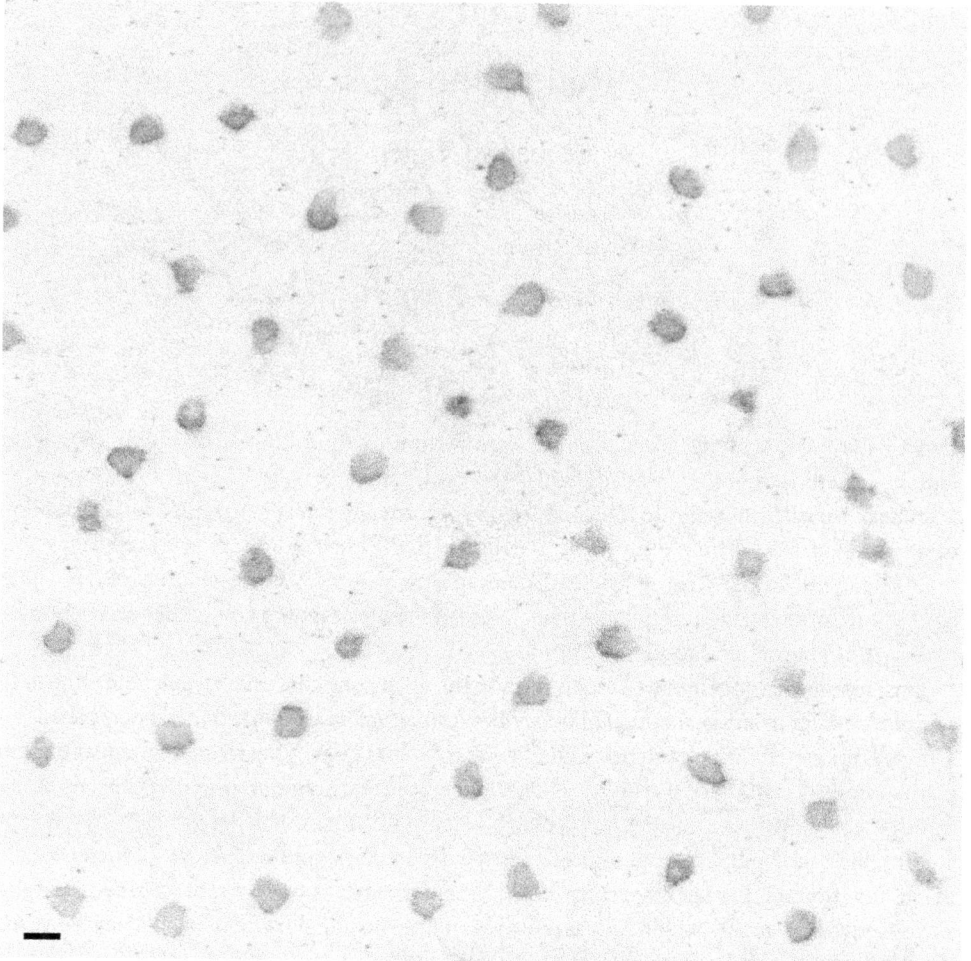

Figure 10.1 Example mosaic of cholinergic amacrine cells from the inner nuclear layer of a postnatal day (P)3 rat retina. Cell bodies, clearly labelled by choline acetyltransferase, are regularly distributed across the sample area. Scale bar is 10 μm.

10.2.1 *Interactions between homotypic (mosaic) cells*

Given that many types of retinal neurons form mosaics during development, it is plausible to think that interactions between cells of different types may influence the formation of retinal mosaics. However, several lines of evidence outlined below indicate that in most cases interactions between different cell types do not guide the development of retinal cells. Instead, it seems that interactions between cells of the same type (homotypic interactions) are sufficient to guide mosaic formation.

First, computer simulations have demonstrated that a spatial distribution of retinal cells can be replicated by assuming that each cell is surrounded by an 'exclusion zone'. In the model, neurons are positioned randomly into the array, subject only to the constraint that a cell cannot be placed within another cell's exclusion zone. This model is known as the d_{min} model, where d_{min} is the parameter that specifies the diameter of the circular exclusion zone surrounding the cell body (Galli-Resta *et al.*, 1997). A limitation of this model is that although it shows some form of exclusion zone acting on homotypic neurons is sufficient to generate mosaics, it does not suggest candidate biological mechanisms that might generate an exclusion zone. However, it does suggest that the mechanism can be fairly local, rather than long-ranging.

Experimental evidence against heterotypic interactions in forming retinal mosaics comes from cross-correlation studies. Rockhill *et al.* (2000) examined six different cell types in rabbit retina and tested for the existence of spatial correlation between pairs of mosaics. In each case, the cross-correlation analysis indicated that the positioning of one type of neuron was spatially independent of another type of neuron. To date, only one positive cross-correlation between cell types, blue cones and blue-cone bipolar cells in the monkey, has been reported (Kouyama and Marshak, 1997). Likewise, only one negative cross-correlation between two cell types, horizontal and short-wavelength cones, has been reported (Ahnelt *et al.*, 2000). Hence, cross-correlations between cell types are likely to be rare exceptions to the general principle of spatial independence between different classes of neuron.

In addition to the computer modelling and analysis of retinal mosaics, direct experimental manipulations have been performed to test for heterotypic interactions in mosaic formation. For example, since the cholinergic amacrine cells and retinal ganglion cells (RGCs) are synaptic partners, it might be possible that these two cell types influence each other during mosaic formation. However, spatial organization of cholinergic amacrine cells in the developing rodent retina was unaffected by either complete loss of RGCs or by doubling the density of RGCs (Galli-Resta, 2000). Likewise, in the case of the horizontal cells the loss of photoreceptor synaptic input, or the presence of a higher than normal cell density around these cells, did not alter their spatial regularity (Raven and Reese, 2003; Rossi *et al.*, 2003). Taken together, these analytical and experimental results suggest that whatever mechanisms are involved in mosaic formation, heterotypic interactions are unlikely to play a major role.

10.2.2 Cell fate

All of the different types of retinal cells originate from one population of postmitotic cells, with a stereotypical sequence in the birthdates of the different types of retinal neuron (see Chapter 3 and Chapter 5). For example, RGCs are born first, whereas bipolar cells are born among the latest (Livesey and Cepko, 2001). Although the fate of an individual postmitotic cell becomes restricted during development (Cepko *et al.*, 1996), it has been suggested that a cell's fate to become a certain type of neuron is influenced by environmental factors, perhaps from neighbouring cells. In analogy to the lateral inhibition process active in *Drosophila*, whereby R8 photoreceptors inhibit neighbouring cells to take the R8 fate, lateral inhibition has been proposed for different types of photoreceptor both in primates and goldfish (Wikler and Rakic, 1991; Stenkamp *et al.*, 1997). It has also been suggested that RGCs secrete factors that inhibit neighbouring postmitotic cells from also becoming RGCs (Waid and McLoon, 1998). Furthermore, a diffusible signal acting as an inhibitory cue in cell fate processes is suggested to be responsible for the spatial arrangement of tyrosine hydroxylase- and 5-hydroxytryptamine-positive cells (Tyler *et al.*, 2005). However, the chemical identity of such inhibitory factors has not yet been determined. In contrast, much is known about the molecular signals involved in the specification of the different classes of photoreceptor in each ommatidium of the *Drosophila* eye (Freeman, 1997). Given the similarities between vertebrate and invertebrate eye development (Jarman, 2000), mechanisms observed in *Drosophila* may well inform future vertebrate studies.

Cell fate mechanisms are likely to be implicated in generating the correct density of each type of neuron, but whether they contribute to the ultimate spatial positioning of cells is unclear. In principle, mutual inhibition of neighbouring cells from adopting the same fate can contribute to spatial order in a population of neurons, as verified by computer models (Eglen and Willshaw, 2002). These theoretical arguments are supported by findings from RGCs in chick, which are seen to be regularly distributed early in development, before migration to their final layer (McCabe *et al.*, 1999). However, since cell fate mechanisms occur relatively early in development, many other subsequent mechanisms (mentioned below) may interfere with any order generated by cell fate mechanisms. This seems to be the case at least for cholinergic amacrine cells; as they migrate to their destination layer, they do not observe minimal spacing constraints (Galli-Resta *et al.*, 1997). This would suggest that if any minimal spacing constraints were created among cholinergic amacrines by early cell fate mechanisms, it is subsequently lost once cells migrate away from the ventricular zone. In contrast, cell fate mechanisms may be sufficient to create ordering in *Drosophila*, since cells do not migrate subsequent to the cell fate decision process.

10.2.3 *Radial and lateral migration*

The lack of good biochemical markers for identifying most types of retinal neuron early in development means that the extent to which cell fate mechanisms can aid mosaic formation is still largely unknown. However, even after cell fate specification, most types of retinal neuron need to migrate away from the ventricular zone into their final destination layer (described in Chapter 4). Depending on the extent of this movement, the initial ordering imposed by any cell fate mechanisms may be destroyed by subsequent migration. Hence, later acting mechanisms must also be involved.

Lineage tracing techniques were used to study the migration of groups of neurons that descended from individual retinal progenitors. These studies found that the cells originated from individual progenitors were organized into columns spanning across the depth of the retina, from which it was suggested that retinal cells migrate predominantly in a radial fashion from the ventricular zone (Turner and Cepko, 1987; Holt *et al.*, 1988). Occasionally, labelled cells were found outside of the radial columns, but since they occurred rarely these were thought to be artefactual or ectopic (Reese and Galli-Resta, 2002). However, by using an X-inactivation transgenic mouse, 50% of all retinal projectors were labelled with the

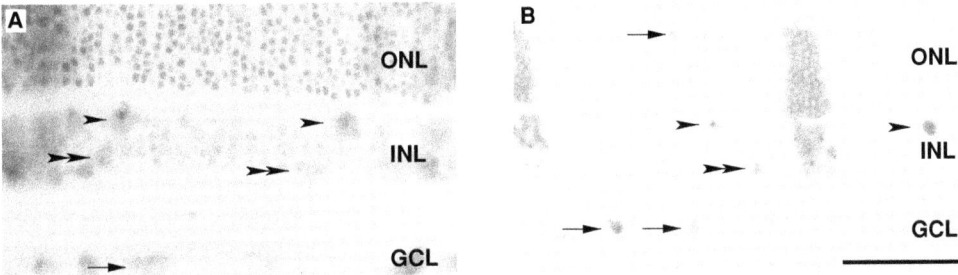

Figure 10.2 Lateral migration of certain classes of retinal neuron away from their clonal column of origin. (A) Retinal section from a female hemizygous transgenic mouse showing labelled cells arranged into columns passing through both the inner and outer nuclear layers (INL and ONL). Outside of these columns, tangentially dispersed cells are observed (indicated by arrows). These cells are horizontal cells (single arrowhead), amacrine cells (double arrowhead) and a ganglion cell (arrow) in the ganglion cell layer (GCL). (B) Retinal section from chimeric mouse (produced by blastocyst injection of embryonic stem cells) whereby a smaller fraction of cells is labelled, thus allowing easier visualization of the clonal columns. Again, tangentially displaced horizontal, amacrine, ganglion cells are observed, together with a blue cone in the ONL (arrow). Scale bar is 50 μm in (A) and 75 μm in (B). Modified from Reese and Galli-Resta (2002) with permission from *Prog. Retin. Eye Research*.

LacZ reporter and hence half of the subsequently generated retinal cells were labelled. In common with the earlier studies, most labelled cells were found in radial columns. However, in addition, many cells were observed outside of these columns, suggesting that some cells moved laterally away from their radial column of origin (Figure 10.2; Reese *et al.*, 1995).

By studying the identity of the neurons that moved, it was found that only certain classes of retinal neuron (cone photoreceptors, horizontal, amacrine and ganglion cells) moved radially, but that for those classes, all cells in each class moved laterally (Reese *et al.*, 1999). The extent of tangential movement away from the nearest column was around 45 to 150 μm. Furthermore, the cell classes that move radially are also the classes whose mosaics are the most regular. From these results, it is suggested that tangential migration of cells away from their radial clone of origin aids mosaic formation. The time at which cells undergo tangential migration has not been precisely determined for most cell types. In the case of mouse horizontal cells, tangential migration occurs around postnatal day (P)1 to P5, which is about the same time as when horizontal cells change from a radial to a horizontal morphology. This correlation between morphological changes and tangential migration hints at a possible mechanism underlying the lateral movements (see Section 10.2.5).

10.2.4 Cell death

Cell death is a process found throughout the developing nervous system, and the retina is no exception. For example, it has been estimated that across a range of mammalian species, 50% to 90% of RGCs that are born will die during development (Finlay and Pallas, 1989). This retinal cell death has largely been implicated in the refinement of the retinal projections to their targets (O'Leary et al., 1986; but see Chapter 11 for discussion of other roles for cell death). However, more recently cell death has been suggested to be involved in the formation of retinal mosaics. If a group of retinal neurons form an irregular mosaic early in development, cell death might kill those neurons that are too close to one another. This would increase the size of the exclusion zone around a cell and thus improve regularity (Galli-Resta, 1998; Cook and Chalupa, 2000). This cell death might be mediated by cells competing with each other for trophic support or contacts from their afferents (Wässle and Riemann, 1978; Kirby and Steineke, 1996), or might even be triggered by neighbouring cells. Part of the difficulty of understanding the impact of cell death is that currently it is very difficult to directly observe its effects; instead we have to rely on indirect observations (after cell death has occurred) to infer what role cell death is playing.

Two recent studies have implicated a role of cell death in the formation of retinal mosaics. First, in cat retina, the mosaic of alpha RGCs increases its regularity during the first postnatal month as the density of alpha RGCs drops by around 20%. This death seems normally driven by mechanisms related to cell spacing, since when spiking activity in the RGCs is blocked the magnitude of cell death is not altered, but mosaic regularity is not improved (Jeyarasasingam et al., 1998). This suggests that the activity patterns of alpha cells influence which cells die during development in a way that improves mosaic regularity. Second, the Bcl-2 overexpressing mouse was used to study the influence of cell death upon mosaic formation. *Bcl-2* is an anti-apoptotic gene, and its overexpression inhibits naturally occurring cell death, leading to an increase in density of most retinal cell types (Strettoi and Volpini, 2002). Out of all neuronal types, dopaminergic amacrine cells showed the largest (tenfold) increase in density compared with wild-type retinas. In the Bcl-2 mouse, the dopaminergic amacrine cells were less regular than in wild type, where many close pairs of dopaminergic cells were often observed (Raven et al., 2003). Since the Bcl-2 mouse inhibits naturally occurring cell death, the suggestion is that cell death would normally remove cells that are positioned too close to neighbouring cells of the same type. These two studies differ in the magnitude of cell death observed (20% versus 90%) in a population, and computer-modelling studies suggest that 20% cell death is insufficient by itself to transform irregular into highly regular arrays (Eglen and Willshaw, 2002). By contrast, the magnitude of cell death observed in the dopaminergic amacrine cells is very large compared with the mild increase in regularity in wild-type animals, suggesting that the choice of which cells to eliminate need not be very selective.

Retinal cell death is also unlikely to be a universal mechanism by which mosaics are generated. In the developing rat retina, around 20% of cholinergic amacrine cells die from P4 to P12, without any increase in mosaic regularity (Galli-Resta and Novelli, 2000). Also, cell death was suggested to be too small to account for the creation of a spatial dependency between blue cones and blue-cone bipolar cells (Kouyama and Marshak, 1997). It is therefore likely that cell death is just one of several mechanisms used to create a regular distribution of neurons. Modelling work suggests for example that it can be used after early cell fate events to produce regular mosaics (Eglen and Willshaw, 2002).

All these studies refer to late waves of cell death, eliminating cells once they have been generated and positioned. Cell death also occurs earlier in retinal development, sculpting the pool of progenitor cells that are available (de la Rosa and de Pablo, 2000), or affecting single cell populations at the same time as new cells are also formed or migrating. An early phase of cell death has been recently shown to contribute to mosaic formation: blocking purinergic receptors in the retina of newborn rodents at P1 to P4 inhibits an early phase of cholinergic cell death and leads to typically a 30% increase in the density of cholinergic cells (Resta *et al.*, 2005). In the treated retinas, many pairs of nearby cholinergic cells are observed, a feature absent from control retinas. Considering that cholinergic cells are thought to be sources themselves of extracellular adenosine triphosphate (e-ATP), and that e-ATP does not diffuse far beyond 50 µm (Newman 2001), it has been suggested that cholinergic cells could be releasing e-ATP in a local fashion to cause close-neighbouring cholinergic cells to die (Resta *et al.*, 2005).

10.2.5 Dendritic interactions

So far we have seen that several mechanisms are implicated in the formation of retinal mosaics. Out of these mechanisms, lateral cell migration seems to play a major role in rearranging cells into a regular array. Computer modelling has been instructive to help our understanding of how mosaics might form. For example, the d_{min} model has shown that interactions need only be between cells of the same type to produce regular arrays similar to those observed experimentally. Furthermore, the d_{min} model also suggests that an exclusion zone mechanism local to each cell is sufficient to generate global order in the array. Following up on a hypothesis suggested by Wässle and Riemann (1978), another computer model suggested that interactions between dendrites of neighbouring cells might cause cells to repel each other and rearrange the cell bodies from a random into a regular array (Eglen *et al.*, 2000). What evidence is there for a role of dendritic interactions driving lateral cell migration and hence mosaic formation?

Two experimental studies provide indirect evidence in favour of dendritic interactions. First, horizontal cells in the mouse migrate to their destination layer, the inner nuclear layer (INL), by P1. At this stage they have a radial morphology. Within the next few days, the cells then displace laterally within their layer at the same time as the cell's morphology changes from radial to horizontal, forming an initial dendritic tree (Reese *et al.*, 1999). Second, when cholinergic cells first arrive in their destination layer, either the INL or ganglion cell layer (GCL), at around embryonic day (E)21, the cell bodies are irregularly spaced and there are many holes in the dendritic network. Soon afterwards (P0) the cell bodies form regular arrays at the same time as the network of dendrites becomes more complete (Galli-Resta *et al.*, 2002).

These findings correlate the emergence of dendritic trees with the formation of retinal mosaics, but do not directly show a role for dendrites upon mosaic formation. However,

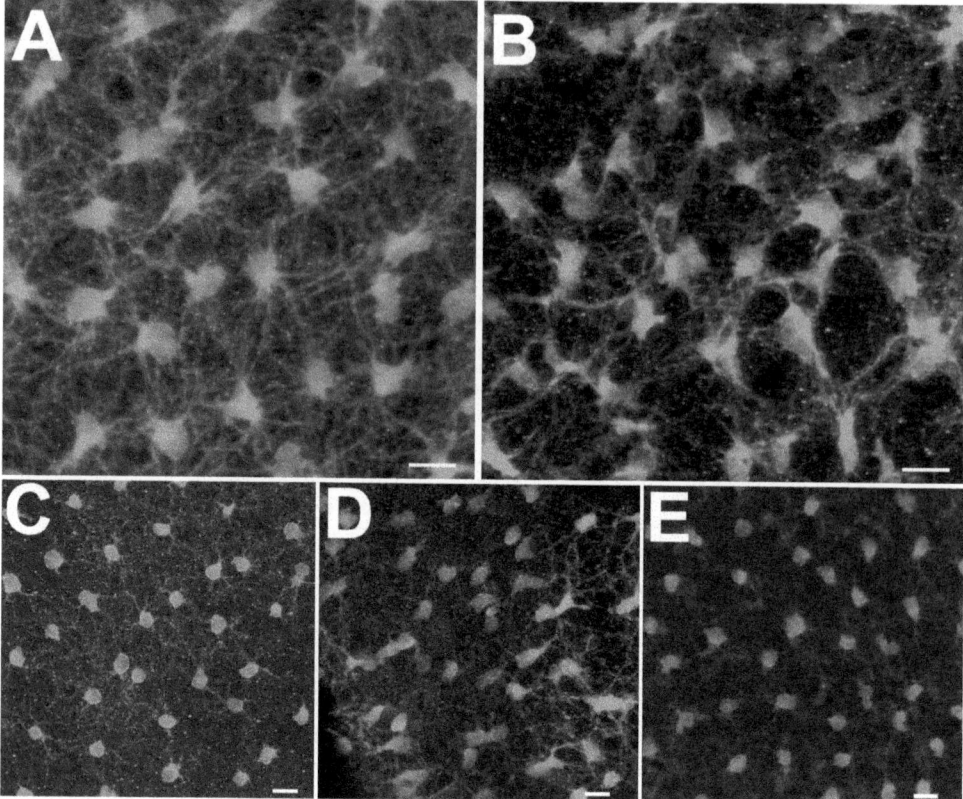

Figure 10.3 Dependence of dendritic network upon mosaic formation. (A) At P0, cholinergic amacrine cell bodies in the rat are regularly spaced, and their dendrites interconnect to form a network. (B) A few days earlier, at E21, the dendritic network is poor and the somas are not so regularly positioned. (C, D, E) State of the network of cholinergic amacrine cells before (C), during (D) and after (E) disruptions of the dendritic network by treatments that affect the microtubule-associated proteins within the dendrites. During the treatment, the regular arrangement of neurons collapses; after the treatment wears off (a few days later), a regular mosaic returns. Scale bars: 10 μm. Figure reproduced from Galli-Resta (2002) with permission from *Trends Neurosci*. For colour version, see Plate 9.

Galli-Resta *et al.* (2002) provided direct evidence by manipulating the dendritic trees during development. Microtubules form the scaffold of both axonal and dendritic processes. Selective perturbation of the microtubules in the cholinergic and horizontal cells, either pharmacologically or with antisense oligonucleotides, disrupted the dendritic network leading to a collapse in the regular spacing of cell bodies within a layer and also to cells moving out of their normal layer (Figure 10.3; Galli-Resta *et al.*, 2002). Remarkably, these effects were reversible, and after around two days, when the experimental treatments had worn off, both the dendritic network and regular mosaic returned. Similar effects were found for both cholinergic and horizontal cells, suggesting this mechanism is not specific to one cell type.

To explain these findings Galli-Resta (2002) has proposed a micromechanical hypothesis of mosaic assembly. As neurons migrate to their destination layer they begin to form connections with dendrites from other neurons of the same type, creating a mechanical network. The microtubules in the dendrites provide a rigid skeleton that counteracts an elastic component to keep the net together. After the net is first formed, cells within the network are positioned at different depths in the retina, but still respect the minimal distance constraints (in 3-d) to neighbouring cells. Shortly afterwards, the cells move such that they are all positioned in the same layer, again respecting minimal distance constraints. From here, the network settles at equilibrium into a regular array of cells. Disruption of the microtubules causes local disturbances in the network, from which cells lose their regular spacing.

This model of mosaic assembly accounts for the results observed disrupting microtubules in the cholinergic and horizontal cells, however, it still requires further testing. For example, what are the mechanisms by which neighbouring neurons of the same type can recognize each other? One requirement of the model is that homotypic neurons must be able to create a network via their dendrites. While dendritic nets are observed in development for the horizontal cells and cholinergic amacrine cells, other types of retinal neuron have little dendritic contact with neighbouring neurons of the same type. For example, beta RGCs form few, if any, contacts with neighbouring beta RGCs, instead forming contacts with presynaptic amacrine and bipolar cells. It is thus to be seen whether this mechanical model of mosaic assembly can account for the regular arrangement of different types of RGCs.

10.3 Mathematical methods for quantifying regularity of retinal mosaics

Once a group of neurons has been labelled by using a specific chemical marker, visual inspection of the tissue may already tell us that the cells are regularly ordered. However, quantitative methods are normally required to evaluate the degree of regularity in retinal mosaics, and to allow comparisons between mosaics from different types of retinal neuron. Several methods have been developed, which are described in this section.

The earliest, and still dominant, technique for quantifying the regularity in the spatial distribution of cells has been the regularity index (RI), based upon finding the distance to nearest neighbours (Wässle and Riemann, 1978; Cook, 1996). For each neuron, we measure the distance to the nearest neighbouring neuron of the same type (Figure 10.4a,b). We then define the RI as the mean of these nearest-neighbour distances divided by their standard deviation. The higher the RI, the more regular the population is. As a rough guideline, an RI around 2 or less typically indicates that the mosaic is randomly arranged. Higher values of the RI indicate that the mosaic is non-randomly arranged. Regularity index values in the range 3 to 9 are typical for regular retinal mosaics from different species.

The RI measure is dominant today because of its relative simplicity to calculate. However, since it measures only the nearest-neighbour distance, it clearly does not capture all the spatial information in a mosaic. For example, a group of cells equally spaced along a (one-dimensional) line will have the same RI as a group of cells positioned in a regular (two-dimensional) grid. Other methods have therefore been developed to capture the relative

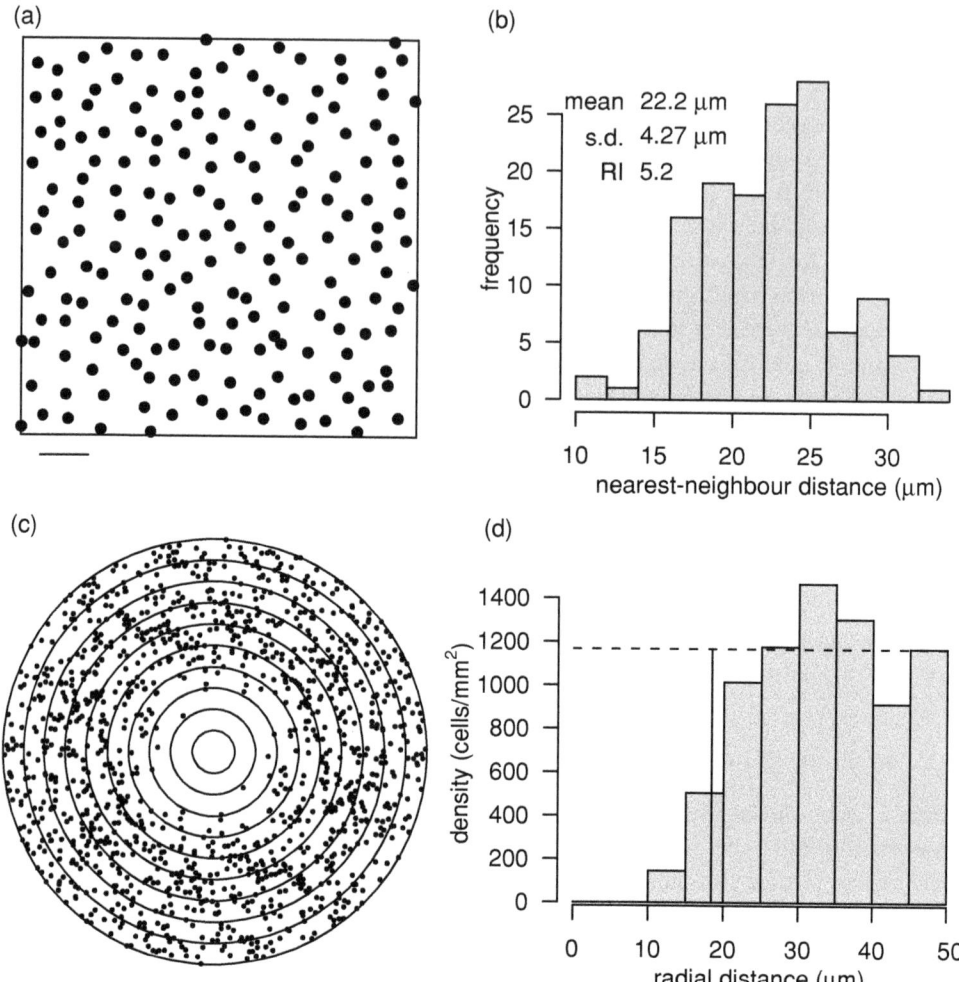

Figure 10.4 Example of a population of cells forming a retinal mosaic. (a) Example mosaic of horizontal cell somas from adult mouse retina. Cell bodies (10 μm diameter) are drawn to scale. Scale bar: 50 μm. (b) Nearest-neighbour histogram computed by finding the distance from each cell in (a) to its nearest neighbour. The regularity index here is 5.2. (c) Autocorrelation plot of the field shown in (a). Each annulus is 10 μm wide. (d) Density recovery profile of the field. Each histogram bar shows the density of cells in the corresponding annulus from the autocorrelation in (c). If cells were randomly positioned, the profile would be flat (shown by dotted line). The 'well' in the histogram out to around 20 μm (vertical line) indicates the size of the exclusion zone for this field. See Section 10.3 for details of the analysis methods used for determining mosaic regularity.

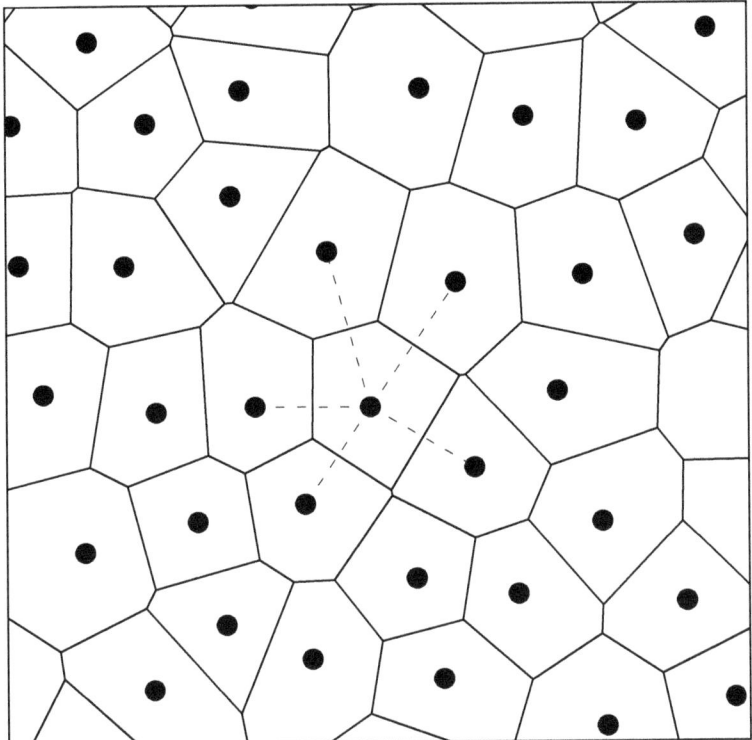

Figure 10.5 Voronoi tessellation of a population of cells. Each retinal cell from a simulated population has been drawn as a circle; each cell is surrounded by its Voronoi polygon, denoting all points of space that are closest to that cell. For one central cell, dashed lines indicate the Delaunay segments to its neighbouring cells.

positioning of many neighbouring neurons. One popular approach has been to study the autocorrelation plot (Figure 10.4c), from which various measures can be calculated (Figure 10.4d; Rodieck, 1991). The autocorrelation plot is created by positioning each neuron at the centre of the plot, and then plotting the relative position of all other neurons. A common feature of the autocorrelation plot for retinal neurons is that there is a central 'exclusion zone' where few, if any, cells can be found. This indicates that no two neurons tend to come closer than some minimal distance from each other. Furthermore, this minimal distance is often much larger than the relatively small distance that is imposed by 'steric hindrance', since two somas cannot overlap in the same layer. A cross-correlation plot is similar to the autocorrelation plot (Figure 10.4c) except that for each cell of type A we plot the relative position of type B cells (Rodieck, 1991).

More recently, measures based on the Voronoi tessellation have been used to evaluate mosaic regularity (Figure 10.5). The Voronoi tessellation divides the retinal surface into non-overlapping polygons, with one retinal neuron inside each polygon; the polygon encloses all points in the plane that are closest to that cell. Equivalently, the Delaunay triangulation is formed by drawing lines between neurons that share a polygon edge; Figure 10.5 shows all of the Delaunay segment lines for one central cell. Various measures taken from the Voronoi tessellation, such as the distribution of polygon areas, have proved to be useful in evaluating spatial regularity of retinal mosaics, especially when comparing real mosaics with those generated by computer simulations (Galli-Resta *et al.*, 1997; Zhan and Troy, 2000).

By using these quantitative techniques on mosaics from both developing and adult tissue, we know that different neuronal types have varying degrees of regularity. Some cell types are therefore clearly more regular than other cell types. These differences in regularity may reflect the spatial requirements of different cell types. For example, beta RGCs are highly regular (Wässle *et al.*, 1981), as might be expected since these cells are responsible for detailed vision. On the other hand, dopaminergic amacrine cells are much less organized (although still non-random); their relative lack of spatial precision may reflect the modulatory role that dopamine plays in retinal processing (Raven *et al.*, 2003). Furthermore, Martin *et al.* (2000) found that regularity of the mosaic of short-wavelength cone photoreceptors varied from species to species, which might reflect the varying functional demands on the same neuronal type in different species.

10.4 Concluding remarks

Retinal neurons form regular arrays within each layer of the retina. The degree of regularity varies from type to type, possibly reflecting the functional requirements of each neuronal type in visual processing. We have outlined several developmental mechanisms that generate spatial patterning of neurons. First, cell fate is an early-acting mechanism that could prevent cells of the same type from being positioned too close to each other. Likewise, if cells of the same type are positioned close to each other, cell death may remove some of them to improve mosaic regularity. However, the dominant mechanism currently thought to produce mosaics is the lateral movement of cells upon their arrival in the destination layer. This lateral movement is mediated by the dendritic interactions, possibly creating a mechanical network of cells that gradually settles to equilibrium. The lack of spatial correlations between cells of different types indicates that heterotypic interactions are not involved in mosaic formation.

In the introduction to this chapter we suggested that retinal mosaics may be needed to ensure that the visual world is uniformly sampled, leaving no holes in visual space. However, another argument in favour of regular mosaics is that it may help the subsequent development of the retina. So far we have shown that after neurons of a particular type are born, they can migrate to their destination layer and form regular arrays independently of any other cell type. Once each type of neuron is arranged in a regular array, the next stage of development is for cells to make appropriate contacts with cells in other layers. For example, the RGCs will need to contact the correct number of amacrine and bipolar cells. Since the amacrine and bipolar cells are already positioned regularly across the INL, the

RGCs do not need to search far to find the appropriate contacts. This searching for contacts could be mediated by filopodial extensions from RGCs (Wong and Wong, 2000; Chapter 12). If, however, the amacrine and bipolar cells were irregularly organized, the RGCs would presumably have further to search for contacts. A regular arrangement of neurons would therefore reduce dendritic wiring lengths and ensure rapid and homogeneous wiring of retinal circuitry.

References

Ahnelt, P. K., Fernández, E., Martinez, O., Bolea, J. A. and Kübber-Heiss, A. (2000). Irregular S-cone mosaics in felid retinae. Spatial interaction with axonless horizontal cells, revealed by cross correlation. *J. Opt. Soc. Am. A Opt. Image Sci. Vis.*, **17**, 580–8.

Cepko, C. L., Austin, C. P., Yang, X., Alexiades, M. and Ezzeddine, D. (1996). Cell fate determination in the vertebrate retina. *Proc. Natl. Acad. Sci. U. S. A.*, **93**, 589–95.

Cook, J. E. (1996). Spatial properties of retinal mosaics: an empirical evaluation of some existing measures. *Vis. Neurosci.*, **13**, 15–30.

Cook, J. E. and Chalupa, L. M. (2000). Retinal mosaics: new insights into an old concept. *Trends Neurosci.*, **23**, 26–34.

de la Rosa, E. J. and de Pablo, F. (2000). Cell death in early neural development: beyond the neurotrophic theory. *Trends Neurosci.*, **23**, 454–8.

Eglen, S. J. and Willshaw, D. J. (2002). Influence of cell fate mechanisms upon retinal mosaic formation: a modeling study. *Development*, **129**, 5399–408.

Eglen, S. J., van Ooyen, A. and Willshaw, D. J. (2000). Lateral cell movement driven by dendritic interactions is sufficient to form retinal mosaics. *Network: Comput. Neural Syst.*, **11**, 103–18.

Finlay, B. L. and Pallas, S. L. (1989). Control of cell number in the developing mammalian visual system. *Prog. Neurobiol.*, **32**, 207–34.

Freeman, M. (1997). Cell determination strategies in the *Drosophila* eye. *Development*, **124**, 261–70.

Galli-Resta, L. (1998). Patterning the vertebrate retina: the early appearance of retinal mosaics. *Semin. Cell Dev. Biol.*, **9**, 279–84.

Galli-Resta, L. (2000). Local, possibly contact-mediated signaling restricted to homotypic neurons controls the regular spacing of cells within the cholinergic arrays in the developing rodent retina. *Development*, **127**, 1509–16.

Galli-Resta, L. (2002). Putting neurons in the right places: local interactions in the genesis of retinal architecture. *Trends Neurosci.*, **25**, 638–43.

Galli-Resta, L. and Novelli, E. (2000). The effects of natural cell loss on the regularity of the retinal cholinergic arrays. *J. Neurosci.*, **20**(RC60), 1–5.

Galli-Resta, L., Resta, G., Tan, S. S. and Reese, B. E. (1997). Mosaics of Islet-1-expressing amacrine cells assembled by short-range cellular interactions. *J. Neurosci.*, **17**, 7831–8.

Galli-Resta, L., Novelli, E. and Viegi, A. (2002). Dynamic microtubule-dependent interactions position homotypic neurones in regular monolayered arrays during retinal development. *Development*, **129**, 3803–14.

Holt, C. E., Bertsch, T. W., Ellis, H. M. and Harris, W. A. (1988). Cellular determination in the Xenopus retina is independent of lineage and birth date. *Neuron*, **1**, 15–26.

Jarman, A. P. (2000). Developmental genetics: vertebrates and insects see eye to eye. *Curr. Biol.*, **10**, R857–9.
Jeyarasasingam, G., Snider, C. J., Ratto, G. M. and Chalupa, L. M. (1998). Activity-regulated cell death contributes to the formation of on and off alpha ganglion cell mosaics. *J. Comp. Neurol.*, **394**, 335–43.
Kirby, M. A. and Steineke, T. C. (1996). Morphogenesis of retinal ganglion cells: a model of dendritic, mosaic, and foveal development. *Perspect. Dev. Neurobiol.*, **3**, 177–94.
Kouyama, N. and Marshak, D. W. (1997). The topographical relationship between two neuronal mosaics in the short wavelength-sensitive system of the primate retina. *Vis. Neurosci.*, **14**, 159–67.
Livesey, F. J. and Cepko, C. L. (2001). Vertebrate neural cell-fate determination: lessons from the retina. *Nat. Rev. Neurosci.*, **2**, 109–18.
Martin, P. R., Grunert, U. and Chan, T. L. (2000). Spatial order in short-wavelength-sensitive cone photoreceptors: a comparative study of the primate retina. *J. Opt. Soc. Am. A Opt. Image Sci. Vis.*, **17**, 557–67.
McCabe, K. L., Gunther, E. C. and Reh, T. A. (1999). The development of the pattern of retinal ganglion cells in the chick retina: mechanisms that control differentiation. *Development*, **126**, 5713–24.
Newman, E. A. (2001). Propagation of intercellular calcium waves in retinal astrocytes and Müller cells. *J. Neurosci.*, **21**, 2215–23.
O'Leary, D. D. M., Fawcett, J. W. and Cowan, W. M. (1986). Topographic targeting errors in the retinocollicular projection and their elimination by ganglion cell death. *J. Neurosci.*, **6**, 3692–705.
Raven, M. A., Eglen, S. J., Ohab, J. J. and Reese, B. E. (2003). Determinants of the exclusion zone in dopaminergic amacrine cell mosaics. *J. Comp. Neurol.*, **461**, 123–36.
Raven, M. A. and Reese, B. E. (2003). Mosaic regularity of horizontal cells in the mouse retina is independent of cone photoreceptor innervation. *Invest. Opthalmol. Vis. Sci.*, **44**, 965–73.
Reese, B. E. and Galli-Resta, L. (2002). The role of tangential dispersion in retinal mosaic formation. *Prog. Retin. Eye Res.*, **21**, 153–68.
Reese, B. E., Harvey, A. R. and Tan, S. S. (1995). Radial and tangential dispersion patterns in the mouse retina are cell-class specific. *Proc. Natl. Acad. Sci. U. S. A.*, **92**, 2494–8.
Reese, B. E., Necessary, B. D., Tam, P. P. L., Faulkner-Jones, B. and Tan, S. S. (1999). Clonal expansion and cell dispersion in the developing mouse retina. *Eur. J. Neurosci.*, **11**, 2965–78.
Resta, V., Novelli, E., Di Virgilio, F. and Galli-Resta, L. (2005). Neuronal death induced by endogenous extracellular ATP in retinal cholinergic neuron density control. *Development*, **132**, 2873–82.
Rockhill, R. L., Euler, T. and Masland, R. H. (2000). Spatial order within but not between types of retinal neurons. *Proc. Natl. Acad. Sci. U. S. A.*, **97**, 2303–7.
Rodieck, R. W. (1991). The density recovery profile: a method for the analysis of points in the plane applicable to retinal studies. *Vis. Neurosci.*, **6**, 95–111.
Rossi, C., Strettoi, E. and Galli-Resta, L. (2003). The spatial order of horizontal cells is not affected by massive alterations in the organization of other retinal cells. *J. Neurosci.*, **23**, 9924–8.
Stenkamp, D. L., Barthel, L. K. and Raymond, P. A. (1997). Spatiotemporal coordination of rod and cone photoreceptor differentiation in goldfish retina. *J. Comp. Neurol.*, **382**, 272–84.

Strettoi, E. and Volpini, M. (2002). Retinal organization in the Bcl-2-over-expressing transgenic mouse. *J. Comp. Neurol.*, **446**, 1–10.

Turner, D. L. and Cepko, C. L. (1987). A common progenitor for neurons and glia persists in rat retina late in development. *Nature*, **328**, 131–6.

Tyler, M. J., Carney, L. H. and Cameron, D. A. (2005). Control of cellular pattern formation in the vertebrate inner retina by homotypic regulation of cell-fate decisions. *J. Neurosci.*, **25**, 4565–76.

Waid, D. K. and McLoon, S. C. (1998). Ganglion cells influence the fate of dividing retinal cells in culture. *Development*, **125**, 1059–66.

Wässle, H. and Riemann, H. J. (1978). The mosaic of nerve cells in the mammalian retina. *Proc. R. Soc. London B*, **200**, 441–61.

Wässle, H., Boycott, B. B. and Illing, R. B. (1981). Morphology and mosaic of on-beta and off-beta cells in the cat retina and some functional considerations. *Proc. R. Soc. London B*, **212**, 177–95.

Wikler, K. C. and Rakic, P. (1991). Relation of an array of early-differentiating cones to the photoreceptor mosaic in the primate retina. *Nature*, **351**, 397–400.

Wong, W. T. and Wong, R. O. L. (2000). Rapid dendritic movements during synapse formation and rearrangement. *Curr. Opin. Neurobiol.*, **10**, 118–24.

Zhan, X. J. and Troy, J. B. (2000). Modeling cat retinal beta-cell arrays. *Vis. Neurosci.*, **17**, 23–39.

11

Programmed cell death

Rafael Linden
Instituto de Biofísica da UFRJ, Rio de Janeiro, Brazil

Benjamin E. Reese
University of California at Santa Barbara, USA

11.1 Introduction

Interest in programmed cell death (PCD) emerged over a century ago (reviewed in Clarke and Clarke, 1996), and such naturally occurring cell death in the developing nervous system has been extensively documented (Oppenheim, 1991 for review). More recently, the concept of PCD has been the subject of some controversy mainly due to the overwhelming interest in one of its forms, apoptosis (Sloviter, 2002). For the purpose of this chapter, PCD is defined simply as a sequence of events based on cellular metabolism that leads to cell destruction (Lockshin and Zakeri, 2001; Guimarães and Linden, 2004), without commitment to particular morphological types.

Programmed cell death has been identified using a variety of techniques, though each of them is prone to errors when estimating the magnitude of cell loss. Estimating the size of the population based on counts of axons in developing nerves or tracts may be confounded by the simultaneous occurrence of both cell death and axonal ingrowth, and by the transient contaminating presence of other axonal populations. Estimates based on cell counts may be influenced by the continuous migration of differentiating cells into spatially delimited cell populations, as well as by the inclusion of other types of cells that are not so readily discriminable at earlier developmental stages. And while great progress has been made in understanding the molecular mechanisms of apoptosis in the last decade, multiple alternative pathways of PCD add a further degree of complexity in understanding developmental cell death and estimating its magnitude. Even with an informed estimate for the clearance rate of such dying profiles, the estimated magnitude of PCD will be flawed when relying on markers for a single apoptotic pathway. The combination of several methods provides the most compelling evidence for PCD, but an accurate determination of its magnitude has remained largely elusive.

Two major roles have been ascribed to developmental cell death in the nervous system: the quantitative matching of interconnecting cell populations developing separately, and the correction of topographical and other targeting errors in neural pathways. While much of this work has been conducted within the retinofugal pathway, little concrete evidence for

Retinal Development, ed. Evelyne Sernagor, Stephen Eglen, Bill Harris and Rachel Wong.
Published by Cambridge University Press. © Cambridge University Press 2006.

the former exists therein, while the latter is frequently a misinterpreted example of axonal retraction or remodelling. Where targeting errors have been shown to be eliminated by cell death, their magnitude is modest, playing a minor role in the formation of the mature retinal architecture, its circuitry and the projection patterns of its optic axons.

The aims of this chapter are, first, to overview the evidence for the occurrence and extent of cell death among various cell classes of the vertebrate retina. As will be seen, evidence for all but the population of retinal ganglion cells is scant, and most conclusions must be provisional at best. Second, the chapter will review the mechanisms of retinal cell death at the cellular and molecular levels, for which the past decade has seen tremendous progress. As these mechanisms mediating cell death in a variety of animal models for retinal disease and degeneration have recently been reviewed (e.g. Pacione et al., 2003), the present chapter will focus almost exclusively on the developing mammalian retina. Third, the intercellular interactions modulating cell death will be discussed, highlighting recent developments that take into consideration those interactions within the histotypical environment of the developing retina. Finally, the chapter will consider the consequences of PCD upon the development of retinal architecture and connectivity, where claims for a functional role for cell death have generally exceeded the published evidence.

11.2 Anatomy of programmed cell death in the developing retina

In the retina, early accounts of PCD have been associated both with morphogenetic and with histogenetic events. The former has been related to the formation of the optic fissure (Silver and Hughes, 1973) and to the penetration of optic axons into the optic stalk (Ulshafer and Clavert, 1979; Cuadros and Rios, 1988; see Chapter 8), but not the formation of the fovea (see Chapter 7), whereas the latter has been examined in the context of the various retinal cell populations, following the early descriptions of developmental cell death among retinal ganglion cells (Hughes and LaVelle, 1975). Of the histogenetic forms of retinal cell death, that occurring in the retinal ganglion cells has received the most attention because this population allows multiple techniques for the detection, as well as for estimating the amount, of naturally occurring cell death. Furthermore, it provides an anatomically discrete population that can be readily separated from its target tissue and independently manipulated from its immediately neighbouring cells, permitting direct testing of the intercellular mechanisms controlling cell survival.

11.2.1 Evidence of cell loss

Counts of axons within the optic nerve at progressive developmental stages have suggested massive retinal ganglion cell death (see Provis and Penfold, 1988, for review). In the chick, for example, nearly four million axons are present at the peak, dropping to about two and a half million ten days later, and remaining at this level thereafter (Rager and Rager, 1978). Such estimates may be contaminated during early development by the presence of retino-retinal fibres (Bunt and Lund, 1981) as well as centrifugal fibres (Reese and Geller, 1995), and also by optic axons that bifurcate within the nerve (Dunlop, 1998), since, in some species, the number of axons present is far in excess of the number of retinal ganglion cells that can be labelled following tracer injections in the brain (Braekevelt et al., 1986; Dunlop and Beazley, 1987). Yet in other species, such as the rat, these are minor factors as estimates of the total number of retrogradely labelled ganglion cells are reasonable replications of the axon counts (Lam et al., 1982; Potts et al., 1982), supporting the contention that large numbers of ganglion cells die.

Still, the simple reduction in the number of axons, or ganglion cells that can be retrogradely labelled from their target visual nuclei, could be interpreted as transformation of a subset of retinal ganglion cells into another type of cell by retraction of their axons, for instance, to become displaced amacrine cells (Hinds and Hinds, 1983). A direct test of this hypothesis, however, failed to validate this interpretation (Perry et al., 1983). Further consistent with the hypothesis of developmental cell death, degenerating profiles were readily observed within the ganglion cell layer (GCL) (Sengelaub and Finlay, 1982). In some species, the time course for pyknotic profiles in the GCL does not coincide with the period of axonal elimination or ganglion cell reduction because a later, though overlapping, wave of cell death amongst displaced amacrine cells also occurs (Cusato et al., 2001). Still, the fact that dying cells were observed within the GCL when the total population was declining has generally been regarded as supportive evidence that the decline is in fact due to PCD. Obtaining an accurate estimate of the amount of this cell death, however, remains elusive.

11.2.2 The magnitude of ganglion cell loss

Estimates of the size of the retinal ganglion cell population, based either on axonal counts or on counts of retrogradely labelled neurons at progressive stages of development, still suffer from the fact that the period of axonal addition is believed to overlap with the period of ganglion cell loss. Thus, the peak number of retinal ganglion cells will underestimate the true size of the total population produced. Others have attempted to follow individual cohorts of ganglion cells to determine the proportion that either lives or dies, yet conspicuously different results have been obtained with this approach. Comparing the rate of cell death in individual cohorts of neurons in the GCL of the rat retina with the period of ganglion cell genesis led to an estimate that as many as 90% of the ganglion cells die during development (Galli-Resta and Ensini, 1996). On the other hand, a somewhat similar approach in the mouse retina led to a rather different conclusion, that naturally occurring ganglion cell death amounts to around 50% (Farah and Easter, 2005). If indeed the earliest cohort of ganglion cells dies only after birth (Farah and Easter, 2005), and assuming no later-generated cohorts die before this cohort, then optic axon counts may provide a reasonable upper limit on the number of generated retinal ganglion cells, and those counts in the mouse suggest that 61% to 70% of the initial population is lost, depending upon the strain (Strom and Williams, 1998). Whether greater numbers of optic axons are present prior to birth, as has been found in the rat's optic nerve (Crespo *et al.*, 1985), suggestive of a ~65% loss therein, remains to be seen.

This estimate for the mouse approximates the 50% to 74% values described for various other mammals (see Dreher and Robinson, 1988, for review). In the majority of these species, the wave of this cell death comes at approximately the same developmental stage, at about three-quarters of the duration of the period between conception and eye opening (Dreher and Robinson, 1988). Little more definitive can be said for the retinal ganglion cell population at this stage, leaving one with the conservative view that perhaps one or two retinal ganglion cells are lost for every one that survives. This loss coincides with estimates for naturally occurring cell death in other populations (Clarke, 1985), and was rapidly incorporated into the growing opinion that developing neuronal populations are overproduced by a factor of one to ensure a sufficient excess in the absence of knowing the ultimate size of a target structure (Clarke, 1985). How general, then, is this scale of overproduction followed by PCD amongst the other types of neuron within the developing retina?

11.2.3 Cell loss amongst other retinal populations

Other retinal neurons must surely undergo PCD, because pyknotic profiles have also been demonstrated within the inner (INL) and outer nuclear layers (ONL) during development (e.g. Young, 1984; Robinson, 1988). Estimating the amount of cell death amongst individual retinal cell populations other than ganglion cells is technically more difficult, because those populations cannot be selectively counted nor labelled via their axonal projections, and because known markers of individual retinal cell types are developmentally regulated (e.g. Pow *et al.*, 1994), so that changes in the number of immunopositive cells as a function of development may reflect events unrelated to cell death. Nevertheless, the relative frequency of dying cells, based on TUNEL-positive profiles, in the different layers of the retina has raised the possibility that naturally occurring cell death may affect different cell types to markedly different extents. In the chick, virtually none were found within the developing ONL (Cook *et al.*, 1998), while their frequency in the mouse, rat, rabbit, cat, ferret and human retina, expressed as TUNEL-positive cells per linear unit of a retinal section, is about one-tenth of that within the INL, occurring largely after the wave of cell death in the INL has finished (Maslim *et al.*, 1997; Georges *et al.*, 1999; Johnson *et al.*, 1999; Mervin and Stone, 2002). These data suggest that TUNEL-positive cell death modulates these two layers at conspicuously different rates. Other studies in the rabbit and quokka retina, however, have reported comparable amounts of dying cells in the INL and ONL that occur simultaneously, based on the frequency of pyknotic profiles labelled with aniline dyes (Robinson, 1988; Harman *et al.*, 1989). Given that multiple cell-execution pathways exist, and TUNEL-staining detects only one of them (considered in the next section), one cannot yet be sure whether naturally occurring cell death affects a large number of photoreceptors.

To date, there are no clear demonstrations of naturally occurring cell death amongst cone photoreceptors, and no documented instances for an overproduction of cones has been made. The fact that no pyknotic profiles were detected in the ONL of the chick retina, a cone-dominant retina (Cook et al., 1998), may indicate that this population is immune to naturally occurring cell death. Rods, on the other hand, clearly undergo some degree of PCD, given the positioning of pyknoses in the inner (cone-free) portion of the ONL (Williams et al., 1990; Johnson et al., 1999), but no estimates of total rod loss have been provided.

Within the INL, pyknotic nuclei are found amongst amacrine and bipolar cells discriminated during the later stages of retinal differentiation on the basis of morphological and positional criteria (Young, 1984). There is also a population of rod photoreceptors that becomes separated from the ONL as the outer plexiform layer (OPL) forms, and these cells are thought to either migrate back across the differentiating OPL or undergo cell death (Spira et al., 1984; Young, 1984). No direct evidence for naturally occurring cell death amongst the horizontal cells has been found; on the contrary, counts of immunolabelled horizontal cells throughout postnatal development show no variation across age, in two different strains of mouse, despite a twofold variation between the strains (Raven et al., 2005). Those results suggest that the initial specification of cellular fate is entirely responsible for regulating horizontal cell number (see Chapter 5). Indirect evidence exists, however, showing that horizontal cells survive via an autocrine mechanism mediated via nerve growth factor (NGF)-TrkA signalling (Karlsson et al., 2001). *Bcl-2*-overexpressing mice contain nearly a 20% increase in the total number of horizontal cells relative to wild-type mice, taken as an indication that this population is normally overproduced and only partially protected by the anti-apoptotic action of Bcl-2 (Strettoi and Volpini, 2002). Dopaminergic amacrine cells have also not been shown to be overproduced during normal development, but show a ninefold increase in the *bcl-2*-overexpressing retina. By contrast, cholinergic amacrine cells are overproduced by about 20% during development (Galli-Resta and Novelli, 2000; Resta et al., 2005), but are not affected by *bcl-2* overexpression (Strettoi and Volpini, 2002). Finally, rod bipolar cells have not been shown to undergo an initial overproduction, but are increased by 32% in the *bcl-2*-overexpressing retina (Strettoi and Volpini, 2002). The limitations of using *bcl-2* overexpression to infer normal cell loss will be discussed in Section 11.3.2.

As indicated above, pyknotic profiles are not uncommon in the INL, yet, to date, no identified population of neurons has been shown to undergo conspicuous cell loss. Estimates for the clearance of pyknotic profiles have, over the years, ranged from three hours to one day, and recently, real-time imaging of nuclear fragmentation in dying cells within the GCL indicates less than one hour from initial detection to clearance (Cellerino et al., 2000). Assuming similarly rapid kinetics within the developing INL, it is surprising that no other populations of identified INL cells have been shown to undergo PCD. As better markers for particular types of immature neurons become available, more populations may be revealed to exhibit overproduction comparable to the ganglion cell population. Alternatively, it may mean that the majority of these pyknotic profiles are proliferating cells, and that the extent of overproduction observed in the population of retinal ganglion cells turns out to be the exception, rather than rule, for retinal neurons.

11.2.4 Loss of proliferating cells

Evidence for naturally occurring death of proliferating retinal cells has been obtained mainly in the chick and, although the magnitude of the cell loss in this population is not known, the frequency of pyknotic profiles in the neuroblastic layer at early stages (prior to all but ganglion cell genesis) is of the same order of magnitude as that for postmitotic ganglion cells (de la Rosa and de Pablo, 2000). Natural cell death amongst undifferentiated postmitotic cells has not been documented (unless protein synthesis is interrupted – Rehen et al., 1996), leading to the suggestion that retinal cells undergo two distinct waves of naturally occurring cell death, namely at late proliferating stages, and then as they differentiate, forming their afferent and efferent connections (de la Rosa and de Pablo, 2000). That former wave may then account for many of the pyknotic cells observed in the outer portion of the INL when it still contains proliferating cells; if it does, it would suggest that cell death within proliferating cells occurs near S-phase, rather than surrounding M-phase, given the sparseness of pyknoses in the ONL at these same times. But it still cannot account for much of the pyknoses in the INL, occurring as they do in the innermost portion (the developing amacrine cell layer), which is vitreal to the S-phase zone, or in the outermost portion (the developing bipolar and horizontal cell layer) after neurogenesis has ceased (see Chapter 13).

11.3 Cellular and molecular biology of programmed cell death in the developing retina

11.3.1 The signature of dying cells

Apoptosis, defined by ultrastructural criteria, was originally proposed as the singular correlate of PCD. This form of cell death presents as a combination of both nuclear and cytoplasmic condensation, accompanied by the relative preservation of cytoplasmic organelles. Blebbing of the plasma membrane leads the cell to break up into pieces known as apoptotic bodies, often containing round remnants of condensed nuclear chromatin. Apoptotic bodies are quickly removed either by mononuclear phagocytes or by neighbouring cells, especially in epithelial tissues (Kerr et al., 1972).

Electron microscopy is, however, impractical for large-scale quantitative studies of PCD, and the specific pattern of internucleosomal DNA cleavage, detected by agarose gel electrophoresis of DNA extracted from apoptotic cells (Wyllie, 1980), ignores the spatial distribution of degenerating cells. This stimulated the widespread acceptance of pyknotic, condensed chromatin as a marker of cell death at the light microscopic level. Indeed, since the early studies of morphogenesis of the optic fissure, several quantitative studies of both normal developing tissue, and of degenerate retina following lesions were undertaken, largely based on counts of pyknotic profiles stained with basic aniline dyes.

The pattern of pyknosis and chromatin breakdown is consistent with the hypothesis that PCD in the retina occurs by apoptosis (Kerr *et al.*, 1972). Still, early after the debut of apoptosis, some investigators loosely employed the term 'necrosis' to refer to the presence of pyknotic profiles in developing retinal tissue (Silver and Hughes, 1973). The TUNEL technique to detect DNA strand breaks deemed to be typical of apoptosis (Gavrieli *et al.*, 1992) allowed for large-scale quantification of this form of cell death, and recent studies of retinal PCD have often, and perhaps hastily (see below), relied exclusively upon the detection of TUNEL-positive apoptotic profiles.

The disassembly of cellular components is attributed mainly to the activity of cysteine-aspartate proteases of the caspase family (Cryns and Yuan, 1998). Induction of DNA strand breaks, for example, was traced to activation of the so-called caspase-activated endonuclease, after caspase-3-catalyzed hydrolysis of an inhibitor (Enari *et al.*, 1998). Thus, the TUNEL technique is useful to detect apoptosis dependent on caspase-3 and, in fact, antibodies to the activated form of this protease label populations of degenerating retinal profiles that overlap with populations of TUNEL-positive profiles (Guimarães *et al.*, 2003). Likewise, internucleosomal fragmentation of DNA typical of apoptosis has been detected during the period of naturally occurring cell death in the GCL (Wong *et al.*, 1994), while deletion of either caspase-3 (CPP32) or Apaf-1, a component of the apoptosome necessary for the activation of caspase-9 and -3 following release of mitochondrial cytochrome c, results in a thickening of the embryonic mouse retina (Cecconi *et al.*, 1998). These data are consistent with the occurrence of caspase-3-mediated apoptosis among developing retinal cells.

11.3.2 The *Bcl-2* gene family

The rate of naturally occurring ganglion cell death is reduced in transgenic mice overexpressing *bcl-2*, and leads to an increased number of optic axons, suggesting that this gene may be involved in the control of developmental cell death among ganglion cells (Martinou *et al.*, 1994; Bonfanti *et al.*, 1996). From fetal life onwards, however, the expression of *bcl-2* within the central nervous system is relatively low compared with a distinct anti-apoptotic member of the same family of genes, *bcl-xL* (Levin *et al.*, 1997). Since anti-apoptotic members of the *Bcl-2* family may substitute for one another upon ectopic expression (Sedlak *et al.*, 1995), the evidence from overexpression is not compelling. In addition, most of the bcl-2 immunoreactivity in the developing retina appears to be concentrated in Müller glial endfeet, rather than in neurons (Chen *et al.*, 1994; Sharma, 2001). In fact, in at least one line of transgenic mice, overexpression of *bcl-2* in Müller cells led to extensive retinal degeneration (Dubois-Dauphin *et al.*, 2000). Although nearly 30% of the retinal ganglion cells degenerate in *bcl-2*-deficient mice, this happens only at a time after the period of naturally occurring cell death (Cellerino *et al.*, 1999). Thus, while *bcl-2* may affect cell death in the developing retina, it is not yet established whether it is relevant for naturally occurring cell death. It follows that estimates of the extent of naturally occurring cell death based on the increased numbers of various types of retinal cells in *bcl-2*-overexpressing transgenic mice (Strettoi and Volpini, 2002) should be taken with caution.

On the other hand, deletion of the pro-apoptotic gene *bax* led to a substantial reduction in ganglion cell death in the mouse retina (e.g. Mosinger Ogilvie *et al.*, 1998), suggesting a role for *bax* upon ganglion cell death in the developing retina. Limited evidence suggests that *bax* also plays a similar role in the embryonic neuroblastic layer and postnatally in the INL (Mosinger Ogilvie *et al.*, 1998; Hahn *et al.*, 2003). Yet the results of overexpression of *bax* are contradictory (Bernard *et al.*, 1998; Eversole-Cire *et al.*, 2002) and, similar to the anti-apoptotic genes of the *Bcl-2* family, pro-apoptotic members of this family may also compensate for one another. Indeed, double-deletion experiments showed that either *bax* or *bak* appear to be sufficient for PCD among the ectopic photoreceptors that disappear during normal retinal development (Hahn *et al.*, 2003). The pro-apoptotic protein bad has also been detected by immunohistochemistry in the developing rat retina (Rickman *et al.*, 1999), and studies in vitro indicated that phosphorylation of bad may be involved in the control of sensitivity to cell death (Campos *et al.*, 2003), but its role in the control of naturally occurring retinal cell death is still unclear.

11.3.3 Transcription and protein synthesis

Early studies have stressed the dependency on protein synthesis as a defining property of PCD. In the retina, inhibitors of protein synthesis blocked ganglion cell death (Rabbachi *et al.*, 1994). This was taken as evidence that ganglion cell death was an 'active, apoptotic cell death' (Rabbachi *et al.*, 1994), although various studies have shown that neither qualification requires protein synthesis (reviewed in Guimarães and Linden, 2004). When present, however, dependency on protein synthesis suggests the need for specific gene expression, and for the activity of transcription factors.

Either mutation or deletion of several transcription factors induces cell death among developing photoreceptors (Becker *et al.*, 1998; Pennesi *et al.*, 2003) and ganglion cells (Gan *et al.*, 1999). The expression of a *fos-lacZ* transgene preceding developmental cell death suggests that c-Fos is involved in retinal PCD (Smeyne *et al.*, 1993). Activation of c-Jun is deemed an essential component of PCD in various postmitotic neuron types, but a xenograft experiment, designed to circumvent the early embryonic lethality of deletion of *c-Jun*, failed to provide evidence that expression of this gene is required for retinal cell death (Herzog *et al.*, 1999).

Indeed, whereas c-Jun-like immunoreactivity has been detected in ganglion cells after optic axon damage (Isenmann and Bähr, 1997; Chiarini *et al.*, 2002), the cytoplasmic location of the protein is inconsistent with a role as a transcription factor. The transcription factors c-Fos, c-Myc and Max, as well as other nuclear proteins such as the bifunctional a purinic-apyrimidinic endonuclease/redox factor Ref-1, were also detected by immunohistochemistry in the cytoplasm of developing retinal cells undergoing induced cell death (Linden and Chiarini, 1999; Linden *et al.*, 1999). Recently it was shown that the nuclear release of certain H1-class histones may be required to trigger the mitochondrial release of cytochrome c in p53-dependent apoptosis induced by DNA damage (Konishi *et al.*, 2003). Nuclear exclusion of the transcription factor Max is an early event following axotomy that is independent of the activation of caspase-3, and precedes the eventual discharge of H1 histones into the cytoplasm (Petrs-Silva et al., 2004). Both the developmental pattern of expression and its loss during cell death (Petrs-Silva et al., 2004), as well as preliminary results of viral vector-mediated overexpression (Petrs-Silva et al, 2005), suggest that Max may have a cytoprotective function in the retina. These data highlight the importance of changes in the subcellular trafficking of nuclear proteins, particularly of transcription factors, for mechanisms of PCD in the developing retina.

Evidence is also available for a role of *Rb* (retinoblastoma) family genes upon the survival of retinal progenitor cells, but not postmitotic cells in the CNS (Slack et al., 1998). Apoptosis detected in the retina of *Rb*-null mice was partially prevented by deletion of the transcription factor *E2F1*, suggesting that the complex pRb/E2F, which is strongly implicated in the control of the cell cycle, may also cooperate in the control of developmental cell death (Saavedra et al., 2002). The pRb protein is a substrate of caspases, and expression of a caspase-resistant form of pRb led to a reduction in the extent of axotomy-induced ganglion cell death in neonatal mice (Chau et al., 2002). These data are consistent with a complex role of pRb in the control of PCD in the developing retina. In turn, the role of the *p53* gene is still uncertain. A remarkable effect of genetic background was found in studies of *p53* deletion in differing mouse strains (Ikeda et al., 1999). Deoxyribonucleic acid damage-induced apoptosis following irradiation is accompanied by increased expression of *p53*, but neither of its usual downstream effectors, $p21^{WAF1/cip1}$ and Bax, were involved (Herzog et al., 2002). In contrast, the Ataxia Teleangiectasia-Mutated protein kinase (Atm) is required upstream of p53 for irradiation-induced cell death (Borges et al., 2004). Thus, a role for *p53* in the developing retina may depend on both cell type and on other simultaneous genetic determinants.

Several other genes have been associated with PCD in the developing retina, such as: *Ap3b1*, encoding the beta3A subunit of the AP-3 adaptor complex, which regulates vesicular trafficking (Feng et al., 2000), and corresponds to the *pearl* mutation that is associated with an altered time course of naturally occurring ganglion cell death (Linden and Pinto, 1985; Williams et al., 1990); *CRB1*, mutations of which are associated with a thickening of the human retina, which may arise from disturbance of naturally occurring cell death (Jacobson et al., 2003); and *Msx2*, overexpression of which induces extensive cell death in the optic vesicle, leading to microphthalmia possibly through deregulated expression of *Bmp4* and/or *Bmp7* genes (Wu et al., 2003). Whereas both the *Ap3b1* (*pearl*) and *CRB1* genes seem to be related specifically to histogenesis of the retina, *Msx2* appears to have a role in morphogenetic cell death.

Current studies have, therefore, provided evidence for the occurrence of TUNEL-positive, caspase-3-mediated, *Bcl-2* family-modulated apoptosis. Nonetheless, this form of cell death cannot be taken as the mechanism of all PCD in tissues (see Guimarães and Linden, 2004, for review), and specifically within the retina, multiple post-translational pathways of PCD have been identified (Guimarães et al., 2003). In contrast with the abundance of markers of apoptosis, only recently has a marker of autophagy been introduced for light microscopy (Munafo and Colombo, 2001). The future introduction of new markers for distinct types of

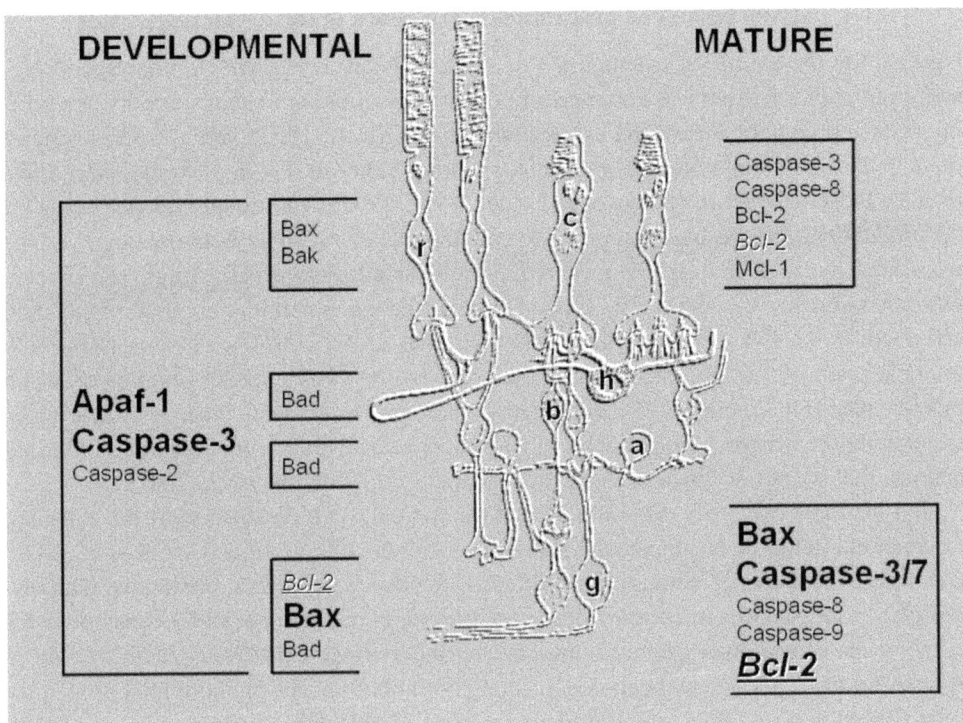

Figure 11.1 Components of the apoptotic execution machinery identified in studies of naturally occurring cell death in developing retina (*left*), as compared with models of retinal degeneration and experimental lesions in mature animals (*right*). A role for some of these components has been described for the tissue in general, without reference to particular retinal cell types (to the left in the diagram). Brackets indicate available data for specific cell types. Larger, bold font indicates stronger evidence. Upright fonts indicate pro-apoptotic components, underlined italics indicate anti-apoptotic roles. Bcl-2 indicates family members, not necessarily the Bcl-2 protein proper. Evidence is clearly stronger for certain molecules that have been tested by various means such as knockout or knock-in, plus pharmacological inhibition, and expression studies. Notice that the evidence is generally more complete for experimental models of mature animals, but rather fragmentary and, in some cases, weak in the developing retina. a, amacrine cell; b, bipolar cell; c, cone; g, ganglion cell; h, horizontal cell; r, rod.

cell death may affect our understanding of both the kinetics and the mechanisms of PCD in the retina. In addition, the assumption that mechanisms involved, for example, in retinal degeneration in adult animals (e.g. Pacione *et al.*, 2003) equally apply to developmental cell death is unwarranted, because clear distinctions exist among mechanisms of retinal cell death depending on either the means of induction or on the stage of differentiation of the retinal cells (Li *et al.*, 2000; Chau *et al.*, 2002; Chiarini *et al.*, 2003; Borges *et al.*, 2004). Thus, a critical appraisal of all these parameters is required to draw a coherent picture of the mechanisms of naturally occurring retinal cell death, in particular the coordinated role of those genes associated with the regulation of developmental cell loss (Figure 11.1).

11.4 Tissue biology of programmed cell death in the developing retina

Upstream to the execution mechanisms of PCD, cells are kept alive by a barrage of signals from their environment, coming from other cells in the immediate vicinity, connecting cells and components of the extracellular matrix (Raff, 1992). Many studies support the hypothesis that competitive interactions among either developing axons or dendrites regulate PCD in the retina (reviewed in Linden, 1987, 1992). Additional evidence exists for the production of molecules that promote ganglion cell survival by both the targets of the retinofugal axons as well as within the retina itself (Schulz et al., 1990; Araujo and Linden, 1993; Ary-Pires et al., 1997). The limited effect of ganglion cell loss upon developmental cell death in the INL (Cusato et al., 2001; Williams et al., 2001) is also consistent with multiple sources of trophic support mediated both by their postsynaptic ganglion cells and by other retinal interneurons. It is noteworthy that massive cell death precedes the development of mature synapses (see Chapter 13), and may include transient interactions of retinal neurons that are not destined to share such associations (Williams et al., 2001).

Non-neuronal cell types may also affect developmental cell death. Despite the abundant evidence that glial cells support the survival of retinal cells in vitro (Garcia et al., 2002 and references therein), there is remarkably little evidence for such a role in vivo (Dubois-Dauphin et al., 2000). In contrast, the retinal pigment epithelium (RPE) is essential for homeostasis of the outer retina (Schraermeyer and Heimann, 1999), including a role in the control of developmental cell death among both photoreceptors and retinal progenitor cells (Sheedlo et al., 1998, 2001; Soderpalm et al., 2000). Effects of the RPE upon both proliferation and cell death have been attributed to melanin-related agents (Jeffery, 1998). However, other genetic differences likely contribute to the control of cell death because, for example, the time courses of naturally occurring cell death in the GCL are similar in congenic wild-type and albino mice (Linden and Pinto, 1985).

11.4.1 Neural activity

Developing ganglion cells in vitro are either killed by blockade of voltage-gated Na^+ channels with tetrodotoxin (Lipton, 1986), or, conversely, protected by the depolarizing agent veratridine (Pereira and Araujo, 1997). In vivo, however, blockade of electrical activity in the retina affected either the pattern (Fawcett et al., 1984) or the timing (Kobayashi, 1993), but not the extent, of ganglion cell death (O'Leary et al., 1986; Scheetz et al., 1995). Thus, evidence for a role of neural activity upon developmental cell death in the retina is limited. It is not known whether the Na^+-dependent currents relevant for ganglion cell survival belong to either the ganglion cells themselves, or to retinal interneurons (Steffen et al., 2003). Indeed, Na^+ channel-dependent action potentials are not required for the release of certain neuroactive substances within the retina (Protti et al., 1997).

11.4.2 Trophic factors

Both neurotrophins and their receptors are expressed in the developing vertebrate retina (see Chapter 6). Interestingly, the prototypical neurotrophic factor NGF induced apoptotic cell death among early postmitotic retinal ganglion cells through its low-affinity receptor p75, and mediated by the p75-interacting zinc finger protein neurotrophin receptor interacting factor (Frade and Barde, 1999). The release of pro-degenerative NGF has been attributed to microglia (Frade and Barde, 1998) and to retinal ganglion cells (Gonzalez-Hoyuela et al., 2001). However, in rat retina the expression of p75 was found in Müller glial processes, and not in ganglion cells (Ding et al., 2001). Also in contrast with the pro-degenerative effect of NGF upon early developing ganglion cells, there is a neuroprotective role of NGF upon both amacrine and horizontal cells at later stages of chick retinal development (Karlsson et al., 2001).

Neurotrophin-3 (NT-3) protects chick retinal ganglion cells from degeneration in vitro (de la Rosa et al., 1994), while antibodies to NT-3 reduce the number of surviving optic axons (Bovolenta et al., 1996). Evidence regarding the TrkB ligands, brain-derived neurotrophic factor (BDNF) and NT-4, is more complex. Both protect ganglion cells and cells of the INL through the high-affinity receptor TrkB (Ma et al., 1998; Cui and Harvey, 2000; Cusato et al., 2002). Nonetheless, overexpression of BDNF in the superior colliculus changed the pattern but did not prevent developmental ganglion cell death in rats (Isenmann et al., 1999), and deletion of neither the *BDNF* nor the *TrkB* genes affected the final number of retinal ganglion cells (Cellerino et al., 1997; Rohrer et al., 2001; Pollock et al., 2003). It is not known whether the lack of an effect in knockout mice is due to either a complementary role of other neurotrophins and/or their receptors, or to other compensatory changes during embryogenesis. Brain-derived neurotrophic factor found within the retina is mainly derived from local sources, particularly intraretinal afferents to the ganglion cells, rather than retrogradely transported from the tectal target (Herzog and von Bartheld, 1998). Thus, the evidence for BDNF is the strongest as an intraretinal, afferent-derived trophic factor for retinal ganglion cells. However, the signalling pathways activated by binding of neurotrophins to their receptors in the retina are still poorly understood, and the role of neurotrophins upon naturally occurring retinal cell death is still in dispute.

Regarding other trophic factors, a link has been found between the binding of either retina-derived insulin or insulin-like growth factors (IGFs) to IGF receptors, and protection from caspase-3-mediated developmental cell death, particularly of ganglion cells, through at least two distinct branches of the phosphatidylinositol-3 kinase signalling pathway (Kermer et al., 2000; Barber et al., 2001; Gutierrez-Ospina et al., 2002; Wu et al., 2003). Glial-derived neurotrophic factor protects photoreceptors from cell death in both dissociated and reaggregate cultures of embryonic chick retina (Politi et al., 2001; Rothermel and Layer, 2003). Also, consistent with work in *Msx2* knockout mice, deletion of the bone morphogenetic protein (BMP) receptor BmprIb led to extensive cell death at the end of the neurogenetic period in postnatal mice (Liu et al., 2003). In addition, evidence was reported for neuroprotective modulation of cell death in developing retinal cells by interleukins (IL)-2 and -4 (Sholl-Franco et al., 2001), tumour necrosis factor-α (Diem et al., 2001) and Stromal Cell-Derived Factor-1 (Chalasani et al., 2003), while transforming growth factor-β induced cell death in chick dissociated retinal cell cultures (Schuster et al., 2002). Thus, besides their roles in the immune system and in neuroimmune interactions, cytokines and chemokines may be involved in the control of developmental cell death in the retina.

In contrast, ciliary neurotrophic factor (CNTF) had no effect upon developmental photoreceptor cell death (Fuhrmann et al., 1998; Kirsch et al., 1998). Studies of fibroblast growth factor (FGF) have also provided conflicting results (Yokoyama et al., 1997; Yamada et al., 2001), possibly because of genuine differences in the effects of this growth factor upon the retina of distinct species (reviewed in Hicks, 1998). Clearly, the evidence for the roles of trophic factors upon the retina is still fragmentary, and further work is required to clarify the sets of neurotrophins and other cytokines that control retinal cell survival during development.

11.4.3 Neurotransmitters

Neurotransmitters and neuromodulators strongly modulate developmental cell death in the retina (see Chapter 6 and Linden et al., 2005 for reviews). Excitotoxicity through either the N-methyl-D-aspartate (NMDA) or α-amino-3-hydroxy-5-methyl-4-isoxazolepropionate (AMPA)/kainate receptors is very low in immature retinal tissue (Cui and Harvey, 1995; Haberecht et al., 1997), and systemic blockade of the NMDA receptor changed the pattern but did not affect the overall extent of retinal ganglion cell death in postnatal rats (Bunch and Fawcett, 1993). This all but excludes excitotoxicity as a relevant mechanism of naturally occurring retinal cell death. On the contrary, glutamate plays a predominantly protective role upon developing retinal cells (Reuter and Zilles, 1993; Fix et al., 1995; Nichol et al., 1995). It has been shown that chronic activation of NMDA receptors in vitro induces a BDNF-dependent neuroprotective state in differentiating retinal cells, and that NMDA receptor activation controls PCD of developing retinal neurons in vivo (Rocha et al., 1999; Martins et al., 2005). Developmental regulation of both distinct subunits and of splice variants of the various types of glutamate receptors follows a complex pattern (e.g. Koulen et al., 1996; Grunder et al., 2000; Johansson et al., 2000; Sucher et al., 2003), and divergent pro- and anti-apoptotic signals were found to be simultaneously induced by NMDA (Manabe and Lipton, 2003). The overall data suggest that the outcome of glutamatergic stimulation in the developing retina depends on a balance between pro- and anti-degenerative signal transduction pathways. Current data mostly suggest that glutamate plays a physiologically protective role upon developing retinal cells.

Nitric oxide (NO) produced in the GCL and INL has also been shown to play a neuroprotective role, mediated by cGMP, upon undifferentiated postmitotic retinal cells (Guimarães et al., 2001). This contrasts with an early suggestion that NO mediates cell death in retinal ganglion cells (Nichol et al., 1995). Other studies have failed to link the activity of NO synthase to retinal ganglion cell death in both rat and chick (Patel et al., 1997; Goureau et al., 1999; Guimarães et al., 2001), although NO may induce photoreceptor cell death (Goureau et al., 1999; Ju et al., 2001).

Dopamine exerts a powerful protective effect upon undifferentiated postmitotic cells in the developing retina, mediated by a D1-like receptor, through the downstream activation of cyclic adenosine monophosphate (cAMP)-dependent protein kinase (Varella et al., 1997, 1999). Protection by cAMP appears to be widespread in the nervous system (Silveira and

Linden, 2005), but the downstream targets of protein kinase A involved in neuroprotection have yet to be identified. Notwithstanding, phosphorylation of the pro-apoptotic protein Bad was recently described in the retina following treatment with the adenylyl-cyclase activator forskolin (Campos *et al.*, 2003), and this may be associated with the mechanisms of cAMP-mediated neuroprotection.

Neuropeptides are also widely expressed in the developing retina (see Linden *et al.*, 2005 for review). Both vasoactive intestinal peptide (VIP) and the related peptide pituitary adenylyl cyclase-activating polypeptide (PACAP) protect developing retinal cells in vitro, through regulation of intracellular cAMP (Kaiser and Lipton, 1990; Silveira *et al.*, 2002). It is still unclear whether dopamine, PACAP or VIP plays such a role upon naturally occurring neuronal death in vivo. Purinergic signalling has recently been implicated in the control of the number of cholinergic amacrine cells, dependent upon ATP-P2X receptors (Resta *et al.*, 2005).

These studies suggest that various neurotransmitters and neuromodulators produced by developing retinal cells are involved in the control of sensitivity to cell death. Further studies are needed to assess the contribution of neurotransmitters to the afferent control of naturally occurring neuronal death (Linden, 1994). These shall add to the multiple trophic factors that operate in the complex tissue environment of the developing retina.

11.4.4 Oxygen

Oxygen tension within the developing retina is an additional factor in the control of cell death among both photoreceptors and proliferating cells (Maslim *et al.*, 1997). Oxidative stress induces, while antioxidants may prevent, retinal cell death (Castagne and Clarke, 1996; Giardino *et al.*, 1998; Borges and Linden, 1999). Developmental cell death is, therefore highly dependent on the maintenance of cellular redox status within a relatively narrow range (Stone *et al.*, 1999; Geiger *et al.*, 2002), and probably on the expression and activity of redox-regulated proteins (Chiarini and Linden, 2000; Tanito *et al.*, 2002).

11.4.5 Extracellular matrix

Tissue-specific components of the extracellular matrix (ECM) may also mediate extracellular signals related to the control of developmental retinal cell death. Laminin beta2 chain-deficient mice show increased cell death in the ONL that may be related to defective development of photoreceptor outer segments (Libby *et al.*, 1999). Increased activity of plasminogen activator and plasmin-mediated degradation of laminin has been correlated with increased apoptosis of retinal ganglion cells (Zhang *et al.*, 2003). An optic tectum-derived trophic factor for retinal ganglion cells was identified as a proteoglycan (Huxlin *et al.*, 1995), and enzyme-induced degradation of heparan sulphate modulated cell death within immature retinal tissue (Erlich *et al.*, 2003). Mechanisms of ECM regulation in the retina are still poorly understood, and may include the modulation of growth factors or cytokines (Schonherr and Hausser, 2000).

11.4.6 Gap junctions

The relevance of tissue structure to the control of developmental cell death is further highlighted by the evidence that gap junctions mediate the transfer of pro-apoptotic signals among developing retinal cells (Cusato *et al.*, 2003). A NO donor, a cAMP analogue and an inhibitor of guanylyl cyclase partially prevented the transmission of pro-apoptotic signals across gap junctions (Cusato *et al.*, 2003). It is unknown whether these second messengers act by modulating gap junctional permeability or whether they directly affect the pro-degenerative messages trafficking through the gap junctions.

11.4.7 Microglia

Finally, clearance of apoptotic bodies is an important aspect of the resolution of PCD (de Almeida and Linden, 2005, for review). In the retina, a role for neuroepithelial cells in the removal of cell debris has been described (Garcia-Porrero and Ojeda, 1979), but a more prominent role for macrophages/microglia and Müller cells has been recognized (Hume *et al.*, 1983; Young, 1984; Linden *et al.*, 1986; Egensperger *et al.*, 1996). Microglia not only phagocytose dead cells in the ONL, but may also be actively involved in the induction of cell death in the INL (Frade and Barde, 1998). Thus, resident phagocytes may play complex roles in the regulation of PCD in the retinal tissue.

Trophic support to retinal cells is thus mediated by various families of extracellular signalling molecules, derived from both their post- and presynaptic connecting partners, as well as from glial and pigment epithelial cells. In particular, retinal PCD is affected by typical tissue characteristics, such as gap junctional communication, oxygen tension within the tissue and resident macrophage populations (Figure 11.2).

11.5 Roles of programmed cell death in retinal development

One of the earliest functional proposals for developmental cell death was that of error elimination (Catsicas *et al.*, 1987), with the defining feature of 'error' generally being some characteristic (e.g. somal positioning, axonal projection) thought to be atypical of the mature retina or visual pathway. Whether such 'errors' truly comprise developmental 'accidents', or in fact play some transient functional role, remains to be seen, but one can largely ignore this issue and simply ask whether cell death is responsible for the observed maturational change of the retina or visual pathway (Figure 11.3).

11.5.1 Elimination of anomalous projections

Growing optic axons frequently invade anomalous targets, that is, structures that in maturity are not normally retino-recipient (Figure 11.3a). For instance, there are transient retinal projections to the opposite retina (Bunt and Lund, 1981), the ventrobasal nucleus (Frost, 1986) and the inferior colliculus (Cooper and Cowey, 1990), all of which disappear as development proceeds. Given the degree of overproduction of retinal ganglion cells considered above, it is reasonable to consider that the elimination of these misprojecting axons is brought about

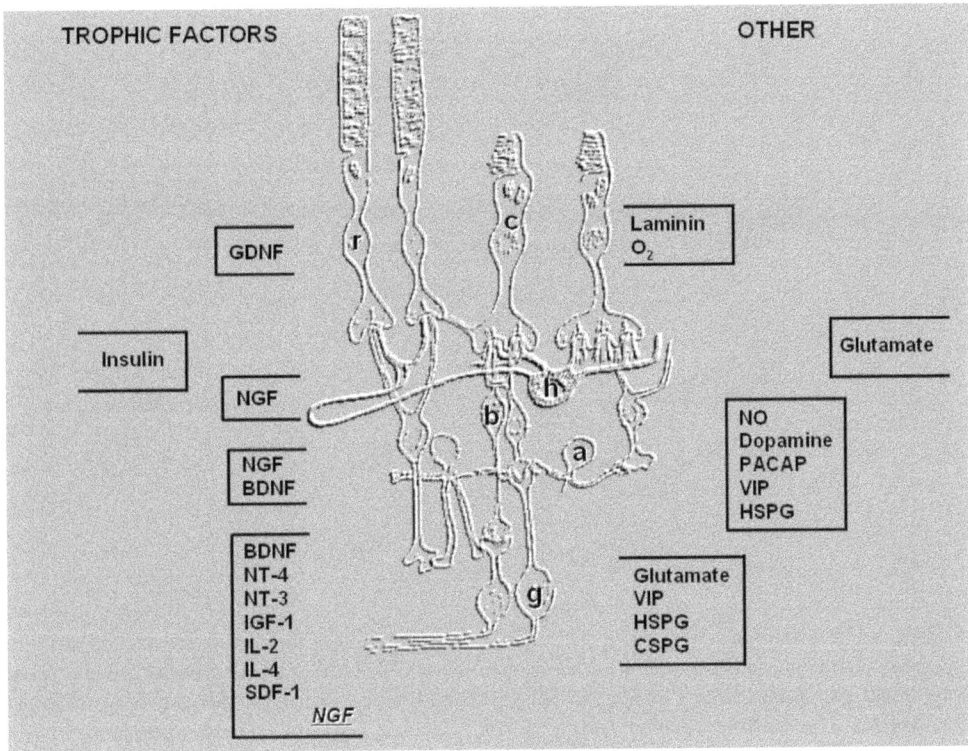

Figure 11.2 Mediators of intercellular control of developmental neuronal death in the retina. The figure contrasts classical trophic factors (to the left) with neurotransmitters and neuromodulators as well as components of the extracellular matrix having trophic actions upon retinal cells (to the right). Underlined italics indicate a pro-apoptotic effect of NGF. Inward brackets (toward the centre) refer to factors with actions on differentiated retinal cell types, box indicates actions upon undifferentiated postmitotic cells and outward brackets indicate effects upon proliferating precursors. a, amacrine cell; b, bipolar cell; c, cone; g, ganglion cell; h, horizontal cell; r, rod; CSPG, chondroitin sulphate proteoglycan; HSPG, heparin sulphate proteoglycan; SDF-1, stromal cell-derived factor 1.

by cell death, particularly since these two events are temporally coincident. However, for none of these instances has cell death been proven to produce the resultant change in axonal projection pattern; rather, axonal retraction appears as likely an explanation, if not more so, given the behaviour of individual developing optic axonal arbors.

11.5.2 Establishment of ocular domains

Retinofugal axons normally terminate within the lateral geniculate nucleus in separate ocular domains, yet during early development, the projections from the two eyes overlap extensively (Linden et al., 1981). The transition from ocular overlap to ocular segregation also coincides with the period of ganglion cell elimination, leading to the suggestion that ganglion cell loss removes those optic axons with terminations in the ocularly incorrect parts of the nucleus (Jeffery, 1984; Figure 11.3b). Labelling of individual axons at different

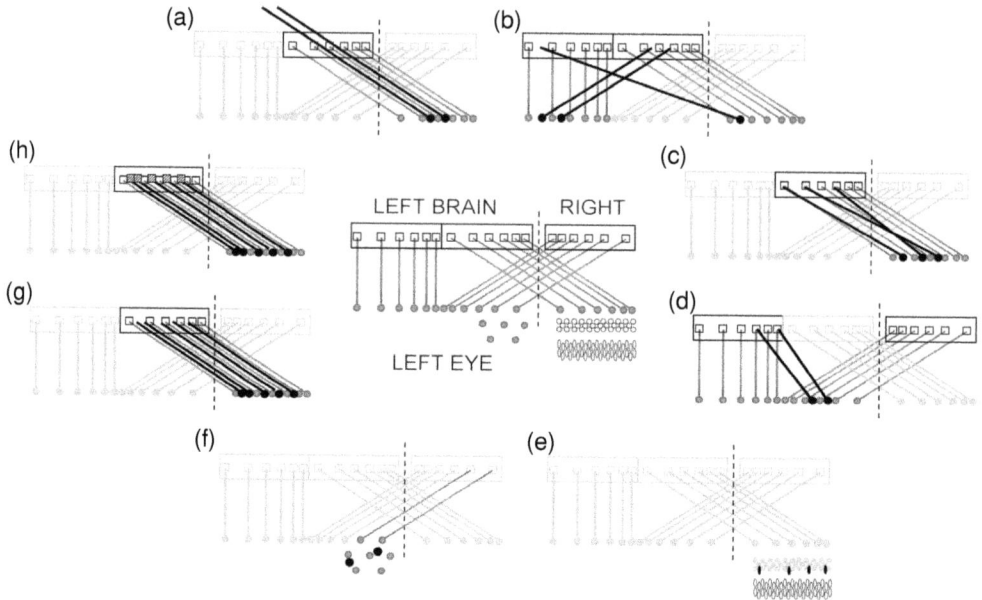

Figure 11.3 The diagrams (a) to (h) depict eight proposed functions for developmental cell death in the retina. The centre diagram represents retinal ganglion cells (circles), both crossed and uncrossed retinofugal projections (lines) and target cells (squares) within primary optic targets in the brain (rectangles). The centro-peripheral density gradient, mosaic organization and decussation line of retinal ganglion cells are schematically represented on the left side. Ocular dominance fields in primary visual targets are represented as boxes side-by-side to the left of the midline (interrupted line). A mosaic of retinal ganglion cells is depicted on the left side, and retinal layers (GCL up) are shown on the right side. (a) to (h), clockwise, show the main features of each one of the eight hypotheses. Cells depicted by black circles and target field cells depicted by shaded squares represent presumptive retinal or target cells eliminated by PCD. Grey symbols represent cells that are to remain throughout development. In each case, elements irrelevant to each hypothesis were dimmed. (a) *Elimination of anomalous projections* is shown as overreaching optic axons. (b) *Establishment of ocular domains* is shown for a binocular target field on the left side of the brain. Ipsilateral and contralateral terminal fields are shown side by side for the sake of clarity. (c) *Refinement of retinotopic mapping* is shown as the elimination of cells with misdirected axons terminating within a single target field. (d) *Sculpting of decussation patterns* is exemplified by the loss of ipsilaterally projecting ganglion cells located in nasal retina. (e) *Elimination of mispositioned neurons* is shown as the loss of photoreceptors located in the INL. (f) *Formation of regular mosaics* is shown as the loss of two cells leaving seven neighbouring ganglion cells regularly distributed. (g) *Creation of centro-peripheral density gradients* is shown for the crossed projection from the right eye to the contralateral target. (h) *Numerical matching* is represented by equivalent neuron loss among both ganglion cells and their target neurons in the brain.

developmental stages, however, revealed a more probable role of axonal remodelling in the progression to ocular segregation (Sretavan and Shatz, 1986). An exuberance of the ipsilateral retinofugal projection into regions of the superior colliculus normally innervated by only the crossed projection has also been documented (Land and Lund, 1979), although in that case the 'error' is more clearly related to the retinotopic organization of the ocular projections, whereas in the thalamus, the error may be largely one of respecting ocular domains (Reese, 1986). Here too, the loss of this exuberant projection coincides with the period of retinal ganglion cell loss (Martin et al., 1983; Insausti et al., 1984), and while some manipulations preserve both cells that would have been eliminated as well as projections to incorrect loci in the colliculus (Fawcett et al., 1984; Isenmann et al., 1999), other studies show that the two features can be dissociated (Yakura et al., 2002). The latter study in particular would suggest that cell death is not critical for the sculpting of the uncrossed retino-collicular termination pattern.

11.5.3 Refinement of retinotopic mapping

The formation of the retinotopic map within target visual structures undergoes a progressive refinement during development, much of this coincident with the period of naturally occurring ganglion cell loss (Figure 11.3c). Some of this appears to be brought about by a remodelling of axonal arbors within the colliculus (Simon and O'Leary, 1992), but there exist independent data supporting an elimination of cells that give rise to topographically incorrect terminations (O'Leary et al., 1986). The topographic targeting errors made are relatively few (Yhip and Kirby, 1990) and their mapping-error is modest rather than extensive (Marotte, 1993), which is surprising, given the widespread nature of immature axonal arbors across the surface of the superior colliculus (Simon and O'Leary, 1992). The capacity of these axonal arbors for remodelling within the colliculus renders the fact that there is any cell loss related to targeting all the more surprising. Clearly, the retinotopic map is not sculpted from an initially indiscriminate population of innervating axons by a process of selective cell death. While cell death may eliminate some retinotopic errors, the amount would appear modest relative to the magnitude of naturally occurring cell death.

11.5.4 Sculpting of decussation patterns

The decussation pattern of retinal ganglion cells has also been suggested to emerge from a process of selective cell death (Figure 11.3d). The uncrossed visual pathway, which originates near-exclusively from the temporal retina in maturity, is comprised of ipsilaterally projecting retinal ganglion cells in both the nasal as well as temporal retina during early development (Insausti et al, 1984; Jeffery, 1984). During the period when the ganglion cell population is being reduced by a magnitude of cell death estimated to be on the order of 50%, there is a disproportionate reduction in this uncrossed nasal projection, on the order of >95% (Jeffery, 1984). This elimination of nasal ganglion cells with uncrossed optic axons has been detected in all mammals examined with the exception of primates (Chalupa and Lia, 1991), but, in each case, the difference in density between nasal and temporal retina is still prominent prior to cell death (Jeffery, 1984; Colello and Guillery, 1990; Thompson and Morgan, 1993).

The complementary border, that defining the temporal limits of the crossed projection, has been more difficult to evaluate, both because it is cell-type unique and because, in some species, it extends to the far temporal limit of the retina, completely overlapping the region giving rise to the uncrossed projection in maturity. For cell types that form a classic partial decussation at the optic chiasm (e.g. beta cells in the carnivore retina), they appear to establish this pattern from the time of pathway formation (Baker and Reese, 1993; Reese *et al.*, 1994), before the period of naturally occurring ganglion cell loss (Henderson *et al.*, 1988). The mature decussation pattern of alpha ganglion cells, by contrast, has been said to depend upon selective cell loss during development in the cat's retina (Leventhal *et al.*, 1988), although no support for this hypothesis was found in the developing ferret (Reese and Urich, 1994). Decussation patterns, then, are created primarily at the time of axonal invasion at the optic chiasm (see Chapter 8); cell death subsequently eliminates the minority of cells that appear to misproject during those earlier stages, particularly those comprising the uncrossed projection from the nasal retina, which may be as little as 2.5% of the size of the crossed projection from this same region of retina during development (Lam *et al.*, 1982; Jeffery, 1984).

11.5.5 Elimination of mispositioned neurons

As retinal neurons migrate to their appropriate laminar position within or beyond the developing neuroblastic layer, some appear to become mispositioned with respect to retinal depth. The evidence for this is greatest amongst photoreceptors, a number of which become subsequently detectable on the 'wrong' side of the OPL as the latter differentiates. As such, they must be inappropriately situated to produce a basally directed terminal reaching that plexiform layer. These cells subsequently die (though some may migrate back to the ONL), no longer being detected in the mature retina (Spira *et al.*, 1984; Young, 1984; Figure 11.3e). This example may be unique, as other cell types have been found at atypical depths that are retained into adulthood (e.g. horizontal cells in the GCL; ganglion cells in the INL; amacrine cells in the GCL – Drager and Olsen, 1980; Silveira *et al.*, 1989; Cook and Becker, 1991; Wassle *et al.*, 2000; Eglen *et al.*, 2003). The discriminating feature would appear to be their ability to extend processes towards the processes of other cells likely to be the source of trophic support: whereas those other cases all give rise to processes extending into their 'normal' plexiform layer, the ectopic photoreceptors do not form synaptic contacts (Spira *et al.*, 1984).

11.5.6 Formation of regular mosaics

The regularity in the mosaic patterning of particular types of retinal nerve cells could, in principle, arise from selective cell death (Linden, 1987; Cook and Chalupa, 2000; Eglen and Willshaw, 2002; Figure 11.3f), yet evidence for such a role is limited (see Chapter 10). In the cat retina, indirect evidence is consistent with a role for alpha cell mosaics (Jeyarasasingam *et al.*, 1998). In mice, mosaic regularity does not change for the cholinergic amacrine cells during the period when their numbers decline to mature levels (Galli-Resta and Novelli, 2000), while horizontal cell mosaic regularity improves during the period when their numbers are not changing (Raven *et al.*, 2005). In the *bcl-2* overexpressing mouse, containing a surplus of dopaminergic amacrine cells, the mosaic is appreciably less regular than in the wild-type mouse (Raven *et al.*, 2003). Indeed, the mosaic in these transgenic retinas is statistically indistinguishable from a random distribution. The modest regularity of this mosaic in the wild-type retina could therefore be sculpted entirely through a process of cell death, but the above provisos concerning *bcl-2* overexpression should be kept in mind.

11.5.7 Creation of centro-peripheral density gradients

The variation in cellular density across the mature retina has been suggested to arise as a consequence of spatially selective cell death (Figure 11.3g). During early development, the distribution of retinal ganglion cells is relatively flat across the surface of the retina; the prominent centro-peripheral density differences characteristic of the mature retina emerge only after cell death is complete (Stone *et al.*, 1982; Robinson *et al.*, 1989). The fact that dying cells have been detected more frequently within the GCL in the periphery, relative to the centre, further advanced this hypothesis (Sengelaub and Finlay, 1982), although other species failed to provide supporting evidence (Henderson *et al.*, 1988; Wikler *et al.*, 1989). Yet other studies showed that the emergence of the centro-peripheral density difference paralleled the increase in retinal areal growth better than it did the period of naturally occurring cell death (Lia *et al.*, 1987; Robinson *et al.*, 1989). A consideration of the spatio-temporal distribution of dying cells in the primate's GCL and INL also showed no correlation with the emergence of the foveal depression (Georges *et al.*, 1999; see Chapter 7). These data indicate that cell death makes, at best, only a modest contribution to the formation of such regional specializations.

11.5.8 Numerical matching

The evidence that ganglion cell survival is dependent upon both targets and afferents is consistent with the matching of spatial (or retinotopic) patterns of dying ganglion cells and their retino-recipient target structures (Cunningham *et al.*, 1981). This led to the notion that interconnected structures each undergo cell death in order to establish a convergence ratio for a given system, often called 'system-matching' or 'numerical matching' (Figure 11.3h). In the retina or visual pathway, experimental manipulation of the size of the target has not been shown to produce a proportional change in the size of the afferents, unlike in other systems (Skeen *et al.*, 1986; Tanaka and Landmesser, 1986; Herrup and Sunter, 1987; see also Linden and Renteria, 1988; Serfaty *et al.*, 1990). An alternative approach has been to determine the correlation between pre- and postsynaptic neuronal populations in maturity, particularly in neuronal systems in which genetic diversity has also been shown to control neuron number. A lack of correlation between neuronal populations comprising retinal ganglion cells and dorsal lateral geniculate neurons was reported (Seecharan *et al.*, 2003), but because the majority of ganglion cells are thought not to innervate the dorsal lateral geniculate nucleus in rodent retina (Martin, 1986), a correlation between these synaptically connected populations is still uncertain. Beyond the simple demonstration that the ganglion cell population and its targets are correspondingly larger or smaller when the other is increased or decreased, there is no convincing evidence for quantitative matching at work. Within the retina itself, no evidence for quantitative matching exists, while the number of counter-examples is conspicuous (Williams et al., 2001; Strettoi and Volpini, 2002; Raven and Reese, 2003), perhaps because most retinal cell types receive multiple sources of innervation. Thus, there is still no compelling evidence that cell death serves quantitative matching of interconnecting retinal cell populations.

The roles of cell death in the sculpting of major traits of the visual system therefore remain unclear. Still, the caveats raised about the assessment of the magnitude of cell death in specific cell populations undermine the conclusive dismissal of either the error correction or the numerical matching hypotheses on the basis of the available data. Either of those hypotheses would be consistent with the strong evidence that the survival of developing neurons is regulated by trophic interactions between cells.

11.6 Concluding remarks

Retinal cell population dynamics are the result of a balance between rates of production (see Chapters 3 and 5) and rates of cell death. The amount of cell death among the various retinal cell types is, however, still disputed. Perhaps the firmest conclusion to be drawn at this stage is that the magnitude of cell death amongst the retinal ganglion cell population is more likely to prove the exception, rather than the rule. Among distinct mechanisms of PCD revealed by experimental studies of various cell and tissue models, caspase-mediated apoptosis modulated by the Bcl-2 family of proteins has been firmly established as one mode of cell death in the retina, but probably not the only one and perhaps not even the most common during development. Upstream, cell death is subject to modulation by several families of extracellular modulators, including distinct neurotrophic factors, cytokines, neuropeptides and neurotransmitters, as well as components of the ECM, intercellular communication through gap junctions and general tissue factors such as the oxygen tension and redox status. The consequences of cell death for shaping retinal populations are also controversial. Whereas the death of individual cells can be explained as the result of a failure to secure either appropriate or sufficient trophic support from connecting partners, little evidence is as yet available to support a major role for cell death in the elimination of either system errors or numerical disparity between retinal and either target or afferent cell populations. Thus, the major questions regarding the amount, mechanisms and roles of developmental cell death in the retina remain outstanding, preventing a greater synthesis of the literature.

The implications of understanding developmental cell death extend beyond embryology. Studies of the developing retina have both benefited from and contributed to the understanding of cell death in the CNS in general, including neurodegenerative diseases and, in particular, retinal dystrophies. Nevertheless, current studies indicate that generalization from single models is not warranted. Many issues of developmental cell death therefore remain provisional for the vertebrate retina. Notwithstanding the differences between immature and adult retinas, mechanisms operating upon the embryonic tissue may apply to both the maintenance of the normal structure as well as to pathological cell death in various retinal dystrophies. In particular, given the early expression of several genes that are mutated in certain retinal degenerations, the study of developmental cell death may provide insight on the mechanisms of cell demise in retinal pathologies. An example is the *rd* mouse, in which the massive photoreceptor degeneration starts at a relatively early stage of photoreceptor outer segment differentiation, widely used as a model for the understanding of *retinitis pigmentosa*. Understanding the balance between proliferation and cell death in the immature retina may also clarify the pathogenesis of retinoblastoma. In the latter case, the coexistence of proliferating retinal cells with differentiated retinal neurons and glia is likely to have a profound impact upon the fate of tumour cells, and models of developing retinal tissue may contribute to the understanding of the interaction among cells at various stages of development in the control of both the cell cycle and PCD.

Future studies are likely to unravel the intricate network of interactions that control the mechanisms of PCD in the developing retina. The introduction of novel early markers of cell differentiation, as well as reliable markers of alternative modes of cell death, should contribute to establish the rates and magnitude of cell death for distinct retinal cell populations. Progress in the design of selective receptor antagonists and further analysis of the expression and function of signalling molecules and membrane receptors should expand our understanding of the intercellular interactions involved in the regulation of cell death. Finally, in depth examination of retinal cell population dynamics should contribute to solve the controversies about the roles of cell death in retinal development.

References

Araujo, E. G. and Linden, R. (1993). Trophic factors produced by retinal cells increase the survival of retinal ganglion cells in vitro. *Eur. J. Neurosci.*, **5**, 1181–8.

Ary-Pires, R., Nakatani, M., Rehen, S. K. and Linden, R. (1997). Developmentally regulated release of intra-retinal neurotrophic factors *in vitro*. *Int. J. Dev. Neurosci.*, **15**, 239–57.

Baker, G. E. and Reese, B. E. (1993). The chiasmatic course of temporal retinal axons during development. *J. Comp. Neurol.*, **330**, 95–104.

Barber, A. J., Nakamura, M., Wolpert, E. B. *et al.* (2001). Insulin rescues retinal neurons from apoptosis by a phosphatidylinositol 3-kinase/Akt-mediated mechanism that reduces the activation of caspase-3. *J. Biol. Chem.*, **276**, 32 814–21.

Becker, T. S., Burgess, S. M., Amsterdam, A. H., Allende, M. L. and Hopkins, N. (1998). *not really finished* is crucial for development of the zebrafish outer retina and encodes a transcription factor highly homologous to human Nuclear Respiratory Factor-1 and avian Initiation Binding Repressor. *Development*, **125**, 4369–78.

Bernard, R., Dieni, S., Rees, S. and Bernard, O. (1998). Physiological and induced neuronal death are not affected in NSE-bax transgenic mice. *J. Neurosci. Res.*, **52**, 247–59.

Bonfanti, L., Strettoi, E., Chierzi, S. *et al.* (1996). Protection of retinal ganglion cells from natural and axotomy-induced cell death in neonatal transgenic mice over-expressing bcl-2. *J. Neurosci.*, **16**, 4186–94.

Borges, H. L. and Linden, R. (1999). Gamma irradiation leads to two waves of apoptosis in distinct cell populations of the retina of new-born rats. *J. Cell Sci.*, **112**, 4315–24.

Borges, H. L., Chao, C., Xu, Y., Linden, R. and Wang, J. Y. J. (2004). Radiation-induced apoptosis in developing mouse retina exhibits dose-dependent requirement for ATM phosphorylation of p53. *Cell Death Differ.*, **11**, 494–502.

Bovolenta, P., Frade, J. M., Marti, E. *et al.* (1996). Neurotrophin-3 antibodies disrupt the normal development of the chick retina. *J. Neurosci.*, **16**, 4402–10.

Braekevelt, C. R., Beazley, S. D., Dunlop, S. A. and Darby, J. E. (1986). Numbers of axons in the optic nerve and of retinal ganglion cells during development in the marsupial *Setonix brachyurus*. *Dev. Brain Res.*, **25**, 117–25.

Bunch, S. T. and Fawcett, J. W. (1993). NMDA receptor blockade alters the topography of naturally occurring ganglion cell death in the rat retina. *Dev. Biol.*, **160**, 434–42.

Bunt, S. M. and Lund, R. D. (1981). Development of a transient retino-retinal pathway in hooded and albino rats. *Brain Res.*, **211**, 399–404.

Campos, C. B. L., Bédard, P. A. and Linden, R. (2003). Selective involvement of the PI3K/PKB/Bad pathway in retinal cell death. *J. Neurobiol.*, **56**, 171–7.

Castagne, V. and Clarke, P. G. (1996). Axotomy-induced retinal ganglion cell death in development: its time-course and its diminution by antioxidants. *Proc. R. Soc. London B. Biol. Sci.*, **263**, 1193–7.

Catsicas, S. and Thanos, S. and Clarke, P. G. H. (1987). Major role for neuronal death during brain development: refinement of topographical connections. *Proc. Natl. Acad. Sci. U. S. A.*, **84**, 8165–8.

Cecconi, F., Alvarez-Bolado, G., Meyer, B. I., Roth, K. A. and Gruss, P. (1998). Apaf1 (CED-4 homolog) regulates programmed cell death in mammalian development. *Cell*, **94**, 727–37.

Cellerino, A., Carroll, P., Thoenen, H. and Barde, Y. A. (1997). Reduced size of retinal ganglion cell axons and hypomyelination in mice lacking brain-derived neurotrophic factor. *Mol. Cell. Neurosci.*, **9**, 397–408.

Cellerino, A., Michaelidis, T., Barski, J. J. *et al.* (1999). Retinal ganglion cell loss after the period of naturally occurring cell death in bcl-2−/− mice. *NeuroReport*, **10**, 1091–5.

Cellerino, A., Galli-Resta, L. and Colombaioni, L. (2000). The dynamics of neuronal death: a time-lapse study in the retina. *J. Neurosci.*, **20**: RC92 (1–5).

Chalasani, S. H., Baribaud, F., Coughlan, C. M. *et al.* (2003). The chemokine stromal cell-derived factor-1 promotes the survival of embryonic retinal ganglion cells. *J. Neurosci.*, **23**, 4601–12.

Chalupa, L. M. and Lia, B. (1991). The nasotemporal division of retinal ganglion cells with crossed and uncrossed projections in the fetal rhesus monkey. *J. Neurosci.*, **11**, 191–202.

Chau, B. N., Borges, H. L., Chen, T. T. *et al.* (2002). Signal-dependent protection from apoptosis in mice expressing caspase-resistant Rb. *Nat. Cell. Biol.*, **4**, 757–65.

Chen, S. T., Garey, L. J. and Jen, L. S. (1994). Bcl-2 proto-oncogene protein immunoreactivity in normally developing and axotomised rat retinae. *Neurosci. Lett.*, **172**, 11–14.

Chiarini, L. B. and Linden, R. (2000). Tissue biology of apoptosis. Ref-1 and cell differentiation in the developing retina. *Ann. New York Acad. Sci.*, **926**, 64–78.

Chiarini, L. B., Freitas, F. G., Leal-Ferreira, M. L., Tolkovsky, A. M. and Linden, R. (2002). Cytoplasmic c-Jun N-terminal immunoreactivity: a hallmark of retinal apoptosis. *Cell. Mol. Neurobiol.*, **22**, 711–26.

Chiarini, L. B., Leal-Ferreira, M. L., de Freitas, F. G. and Linden, R. (2003). Changing sensitivity to cell death during development of retinal photoreceptors. *J. Neurosci. Res.*, **74**, 875–83.

Clarke, P. G. H. (1985). Neuronal death in the development of the vertebrate nervous system. *Trends Neurosci.*, **8**, 345–9.

Clarke, P. G. H. and Clarke, S. (1996). Nineteenth century research on naturally occurring cell death and related phenomena. *Anat. Embryol. (Berl.)*, **193**, 81–99.

Colello, R. J. and Guillery, R. W. (1990). The early development of retinal ganglion cells with uncrossed axons in the mouse: retinal position and axonal course. *Development*, **108**, 515–23.

Cook, B., Portera-Cailliau, C. and Adler, R. (1998). Developmental neuronal death is not a universal phenomenon among cell types in the chick embryo retina. *J. Comp. Neurol.*, **396**, 12–19.

Cook, J. E. and Becker, D. L. (1991). Regular mosaics of large displaced and non-displaced ganglion cells in the retina of a cichlid fish. *J. Comp. Neurol.*, **306**, 668–84.

Cook, J. E. and Chalupa, L. M. (2000). Retinal mosaics: new insights into an old concept. *Trends Neurosci.*, **23**, 26–34.

Cooper, A. M. and Cowey, A. (1990). Development and retraction of a crossed retinal projection to the inferior colliculus in neonatal pigmented rats. *Neuroscience*, **35**, 335–44.

Crespo, D., O'Leary, D. D. and Cowan, W. M. (1985). Changes in the numbers of optic nerve fibers during late prenatal and postnatal development in the albino rat. *Brain Res.*, **351**, 129–34.

Cryns, V. and Yuan, J. (1998). Proteases to die for. *Genes Dev.*, **12**, 1551–70.

Cuadros, M. A. and Rios, A. (1988). Spatial and temporal correlation between early nerve fiber growth and neuroepithelial cell death in the chick embryo retina. *Anat. Embryol. (Berl.)*, **178**, 543–51.

Cui, Q. and Harvey, A. R. (1995). At least two mechanisms are involved in the death of retinal ganglion cells following target ablation in neonatal rats. *J. Neurosci.*, **15**, 8143–55.

Cui, Q. and Harvey, A. R. (2000). NT-4/5 reduces cell death in inner nuclear as well as ganglion cell layers in neonatal rat retina. *NeuroReport*, **11**, 3921–4.

Cunningham, T. J., Mohler, I. M. and Giordano, D. L. (1981). Naturally occurring neuron death in the ganglion cell layer of the neonatal rat: morphology and evidence for regional correspondence with neuron death in superior colliculus. *Brain Res.*, **254**, 203–15.

Cusato, K., Stagg, S. B. and Reese, B. E. (2001). Two phases of increased cell death in the inner retina following early elimination of the ganglion cell population. *J. Comp. Neurol.*, **439**, 440–9.

Cusato, K., Bosco, A., Linden, R. and Reese, B. E. (2002). Cell death in the inner nuclear layer of the retina is modulated by BDNF. *Dev. Brain Res.*, **139**, 325–30.

Cusato, K., Bosco, A., Rozental, R. *et al.* (2003). Gap junctions mediate bystander cell death in developing mammalian retina. *J. Neurosci.*, **23**, 6413–22.

de Almeida, C. J. and Linden, R. (2005). Phagocytosis of apoptotic cells: a matter of balance. *Cell. Mol. Life Sci.*, **62**, 1532–46.

de la Rosa and E. J. and de Pablo, F. (2000). Cell death in early neural development: beyond the neurotrophic theory. *Trends Neurosci.*, **23**, 454–8.

de la Rosa, E. J., Arribas, A., Frade, J. M. and Rodriguez-Tebar, A. (1994). Role of neurotrophins in the control of neural development: neurotrophin-3 promotes both neuron differentiation and survival of cultured chick retinal cells. *Neuroscience*, **58**, 347–52.

Diem, R., Meyer, R., Weishaupt, J. H. and Bahr, M. (2001). Reduction of potassium currents and phosphatidylinositol 3-kinase-dependent AKT phosphorylation by tumor necrosis factor-(alpha) rescues axotomized retinal ganglion cells from retrograde cell death in vivo. *J. Neurosci.*, **21**, 2058–66.

Ding, J., Hu, B., Tang, L. S. and Yip, H. K. (2001). Study of the role of the low-affinity neurotrophin receptor p75 in naturally occurring cell death during development of the rat retina. *Dev. Neurosci.*, **23**, 390–8.

Drager, U. C. and Olsen, J. F. (1980). Origins of crossed and uncrossed retinal projections in pigmented and albino mice. *J. Comp. Neurol.*, **191**, 383–412.

Dreher, B. and Robinson, S. R. (1988). Development of the retinofugal pathway in birds and mammals: evidence for a common 'timetable'. *Brain Behav. Evol.*, **31**, 369–90.

Dubois-Dauphin, M., Poitry-Yamate, C., de Bilbao, F., *et al.* (2000). Early postnatal Müller cell death leads to retinal but not optic nerve degeneration in NSE-Hu-Bcl-2 transgenic mice. *Neuroscience*, **95**, 9–21.

Dunlop, S. A. (1998). Transient axonal side branches in the developing mammalian optic nerve. *Cell Tissue Res.*, **291**, 43–56.

Dunlop, S. A. and Beazley, L. D. (1987). Cell death in the developing retinal ganglion cell layer of the wallaby *Setonix brachyurus*. *J. Comp. Neurol.*, **264**, 14–23.

Egensperger, R., Maslim, J., Bisti, S., Hollander, H. and Stone, J. (1996). Fate of DNA from retinal cells dying during development: uptake by microglia and macroglia (Müller cells). *Dev. Brain Res.*, **97**, 1–8.

Eglen, S. J. and Willshaw, D. J. (2002). Influence of cell fate mechanisms upon retinal mosaic formation: a modeling study. *Development*, **129**, 5399–408.

Eglen, S. J., Raven, M. A., Tamrazian, E. and Reese, B. E. (2003). Dopaminergic amacrine cells in the inner nuclear layer and ganglion cell layer comprise a single functional retinal mosaic. *J. Comp. Neurol.*, **466**, 343–55.

Ehrlich, R. B., Werneck, C. C., Mourao, P. A. and Linden, R. (2003). Major glycosaminoglycan species in the developing retina: synthesis, tissue distribution and effects upon cell death. *Exp. Eye Res.*, **77**, 157–65.

Enari, M., Sakahira, H., Yokoyama, H. *et al.* (1998). A caspase-activated DNase that degrades DNA during apoptosis, and its inhibitor ICAD. *Nature*, **391**, 43–50.

Eversole-Cire, P., Chen, J. and Simon, M. I. (2002). Bax is not the heterodimerization partner necessary for sustained anti-photoreceptor-cell-death activity of Bcl-2. *Invest. Ophthalmol. Vis. Sci.*, **43**, 1636–44.

Farah, M. H. and Easter, S. S. (2005). Cell birth and death in the mouse retinal ganglion cell layer. *J. Comp. Neurol.*, **489**, 120–34.

Fawcett, J. W., O'Leary, D. D. and Cowan, W. M. (1984). Activity and the control of ganglion cell death in the rat retina. *Proc. Natl. Acad. Sci. U. S. A.*, **81**, 5589–93.

Feng, L., Rigatti, B. W., Novak, E. K., Gorin, M. B. and Swank, R. T. (2000). Genomic structure of the mouse *Ap3b1* gene in normal and pearl mice. *Genomics.*, **69**, 370–9.

Fix, A. S., Horn, J. W., Hall, R. L., Johnson, J. A. and Tizzano, J. P. (1995). Progressive retinal toxicity in neonatal rats treated with D,L-2-amino-3-phosphonopropionate (D,L-AP3). *Vet. Pathol.*, **32**, 521–31.

Frade, J. M. and Barde, Y. A. (1998). Microglia-derived nerve growth factor causes cell death in the developing retina. *Neuron*, **20**, 35–41.

Frade, J. M. and Barde, Y. A. (1999). Genetic evidence for cell death mediated by nerve growth factor and the neurotrophin receptor p75 in the developing mouse retina and spinal cord. *Development*, **126**, 683–90.

Frost, D. O. (1986). Development of anomalous retinal projections to nonvisual thalamic nuclei in Syrian hamsters: a quantitative study. *J. Comp. Neurol.*, **252**, 95–105.

Fuhrmann, S., Heller, S., Rohrer, H. and Hofmann, H. D. (1998). A transient role for ciliary neurotrophic factor in chick photoreceptor development. *J. Neurobiol.*, **37**, 672–83.

Galli-Resta, L. and Ensini, M. (1996). An intrinsic time limit between genesis and death of individual neurons in the developing retinal ganglion cell layer. *J. Neurosci.*, **16**, 2318–24.

Galli-Resta, L. and Novelli, E. (2000). The effects of natural cell loss on the regularity of the retinal cholinergic arrays. *J. Neurosci.*, **20**, RC60.

Gan, L., Wang, S. W., Huang, Z. and Klein, W. H. (1999). POU domain factor Brn-3b is essential for retinal ganglion cell differentiation and survival but not for initial cell fate specification. *Dev. Biol.*, **210**, 469–80.

Garcia, M., Forster, V., Hicks, D. and Vecino, E. (2002). Effects of Müller glia on cell survival and neuritogenesis in adult porcine retina in vitro. *Invest. Ophthalmol. Vis. Sci.*, **43**, 3735–43.

Garcia-Porrero, J. A. and Ojeda, J. L. (1979). Cell death and phagocytosis in the neuroepithelium of the developing retina. A TEM and SEM study. *Experientia*, **35**, 375–376.

Gavrieli, Y. and Sherman, Y., Ben-Sasson, S. A. and (1992). Identification of programmed cell death in situ via specific labeling of nuclear DNA fragmentation. *J. Cell Biol.*, **119**, 493–501.

Geiger, L. K., Kortuem, K. R., Alexejun, C. and Levin, L. A. (2002). Reduced redox state allows prolonged survival of axotomized neonatal retinal ganglion cells. *Neuroscience*, **109**, 635–42.

Georges, P., Madigan, M. C. and Provis, J. M. (1999). Apoptosis during development of the human retina: relationship to foveal development and retinal synaptogenesis. *J. Comp. Neurol.*, **413**, 198–208.

Giardino, I., Fard, A. K., Hatchell, D. L. and Brownlee, M. (1998). Aminoguanidine inhibits reactive oxygen species formation, lipid peroxidation, and oxidant-induced apoptosis. *Diabetes*, **47**, 1114–20.

Gonzalez-Hoyuela, M., Barbas, J. A. and Rodriguez-Tebar, A. (2001). The autoregulation of retinal ganglion cell number. *Development*, **128**, 117–24.

Goureau, O., Regnier-Ricard, F., Desire, L. and Courtois, Y. (1999). Role of nitric oxide in photoreceptor survival in embryonic chick retinal cell culture. *J. Neurosci. Res.*, **55**, 423–31.

Grunder, T., Kohler, K., Kaletta, A. and Guenther, E. (2000). The distribution and developmental regulation of NMDA receptor subunit proteins in the outer and inner retina of the rat. *J. Neurobiol.*, **44**, 333–42.

Guimarães, C. A. and Linden, R. (2004). Programmed cell deaths: apoptosis and alternative deathstyles. *Eur. J. Biochem.*, **271**, 1638–50.

Guimarães, C. A., Assreuy, J. and Linden, R. (2001). Paracrine anti-apoptotic function of nitric oxide in developing retina. *J. Neurochem.*, **76**, 1233–41.

Guimarães, C. A., Benchimol, M., Amarante-Mendes, G. and Linden, R. (2003). Alternative programs of cell death in retinal tissue. *J. Biol. Chem.*, **278**, 41 938–46.

Gutierrez-Ospina, G., Gutierrez, de la Barrera, A., Larriva, J. and Giordano, M. (2002). Insulin-like growth factor I partly prevents axon elimination in the neonate rat optic nerve. *Neurosci. Lett.*, **325**, 207–10.

Haberecht, M. F., Mitchell, C. K., Lo, G. J. and Redburn, D. A. (1997). N-methyl-D-aspartate-mediated glutamate toxicity in the developing rabbit retina. *J. Neurosci. Res.*, **47**, 416–26.

Hahn, P., Lindsten, T., Ying, G. S. *et al.* (2003). Proapoptotic bcl-2 family members, Bax and Bak, are essential for developmental photoreceptor apoptosis. *Invest. Ophthalmol. Vis. Sci.*, **44**, 3598–605.

Harman, A. M., Snell, L. L. and Beazley, S. D. (1989). Cell death in the inner and outer nuclear layers of the developing retina in the wallaby *Setonix brachyurus* (Quokka). *J. Comp. Neurol.*, **289**, 1–10.

Henderson, Z., Finlay, B. L. and Wikler, K. C. (1988). Development of ganglion cell topography in ferret retina. *J. Neurosci.*, **8**, 1194–205.

Herrup, K. and Sunter, K. (1987). Numerical matching during cerebellar development: quantitative analysis of granule cell death in *staggerer* mouse chimeras. *J. Neurosci.*, **7**, 829–36.

Herzog, K. H. and von Bartheld, C. S. (1998). Contributions of the optic tectum and the retina as sources of brain-derived neurotrophic factor for retinal ganglion cells in the chick embryo. *J. Neurosci.*, **18**, 2891–906.

Herzog, K. H., Chen, S. C. and Morgan, J. I. (1999). c-jun is dispensable for developmental cell death and axogenesis in the retina. *J. Neurosci.*, **19**, 4349–59.

Herzog, K. H., Braun, J. S., Han, S. H. and Morgan, J. I. (2002). Differential post-transcriptional regulation of p21WAF1/Cip1 levels in the developing nervous system following gamma-irradiation. *Eur. J. Neurosci.*, **15**, 627–36.

Hicks, D. (1998). Putative functions of fibroblast growth factors in retinal development, maturation and survival. *Semin. Cell Dev. Biol.*, **9**, 263–9.

Hinds, J. W. and Hinds, P. L. (1983). Development of retinal amacrine cells in the mouse embryo: evidence for two modes of formation. *J. Comp. Neurol.*, **213**, 1–23.

Hughes, W. F. and LaVelle, A. (1975). The effects of early tectal lesions on development in the retinal ganglion cell layer of chick embryos. *J. Comp. Neurol.*, **163**, 265–83.

Hume, D. A., Perry, V. H. and Gordon, S. (1983). Immunohistochemical localization of a macrophage-specific antigen in developing mouse retina: phagocytosis of dying neurons and differentiation of microglial cells to form a regular array in the plexiform layers. *J. Cell Biol.*, **97**, 253–7.

Huxlin, K. R., Carr, R., Schulz, M., Sefton, A. J. and Bennett, M. R. (1995). Trophic effect of collicular proteoglycan on neonatal rat retinal ganglion cells in situ. *Dev. Brain Res.*, **84**, 77–88.

Ikeda, S., Hawes, N. L., Chang, B. *et al.* (1999). Severe ocular abnormalities in C57BL/6 but not in 129/Sv p53-deficient mice. *Invest. Ophthalmol. Vis. Sci.*, **40**, 1874–8.

Insausti, R., Blakemore, C. and Cowan, W. M. (1984). Ganglion cell death during development of ipsilateral retino-collicular projection in golden hamster. *Nature*, **308**, 362–5.

Isenmann, S. and Bähr, M. (1997). Expression of c-Jun protein in degenerating retinal ganglion cells after optic nerve lesion in the rat. *Exp. Neurol.*, **147**, 28–36.

Isenmann, S., Cellerino, A., Gravel, C. and Bähr, M. (1999). Excess target-derived brain-derived neurotrophic factor preserves the transient uncrossed retinal projection to the superior colliculus. *Mol. Cell. Neurosci.*, **14**, 52–65.

Jacobson, S. G., Cideciyan, A. V., Aleman, T. S. et al. (2003). Crumbs homolog 1 (CRB1) mutations result in a thick human retina with abnormal lamination. *Hum. Mol. Genet.*, **12**, 1073–8.

Jeffery, G. (1984). Retinal ganglion cell death and terminal field retraction in the developing rodent visual system. *Dev. Brain Res.*, **13**, 81–96.

Jeffery, G. (1998). The retinal pigment epithelium as a developmental regulator of the neural retina. *Eye*, **12**, 499–503.

Jeyarasasingam, G., Snider, C. J., Ratto, G. M. and Chalupa, L. M. (1998). Activity-regulated cell death contributes to the formation of ON and OFF alpha ganglion cell mosaics. *J. Comp. Neurol.*, **394**, 335–43.

Johansson, K., Bruun, A., Torngren, M. and Ehinger, B. (2000). Development of glutamate receptor subunit 2 immunoreactivity in postnatal rat retina. *Vis. Neurosci.*, **17**, 737–42.

Johnson, P. T., Williams, R. R., Cusato, K. and Reese, B. E. (1999). Rods and cones project to the inner plexiform layer during development. *J. Comp. Neurol.*, **414**, 1–12.

Ju, W. K., Chung, I. W., Kim, K. Y. et al. (2001). Sodium nitroprusside selectively induces apoptotic cell death in the outer retina of the rat. *NeuroReport*, **12**, 4075–9.

Kaiser, P. K. and Lipton, S. A. (1990). VIP-mediated increase in cAMP prevents tetrodotoxin-induced retinal ganglion cell death in vitro. *Neuron*, **5**, 373–81.

Karlsson, M., Mayordomo, R., Reichardt, L. F. et al. (2001). Nerve growth factor is expressed by post-mitotic avian retinal horizontal cells and supports their survival during development in an autocrine mode of action. *Development*, **128**, 471–9.

Kermer, P., Klocker, N., Labes, M. and Bähr, M. (2000). Insulin-like growth factor-I protects axotomized rat retinal ganglion cells from secondary death via PI3-K-dependent Akt phosphorylation and inhibition of caspase-3 in vivo. *J. Neurosci.*, **20**, 2–8.

Kerr, J. F. R., Wyllie, A. H. and Currie, A. R. (1972). Apoptosis: a basic biological phenomenon with wide-ranging implications in tissue kinetics. *Br. J. Cancer*, **26**, 239–57.

Kirsch, M., Schulz-Key, S., Wiese, A., Fuhrmann, S. and Hofmann, H. (1998). Ciliary neurotrophic factor blocks rod photoreceptor differentiation from post-mitotic precursor cells in vitro. *Cell Tissue Res.*, **291**, 207–16.

Kobayashi, T. (1993). Delay of ganglion cell death by tetrodotoxin during retinal development in chick embryos. *Neurosci. Res.*, **16**, 187–94.

Konishi, A., Shimizu, S., Hirota, J. et al. (2003). Involvement of histone H1.2 in apoptosis induced by DNA double-strand breaks. *Cell*, **114**, 673–88.

Koulen, P., Malitschek, B., Kuhn, R., Wassle, H. and Brandstatter, J. H. (1996). Group II and group III metabotropic glutamate receptors in the rat retina: distributions and developmental expression patterns. *Eur. J. Neurosci.*, **8**, 2177–87.

Lam, K., Sefton, A. J. and Bennett, M. R. (1982). Loss of axons from the optic nerve of the rat during early postnatal development. *Dev. Brain Res.*, **3**, 487–91.

Land, P. W. and Lund, R. D. (1979). Development of the rat's uncrossed retinotectal pathway and its relation to plasticity studies. *Science*, **205**, 698–700.

Leventhal, A. G., Schall, J. D., Ault, S. J., Provis, J. M. and Vitek, D. J. (1988). Class-specific cell death shapes the distribution and pattern of central projection of cat retinal ganglion cells. *J. Neurosci.*, **8**, 2011–27.

Levin, L. A., Schlamp, C. L., Spieldoch, R. L., Geszvain, K. M. and Nickells, R. W. (1997). Identification of the bcl-2 family of genes in the rat retina. *Invest. Ophthalmol. Vis. Sci.*, **38**, 2545–53.

Li, Y., Schlamp, C. L., Poulsen, K. P. and Nickells, R. W. (2000). Bax-dependent and independent pathways of retinal ganglion cell death induced by different damaging stimuli. *Exp. Eye Res.*, **71**, 209–213.

Lia, B., Williams, R. W. and Chalupa, L. M. (1987). Formation of retinal ganglion cell topography during prenatal development. *Science*, **236**, 848–51.

Libby, R. T., Lavallee, C. R., Balkema, G. W., Brunken, W. J. and Hunter, D. D. (1999). Disruption of laminin beta2 chain production causes alterations in morphology and function in the CNS. *J. Neurosci.*, **19**, 9399–411.

Linden, D. C., Guillery, R. W. and Cucchiaro, J. (1981). The dorsal lateral geniculate nucleus of the normal ferret and its postnatal development. *J. Comp. Neurol.*, **203**, 189–211.

Linden, R. (1987). Competitive interactions and regulation of developmental neuronal death in the retina. In *Developmental Neurobiology of Mammals* ed. C. Chagas and R. Linden. The Vatican: Pontifical Academy of Sciences, pp. 109–40.

Linden, R. (1992). Dendritic competition: a principle of retinal development. In *The Visual System: From Genesis to Maturity*, ed. R. Lent. Boston: Birkhauser, pp. 86–103

Linden, R. (1994). The survival of developing neurons: a review of afferent control. *Neuroscience*, **58**, 671–82.

Linden, R. and Chiarini, L. B. (1999). Nuclear exclusion of transcription factors in retinal apoptosis. *Braz. J. Med. Biol. Res.*, **32**, 813–20.

Linden, R. and Pinto, L. H. (1985). Developmental genetics of the retina: evidence that the pearl mutation in the mouse affects the time course of natural neuronal death in the ganglion cell layer. *Exp. Brain Res.*, **60**, 79–86.

Linden, R. and Renteria, A. S. (1988). Afferent control of neuron numbers in the developing brain. *Dev. Brain Res.*, **44**, 291–5.

Linden, R., Cavalcante, L. A. and Barradas, P. C. (1986). Mononuclear phagocytes in the retina of developing rats. *Histochemistry*, **85**, 335–40.

Linden, R., Rehen, S. K. and Chiarini, L. B. (1999). Apoptosis in developing retinal tissue. *Prog. Retin. Eye Res.*, **18**, 133–65.

Linden, R., Martins, R. A. and Silveira, M. S. (2005). Control of programmed cell death by neurotransmitters and neuropeptides in the developing mammalian retina. *Prog. Retin. Eye Res.*, **24**, 457–91.

Lipton, S. A. (1986). Blockade of electrical activity promotes the death of mammalian retinal ganglion cells in culture. *Proc. Natl. Acad. Sci. U. S. A.*, **83**, 9774–8.

Liu, J., Wilson, S. and Reh, T. (2003). BMP receptor 1b is required for axon guidance and cell survival in the developing retina. *Dev. Biol.*, **256**, 34–48.

Lockshin, R. A. and Zakeri, Z. (2001). Programmed cell death and apoptosis: origins of the theory. *Nat. Rev. Mol. Cell. Biol.*, **2**, 545–50.

Ma, Y. T., Hsieh, T., Forbes, M. E., Johnson, J. E. and Frost, D. O. (1998). BDNF injected into the superior colliculus reduces developmental retinal ganglion cell death. *J. Neurosci.*, **18**, 2097–107.

Manabe, S. and Lipton, S. A. (2003). Divergent NMDA signals leading to proapoptotic and antiapoptotic pathways in the rat retina. *Invest. Ophthalmol. Vis. Sci.*, **44**, 385–92.

Marotte, L. R. (1993). Location of retinal ganglion cells contributing to the early imprecision in the retinotopic order of the developing projection to the superior colliculus of the wallaby (*Macropus eugenii*). *J. Comp. Neurol.*, **331**, 1–13.

Martin, P. R. (1986). The projection of different retinal ganglion cell classes to the dorsal lateral geniculate nucleus in the hooded rat. *Exp. Brain Res.*, **62**, 77–88.

Martin, P. R., Sefton, A. J. and Dreher, B. (1983). The retinal location and fate of ganglion cells which project to the ipsilateral superior colliculus in neonatal albino and hooded rats. *Neurosci. Lett.*, **41**, 219–26.

Martinou, J. C., Dubois-Dauphin, M., Staple, J. K. *et al.* (1994). Overexpression of BCL-2 in transgenic mice protects neurons from naturally occurring cell death and experimental ischemia. *Neuron*, **13**, 1017–30.

Martins, R. A. P., Silveira, M. S., Curado, M. R., Police, A. I. and Linden, R. (2005). NMDA receptor activation modulates programmed cell death during early post-natal retinal development: a BDNF-dependent mechanism. *J. Neurochem.*, **95**, 244–53.

Maslim, J., Valter, K., Egensperger, R., Hollander, H. and Stone, J. (1997). Tissue oxygen during a critical developmental period controls the death and survival of photoreceptors. *Invest. Ophthalmol. Vis. Sci.*, **38**, 1667–77.

Mervin, K. and Stone, J. (2002). Developmental death of photoreceptors in the C57BL/6J mouse: association with retinal function and self-protection. *Exp. Eye Res.*, **75**, 703–13.

Mosinger Ogilvie, J., Deckwerth, T. L., Knudson, C. M. and Korsmeyer, S. J. (1998). Suppression of developmental retinal cell death but not of photoreceptor degeneration in Bax-deficient mice. *Invest. Ophthalmol. Vis. Sci.*, **39**, 1713–20.

Munafo, D. B. and Colombo, M. I. (2001). A novel assay to study autophagy: regulation of autophagosome vacuole size by amino acid deprivation. *J. Cell Sci.*, **114**, 3619–29.

Nichol, K. A., Schulz, M. W. and Bennett, M. R. (1995). Nitric oxide-mediated death of cultured neonatal retinal ganglion cells: neuroprotective properties of glutamate and chondroitin sulfate proteoglycan. *Brain Res.*, **697**, 1–16.

O'Leary, D. D., Fawcett, J. W. and Cowan, W. M. (1986). Topographic targeting errors in the retinocollicular projection and their elimination by selective ganglion cell death. *J. Neurosci.*, **6**, 3692–705.

Oppenheim, R. W. (1991). Cell death during development of the nervous system. *Annu. Rev. Neurosci.*, **14**, 453–501.

Pacione, L. R., Szego, M. J., Ikeda, S., Nishina, P. M. and McInnes, R. R. (2003). Progress toward understanding the genetic and biochemical mechanisms of inherited photoreceptor degenerations. *Annu. Rev. Neurosci.*, **26**, 657–700.

Patel, J. I., Gentleman, S. M., Jen, L. S. and Garey, L. J. (1997). Nitric oxide synthase in developing retinae and after optic tract section. *Brain Res.*, **761**, 156–60.

Pennesi, M. E., Cho, J. H., Yang, Z. *et al.* (2003). BETA2/NeuroD1 null mice: a new model for transcription factor-dependent photoreceptor degeneration. *J. Neurosci.*, **23**, 453–61.

Pereira, S. P. and Araujo, E. G. (1997). Veratridine increases the survival of retinal ganglion cells in vitro. *Braz. J. Med. Biol. Res.*, **30**, 1467–70.

Perry, V. H., Henderson, Z. and Linden, R. (1983). Postnatal changes in retinal ganglion cell and optic axon populations in the pigmented rat. *J. Comp. Neurol.*, **219**, 356–70.

Petrs-Silva, H., Freitas, F. G., Linden, R. and Chiarini, L. B. (2004). Early nuclear exclusion of the transcription factor Max is associated with retinal ganglion cell death independent of caspase activity. *J. Cell. Physiol.*, **198**, 179–87.

Petrs-Silva, H., Chiodo, V., Chiarini, L. B., Hauswirth, W. W. and Linden, R. (2005). Modulation of the expression of the transcription factor Max in rat retinal ganglion cells by a recombinant adeno-associated viral vector. *Braz. J. Med. Biol. Res.*, **38**, 375–9.

Politi, L. E., Rotstein, N. P. and Carri, N. G. (2001). Effect of GDNF on neuroblast proliferation and photoreceptor survival: additive protection with docosahexaenoic acid. *Invest. Ophthalmol. Vis. Sci.*, **42**, 3008–15.

Pollock, G. S., Robichon, R., Boyd, K. A. *et al.* (2003). TrkB receptor signaling regulates developmental death dynamics, but not final number, of retinal ganglion cells. *J. Neurosci.*, **23**, 10 137–45.

Potts, R. A., Dreher, B. and Bennett, M. R. (1982). The loss of ganglion cells in the developing retina of the rat. *Dev. Brain Res.*, **3**, 481–6.

Pow, D. V., Crook, D. K. and Wong, R. O. (1994). Early appearance and transient expression of putative amino acid neurotransmitters and related molecules in the developing rabbit retina: an immunocytochemical study. *Vis. Neurosci.*, **11**, 1115–34.

Protti, D. A., Gerschenfeld, H. M. and Llano, I. (1997). GABAergic and glycinergic IPSCs in ganglion cells of rat retinal slices. *J. Neurosci.*, **17**, 6075–85.

Provis, J. M. and Penfold, P. L. (1988). Cell death and the elimination of retinal axons during development. *Prog. Neurobiol.*, **31**, 331–47.

Rabacchi, S. A., Bonfanti, L., Liu, X. H. and Maffei, L. (1994). Apoptotic cell death induced by optic nerve lesion in the neonatal rat. *J. Neurosci.*, **14**, 5292–301.

Raff, M. C. (1992). Social controls on cell survival and cell death. *Nature*, **356**, 397–400.

Rager, G. and Rager, U. (1978). Systems-matching by degeneration. I. A quantitative electron microscopic study of the generation and degeneration of retinal ganglion cells in the chicken. *Exp. Brain Res.*, **33**, 65–78.

Raven, M. A. and Reese, B. E. (2003). Mosaic regularity of horizontal cells in the mouse retina is independent of cone photoreceptor innervation. *Invest. Ophthalmol. Vis. Sci.*, **44**, 965–73.

Raven, M. A., Eglen, S. J., Ohab, J. J. and Reese, B. E. (2003). Determinants of the exclusion zone in dopaminergic amacrine cell mosaics. *J. Comp. Neurol.*, **461**, 123–36.

Raven, M. A., Stagg, S. B., Nassar, H. and Reese, B. E. (2005). Developmental improvement in the regularity and packing of mouse horizontal cells: implications for mechanisms underlying mosaic pattern formation. *Vis. Neurosci.*, **22**, 569–73.

Reese, B. E. (1986). The topography of expanded uncrossed retinal projections following neonatal enucleation of one eye: differing effects in dorsal lateral geniculate nucleus and superior colliculus. *J. Comp. Neurol.*, **250**, 8–32.

Reese, B. E. and Geller, S. F. (1995). Precocious invasion of the optic stalk by transient retinopetal axons. *J. Comp. Neurol.*, **353**, 572–84.

Reese, B. E. and Urich, J. L. (1994). Does early enucleation affect the decussation pattern of alpha cells in the ferret? *Vis. Neurosci.*, **11**, 447–54.

Reese, B. E., Thompson, W. F. and Peduzzi, J. D. (1994). Birthdates of neurons in the retinal ganglion cell layer in the ferret. *J. Comp. Neurol.*, **41**, 464–75.

Rehen, S. K., Varella, M. H., Freitas, F. G., Moraes, M. O. and Linden, R. (1996). Contrasting effects of protein synthesis inhibition and of cyclic AMP on apoptosis in the developing retina. *Development*, **122**, 1439–48.

Resta, V., Novelli, E., Di Virgilio, F. and Galli-Resta, L. (2005). Neuronal death induced by endogenous extracellular ATP in retinal cholinergic neuron density control. *Development*, **132**, 2873–82.

Reuter, G. and Zilles, K. (1993). Reduction of naturally-occurring cell death by kainic acid in the retina of chicken embryos. *Anat. Anz.*, **175**, 243–51.

Rickman, D. W., Nacke, R. E. and Rickman, C. B. (1999). Characterization of the cell death promoter, Bad, in the developing rat retina and forebrain. *Dev. Brain Res.*, **115**, 41–7.

Robinson, S. R. (1988). Cell death in the inner and outer nuclear layers of the developing cat retina. *J. Comp. Neurol.*, **267**, 507–15.

Robinson, S. R., Dreher, B. and McCall, M. J. (1989). Nonuniform retinal expansion during the formation of the rabbit's visual streak: implications for the ontogeny of mammalian retinal topography. *Vis. Neurosci.*, **2**, 201–19.

Rocha, M., Martins, R. A. P. and Linden, R. (1999). Activation of NMDA receptors protects against glutamate neurotoxicity in the retina: evidence for the involvement of neurotrophins. *Brain Res.*, **827**, 79–92.

Rohrer, B., LaVail, M. M., Jones, K. R. and Reichardt, L. F. (2001). Neurotrophin receptor TrkB activation is not required for the postnatal survival of retinal ganglion cells in vivo. *Exp. Neurol.*, **172**, 81–91.

Rothermel, A. and Layer, P. G. (2003). GDNF regulates chicken rod photoreceptor development and survival in reaggregated histotypic retinal spheres. *Invest. Ophthalmol. Vis. Sci.*, **44**, 2221–8.

Saavedra, H. I., Wu, L., de Bruin, A. *et al.* (2002). Specificity of E2F1, E2F2, and E2F3 in mediating phenotypes induced by loss of Rb. *Cell Growth Differ.*, **13**, 215–25.

Scheetz, A. J., Williams, R. W. and Dubin, M. W. (1995). Severity of ganglion cell death during early postnatal development is modulated by both neuronal activity and binocular competition. *Vis. Neurosci.*, **12**, 605–10.

Schonherr, E. and Hausser, H. J. (2000). Extracellular matrix and cytokines: a functional unit. *Dev. Immunol.*, **7**, 89–101.

Schraermeyer, U. and Heimann, K. (1999). Current understanding on the role of retinal pigment epithelium and its pigmentation. *Pigment Cell Res.*, **12**, 219–36.

Schulz, M., Raju, T., Ralston, G. and Bennett, M. R. (1990). A retinal ganglion cell neurotrophic factor purified from the superior colliculus. *J. Neurochem.*, **55**, 832–41.

Schuster, N., Dunker, N. and Krieglstein, K. (2002). Transforming growth factor-beta induced cell death in the developing chick retina is mediated via activation of c-jun N-terminal kinase and downregulation of the anti-apoptotic protein Bcl-X(L). *Neurosci. Lett.*, **330**, 239–42.

Sedlak, T. W., Oltvai, Z. N., Yang, E. *et al.* (1995). Multiple Bcl-2 family members demonstrate selective dimerizations with Bax. *Proc. Natl. Acad. Sci. U. S. A.*, **92**, 7834–8.

Seecharan, D. J., Kulkarni, A. L., Lu, L., Rosen, G. D. and Williams, R. W. (2003). Genetic control of interconnected neuronal populations in the mouse primary visual system. *J. Neurosci.*, **23**, 11179–88.

Sengelaub, D. R. and Finlay, B. L. (1982). Cell death in the mammalian visual system during normal development. I. Retinal ganglion cells, *J. Comp. Neurol.*, **204**, 311–17.

Serfaty, C. A., Reese, B. E. and Linden, R. (1990). Cell death and interocular interactions among retinofugal axons: lack of binocularly matched specificity. *Dev. Brain Res.*, **56**, 198–204.

Sharma, R. K. (2001). Expression of Bcl-2 during the development of rabbit retina. *Curr. Eye Res.*, **22**, 208–14.

Sheedlo, H. J., Brun-Zinkernagel, A. M., Oakford, L. X. and Roque, R. S. (2001). Rat retinal progenitor cells and a retinal pigment epithelial factor. *Brain Res. Dev. Brain Res.*, **127**, 185–7.

Sheedlo, H. J., Nelson, T. H., Lin, N. *et al.* (1998). RPE secreted proteins and antibody influence photoreceptor cell survival and maturation. *Brain Res. Dev. Brain Res.*, **107**, 57–69.

Sholl-Franco, A., Figueiredo, K. G. and de Araujo, E. G. (2001). Interleukin-2 and interleukin-4 increase the survival of retinal ganglion cells in culture. *NeuroReport*, **12**, 109–12.

Silveira, L. C., Yamada, E. S. and Picanco-Diniz, C. W. (1989). Displaced horizontal cells and biplexiform horizontal cells in the mammalian retina. *Vis. Neurosci.*, **3**, 483–8.

Silveira, M. S. and Linden, R. (2005). Neuroprotection by cAMP. In *Brain Repair*, ed. M. Bähr. New York: Springer, pp. 164–76.

Silveira, M. S., Costa, M. R., Bozza, M. and Linden, R. (2002). Pituitary adenylyl cyclase-activating polypeptide prevents induced cell death in retinal tissue through activation of cyclic AMP-dependent protein kinase. *J. Biol. Chem.*, **277**, 16 075–80.

Silver, J. and Hughes, A. F. (1973). The role of cell death during morphogenesis of the mammalian eye. *J. Morphol.*, **140**, 159–70.

Simon, D. K. and O'Leary, D. D. (1992). Influence of position along the medial-lateral axis of the superior colliculus on the topographic targeting and survival of retinal axons. *Dev. Brain Res.*, **69**, 167–72.

Skeen, L. C., Due, B. R. and Douglas, F. E. (1986). Neonatal sensory deprivation reduces tufted cell number in mouse olfactory bulbs. *Neurosci. Lett.*, **63**, 5–10.

Slack, R. S., El-Bizri, H., Wong, J., Belliveau, D. J. and Miller, F. D. (1998). A critical temporal requirement for the retinoblastoma protein family during neuronal determination. *J. Cell Biol.*, **140**, 1497–509.

Sloviter, R. S. (2002). Apoptosis: a guide for the perplexed. *Trends Pharmacol. Sci.*, **23**, 19–24.

Smeyne, R. J., Vendrell, M., Hayward, M. *et al.* (1993). Continuous c-fos expression precedes programmed cell death *in vivo*. *Nature*, **363**, 166–9.

Soderpalm, A. K., Fox, D. A., Karlsson, J. O. and van Veen, T. (2000). Retinoic acid produces rod photoreceptor selective apoptosis in developing mammalian retina. *Invest. Ophthalmol. Vis. Sci.*, **41**, 937–47.

Spira, A., Hudy, S. and Hannah, R. (1984). Ectopic photoreceptor cells and cell death in the developing rat retina. *Anat. Embryol. (Berl.)*, **169**, 293–301.

Sretavan, D. W. and Shatz, C. J. (1986). Prenatal development of retinal ganglion cell axons: segregation into eye-specific layers within the cat's lateral geniculate nucleus. *J. Neurosci.*, **6**, 234–51.

Steffen, M. A., Seay, C. A., Amini, B. *et al.* (2003). Spontaneous activity of dopaminergic retinal neurons. *Biophys. J.*, **85**, 2158–69.

Stone, J., Rapaport, D. H., Williams, R. W. and Chalupa, L. (1982). Uniformity of cell distribution in the ganglion cell layer of prenatal cat retina: implications for mechanisms of retinal development. *Dev. Brain Res.*, **2**, 231–42.

Stone, J., Maslim, J., Valter-Kocsi, K. (1999). Mechanisms of photoreceptor death and survival in mammalian retina. *Prog. Retin. Eye Res.*, **18**, 689–735.

Strettoi, E. and Volpini, M. (2002). Retinal organization in the bcl-2-over-expressing transgenic mouse. *J. Comp. Neurol.*, **446**, 1–10.

Strom, R. C. and Williams, R. W. (1998). Cell production and cell death in the generation of variation in neuron number. *J. Neurosci.*, **18**, 9948–53.

Sucher, N. J., Kohler, K., Tenneti, L. *et al.* (2003). N-methyl-D-aspartate receptor subunit NR3A in the retina: developmental expression, cellular localization, and functional aspects. *Invest. Ophthalmol. Vis. Sci.*, **44**, 4451–6.

Tanaka, H. and Landmesser, L. T. (1986). Cell death of lumbosacral motoneurons in chick, quail, and chick–quail chimera embryos: a test of the quantitative matching hypothesis of neuronal cell death. *J. Neurosci.*, **6**, 2889–99.

Tanito, M., Masutani, H., Nakamura, H., Ohira, A. and Yodoi, J. (2002). Cytoprotective effect of thioredoxin against retinal photic injury in mice. *Invest. Ophthalmol. Vis. Sci.*, **43**, 1162–7.

Thompson, I. D. and Morgan, J. E. (1993). The development of retinal ganglion cell decussation patterns in postnatal pigmented and albino ferrets. *Eur. J. Neurosci.*, **5**, 341–56.

Ulshafer, R. J. and Clavert, A. (1979). Cell death and optic fiber penetration in the optic stalk of the chick. *J. Morphol.*, **162**, 67–76.

Varella, M. H., Correa, D. F., Campos, C. B. L., Chiarini, L. B. and Linden, R. (1997). Protein kinases selectively modulate apoptosis in the developing retina in vitro. *Neurochem. Int.*, **31**, 217–27.

Varella, M. H., de Mello, F. G. and Linden, R. (1999). Evidence for an antiapoptotic role of dopamine in developing retinal tissue. *J. Neurochem.*, **73**, 485–92.

Wässle, H., Dacey, D. M., Haun, T. *et al.* (2000). The mosaic of horizontal cells in the macaque monkey retina: with a comment on biplexiform ganglion cells. *Vis. Neurosci.*, **17**, 591–608.

Wikler, K. C., Perez, C. and Finlay, B. L. (1989). Duration of retinogenesis: its relationship to retinal organization in two cricetine rodents. *J. Comp. Neurol.*, **285**, 157–76.

Williams, M. A., Piñòn, L. G. P., Linden, R. and Pinto, L. H. (1990). The pearl mutation accelerates the schedule of natural cell death in the early postnatal retina. *Exp. Brain Res.*, **82**, 393–400.

Williams, R. R., Cusato, K., Raven, M. A. and Reese, B. E. (2001). Organization of the inner retina following early elimination of the retinal ganglion cell population: effects on cell numbers and stratification patterns. *Vis. Neurosci.*, **18**, 233–44.

Wong, P., Smith, S. B., Bora, N. and Gentleman, S. (1994). The use of C0t-1 probe DNA for the detection of low levels of DNA fragmentation. *Biochem. Cell Biol.*, **72**, 649–53.

Wu, L. Y., Li, M., Hinton, D. R. *et al.* (2003). Microphthalmia resulting from MSX2-induced apoptosis in the optic vesicle. *Invest. Ophthalmol. Vis. Sci.*, **44**, 2404–12.

Wyllie, A. H. (1980). Glucocorticoid-induced thymocyte apoptosis is associated with endogenous endonuclease activation. *Nature*, **284**, 555–6.

Yakura, T., Fukuda, Y. and Sawai, H. (2002). Effect of Bcl-2 over-expression on establishment of ipsilateral retinocollicular projection in mice. *Neuroscience*, **110**, 667–73.

Yamada, H., Yamada, E., Ando, A. *et al.* (2001). Fibroblast growth factor-2 decreases hyperoxia-induced photoreceptor cell death in mice. *Am. J. Pathol.*, **159**, 1113–20.

Yhip, J. P. and Kirby, M. A. (1990). Topographic organization of the retinocollicular projection in the neonatal rat. *Vis. Neurosci.*, **4**, 313–29.

Yokoyama, Y., Ozawa, S., Seyama, Y. *et al.* (1997). Enhancement of apoptosis in developing chick neural retina cells by basic fibroblast growth factor. *J. Neurochem.*, **68**, 2212–15.

Young, R. W. (1984). Cell death during differentiation of the retina in the mouse. *J. Comp. Neurol.*, **229**, 362–73.

Zhang, X., Chaudhry, A. and Chintala, S. K. (2003). Inhibition of plasminogen activation protects against ganglion cell loss in a mouse model of retinal damage. *Mol. Vis.*, **9**, 238–48.

12

Dendritic growth

Jeff Mumm

Washington University School of Medicine, St. Louis, USA
(Currently at Luminomics, St. Louis, USA)

Christian Lohmann

Max-Planck-Institute of Neurobiology, Germany

12.1 Introduction

Retinal neuron arbors are organized in relation to three central functions. (1) Outgrowth is regulated in the lateral dimension to delimit receptive-field size, a property linked to spatial acuity. (2) Interactions between individual neuronal subtypes are coordinated with respect to neuritic overlap to promote complete coverage, or tiling, of the retina, thus assuring that distinct functions have representation over the entire area of the retina (see Chapter 10). (3) Interactions between pre- and postsynaptic partners are organized in the vertical dimension such that functionally discrete circuits are physically isolated within the synaptic neuropil. For instance, during development of the inner plexiform layer (IPL) connections between subsets of bipolar, amacrine and retinal ganglion cells come to be arranged in a laminar fashion, sometimes occupying single strata within a multilayered array of concentric circuits (Figure 12.1).

In this chapter the current state of understanding regarding the structural development of retinal neuron arbors is discussed: from mechanisms that impact individual neuronal morphologies to those that orchestrate interactions between synaptic partners. In the first section, issues concerning initial neurite extension are discussed. These include establishing cellular polarity and compartmentalization of neurites into the axon and dendrites. Section two focuses on the establishment of dendritic territory and interactions that influence receptive-field size. The last section deals with the process of sublamination, whereby individual neuritic arbors resolve into monostratified, multistratified, or diffuse (non-stratified) configurations within the IPL. The chapter concludes with a discussion of some of the outstanding questions in this field, including a highlight of recently developed techniques that are being applied to further elucidate the fascinating process of how retinal neurons form and how neurite patterning defines visual function.

12.2 Mechanisms regulating initial neurite outgrowth patterns

How do retinal neurons come to assume their unique shapes and what are the cellular and molecular cues that guide this process? Shortly after becoming specified neurons begin

Retinal Development, ed. Evelyne Sernagor, Stephen Eglen, Bill Harris and Rachel Wong.
Published by Cambridge University Press. © Cambridge University Press 2006.

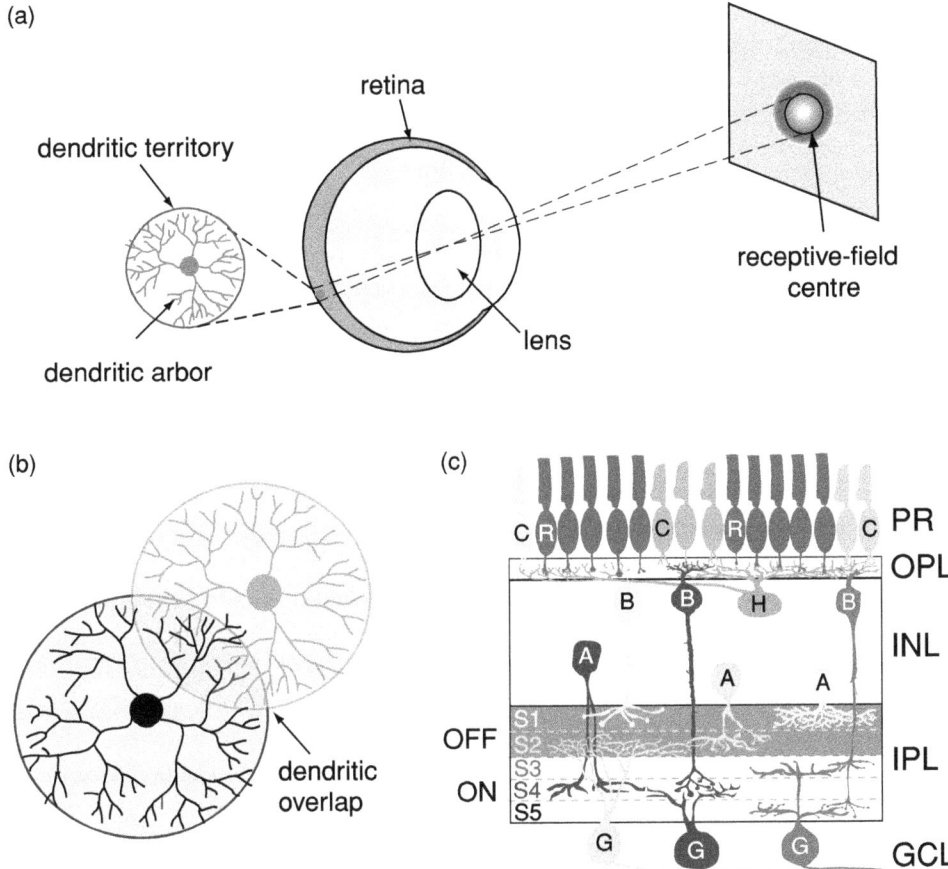

Figure 12.1 Three principal functions of retinal neuritic patterning. (a) The extent of a neuron's denritic territory corresponds directly to a region in visual space, the 'receptive field'. (b) Regions of denritic overlap are organized such that, collectively, discrete neuronal subtypes completely cover, or 'tile', the retina. (c) Neuronal arbors often stratify within the IPL such that they occupy discrete laminar positions. On a gross level, the IPL is split roughly in half such that OFF-responding circuitry is restricted to the outermost region (grey bar) while ON-responding circuitry is confined to an innermost sublayer (white bar). Beyond this, the IPL can be divided into finer sublaminae, typically five can be delineated (S1 to S5). Individual retinal circuits are comprised of connections between specific subtypes of bipolar, amacrine and ganglion cells (corresponding shades of grey) whose arbors co-stratify within a particular sublamina, or sublaminae. A, amacrine cell; B, bipolar cell; C, cone photoreceptor; G, retinal ganglion cell; H, horizontal cell; R, rod photoreceptor; GCL, ganglion cell layer; INL, inner nuclear layer; IPL, inner plexiform layer; OPL, outer plexiform layer; PR, photoreceptor.

to elaborate neurites. Through the process of polarization neurons typically establish a single process as the axon with the remainder, or subsequent, processes becoming dendrites (see Horton and Ehlers, 2003, for review). Morphological analyses have shown that retinal ganglion cells (RGCs) extend a single axonal process prior to elaborating dendrites (Hinds and Hinds, 1974; Maslim et al., 1986). Molecular pathways that regulate neuronal polarity in other neural systems have recently been defined (Wodarz, 2002; Shi *et al.*, 2004; Jiang *et al.*, 2005; Yoshimura *et al.*, 2005). It is not yet known whether the same pathways regulate this process in the retina. Until these details are worked out we can ask simply what happens if cellular polarity is disrupted? The molecular identification of zebrafish mutants with severe retinal patterning defects (i.e. disruption of laminar organization) provides possible inroads to these questions (Horne-Badovinac *et al.*, 2001; Wei and Malicki, 2002; Malicki *et al.*, 2003). Of three loci cloned to date, homologues of two (aPKC or protein kinase C type a and membrane-associated guanylate kinase) have been shown to interact with members of the protein complex implicated in neuronal polarity, and the third (N-cadherin) promotes adhesive interactions at the cell surface that underlie compartmentalization of related groups of cells (Hayashi and Carthew, 2004). Thus, it appears that when cellular polarity or specific cell–cell interactions are disrupted, laminar patterning of the retina is lost. Factors that specifically stimulate axonal or dendritic outgrowth in retinal neurons still remain largely undefined. However, cellular and molecular determinants that stimulate general neurite outgrowth in purified cultures of rat RGCs have been revealed.

Members of the bone morphogenetic protein family (BMP2, BMP13, GDF8; growth and differentiation factor 8) and the neurotrophin, brain-derived neurotrophic factor (BDNF), increase neurite length and complexity (branch point number), while BMPs also promote neurite number in RGC cultures (Kerrison *et al.*, 2005). Similarly, an extensive panel of trophic factors, hormones and adhesion molecules was shown to promote neurite extension in RGC cultures grown at clonal densities (Goldberg *et al.*, 2002a). Interestingly, when RGCs were electrically stimulated via multielectrode arrays, responses to exogenous factors were substantially potentiated. The principle caveat of these studies is whether these factors act directly or simply improve neuronal 'health' under culture conditions. In addition, it is difficult to extrapolate precisely how factors that effect neurite growth patterns in culture operate to shape neuronal structure, particularly sublamination patterns, in vivo. Nevertheless, as this system has been developed as a high throughput application, capable of testing hundreds of molecules for similar effects, it should prove useful for identifying new leads (Kerrison *et al.*, 2005). The Reelin–Dab-1 pathway also appears to stimulate outgrowth from newly born RGCs (and amacrines) in the chick retina (Katyal and Godbout, 2004). Interestingly, Dab-1 can be rendered unresponsive to secreted Reelin via alternative splicing that removes critical phosphorylation domains; this form is preferentially expressed in undifferentiated neuroblasts. Expression of the fully functional form of Dab-1 is limited to differentiating cells and thereby promotes Reelin-stimulated outgrowth. Details akin to this level of regulation will no doubt become more evident as our understanding of neurite promoting molecular factors increases.

Insights into *cellular* mechanisms influencing axonal and dendritic growth have also come from studies utilizing purified rat RGC cultures. Goldberg *et al.* (2002b) demonstrated that cultured RGCs, much like RGCs in vivo, display an age-related bias in the rate of growth of axonal versus dendritic processes; embryonic cells (E20) preferentially grow axons, while postnatal cells (P8) slow axonal growth and switch to extending dendrites primarily. The data suggest the presence of an extrinsic contact-mediated signal responsible for shifting ganglion cell outgrowth from an axonal to a dendritic mode. Interestingly, this signal does not come from the axonal target zone but instead from direct contact with amacrine cells within the retina. Radial glia have also been implicated in providing localized signals that preferentially promote and/or sustain axonal (glial endfeet) or dendritic (glial somata) outgrowth of RGCs in the chicken retina (Bauch et al., 1998; Stier and Schlosshauer, 1998). These studies implicate a contact-dependent mechanism rather than the production of exogenous factors in regulating neurite extension. Thus, assigning discrete roles for locally secreted factors awaits further elucidation.

Signalling in the target area of RGC axons can also affect the growth and patterning of RGC dendrites. In developing frogs, BDNF is endogenously released in the retina as well as the tectum where RGC axons make synapses. Furthermore, RGCs express the neurotrophin receptor (TrkB) that preferentially binds BDNF. Experimentally increasing BDNF levels in the tectum results in more complex RGC dendritic and axonal arborizations (Lom and Cohen-Cory, 1999; Lom et al., 2002). Surprisingly, increasing retinal BDNF levels has the opposite effect on RGC dendrites: they develop fewer primary dendrites and dendritic branches. In contrast, axonal complexity is not affected by increasing retinal BDNF levels. Together, these studies show that the development of RGC dendrites is not only regulated by intraretinal signalling, but also by neuronal interactions in the brain. In addition, they reveal that the effect of individual molecular factors is dependent on the site of action.

In contrast to the detailed studies on the development of RGC morphology, comparatively little is known about factors that shape the arborizations of the retina's interneurons. Amacrine, horizontal and bipolar cell studies have largely been limited to morphological analyses using techniques such as immunocytochemistry, Golgi staining and electron microscopy. Although instrumental in defining numerous mature morphological neuronal subtypes the static nature of these approaches makes it difficult to piece together a developmental series of events. For instance, Golgi staining of presumptive amacrines in the chick retina suggests that a 'multipodial' subpopulation extends neurites in all directions while migrating towards the nascent IPL (Prada et al., 1987). A second 'smooth' population is believed to refrain from extending neurites until after reaching the IPL. Similar nascent 'amacrine' morphologies have been observed in serial electron microscope reconstructions of the embryonic mouse retina (Hinds and Hinds, 1978, 1983). Thus, the arbors of particular subtypes of retinal interneurons may undergo significant remodelling during development, the principle caveat of these studies being the inability to follow the cells displaying these early morphologies through time in order to verify their cellular identity. Nevertheless, careful observations have resulted in a generalized scheme for the morphological development of the major neuronal subtypes of the retina (Figure 12.2).

12.3 Mechanisms regulating neuritic field size

After neurite outgrowth is initiated, mechanisms become active to constrain it. These constraints determine the specific structures of mature retinal neurons, which underlie their

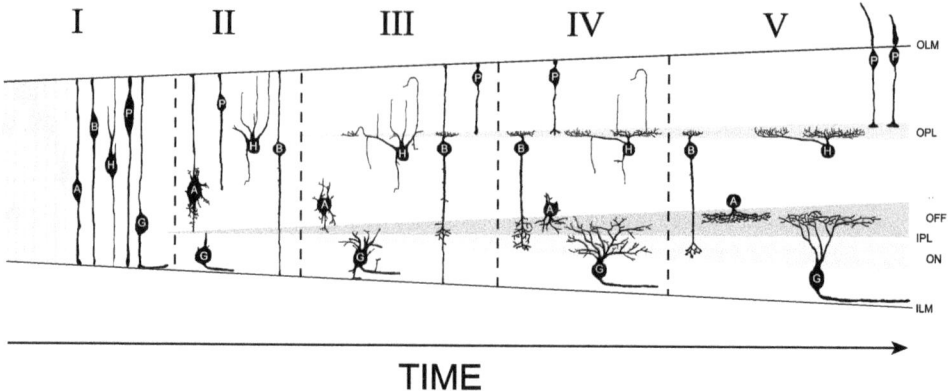

Figure 12.2 Morphological differentiation of major retinal neuron types. Depicted here is a stereotyped series of steps in the development of the five archetypal neuronal types of the vertebrate retina. An attempt has been made to correlate morphological stages of development between cell types with the understanding that this represents a gross oversimplification due to neuronal subtype, and species-specific variations. A series of schematized stages denotes major developmental events. (I) Neuroblastic – the ganglion cell is the first to begin the differentiation process, RGC axon outgrowth is believed to precede release from the limiting membranes. (II) Initial neurite outgrowth – neurons begin to extend neurites within the retina as they become postmitotic and lose connections to the limiting membranes, inner plexiform layer (IPL) formation commences via ingrowth of ganglion and amacrine cells followed shortly thereafter by synaptogenesis. (III) Neurite exploration – neurons extend many neuritic branches within areas that they will eventually retract from, outer plexiform layer (OPL) formation commences via ingrowth of photoreceptor, horizontal and bipolar cells. (IV) Neurite elaboration – neurons begin to establish neuritic territories within each of the synaptic neuropils. (V) Neurite stratification – neurons establish sublaminated neurite patterns, most notably in regards to ON and OFF strata formation. A, amacrine cell; B, bipolar cell; G, retinal ganglion cell; H, horizontal cell; P, photoreceptor cell; ILM, inner limiting membrane; OLM, outer limiting membrane. This figure was adapted from a drawing by Josh Morgan.

respective functions. In this section, how the arborizations of neurites are shaped in the lateral dimension will be addressed. The lateral extent and complexity of an individual neuron's dendritic tree determines the proportion of the visual scene that can contribute to the activity of the neuron, its 'receptive field' (Hartline, 1940).

In the 1980s Wässle and colleagues discovered that the dendrites of RGCs are not randomly distributed within the horizontal plane of the retina, but obey certain borders and form territories (Wässle et al., 1981). The overlap between dendritic trees of neighbouring RGCs is surprisingly constant between cells of the same subtype. In contrast, the overlap of the dendritic arborizations of cells of different subtypes (e.g. ON or OFF RGCs) appear to be independent. This type of dendritic patterning ensures that each population of functionally distinct RGC subtype samples the visual field evenly throughout the entire area of the retina.

How is the patterned organization of RGC dendrite territories achieved during development? Wässle et al. (1981) showed that in the mature retina the area of the dendritic arborization of each RGC is irregular. The extension of the dendrites is approximated by the so-called 'Dirichlet domains', the areas around each cell body that are restricted by the distance to the nearest neighbour. This indicates that the immediate neighbours of RGCs influence the extent of their dendritic outgrowth. In addition, replacing the dendritic territories with their mirror image produces an incomplete coverage of the area. Together, these observations suggested that local interactions between neighbouring RGCs of the same subtype shape their dendrites during development and yield the complete and uniform coverage that is necessary for vision at maturity.

Furthermore, lesion experiments supported this view; depletion of RGCs in one spot of the retina during a critical period of development induced sprouting of dendrites in the RGCs adjacent to the lesion site. In fact, dendrites grew towards the site where neighbouring RGCs were absent, indicating that interactions with the local environment regulate dendritic growth on the level of single dendrites (Figure 12.3a; Perry and Linden, 1982; Eysel et al., 1985; Deplano et al., 1994).

Depletion experiments not only suggested that regulation of dendritic extension is local, but also that some form of interaction between neighbouring RGCs may shape their dendrites. However, whether this interaction is direct, e.g. by contact-mediated inhibition of dendritic growth, or indirect, by competition for synaptic input or growth factors from presynaptic neurons, has been debated. Troilo et al. (1996) used two different strategies to decrease the density of RGCs in retinas from developing chicks. First, they increased eye growth by inducing visual form deprivation. This manipulation resulted in a decreased density of RGCs as well as their presynaptic neurons. Second, they induced a selective reduction of RGC density without affecting the number of presynaptic neurons by a partial optic nerve section. They observed that both types of experiments resulted in larger dendritic fields; however, only the latter manipulation caused an additional increase in dendritic branching. These results suggest that direct interactions – possibly contact-mediated inhibition – between neighbouring RGCs restrict the extension of dendritic fields, because the increase of dendritic area is independent of the density of presynaptic inputs. Yet, arborization complexity seems to be dependent on presynaptic inputs, since only the experiment that increased the ratio between presynaptic neurons and postsynaptic RGCs facilitated dendritic branching. Interestingly, a recent study in horizontal cells demonstrated a very similar dichotomy in the regulation of dendritic field size versus dendritic branching. In mouse horizontal cells the amount of dendritic branching depends on the presence of afferent cells, however, dendritic field size is regulated by neighbour interactions (Reese et al., 2005).

The notion that contact-mediated inhibition of dendritic growth regulates the extension of RGC dendrites is supported by the presence of dendro-dendritic contacts between RGCs of the same type in developing ferret retinas (Lohmann and Wong, 2001). Virtually all terminal dendrites of ferret alpha-RGCs form contacts with dendrites of neighbouring RGCs of the same type (Figure 12.3b). In addition, dendrites extend only a few micrometres beyond a point of contact, but tens of micrometres after a point where they cross a dendrite of a neighbouring RGC but fail to make a contact. It is not clear whether RGCs indeed signal

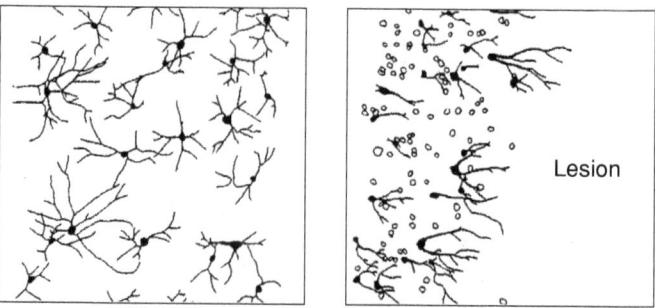
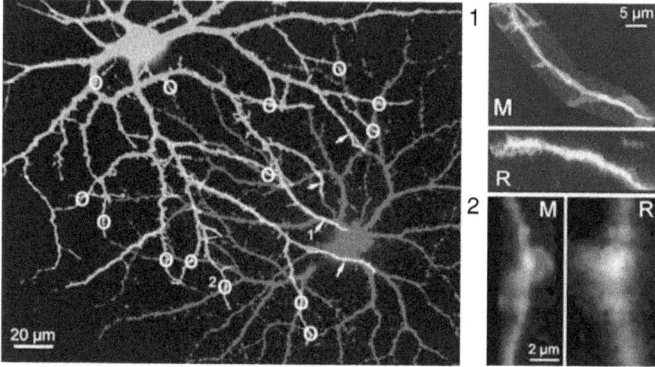
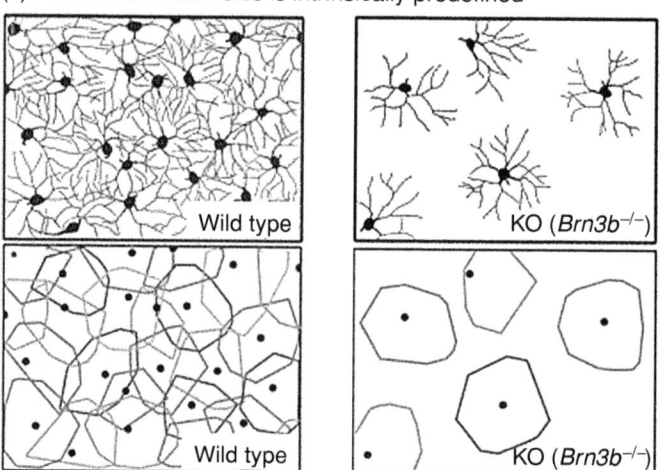

Figure 12.3 Dendritic territories are shaped by interactions between neighbouring RGCs of the same type. (a) Population of rat RGCs stained by injections of horseradish peroxidase into the optic tract in a control retina (*left*) and in a lesioned retina (*right*). Note that dendrites of those RGCs that are located close to the lesion site are directed towards the depleted area. (b) Pair of OFF-α-RGCs in a ferret retinal whole mount. Dendritic contacts are circled in white. Fasciculation of dendrites are marked by arrows. High magnification (m) and 90 degree rotation (R) views are shown for two contacts (1,2). (c) Camera lucida drawings (*top*) and dendritic fields (*bottom*) of mouse RGCs from a wild-type (*left*) and a knockout (Ko) retina ($Brn3b^{-/-}$) (*right*). Note that although the density of neurons is strongly reduced in the knockout retina, the extent of the dendritic area is comparable to those in the wild-type retina. For colour version, see Plate 10.

via the dendritic appositions. If so, contact-mediated inhibition may occur via homophilic interactions of adhesion molecules, such as cadherins (Inuzuka *et al.*, 1991) or protocadherins (Gao *et al.*, 2000), or it may involve Notch–Delta signalling (Berezovska *et al.*, 1999; Sestan *et al.*, 1999; Redmond *et al.*, 2000) or gap junctions (Vaney, 1990; Penn *et al.*, 1994). Interestingly, theoretical (Eglen *et al.*, 2000) as well as experimental (Galli-Resta *et al.*, 2002) evidence suggests that dendro-dendritic interactions may not only determine the shape of RGC dendrites, but also the position of their somata within 'mosaics', orderly arrays of retinal neuron cell bodies (see Chapter 10).

It is important to note, however, that contact inhibition is not the only mechanism that constrains dendritic growth. Intrinsic growth restriction mechanisms are also important. In two mutant mouse lines in which the number of RGCs is reduced ($Brn3b^{-/-}$ or $Math5^{-/-}$ mice) most, if not all, RGC types are present, albeit at a much reduced density (20% of wild type in $Brn3b^{-/-}$, 5% of wild type in $Math5^{-/-}$ mice). Surprisingly, in these mice RGC dendritic areas are not significantly different from those in wild-type animals (Figure 12.3c; Lin *et al.*, 2004). This finding suggests that dendrites may grow until they reach an intrinsically determined maximal length, but, if neighbouring neurons are present, local inhibitory interactions determine the final shape of the dendritic tree. Whether homotypic neighbour interactions or intrinsic factors determine the area of the dendritic field may also depend on the cell type. For example, the extent of the dendritic arborization of horizontal cells depends on the distance from the nearest neighbour (Reese *et al.*, 2005). In contrast, the dendritic field size of cholinergic amacrine cells in mice is not changed in the absence of neighbours of the same type (Farajian *et al.*, 2004). Thus, extrinsic and intrinsic factors cooperate to form the lateral extent of the dendritic arborization and, consequently, the receptive field size of retinal neurons. However, the relative contribution of each factor may differ between different retinal cell types.

12.4 Mechanisms regulating structural lamination of the IPL

A relationship between form and function in the retina was elegantly revealed upon discovery of the morphological correlate of ON- and OFF-centre responding retinal circuits (Kuffler, 1953): ON-pathway cells restrict their arbors to the inner portion of the IPL (nearest the RGC layer), OFF-pathway cells occupy the outer region (Famiglietti and Kolb, 1976; Famiglietti *et al.*, 1977; Stell *et al.*, 1977; Nelson *et al.*, 1978). Later it was shown that ON- and OFF-responding cells (ON–OFF cells) ramify in both regions (Ammermuller and Kolb, 1995). ON and OFF streams are responsible for imparting contrast sensitivity to the retina; ON-centre paths delineate lighter than background stimuli (or light onset) and OFF-centre detect darker than background signals (or light offset, Hartline, 1938). Moreover, the ON and OFF regions of the IPL are each composed of several finer subcircuit stratifications, or sublaminae, that are believed to represent unique temporal and spatial aspects of the visual scene (Lettvin *et al.*, 1959; Roska and Werblin, 2001, Figure 12.1). Thus, retinal neurons that innervate the IPL often display highly laminated arborizations such that their neurites are restricted to either ON or OFF sublayers and/or contribute to individual, or specific subsets of, sublaminae. Defining distinct signalling roles for these laminar circuits, and the neuronal subpopulations that comprise them, would serve to advance our understanding of the organizational principles of retinal networks.

In most species, five prominent sublaminae are identifiable in the IPL, termed S1 to S5 (Cajal, 1972). Sublaminae S1 and S2 reside in the OFF region while S3 to S5 occupy the ON region. Generally though, the sublamination patterns of retinal neurons in non-mammalian species are more complex than in mammals. In agreement with this, eight to ten IPL sublaminae can be resolved in the chick (Naito and Chen, 2004; Mumm *et al.*, 2005) and the salamander (Pang *et al.*, 2002). Highly laminated circuitry of this type provides a tenable morphological correlate to synaptic specificity in the CNS (Sanes and Yamagata, 1999). Thus, experiments aimed at determining the mechanisms that regulate the process of neuritic sublamination in the retina provide a glimpse into how complex circuits form and provide possible insight into how synaptic specificity is achieved. In this section, the process of IPL sublamination, and potential regulatory mechanisms will be discussed.

12.4.1 Early synaptogenic events and dynamic dendritic motility

How do neurons go about defining synaptic relationships that ultimately serve to shape and stabilize neuritic structure? In terms of establishing contact with appropriate synaptic partners, neurite outgrowth patterns can be divided into two simplified models: (1) neurites could innervate target zones in a highly directed manner, suggesting that mature circuit architectures result from an intrinsic developmental strategy linking cell fate identities directly to final neuritic morphologies; or (2) neurites might rapidly extend and retract within target zones, suggesting that exploratory 'sampling' is utilized to select appropriate connections from a pool of possible synaptic partnerships. While it has been acknowledged for some time that RGCs undergo dendritic remodelling during development, only recently has time-lapse microscopy revealed the dynamic nature of this process. This phenomenon of rapid dendritic 'filopodial' motility was first described in developing hippocampal neurons (Dailey and Smith, 1996), but also occurs in many other neuronal types, including RGCs (Wong and Wong, 2000). For example, dendritic filopodia of embryonic chick RGCs extend and retract at rates of minutes and even seconds (Figure 12.4a; Wong *et al.*, 2000). Filopodial motility does not, however, appear to contribute significantly to overall growth of the dendritic arborization; instead, growth and retraction are very well balanced. Therefore, this phenomenon appears to result in subtle changes to the dendritic branching pattern. This suggests that RGCs utilize filopodial motility to increase the volume of the area explored by the dendrite; effectively combing the immediate microenvironment for cues from surrounding cells. In accordance with this theory, the vast majority of filopodial extensions are transient; presumably only those encountering meaningful extracellular signals are stabilized. Thus, at the level of circuit formation rapid filopodial motility may serve to make synaptogenic contacts, to pull those contacts to the dendritic shaft and/or to select among

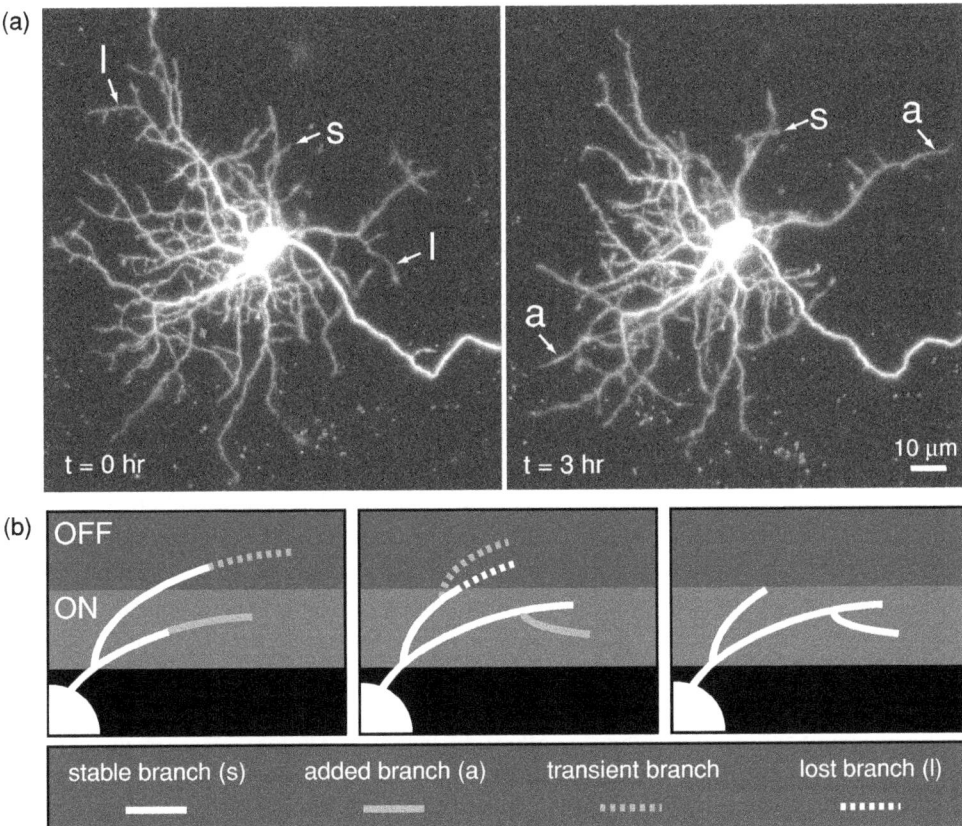

Figure 12.4 Structural plasticity and the development of stratified dendrites. (a) Confocal images of a green fluorescent protein-labelled RGC in a retinal whole mount from a developing mouse (P8). During a 3 hour period new dendritic branches are added (a), some are lost (l), whereas others remain stable (s). (b) Model for the development of stratified dendrites by dendritic motility and selective stabilization of dendrites in appropriate sublaminae (in this example the ON sublamina).

competing inputs by withdrawing from inappropriate contacts. This would require RGCs to be capable of determining an appropriate synapse from an inappropriate synapse and/or some level of network coordination such that Hebbian-type principles could apply. What signals have been shown to regulate dendritic motility? Interestingly, synaptic transmission can affect filopodial extension and retraction rates of immature chick RGCs, as can the relative activity of cytoskeletal regulators such as Rac and Rho (Wong et al., 2000; Wong and Wong, 2001). The developmental profile of dendritic motility, the patterns of its regulation and the occurrence of synapses on dendritic filopodia (Fiala et al., 1998) strongly suggest that filopodial motility plays an important role in synapse formation in the retina.

Thus, it appears that RGC dendrites actively search the environment for specific extracellular signals that shape the growing arbor by regulating synaptic partnerships. Indeed, the selective loss of dendritic branches from 'inappropriate' areas is currently hypothesized to underlie the formation of ON and OFF stratification patterns of mammalian RGCs (Figure 12.4b). It may be useful at this point to step back and ask what coordinated patterns of neurite outgrowth between retinal subpopulations could result in the formation of a stratified network made up of multiple sublaminae?

12.4.2 Sublamination: possible strategies

Laminated circuits could arise via several non-exclusive strategies (Figure 12.5): (a) sequential layering, successive strata stacking to either side of preceding networks; (b) remodelling, sublamination being the result of eliminating neurites from inappropriate regions; (c) targeting, neurons extending laterally stratified processes after encountering localized molecular guidance signals, or, (d) contact-dependent guidance signals; (e) hierarchical, laminar organization following refinement of afferent inputs or efferent target fields. Each of these strategies makes certain predictions about how retinal development would proceed and/or how neurons would behave. For instance, sequential layering would suggest that later-born cells would be limited to outer layers of the neuropil or that one side of the IPL would need to arise prior to the other. Data have indeed suggested that sublaminated amacrine and bipolar cells are evident in the OFF region prior to the ON region of the nascent IPL (Okada *et al.*, 1994; Crooks *et al.*, 1995; Gunhan-Agar *et al.*, 2000; Sherry *et al.*, 2003; Drenhaus *et al.*, 2004).

Somewhat at odds with the hypothesis that OFF layers form before ON are results showing that in postnatal cat retina the vast majority of immature RGCs are either ON- or OFF-responding from the outset; no evidence of one type preceding the other was found (Dubin *et al.*, 1986; Tootle, 1993). These findings suggested that sublamination of RGCs into ON and OFF layers resolves simultaneously, or that RGC dendrites are intrinsically 'hardwired' to grow within specific laminar regions. Support for the latter was obtained in a study of postnatal cat RGC morphology. The data show that at birth the majority of cat RGC dendrites are monostratified within ON or OFF regions, only a small minority of cat RGCs display a diffuse, non-stratified morphology (Dann *et al.*, 1988). Similarly, fetal macaque RGCs are either monostratified or bistratified, no evidence of diffuse patterns was observed (Kirby and Steineke, 1991). However, studies of the morphological development of *embryonic* cat and rat RGCs provide a much different picture; immature RGC arbors are shown to ramify throughout the entire depth of the IPL before becoming stratified (Maslim *et al.*, 1986; Maslim and Stone, 1988). Taken together, these data suggest that immature RGCs are innervated preferentially by either ON or OFF afferents from the outset of synaptogenesis, despite having dendritic arbors that extend throughout the entire depth of the IPL. However, patch-clamp recordings in postnatal ferret retinas show that the majority of immature RGCs respond to both light onset and light offset (Wang *et al.*, 2001). Moreover, subsequent dye filling allowed direct correlation between ON–OFF responses and diffuse dendritic arbors that spanned ON and OFF IPL sublayers. Immature mouse RGCs have also been characterized as predominantly ON–OFF-responding (Tian and Copenhagen, 2003). Thus,

Dendritic growth

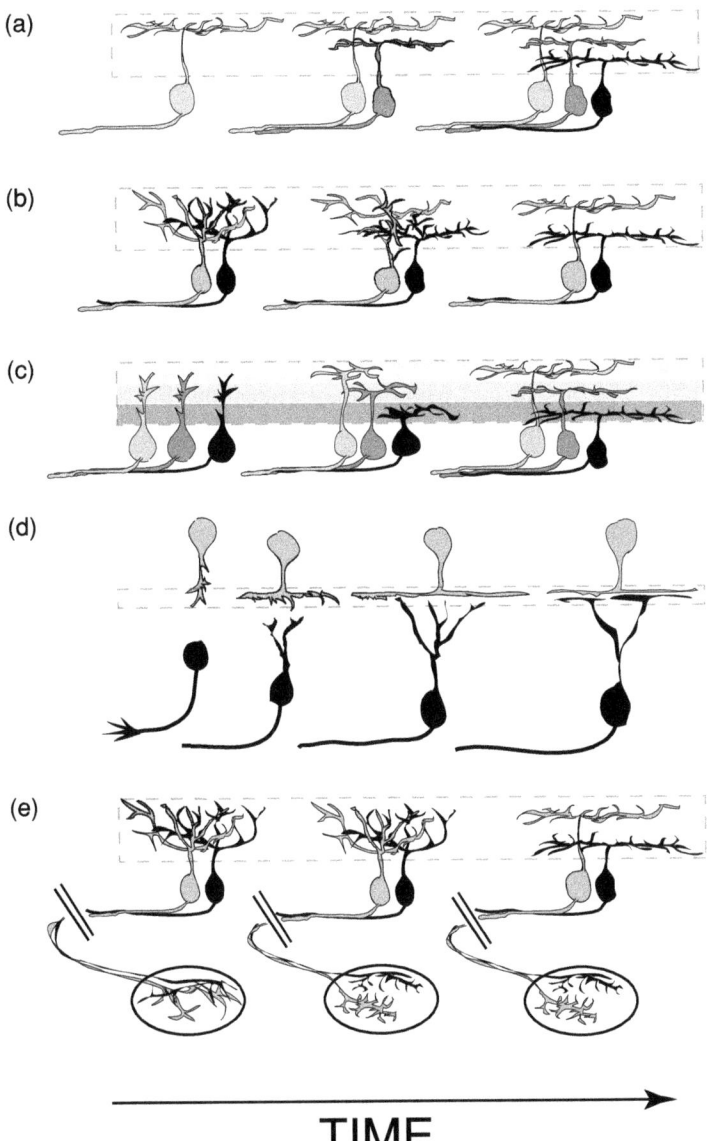

Figure 12.5 Non-exclusive strategies for the formation of laminar circuits. (a) Sequential layering; initial neurons elaborate stratified arbors within the neuropil, subsequent neurons establish strata above or below pre-existing layers. (b) Remodelling; neuritic arbors branch throughout the neuropil initially, subsequent elimination, or selective stabilization, of branches results in stratified patterns at maturity. (c) Targeted (molecular); neurons respond to molecular guidance cues arranged in a laminar fashion such that they elaborate lateral processes only in specific sublaminae. (d) Targeted (cellular); neurons respond to contact-dependent cues presented by guidepost cells and/or synaptic partners such that laminar outgrowth is restricted to specific sublaminae. (e) Hierarchical; neurons organize axonal and dendritic arbors in a serial fashion such that resolving the axonal target field area precedes dendritic rearrangements or vice versa. (a–e) Dashed box indicates neuropil (e.g. inner plexiform layer), all panels proceed left to right from immature to mature states.

the current hypothesis is that the majority of mammalian RGCs progress from an early non-stratified stage – characterized by ON–OFF responses and ramification throughout the depth of the IPL – to a mature stratified stage whereupon they respond as strictly ON *or* OFF cells and dendritic arbors are restricted to corresponding regions of the IPL (Bodnarenko et al., 1995, 1999; Lohmann and Wong, 2001; Stacy and Wong, 2003).

The morphologies of retinal neurons in non-mammalian species are known to be considerably more complex in terms of their stratification patterns (Cajal, 1972; Ammermuller and Kolb, 1995; Wu et al., 2000; Mangrum et al., 2002; Pang et al., 2002; Naito and Chen, 2004). One by-product of this is a veritable explosion in the number of morphological subtypes of retinal neurons encountered in these species. For instance, studies correlating morphology and physiology of retinal neurons in the tiger salamander have found more than 70 amacrine and bipolar cell patterns of stratification (Pang et al., 2002 and Pang et al., 2004, respectively). Interestingly, the physiological data indicate that of the ten IPL sublaminae delineated in the salamander retina; strata 1, 2 and 4 are OFF-responding, 3 and 7 to 10 are ON-responding, and 5 and 6 are ON–OFF. Thus, the physical segregation of ON and OFF channels is not strictly maintained and the overall pattern is not simply one of roughly splitting the IPL into two regions. It will be interesting to determine if the remodelling strategy of overgrowth and retraction, typified for mammalian RGCs, holds for the neuritic maturation of other subtypes of retinal neurons (and/or for different species), and to ascertain functions for the multiple strata found in retinas with such elegantly sublaminated circuitry.

What cellular interrelationships are required to form stratified arborizations in the IPL? Cellular ablation studies have provided insights into this question and to the cellular origins of stratification signals. For instance, RGCs can be induced to degenerate by severing the optic nerve. This approach was used to test whether the stratification patterns of immunolabelled subpopulations of amacrine and bipolar cells depend on the presence of RGCs in rats and ferrets (Gunhan-Agar et al., 2000; Williams et al., 2001, respectively). Despite concomitant effects on the width of the IPL ($\sim$70% the normal width) and amacrine numbers (several subtypes reduced), in the absence of RGC-derived targets all cell types tested formed relatively normal innervation patterns in the IPL. Similarly, in a zebrafish mutant in which RGCs never develop (*lakritz*), the bistratification pattern of a transgenically defined amacrine cell subpopulation was found to be nearly normal at maturity; showing only circumscribed areas of disruption (Kay et al., 2004). Time-lapse analysis revealed more pronounced abnormalities during the initial phases of IPL formation in *lakritz*, but the majority of these anomalies appear to resolve over time. Thus, the development of stratified arbor morphologies of bipolar and amacrine cells does not appear to depend on the presence of RGC dendrites. Interestingly, in those regions displaying aberrant amacrine cell stratification in *lakritz* mutants, a population of ON-stratifying bipolar cell axons (PKCβ1-immunoreactive) appeared to shadow the ectopic laminar patterns of the amacrine cells. This finding suggests that distinct bipolar and amacrine subpopulations are either interdependent regarding sublamination or that bipolar axons seek out cues from amacrine cells. In addition, it has been shown in several systems that cholinergic amacrine cells stratify prior to most RGCs and bipolar cells (Reese et al., 2001; Gunhan et al., 2002; Drenhaus et al., 2003; Stacy and Wong, 2003). Thus, cholinergic amacrines may be positioned upstream, in a hierarchical sense, providing guidance signals for the bipolar and RGC subtypes that subsequently ramify within the sublaminae that they establish (Bansal et al., 2000). In contrast, excitotoxic ablation of cholinergic amacrines in the postnatal ferret does not disrupt stratification patterns defined by several other neurotransmitter types (Reese et al., 2001). Furthermore, immunotoxin-mediated ablation of cholinergic amacrine cells in postnatal rats does not disrupt bistratification of recoverin-expressing cone bipolar cells (Gunhan et al., 2002). However, whether any of the cell types labelled in these studies actually interact with cholinergic amacrines at the synaptic level is not known. Similar studies performed on definitive sets of synaptic partners are needed to resolve this issue further.

12.4.3 Role of molecular guidance cues

How might molecules be used to specify retinal stratification patterns? One of the most attractive models for how neural circuits form involves unique molecular adhesion profiles that are shared between synaptic partners. Differential adhesion would thereby provide the initial stimulus to stabilize neuritic contacts. Interestingly, a number of cell–cell adhesion molecules are expressed in stratified patterns in the IPL of the retina (Wohrn et al., 1998; Honjo et al., 2000; Drenhaus et al., 2003, 2004). However, direct evidence for this model has been elusive, possibly due to the fact that disruption of cell–cell adhesion molecules often results in a number of pleiotropic effects. Evidence for general repulsive and/or attractive cues is also lacking. However, the distance between retinal synaptic partners is small enough that contact may result from polarized outgrowth alone without the need for positional gradients.

Despite the difficulties inherent in these studies several molecules have been implicated in IPL formation and patterning: Disruption of Plexin function, a cell surface co-receptor that mediates the repulsive effects of Semaphorins, results in a failure of IPL formation in the chick retina (Ohta et al., 1992). Interestingly, individual plexin subfamily members are expressed in amacrine, bipolar and retinal ganglion cell subsets. However, it is not yet known whether this corresponds to expression within individual strata of the IPL. Beyond its role in organizing retinal patterning in general (Matsunaga et al., 1988; Erdmann et al., 2003; Malicki et al., 2003), N-cadherin may function specifically to promote proper targeting and lamination of retinal neurites. A hypomorphic mutation in zebrafish ($N\text{-}cad^{RW95}$), which retains partial N-cadherin function, has been shown to cause defects in amacrine neurite patterning despite the relatively normal appearance of bipolar and retinal ganglion cell morphologies (Masai et al., 2003). However, expression of a dominant negative form in single cells of the frog retina suggests that N-cadherin functions primarily to promote RGC outgrowth while having lesser effect on other retinal cell types (Riehl et al., 1996). Finally, recent evidence implicates two new members of the immunoglobin superfamily as direct sublamination guidance cues in the chick retina (Yamagata *et al.*, 2002). Sidekicks (1 and 2) have been shown to promote homophilic binding and are expressed in retinal neuron subsets. Like several other immunoglobin family members, Sidekicks are expressed in specific IPL sublaminae suggesting these molecules may serve as laminar adhesion zones. In support of this idea, *sidekick-1* overexpression is correlated with stratification patterns that suggest it acts as a laminar-patterning determinant. This study, therefore, provides the first evidence of a molecular 'matchmaker' involved in patterning IPL circuitry. Proposed loss of function studies in the mouse should help to resolve several outstanding questions regarding the roles Sidekicks play in retinal development.

Finding irrefutable stratification regulators may prove difficult by reverse genetics methods (e.g. knockout mice) since the factors implicated to date are likely to perform multiple developmental functions. In fact, direct cause and effect relationships become increasingly difficult at the single molecule level as the biological process of interest becomes more complex. Circuit formation may rely heavily on mechanistic redundancies that simply render it unsusceptible to the loss of individual molecular components. One approach, which might provide evidence of singularly critical factors, is forward genetics screening in zebrafish. For instance, transgenic zebrafish with fluorescently labelled retinal neuron subtypes (see Chapter 17), or lamina-specific antibodies (Yazulla and Studholme, 2001) could prove particularly useful in identifying mutants that specifically disrupt sublamination patterns in the IPL.

12.4.4 Role of activity

It is clear that not only molecular cues, but also neuronal activity, in particular synaptic signalling, plays a role in the stratification process. Initial evidence for a role of neuronal activity in regulating the vertical extent of RGC dendrites has been obtained by manipulating synaptic signalling in the postnatal cat retina using 2-amino-4-phosphonobutyric acid (APB), an agonist of the metabotropic glutamate receptor mGluR6. In developing ferrets APB blocks all light-induced activity in ON/OFF RGCs (Wang *et al.*, 2001; note that in the adult only ON bipolar cells express the mGluR6 receptor and consequently APB affects only the activity of the ON pathway). Single intraocular injections of APB in newborn cats delay the stratification of ON and OFF RGCs (Bodnarenko and Chalupa, 1993; Bodnarenko *et al.*, 1995) while long-term APB application prevents stratification completely (Deplano *et al.*, 2004; see also Chapter 14). Therefore, synaptic activity – probably partly resulting from visual stimulation – appears to be required for the stratification of RGC dendrites. Interestingly, the maturation of ON and OFF responses of RGCs is also inhibited by APB applications (Bisti *et al.*, 1998; see also Chapter 14). These studies demonstrate the close relationship between the development of function, such as ON or OFF responses, and the structure of dendrites, namely their segregation into the ON or OFF subregions of the IPL.

Surprisingly, impairing mGluR6 signalling in transgenic mice – where synaptic activity onto RGCs is most likely increased – does not affect stratification. For example, the structural development of dendrites is not affected in mGluR6 knockout mice (Tagawa *et al.*, 1999) nor in mice lacking the G-protein subunit, Gαo, required for mGluR6 signalling (Dhingra et al., 2000). These findings may indicate that a certain level of glutamatergic activation from presynaptic bipolar cells is required for the segregation of RGC dendrites, whereas the exact pattern of activity may be less critical.

Besides bipolar cells, which provide glutamatergic inputs onto RGCs, various types of amacrine cells also make synaptic connections with ganglion cells. Amacrine cells are present early in retinal development and it has been shown that cholinergic signalling regulates dendritic stratification. In mutant mice that lack the β2 subunit of the nicotinic acetylcholine receptor stratification is delayed (Bansal et al., 2000). More dramatic is the phenotype of acetylcholine esterase knockout mice: stratification of neurites from various cell types in the IPL is severely altered (Bytyqi et al., 2004). Since not all RGC types establish contacts with processes of cholinergic amacrine cells (Stacy and Wong, 2003), it is not surprising that the disruption of cholinergic signalling affects only a subpopulation of RGCs. Other amacrine cell types may influence the development of RGC dendrites as well. However, it is unknown whether RGC dendrites stratify in the complete absence of any specific subset of amacrine cells. In addition, it has not been investigated whether transmitters that are released by different types of amacrine cells (e.g. γ-aminobutyric acid, glycine, or dopamine) may be involved in RGC dendrite stratification.

The fact that synaptic signalling is required for the refinement of the structure of RGC dendrites raises the question whether spontaneous or light-induced activity or both are involved in this process. In fact, vision appears to be important, since dark rearing for one month after birth results in a higher number of ON/OFF neurons and multistratified RGCs in mice compared with normally raised animals (Tian and Copenhagen, 2003). Also the finding that more dorsal lateral geniculate nucleus (dLGN) neurons respond to both increases and decreases of light after dark rearing may be interpreted as a deficit in intraretinal wiring (Akerman et al., 2002). At least two observations suggest that the mean activity of RGCs and their inputs is reduced during dark rearing. (1) Vision increases the spike activity of ferret dLGN neurons, even before eye opening (Akerman et al., 2002). (2) Dark rearing leads to a reduction of spontaneous synaptic events in RGCs of mice during the first postnatal month (Tian and Copenhagen, 2001). Thus, the results from dark rearing experiments further strengthen the conclusion that a certain amount of synaptic activity is required for the stratification of RGC dendrites and the development of selective ON or OFF response patterns of RGCs.

What are the activity-dependent processes that allow neurons to direct or rearrange their processes into specific sublaminae? As described above, RGC dendrites show a high degree of plasticity, which is regulated by synaptic activity. Over time, this dendritic plasticity may convert unspecifically distributed dendrites into laminated structures by selectively pruning the processes that are located in inappropriate layers and stabilizing those in appropriate areas (Fig. 12.4b). Recently, a mechanism has been described that allows RGCs to selectively stabilize single dendritic branches by transmitter-induced local release of Ca^{2+} from internal stores (Lohmann et al., 2002). Local Ca^{2+} signals may stabilize those dendritic branches that receive synaptic inputs, whereas other branches that receive no or only a few synapses would retract. Such a mechanism could explain the retraction of ectopic dendrites and stabilization of those dendrites that are located within the 'correct' sublamina. However, this model assumes that developing RGCs receive more ON- than OFF-type inputs or vice versa even before the onset of stratification. As Wang *et al.* (2001) point out, this may not necessarily be the case. Furthermore, even if synaptic connectivity is biased before stratification occurs, the question remains what the cues are that help establish this initial bias. Therefore, it is currently not clear whether synaptic activity has an instructive function beyond its permissive role in dendritic remodelling and sublamination of the IPL. Alternatively, differential expression patterns of molecular markers (see p. 256) within the ON and the OFF systems could initiate the segregation of processes into the respective sublaminae.

12.5 Concluding remarks

Despite numerous advances in our understanding of how retinal neurons develop and the relationship between form and function in the retina there remain many outstanding issues: patterns of neurite growth that shape receptive fields and create laminated circuitry are uncharacterized for the majority of retinal neuron subtypes. How the relative roles of neuronal activity and molecular guidance are integrated to resolve patterns of circuitry is not understood. Further complicating these issues is the question of whether insights gained from one system can be applied universally or represent evolutionary divergent and/or cell type-specific solutions to a given problem. What roles do specific retinal subtypes play in coordinating arborizations of neighbouring cells and their partners? Elimination of select neuronal subpopulations and/or manipulation of neurotransmission in the retina have been useful in providing some details in this regard; further refinement of these techniques and identification of knockouts or mutants that lack individual neuronal subtypes will provide valuable tools to elaborate further on this question. Beyond determining the roles specific cell types play in regulating connectivity, inducible ablation techniques should allow functional roles to be assigned to targeted neuronal populations at the level of visual behaviours. The ability to link discrete retinal elements – cellular subsets, defined circuits, specific modes of transmission – to specific developmental and/or visual system functions represents perhaps the ultimate goal in the quest to understand how the retina forms and how it functions. The ability to accurately assess changes over time as a series of causally linked events, be they morphological stages or physiological response patterns, is critical to moving this field forward. Fortunately, recent technological advances have provided modern researchers with the means to view the retina from wholly new perspectives (Megason and Fraser, 2003; Morgan *et al.*, 2005, see also Chapters 15 to 17).

Perhaps most notably, with regard to developmental studies, is the advent of high-resolution, time-lapse imaging of non-invasive cellular reporters, such as green fluorescent protein (GFP). Transgenesis is being used to express GFP-type reporters in select neuronal subsets in many different model systems. High-resolution microscopy, such as confocal and multiphoton, allows more refined resolution of morphological criteria, such as sublamination patterns, from these genetically defined subpopulations. From the standpoint of morphological analyses alone retinal neurons can now be visualized, in vivo, from the time of initial neurite outgrowth until fully mature morphologies are resolved. Thus, early neuritic behaviours and nascent morphological stages can be directly linked to final functional forms. Transgenically defined subsets labelled with fluorescent reporters can also be targeted for electrophysiological characterization, an approach historically utilized in the reverse to elucidate the link between sublamination patterning and response patterns. In addition, an ever-expanding palette of fluorescent reporter colours facilitates simultaneous visualization of separate cellular subpopulations, allowing correlations of differentiation state changes and direct interactions between cells to be monitored over time. Thus, multi-coloured 'contextual imaging' could reveal which specific cell types laminate prior to others, suggesting possible hierarchical orders of dependence, or if partnered populations tend to co-laminate in an interdependent fashion. Fluorescent reporters can also be linked to other proteins in order to reveal discrete subcellular compartments or modified to read out metabolic changes in real time. For instance, reporters that indicate relative levels of neuronal activity and/or that are localized to synapses have been developed. Clearly these and other advances in cellular imaging will provide a wealth of new insights into the formation and function of retinal circuitry in the coming years.

Acknowledgements

We would like to thank Rachel Wong for her mentorship, providing a richly rewarding scientific environment, assistance in the preparation of this manuscript and for all the food; past and present members of the Wong laboratory for valuable insights and discussions; and Josh Morgan for allowing us to abscond and adapt his artistic renderings of the retina. This work was supported by NIH NRSA (JM), DFG (CL), and Schloessmann (CL) fellowships.

References

Akerman, C. J., Smyth, D. and Thompson, I. D. (2002). Visual experience before eye-opening and the development of the retinogeniculate pathway. *Neuron*, **36**, 869–79.

Ammermuller, J. and Kolb, H. (1995). The organization of the turtle retina. I. ON- and OFF-center pathways. *J. Comp. Neurol.*, **358**, 1–34.

Bansal, A., Singer, J. H., Hwang, B. J. *et al.* (2000). Mice lacking specific nicotinic acetylcholine receptor subunits exhibit dramatically altered spontaneous activity patterns and reveal a limited role for retinal waves in forming ON and OFF circuits in the inner retina. *J. Neurosci.*, **20**, 7672–81.

Bauch, H., Stier, H. and Schlosshauer, B. (1998). Axonal versus dendritic outgrowth is differentially affected by radial glia in discrete layers of the retina. *J. Neurosci.*, **18**, 1774–85.

Berezovska, O., McLean, P., Knowles, R. *et al.* (1999). Notch1 inhibits neurite outgrowth in post-mitotic primary neurons. *Neuroscience*, **93**, 433–9.

Bisti, S., Gargini, C. and Chalupa, L. M. (1998). Blockade of glutamate-mediated activity in the developing retina perturbs the functional segregation of ON and OFF pathways. *J. Neurosci.*, **18**, 5019–25.

Bodnarenko, S. R. and Chalupa, L. M. (1993). Stratification of ON and OFF ganglion cell dendrites depends on glutamate-mediated afferent activity in the developing retina. *Nature*, **364**, 144–6.

Bodnarenko, S. R., Jeyarasasingam, G. and Chalupa, L. M. (1995). Development and regulation of dendritic stratification in retinal ganglion cells by glutamate-mediated afferent activity. *J. Neurosci.*, **15**, 7037–45.

Bodnarenko, S. R., Yeung, G., Thomas, L. and McCarthy, M. (1999). The development of retinal ganglion cell dendritic stratification in ferrets. *NeuroReport*, **10**, 2955–9.

Bytyqi, A. H., Lockridge, O., Duysen, E. *et al.* (2004). Impaired formation of the inner retina in an AChE knock-out mouse results in degeneration of all photoreceptors. *Eur. J. Neurosci.*, **20**, 2953–62.

Cajal, S.R. (1972). *The Structure of the Retina*, ed. S. A. Thorpe and M. Glickstein. Springfield: Charles C Thomas, pp. 17–132.

Crooks, J., Okada, M. and Hendrickson, A. E. (1995). Quantitative analysis of synaptogenesis in the inner plexiform layer of macaque monkey fovea. *J. Comp. Neurol.*, **360**, 349–62.

Dailey, M. E. and Smith, S. J. (1996). The dynamics of dendritic structure in developing hippocampal slices. *J. Neurosci.*, **16**, 2983–94.

Dann, J. F., Buhl, E. H. and Peichl, L. (1988). Postnatal dendritic maturation of alpha and beta ganglion cells in cat retina. *J. Neurosci.*, **8**, 1485–99.

Deplano, S., Ratto, G. M. and Bisti, S. (1994). Interplay between the dendritic trees of alpha and beta ganglion cells during the development of the cat retina. *J. Comp. Neurol.*, **342**, 152–60.

Deplano, S., Gargini, C., Maccarone, R., Chalupa, L. M. and Bisti, S. (2004). Long-term treatment of the developing retina with the metabotropic glutamate agonist APB induces long-term changes in the stratification of retinal ganglion cell dendrites. *Dev. Neurosci.*, **26**, 396–405.

Dhingra, A., Lyubarsky, A., Jiang, M. *et al.* (2000). The light response of ON bipolar neurons requires Gαo. *J. Neurosci.*, **20**, 9053–8.

Drenhaus, U., Morino, P. and Veh, R. W. (2003). On the development of the stratification of the inner plexiform layer in the chick retina. *J. Comp. Neurol.*, **460**, 1–12.

Drenhaus, U., Morino, P. and Rager, G. (2004). Expression of axonin-1 in developing amacrine cells in the chick retina. *J. Comp. Neurol.*, **468**, 496–508.

Dubin, M. W., Stark, L. A. and Archer, S. M. (1986). A role for action-potential activity in the development of neural connections in the kitten retinogeniculate pathway. *J. Neurosci.*, **6**, 1021–36.

Eglen, S. J., Van Ooyen, A. and Willshaw, D. J. (2000). Lateral cell movement driven by dendritic interactions is sufficient to form retinal mosaics. *Network: Comput. Neural Syst.*, **11**, 103–18.

Erdmann, B., Kirsch, F. P., Rathjen, F. G. and More, M. I. (2003). N-cadherin is essential for retinal lamination in the zebrafish. *Dev. Dyn.*, **226**, 570–7.

Eysel, U. T., Peichl, L. and Wassle, H. (1985). Dendritic plasticity in the early postnatal feline retina: quantitative characteristics and sensitive period. *J. Comp. Neurol.*, **242**, 134–45.

Famiglietti, E. V. Jr and Kolb, H. (1976). Structural basis for ON- and OFF-center responses in retinal ganglion cells. *Science*, **194**, 193–5.

Famiglietti, E. V. Jr, Kaneko, A. and Tachibana, M. (1977). Neuronal architecture of on and off pathways to ganglion cells in carp retina. *Science*, **198**, 1267–1269.

Farajian, R., Raven, M. A., Cusato, K. and Reese, B. E. (2004). Cellular positioning and dendritic field size of cholinergic amacrine cells are impervious to early ablation of neighboring cells in the mouse retina. *Vis. Neurosci.*, **21**, 13–22.

Fiala, J. C., Feinberg, M., Popov, V. and Harris, K. M. (1998). Synaptogenesis via dendritic filopodia in developing hippocampal area CA1. *J. Neurosci.*, **18**, 8900–11.

Galli-Resta, L., Novelli, E. and Viegi, A. (2002). Dynamic microtubule-dependent interactions position homotypic neurones in regular monolayered arrays during retinal development. *Development*, **129**, 3803–14.

Gao, F. B., Kohwi, M., Brenman, J. E., Jan, L. Y. and Jan, Y. N. (2000). Control of dendritic field formation in Drosophila: the roles of flamingo and competition between homologous neurons. *Neuron*, **28**, 91–101.

Goldberg, J. L., Espinosa, J. S., Xu, Y. *et al.* (2002a). Retinal ganglion cells do not extend axons by default: promotion by neurotrophic signaling and electrical activity. *Neuron*, **33**, 689–702.

Goldberg, J. L., Klassen, M. P., Hua, Y. and Barres, B. (2002b). Amacrine-signaled loss of intrinsic axon growth ability by retinal ganglion cells. *Science*, **296**, 1860–4.

Gunhan, E., Choudary, P. V., Landerholm, T. E. and Chalupa, L. M. (2002). Depletion of cholinergic amacrine cells by a novel immunotoxin does not perturb the formation of segregated on and off cone bipolar cell projections. *J. Neurosci.*, **22**, 2265–73.

Gunhan-Agar, E., Kahn, D. and Chalupa, L. M. (2000). Segregation of on and off bipolar cell axonal arbors in the absence of retinal ganglion cells. *J. Neurosci.*, **20**, 306–14.

Hartline, H. K. (1938). The response of single optic nerve fibers of the vertebrate eye to illumination of the retina. *Am. J. Physiol.*, **121**, 400–15.

Hartline, H. K. (1940). The receptive fields of optic nerve fibers. *Am. J. Physiol.*, **130**, 690–9.

Hayashi, T. and Carthew, R. W. (2004). Surface mechanics mediate pattern formation in the developing retina. *Nature*, **431**, 647–52.

Hinds, J. W. and Hinds, P. L. (1974). Early ganglion cell differentiation in the mouse retina: an electron microscopic analysis utilizing serial sections. *Dev. Biol.*, **37**, 381–416.

Hinds, J. W. and Hinds, P. L. (1978). Early development of amacrine cells in the mouse retina: an electron microscopic serial section analysis. *J. Comp. Neurol.*, **179**, 277–300.

Hinds, J. W. and Hinds, P. L. (1983). Development of retinal amacrine cells in the mouse embryo: evidence for two modes of formation. *J. Comp. Neurol.*, **213**, 1–23.

Honjo, M., Tanihara, H., Suzuki, S. *et al.* (2000). Differential expression of cadherin adhesion receptors in neural retina of the postnatal mouse. *Invest. Opthamol. Vis. Sci.*, **41**, 546–51.

Horne-Badovinac, S., Lin, D., Waldron, S. (2001). Positional cloning of heart and soul reveals multiple roles for PKC lambda in zebrafish organogenesis. *Curr. Biol.*, **11**, 1492–502.

Horton, A. C. and Ehlers, M. D. (2003). Neuronal polarity and trafficking. *Neuron*, **40**, 277–95.

Inuzuka, H., Miyatani, S. and Takeichi, M. (1991). R-cadherin: a novel Ca(2+)-dependent cell–cell adhesion molecule expressed in the retina. *Neuron*, **7**, 69–79.

Jiang, H., Guo, W., Liang, X. and Rao, Y (2005). Both the establishment and the maintenance of neuronal polarity require active mechanisms: critical roles of GSK-3beta and its upstream regulators. *Cell*, **120**, 123–35.

Katyal, S. and Godbout, R. (2004). Alternative splicing modulates Disabled-1 (Dab1) function in the developing chick retina. *EMBO J.*, **23**, 1878–88.

Kay, J. N., Roeser, T., Mumm, J. S. *et al.* (2004). Transient requirement for ganglion cells during assembly of retinal synaptic layers. *Development*, **131**, 1331–42.

Kerrison, J. B., Lewis, R. N., Otteson, D. C. and Zack, D. J. (2005). Bone morphogenetic proteins promote neurite outgrowth in retinal ganglion cells. *Mol. Vis.*, **11**, 208–15.

Kirby, M. A. and Steineke, T. C. (1991). Early dendritic outgrowth of primate retinal ganglion cells. *Vis. Neurosci.*, **7**, 513–30.

Kuffler, S. W. (1953). Discharge patterns and functional organization of mammalian retina. *J. Neurophysiol.*, **16**, 37–68.

Lettvin, J. Y., Maturana, H. R., McCulloch, W. S. and Pitts, W. H. (1959). What the frog's eye tells the frog's brain. *Proc. Inst. Radio Engr.*, **47**, 1940–51.

Lin, B., Wang, S. W. and Masland, R. H. (2004). Retinal ganglion cell type, size, and spacing can be specified independent of homotypic dendritic contacts. *Neuron*, **43**, 475–85.

Lohmann, C. and Wong, R. O. L. (2001). Cell-type specific dendritic contacts between retinal ganglion cells during development. *J. Neurobiol.*, **48**, 150–62.

Lohmann, C., Myhr, K. L. and Wong, R. O. (2002). Transmitter-evoked local calcium release stabilizes developing dendrites. *Nature*, **418**, 177–81.

Lom, B. and Cohen-Cory, S. (1999). Brain-derived neurotrophic factor differentially regulates retinal ganglion cell dendritic and axonal arborization *in vivo*. *J. Neurosci.*, **19**, 9928–38.

Lom, B., Cogen, J., Sanchez, A. L., Vu, T. and Cohen-Cory, S. (2002). Local and target-derived brain-derived neurotrophic factor exert opposing effects on the dendritic arborization of retinal ganglion cells in vivo. *J. Neurosci.*, **22**, 7639–49.

Malicki, J., Jo, H. and Pujic, Z. (2003). Zebrafish N-cadherin, encoded by the glass onion locus, plays an essential role in retinal patterning. *Dev. Biol.*, **259**, 95–108.

Mangrum, W. I., Dowling, J. E. and Cohen, E. D. (2002). A morphological classification of ganglion cells in the zebrafish retina. *Vis. Neurosci.*, **19**, 767–79.

Masai, I., Lele, Z., Yamaguchi, M. *et al.* (2003). N-cadherin mediates retinal lamination, maintenance of forebrain compartments and patterning of retinal neurites. *Development*, **130**, 2479–94.

Maslim, J. and Stone, S. (1988). Time course of stratification of the dendritic fields of ganglion cells in the retina of the cat. *Dev. Brain Res.*, **44**, 87–93.

Maslim, J., Webster, M. and Stone, J. (1986). Stages in the structural differentiation of retinal ganglion cells. *J. Comp. Neurol.*, **254**, 382–402.

Matsunaga, M., Hatta, K. and Takeichi, M. (1988). Role of N-cadherin cell adhesion molecules in the histogenesis of neural retina. *Neuron*, **1**, 289–95.

Megason, S. G. and Fraser, S. E. (2003). Digitizing life at the level of the cell: high-performance laser-scanning microscopy and image analysis for in toto imaging of development. *Mech. Dev.*, **120**, 1407–20

Morgan, J., Huckfeldt, R. and Wong, R. O. L. (2005). Imaging techniques in retinal research. *Exp. Eye Res.*, **80**, 297–306.

Mumm, J. S., Godinho, L., Morgan, J. L. *et al.* (2005). Laminar circuit formation in the vertebrate retina. *Prog. Brain Res.*, **147**, 155–69.

Naito, J. and Chen, Y. (2004). Morphological features of chick retinal ganglion cells. *Anat. Sci. Int.*, **79**, 213–25.

Nelson, R., Famiglietti, E. V. J. and Kolb, H. (1978). Intracellular staining reveals different levels of stratification for on- and off-center ganglion cells in the cat retina. *J. Neurophysiol.* **41**, 472–83

Ohta, K., Takagi, S., Asou, H. and Fujisawa, H. (1992). Involvement of neuronal cell surface molecule B2 in the formation of retinal plexiform layers. *Neuron*, **9**, 151–61.

Okada, M., Erickson, A. and Hendrickson, A. (1994). Light and electron microscopic analysis of synaptic development in Macaca monkey retina as detected by immunocytochemical labeling for the synaptic vesicle protein SV2. *J. Comp. Neurol.*, **339**, 535–58.

Pang, J.-J., Gao, F. and Wu, S. M. (2002). Segregation and integration of visual channels: layer-by-layer computation of ON–OFF signals by amacrine cell dendrites. *J. Neurosci.*, **22**, 4693–701.

Pang, J.-J., Gao, F. and Wu, S. M. (2004). Stratum-by-stratum projection of light response attributes by retinal bipolar cells of *Ambystoma*. *J. Physiol.*, **558**, 249–62.

Penn, A. A., Wong, R. O. and Shatz, C. J. (1994). Neuronal coupling in the developing mammalian retina. *J. Neurosci.*, **14**, 3805–15.

Perry, V. H. and Linden, R. (1982). Evidence for dendritic competition in the developing retina. *Nature*, **297**, 683–5.

Prada, C., Puelles, L., Genis-Galvez, J. M. and Ramirez, G. (1987). Two modes of free migration of amacrine cell neuroblasts in the chick retina. *Anat. Embryol.*, **175**, 281–7.

Redmond, L., Oh, S. R., Hicks, C., Weinmaster, G. and Ghosh, A. (2000). Nuclear Notch1 signaling and the regulation of dendritic development. *Nat. Neurosci.*, **3**, 30–40.

Reese, B. E., Raven, M. A., Giannotti, K. A. and Johnson, P. T. (2001). Development of cholinergic amacrine cell stratification in the ferret retina and the effects of early excitotoxic ablation. *Vis. Neurosci.*, **18**, 559–70.

Reese, B. E., Raven, M. A. and Stagg, S. B. (2005). Afferents and homotypic neighbors regulate horizontal cell morphology, connectivity, and retinal coverage. *J. Neurosci.*, **25**, 2167–75.

Riehl, R., Johnson, K., Bradley, R. *et al.* (1996). Cadherin function is required for axon outgrowth in retinal ganglion cells *in vivo*. *Neuron*, **17**, 837–48.

Roska, B. and Werblin, F. (2001). Vertical interactions across ten parallel, stacked representations in the mammalian retina. *Nature*, **410**, 583–7.

Sanes, J. R. and Yamagata, M. (1999). Formation of lamina-specific synaptic connections. *Curr. Opin. Neurobiol.*, **9**, 79–87.

Sestan, N., Artavanis-Tsakonas, S. and Rakic, P. (1999). Contact-dependent inhibition of cortical neurite growth mediated by notch signaling. *Science*, **286**, 741–6.

Sherry, D. M., Wang, M. M., Bates, J. and Frishman, L. J. (2003). Expression of vesicular glutamate transporter 1 in the mouse retina reveals temporal ordering in development of rod vs. cone and ON vs OFF circuits. *J. Comp. Neurol.*, **465**, 480–98.

Shi, S. H., Cheng, T. Jan, L. Y. and Jan, Y. N. (2004). APC and GSK-3beta are involved in mPar3 targeting to the nascent axon and establishment of neuronal polarity. *Curr. Biol.*, **14**, 2025–32.

Stacy, R. C. and Wong, R. O. L. (2003). Developmental relationship between cholinergic amacrine cell processes and ganglion cell dendrites of the mouse retina. *J. Comp. Neurol.*, **456**, 154–66.

Stell, W. K., Ishida, A. T. and Lightfoot, D. O. (1977). Structural basis for On- and Off-center responses in retinal bipolar cells. *Science*, **198**, 1269–71.

Stier, H. and Schlosshauer, B. (1998). Different cell surface areas of polarized radial glia having opposite effects on axonal outgrowth. *Eur. J. Neurosci.*, **10**, 1000–10.

Tagawa, Y., Sawai, H., Ueda, Y., Tauchi, M. and Nakanishi, S. (1999). Immunohistological studies of metabotropic glutamate receptor subtype 6-deficient mice show no abnormality of retinal cell organization and ganglion cell maturation. *J. Neurosci.*, **19**, 2568–79.

Tian, N. and Copenhagen, D. R. (2001). Visual deprivation alters development of synaptic function in inner retina after eye-opening. *Neuron*, **32**, 439–49.

Tian, N. and Copenhagen, D. R. (2003). Visual stimulation is required for refinement of ON and OFF pathways in postnatal retina. *Neuron*, **39**, 85–96.

Tootle, J. S. (1993). Early postnatal development of visual function in ganglion cells of the cat retina. *J. Neurophysiol.*, **69**, 1645–60.

Troilo, D., Xiong, M., Crowley, J. C. and Finlay, B. L. (1996). Factors controlling the dendritic arborization of retinal ganglion cells. *Vis. Neurosci.*, **13**, 721–33.

Vaney, D. I. (1990). Patterns of neuronal coupling in the retina. *Prog. Retin. Eye Res.*, **13**, 301–55.

Wang, G. Y., Liets, L. C. and Chalupa, L. M. (2001). Unique functional properties of on and off pathways in the developing mammalian retina. *J. Neurosci.*, **21**, 4310–17.

Wässle, H., Peichl, L. and Boycott, B. B. (1981). Dendritic territories of cat retinal ganglion cells. *Nature*, **292**, 344–5.

Wei, X. and Malicki, J. (2002). Nagie oko, encoding a MAGUK-family protein, is essential for cellular patterning of the retina. *Nat. Genet.*, **31**, 150–7.

Williams, R. R., Cusato, K., Raven, M. A. and Reese, B. E. (2001). Organization of the inner retina following early elimination of the retina ganglion cell population: effects on cell numbers and stratification patterns. *Vis. Neurosci.*, **18**, 233–44.

Wodarz, A. (2002). Establishing cell polarity in development. *Nat. Cell Biol.*, **4**, E39–E44.

Wohrn, J. C. P., Puelles, L., Nakagma, S., Takeichi, M. and Redies, C. (1998). Cadherin expression in the retina and retinofugal pathways of the chicken embryo. *J. Comp. Neurol.*, **396**, 20–38.

Wong, W. T. and Wong, R. O. (2000). Rapid dendritic movements during synapse formation and rearrangement. *Curr. Opin. Neurobiol.*, **10**, 118–24.

Wong, W. T. and Wong, R. O. (2001). Changing specificity of neurotransmitter regulation of rapid dendritic re-modeling during synaptogenesis. *Nat. Neurosci.*, **4**, 351–2.

Wong, W. T., Faulkner-Jones, B. E., Sanes, J. R. and Wong, R. O. (2000). Rapid dendritic re-modeling in the developing retina: dependence on neurotransmission and reciprocal regulation by Rac and Rho. *J. Neurosci.*, **20**, 5024–36.

Wu, S. M., Gao, F. and Maple, B. R. (2000). Functional architecture of synapses in the inner retina: segregation of visual signals by stratification of bipolar cell axon terminals. *J. Neurosci.*, **20**, 4462–70.

Yamagata, M., Weiner, J. A. and Sanes, J. R. (2002). Sidekicks: synaptic adhesion molecules that promote lamina-specific connectivity in the retina. *Cell*, **110**, 649–60.

Yazulla, S. and Studholme, K. M. (2001). Neurochemical anatomy of the zebrafish retina as determined by immunocytochemistry. *J. Neurocytol.*, **30**, 551–92.

Yoshimura, T., Kawano, Y., Arimura, N. *et al.* (2005). GSK-3beta regulates phosphorylation of CRMP-2 and neuronal polarity. *Cell*, **120**, 137–49.

13

Synaptogenesis and early neural activity

Evelyne Sernagor
University of Newcastle upon Tyne, Newcastle upon Tyne, UK

13.1 Introduction

Once various cell types have migrated to their final location, they start synthesizing neurotransmitters and extend neurites. At that stage, they are ready to begin forming synaptic connections with other retinal neurons. We have already seen in Chapter 6 that neurotransmitters are present before the formation of functional synapses, before electrical activity can be detected, suggesting that they play a trophic role during retinal development. However, retinal visual processing, the conversion of light into electrical signals and the relaying of these signals to the visual centres of the brain, cannot occur until neurons have established synaptic contacts with each other.

The first part of this chapter describes the formation of synaptic connections in the various retinal layers. In the last decade, important issues regarding the establishment of synaptic connections have been resolved thanks to the advent of genetic engineering and to the development of powerful specific cellular or subcellular markers, some of which are reviewed here. We put a particular emphasis on the formation of synapses between photoreceptors and second-order neurons because photoreceptors are involved in many various types of hereditary retinal degenerations. We review studies using transgenic mouse models because they provide invaluable knowledge about factors influencing photoreceptor synaptogenesis (Farber and Danciger, 1997). Understanding factors that affect synapse formation between photoreceptors and retinal neurons is crucial for reaching better insights into these devastating diseases that often lead to blindness. We will also discuss briefly the formation of gap junctions.

The second part of the chapter reviews the nature of early network-driven activity. Long before visual experience is even possible the retina exhibits spontaneous activity. This activity is correlated between neighbouring ganglion cells, resulting in propagating waves. This spontaneous activity is present during a limited temporal window of development, presumably a 'critical period'. We will review the mechanisms underlying the generation and propagation of these waves and will also discuss how their patterns change with development and what cause their disappearance shortly after birth.

Retinal Development, ed. Evelyne Sernagor, Stephen Eglen, Bill Harris and Rachel Wong.
Published by Cambridge University Press. © Cambridge University Press 2006.

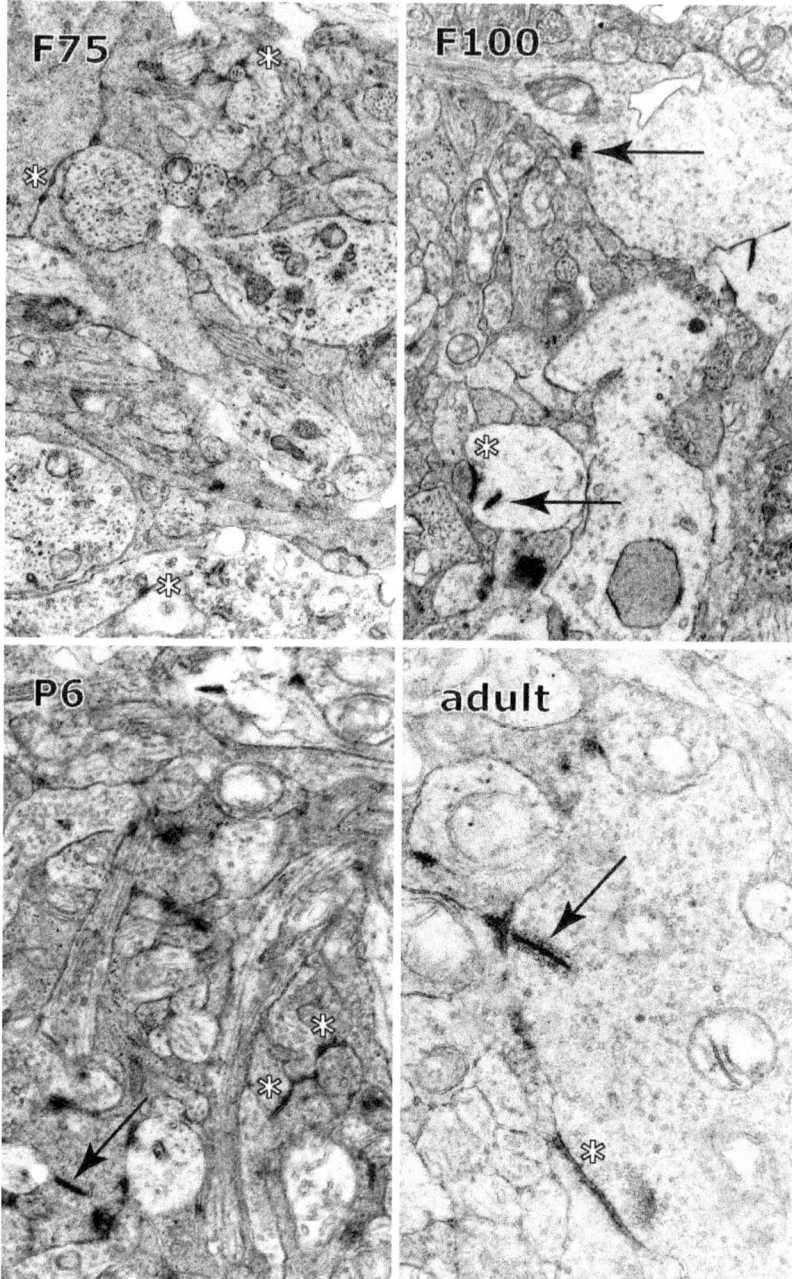

Figure 13.1 Synapse formation in rhesus monkey (*Macaca mulatta*) IPL. The four panels illustrate electron micrographs of the IPL at different developmental stages. Gestation lasts 160 to 170 days. Fetal day (F) 75: the IPL contains processes from retinal ganglion cells (RGCs), ACs and cone BCs. The neuropil exhibits large extracellular spaces and shows no ribbons or even conventional synapses. A few 'spot welds' of primitive synapses occur (indicated by asterisks). F100: neural processes have

13.2 Synapse formation

13.2.1 Chemical synapses

Functional synaptic connections emerge once the plexiform layers are established through the formation of specific contacts between neurites originating from diverse cell types expressing different neurotransmitters (see previous chapter). Synapse formation proceeds in a centrifugal manner, from the inner to the outer retina, and it occurs first between cells that spread information laterally within the plexiform layers (amacrine cells (ACs)) and (horizontal cells (HCs)), followed by vertical connections (through bipolar cells (BCs)).

Ultrastructural studies

Substantial knowledge on the formation of retinal synaptic connections comes from ultrastructural studies (see Robinson, 1991, for a review). Synaptogenesis starts before eye opening, and even before birth, thus long before visual experience is even possible. Synaptic contacts are initially made in the inner plexiform layer (IPL) (Figure 13.1), and then they proceed towards the outer retina where the process continues well after eye opening. The time course of synapse formation in the IPL and in the outer plexiform layer (OPL) suggests important functional issues. First, the presence of early synaptic contacts in the inner retina coincides with the period during which the retina is spontaneously active, indicating the central role of retinal ganglion cells (RGCs) and ACs in generating this activity (this will be discussed in Section 13.3). Second, the fact that synaptic wiring in the outer retina continues after eye opening suggests that the refinement of circuitry in the outer layers is more likely to be influenced by visual experience.

The first connections observed in the IPL are amongst ACs and between ACs and RGCs (cat: Maslim and Stone, 1986; monkey: Nishimura and Rakic, 1985, 1987; mouse: Fisher, 1979; rat: Horsburgh and Sefton, 1987). Even before synaptic vesicles become detectable, apposed pairs of membrane specializations called junctions become apparent throughout the IPL. Nishimura and Rakic (1987) have performed a detailed analysis of the sequence of events occurring during the formation of synaptic contacts between ganglion and ACs in the monkey, showing that synaptic differentiation proceeds from the postsynaptic to the presynaptic side. Indeed, specialized areas of membrane thickenings occur first on the dendrites of RGCs. Then, they become apposed to amacrine processes, which in turn develop their own membrane specializations, including synaptic vesicles (Figure 13.1). These are the

---

Figure 13.1 (*cont.*) expanded (and some now belong to rod BCs) and it is now possible to detect few synaptic vesicles in the richer neuropil. Immature, primordial ribbons (dark structures encircled by vesicles, indicated by arrows) can be seen 'floating' in neural processes, not yet attached to synapses. Conventional synapses are also appearing (asterisk). Postnatal day 6 (P6): the neuropil is much denser and shows mature conventional synaptic contacts (asterisks) (between ACs, RGCs and BCs). Immature ribbon synapses (triads) are also present (arrow). Adult: synaptic contacts are now completely mature, with conventional synapses (asterisk) as well as clearly delineated ribbon triads (arrow). Courtesy of Les Westrum and Anita Hendrickson (Les Westrum sadly passed away shortly after providing the micrographs. We will always remember his kindness and support).

first noticeable conventional synapses, and are followed by the appearance of conventional synapses between ACs. Ribbon synapses, characteristic of BCs, emerge much later, around birth (Figure 13.1), suggesting that functional contacts between BCs and ACs or RGCs do not emerge until late, at postnatal stages. The order of establishment of ribbon synapses is first between a BC and a single RGC or AC (dyads), followed by the formation of triads (Figure 13.1) through the addition of processes from another AC or RGC to dyads. The circuitry in the IPL is then completed by the addition of a feedback loop through amacrine or interplexiform cell processes onto bipolar axons. In primates, ribbon synapses appear after conventional synapses in the peripheral retina while the order is reversed in the fovea (Crooks *et al.*, 1995). Issues related to primates are discussed further in Chapter 7.

In the OPL, the first synapses to appear are ribbon synapses between photoreceptors and HCs (rabbit: McArdle *et al.*, 1977; cat: Maslim and Stone, 1986). Contacts are first established between cones and HCs (mouse: Rich *et al.*, 1997), closely followed by wiring between rods and HCs. Vertical functional connections between photoreceptors and BCs arise later, once again starting with cones, and followed by rods.

Markers for various synaptic release-associated proteins

Synaptic transmission requires several important steps. First, neurotransmitters must be packed into vesicles, and then these vesicles must be docked at the active zone where they fuse with the membrane of the terminal, releasing their content into the synaptic cleft, through the universal mechanism of exocytosis. Exocytosis is Ca^{2+}-dependent, and it is under the control of a vast family of proteins associated either with the membrane of the vesicle or with the membrane of the terminal. Invaluable information can therefore be gathered on the ontogeny of synaptic vesicular release by studying the development of mechanisms controlling vesicular neurotransmitter transport, or by following the developmental expression of synaptic release proteins in different types of retinal synapses.

A recent developmental study of the expression of the vesicular transporter for glutamate, VGLUT1, in the mouse retina reports that vesicular glutamatergic neurotransmission appears in cone circuits before it does in rod circuits, both in the IPL and OPL (Sherry *et al.*, 2003). The vesicular transporter VGLUT1 also appears in OFF circuits before it does in ON circuits in the IPL (see Chapter 14).

The developmental expression of VGAT, the vesicular transporter for γ-aminobutyric acid (GABA) and glycine, precedes that of VGLUT1 by several days in the postnatal rodent retina (Johnson *et al.*, 2003). Indeed, VGAT is already present in the IPL by postnatal day (P)1. On the other hand, VGLUT1 does not appear until several days later (rat: P5 to P7 in the OPL and P7 in the IPL; mouse: P3 in the OPL and P5 in the IPL), reaching its adult pattern by P14. Its expression precedes the formation of mature ribbon synapses (P11 in mouse IPL and P13 in rat IPL), but is already in place when spontaneous excitatory postsynaptic responses begin (P7 in mouse, see Section 13.3). These observations point towards several interesting conclusions. First of all, they suggest that vesicular GABAergic and/or glycinergic neurotransmission at conventional synapses occurs at early developmental stages in the

IPL, before glutamatergic synapses become functional. This is in accordance with several physiological studies demonstrating an important role for GABAergic neurotransmission in the generation of early retinal spontaneous activity (see Section 13.3.3). The second important conclusion is that there is vesicular glutamatergic release in the IPL before the maturation of ribbon synapses. One possible explanation for the presence of this early glutamatergic neurotransmission is that it originates in photoreceptors, known to make transient connections in the IPL at early stages, during the period of spontaneous activity (Johnson et al., 1999). Second, it may originate from axon collaterals of RGCs projecting back into the IPL. Such collaterals have been observed in the immature retina (discussed in Sernagor et al., 2001).

Many proteins involved in synaptic transmission, either at the presynaptic or at the postsynaptic level, have been identified in various parts of the nervous system. Synapse formation can thus be assessed using various approaches to identify these proteins because their expression during development is prone to be involved in the functional maturation of synaptic circuitry. In general, they appear at early stages, long before synapses mature morphologically, which indicates that the establishment of the basic membrane synaptic machinery is an early step during synaptogenesis. The fact that these proteins are present before synapses are functionally formed suggests that they play a role in modulating synaptic maturation, perhaps through trafficking synaptic vesicles to growing neurites.

Synaptogenesis can be followed at the light microscope level using immunocytochemical labelling of synaptophysin or syntaxin, transmembrane proteins found on synaptic vesicles, believed to be involved in neurosecretion. Both are present in the IPL at the time synapse formation starts in the central retina (human: Nag and Wadhwa, 2001). In the outer retina, synaptophysin or syntaxin staining appears once the OPL is fully differentiated (rat: Dhingra et al., 1997; human: Nag and Wadhwa, 2001; mouse: Sharma et al., 2003). The expression of both these proteins follows a central-to-peripheral gradient.

Synaptic vesicle protein 2 (SV2) is another important protein involved in the regulation of Ca^{2+}-stimulated synaptic vesicle exocytosis (Buckley and Kelly, 1985). Three isoforms (SV2A, SV2B, SV2C) encoded by different genes show differential spatial and temporal patterns of expression in the developing and adult mouse retina (Wang et al., 2003). At maturity, SV2A is present in cone terminals and in the IPL, SV2B in ribbon synapses between rods/cones and BCs, and SV2C in starburst ACs and other conventional synapses in the IPL. All three show a central-to-peripheral gradient of developmental expression and are already present before the onset of synaptogenesis. Synaptic vesicle protein 2A shows a transient high level of expression in HCs, at the time when they start forming contacts with cone photoreceptors.

In the last decade, studies on the SNARE complex have advanced our understanding of molecular mechanisms of synaptic release. The SNARE complex is a family of proteins that is essential for the fusion of synaptic vesicles with the terminal plasma membrane (Sollner et al., 1993). According to the SNARE hypothesis, vesicles fuse with the membrane of the terminal through the binding of a VAMP (vesicle-associated membrane protein) with two proteins on the terminal membrane, SNAP-25 (synaptosomal-associated protein of 25 kDa) and syntaxin (Rothman and Warren, 1994). The SNARE complex then triggers additional events that lead to vesicular fusion. Hence, studying the expression of SNARE complex proteins during development provides important insight about mechanisms of synapse formation. Prior to eye opening, cholinergic ACs exhibit transient high levels of SNAP-25 (Brazilian opossum *Monodelphis domestica*: West Greenlee *et al.*, 1998; rat: West Greenlee *et al.*, 2001), but not of syntaxin and VAMP (West Greenlee *et al.*, 2001). As development proceeds, towards eye opening SNAP-25 expression decreases to basal levels. The transient high expression of SNAP-25 temporally correlates with the period during which cholinergic ACs participate in the generation of spontaneous waves of activity (see next section), suggesting that the regulation of SNARE protein expression is important for retinal wiring. Perhaps high levels of SNAP-25 facilitate interactions between acetylcholine-containing synaptic vesicles and amacrine terminal membranes, thereby causing a transient increase in the probability of release. Syntaxin and SNAP-25 expression precedes the expression of VAMP both in the IPL and OPL (West Greenlee *et al.*, 2001), suggesting a role for these two proteins during synaptogenesis.

A detailed study of the chronological appearance of 11 membrane and membrane-associated synaptic proteins in mouse ribbon synapses shows that several synaptic membrane proteins appear before ribbon synapses maturation, and that some are initially found at extra-synaptic locations (Kriegstein and Schmitz, 2003). However, all proteins cluster at ribbon synapses within a short temporal window, between P4 and P5, several days earlier than ribbon synapses mature morphologically, between P7 and P12.

Ribbon synapses contain scaffolds of highly specialized proteins associated with synaptic release at the active zone. Bassoon is one of these structural proteins, and it is found in photoreceptor ribbon synapses (Brandstätter *et al.*, 1999). Bassoon is important for the formation of these synapses during development. Indeed, in mice lacking a functional Bassoon protein, few ribbon synapses develop, and they do not appear normal. Ribbons fail to anchor to the presynaptic active zone, resulting in impaired neurotransmission at these synapses as well as in the formation of ectopic synapses and abnormal dendritic branching in second-order neurons (Dick *et al.*, 2003).

The extracellular matrix influences synapse formation

Many factors undoubtedly play a role in the appropriate wiring of retinal circuits. Laminin, which is part of the extracellular matrix, is one of these factors. Laminins are a complex family of extracellular glycoproteins with α, β and γ chains (Timpl, 1996). Beta2-containing laminins are expressed on the apical surface of the retina and in the OPL and they are important in retinal development, including synapse formation (Libby *et al.*, 1999). Indeed, laminin $\beta2$ chain-deficient mice exhibit abnormal synaptic contacts between rods and second-order neurons while synapses in the IPL appear normal. Triads are rare in these animals, whereas dyads are more common. The most striking malformation is the presence of floating ribbons, presynaptic specializations not apposed to any postsynaptic structure. Another type of laminin, containing the $\alpha4$ chain, is present in the inner retina (Libby *et al.*, 2000), suggesting a role for extracellular matrix components in synapse formation in the IPL as well.

13.2.2 Electrical synapses

Gap junctions allow electrical and metabolic communication between adjacent cells. They are formed by the apposition of two hemichannels or connexons, each made of six large transmembrane proteins called connexins. Connexins exist in many different forms and they are subdivided in four groups, α, β, γ and δ. In the vertebrate retina, connexins 26, 32 and 43 have been identified (in addition to connexin 35 in fish and 36 in mammals). Both the specific cell and connexin types determine particular roles of gap junctions. Extensive networks of gap junctions are present between most cell types in the adult retina, where they are important for the modulation of responses to changing light levels, when network activity shifts from cone pathways in daylight to rod pathways at night.

Gap junctions play an important role during CNS development, including neurogenesis, cell death, pattern formation and synchronization of impulses across neuronal populations during synapse formation (synapse elimination or strengthening). The precise role of gap junctions during these major developmental events and the exact nature of the signals that are transmitted between cells through these channels remain to be elucidated.

It is known that gap junctions are absent during the early stages of synaptogenesis in the monkey IPL, not appearing until the density of chemical synapses has reached mature levels (Nishimura and Rakic, 1985). Immature A and B type HCs in the rabbit retina show homologous coupling (i.e., coupling between cells of the same type) at birth, long before synapse formation is completed, suggesting that direct communication between these cells is essential for OPL maturation (Johnson *et al.*, 2000).

Most of our knowledge of the role of gap junctions during retinal development comes from studies on the developing chick retina (reviewed in Becker *et al.*, 1998). Connexins 26, 32 and 43 are all expressed in the immature chick retina, but their expression changes with development. As a general rule, these connexins are expressed at transient high levels during embryogenesis, while cells proliferate and migrate, and during the period of spontaneous correlated activity (see Section 13.3). Connexin 43 is the first to be detected in the neuroepithelium of the eyecup, followed by connexins 32 and 26 at embryonic day (E)4 to E4.5, when the first RGCs are born. By E5, ganglion cells are connected to few other cells, and coupling increases until E11, when clusters of about a dozen cells are connected, with their soma in the ganglion cell layer (GCL) or in the INL (Becker *et al.*, 2002). High levels of connexin 32 correlate with coupling between RGCs, and it is suggested that this type of communication between ganglion cells is involved in the regulation of new RGC differentiation. Connexin 43 is the only one of the connexins expressed in the pigment epithelium, and it is suggested that it regulates neuronal cell proliferation. Indeed, when its expression is knocked down eyes become smaller than normal (Becker *et al.*, 1998). In support, the rate of cell proliferation is slower in albino animals (Ilia and Jeffery, 1996). A new study reports that ATP, released from the retinal pigment epithelium onto the neural retina via connexin 43 hemichannels, regulates cell proliferation in the neural retina (Pearson *et al.*, 2005). Connexin 43 and 26 levels decrease at the onset of cell death, which in the chick retina is between E9 and E12.

Of greater relevance to the theme of this chapter, is the fact that connexin levels decrease to normal between E14 and E18, during the period of synapse formation. Connexin 26 shows a delimited transient increase (lasting about one day) on dendritic processes of starburst ACs in the OFF sublamina of the IPL during synaptogenesis. As yet, however, its role in synapse formation is not understood. By E14, coupling between RGCs decreases to clusters of three to four cells, all of the same morphological type (Becker *et al.*, 2002).

Although gap junctions do not appear to mediate correlated spontaneous activity in the immature inner retina after the onset of synaptogenesis (see Section 13.3), they seem to mediate vertically propagating spontaneous Ca^{2+} waves that travel throughout the thickness of the chick embryo retina prior to synapse formation (Catsicas *et al.*, 1998). This indicates that gap junctions are important for communication between retinal layers during early development, but the role of these vertical waves remains to be unravelled. Gap junctions have also been recently reported to mediate early lateral waves in the rabbit, prior to synapse formation (Syed *et al.*, 2004a).

13.3 Early synaptic network-driven neural activity

13.3.1 *Spontaneous bursting activity in immature ganglion cells and amacrine cells*

Once conventional synaptic contacts have been established between ACs and RGCs in the IPL, long before vision is even possible, spontaneous network-driven activity begins in the inner retina, occurring first between ACs and RGCs (reviewed in Wong, 1999; Sernagor *et al.*, 2001). Intracellular recordings reveal the presence of spontaneous excitatory and inhibitory synaptic responses in ganglion cells before eye opening in mammals (rat: Rorig and Grantyn, 1993; mouse: Johnson *et al.*, 2003); and in turtle, spontaneous synaptic events are present even at early embryonic stages (Sernagor *et al.*, 2001). In mice, the frequency of these early events increases after eye opening, and this is due to a higher probability of release from presynaptic BCs and ACs (Tian and Copenhagen, 2001).

These early spontaneous synaptic responses give rise to action potentials in RGCs. Indeed, in vitro extracellular recordings from the neonatal rabbit retina first revealed that developing RGCs periodically fire bursts of action potentials in the absence of light stimulation (Masland, 1977). Subsequently, Galli and Maffei (1988) (see also Maffei and Galli-Resta, 1990) succeeded in demonstrating that spontaneous bursting activity is present in vivo in the fetal rat retina. In mammals, these discharge patterns gradually switch from brief periodic bursts to more sustained firing with increasing age (Tootle, 1993; Wong *et al.*, 1993). In comparison, RGCs in turtle exhibit a wide variety of spontaneous discharges throughout development (Sernagor and Grzywacz, 1995; Grzywacz and Sernagor, 2000; Sernagor *et al.*, 2001). Rhythmic bursting activity therefore appears to be a general characteristic of the immature retina. It has been recorded in other parts of the nervous system

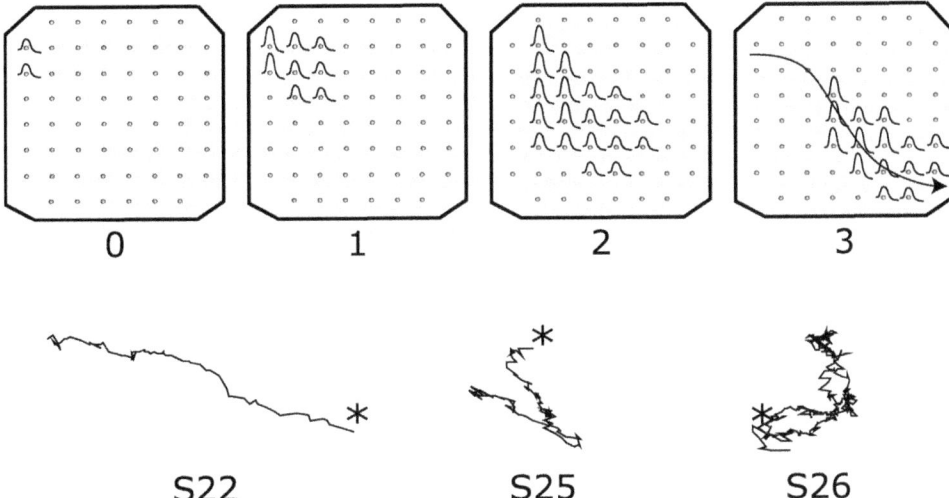

Figure 13.2 Spontaneous waves of activity in the developing retina. The upper panels depict an array of 60 neighbouring extracellular electrodes (each dot represents one electrode) used to record spontaneous bursting activity from the GCL (the retina is placed with the GCL facing down onto the electrodes). The activity (not from real data) on one electrode is represented by a small waveform, and the amount, by the size of the waveform. Each panel represents activity from the same retina at 1 s intervals. At time 0, spontaneous activity is detected on two electrodes in the upper left side of the array. The activity propagates diagonally to the bottom right. The curved arrow in the last panel depicts the trajectory of the wave from start to end. The three lower panels of the figure illustrate the trajectory of the centre of mass of retinal waves recorded at different developmental stages in turtle, with the asterisk representing the point of origin of the activity. At S22 (three weeks before hatching, when waves emerge), the waves propagate in a direct line, illustrated by the relatively straight trajectory. Two weeks later, at S25, the waves are much more winding (and slower). In the particular example depicted here, the wave goes first downward and then turns left and upward. At hatching (S26), waves stop propagating, and become replaced by stationary patches of coactivated cells. Consequently, it is not possible to detect a clear wave trajectory anymore at that time.

(O'Donovan, 1999), suggesting that such activity may serve a fundamental role in neural development.

13.3.2 Retinal waves: how and why?

An important and rather unique feature of this spontaneous bursting activity is that it is temporally correlated between neighbouring cells, as first reported by Maffei and Galli-Resta in 1990. Using a multielectrode array that allowed them to record simultaneously from large populations of RGCs (in cat and ferret), Meister *et al.* (1991) discovered that this bursting activity propagates across the retina as waves (Figure 13.2). These waves are rather slow, occurring on the order of every few minutes; they can originate at any point in the retina and then propagate in random directions and patterns (Meister *et al.*, 1991;

Wong et al., 1993; Feller et al., 1996). Retinal waves propagate not only within the GCL, but involve ACs as well (Figure 13.4) (Wong et al., 1995; Zheng et al., 2004). Correlated spontaneous activity and propagating waves have been recorded electrophysiologically as well as with Ca^{2+} imaging in a wide variety of mammals (mouse, rabbit, ferret, cat), chick and turtle (reviewed in Sernagor et al., 2001), and are therefore likely to be a general feature of the developing vertebrate retina. Recently, it has been shown in the rabbit that waves initiated in the inner retina send retrograde signals to differentiating cells in the ventricular zone where slow Ca^{2+} waves occur in coincidence with the waves in the GCL (Syed et al., 2004a). The exact nature of these signals remains to be elucidated.

Retinal waves have attracted substantial interest since they were discovered more than a decade ago. The reason is that they encode spatiotemporal cues that may provide important information during the wiring of the visual system, when RGC axons establish connectivity with their central targets, or even while intraretinal circuitry is being formed. Bursting patterns appear to be necessary for the strengthening of synaptic transmission (long-term potentiation) between RGCs and their targets in the lateral geniculate nucleus (LGN) (Mooney et al., 1996). Moreover, since neighbouring RGCs project to neighbouring geniculate neurons, temporal correlation between near-neighbours during these retinal waves ensures that neighbouring geniculate cells are coactivated as well. In line with the Hebbian postulate this would result in the refinement of developing retinotopic maps (Crair, 1999; Eglen, 1999; Penn and Shatz, 1999; Wong, 1999). Retinal waves are unlikely to occur simultaneously in both eyes so they have been viewed as inducing coactivation or cooperation between projections originating from the same eye, where a wave will recruit a massive number of RGCs to fire nearly simultaneously, while at the same time causing competition with inputs from the other eye, which is unlikely to manifest such discharges simultaneously. Let us briefly consider the empirical evidence for the role of activity, and retinal waves in particular, in the refinement of retinal projections.

Pharmacological manipulations that eliminate all retinal activity prevent the segregation of eye-specific inputs to the LGN (Shatz and Stryker, 1988; Penn et al., 1998; Huberman et al., 2003), and changing the balance in activity in the two eyes causes an increase in the territory innervated by the more active eye at the expense of the less active eye (Penn et al., 1998; Stellwagen and Shatz, 2002). Such studies demonstrate that retinal activity is required to form segregated eye-specific projections, but they do not address the role of retinal waves per se. In the mammalian retina, waves are mediated in part by cholinergic ACs and nicotinic receptors in the inner retina (see next section), and a number of studies have now exploited this fact to examine the effects of manipulating wave activity on the formation of segregated retinal-LGN inputs as well as retinotopic organization. In mice lacking the $\beta 2$ subunit of the nicotinic receptor ($\beta 2^{-/-}$) retinal waves are not present, although individual RGCs fire spikes at seemingly normal discharge rates (Bansal et al., 2000). Thus, the $\beta 2^{-/-}$ mice have provided an opportunity to assess the role of retinal waves in the development of the retina and retinal projections. These mutants have been found to exhibit a number of abnormalities in the organization of their visual system, including a failure to form normally segregated retinogeniculate projections (Rossi et al., 2001; Muir-Robinson *et al.*, 2002, Torborg *et al.*, 2004) as well as a lack of refined retinotopy in the LGN and the superior colliculus (Grubb *et al.*, 2003; McLaughlin *et al.*, 2003), and unexpectedly, an abnormal spatial segregation of ON- and OFF-centre LGN cells (Grubb *et al.*, 2003). These studies on $\beta 2^{-/-}$ mice would seem to lend support to the notion that retinal waves are essential for the normal segregation of retinogeniculate projections, as well as a number of other salient properties of the visual system. By contrast, in newborn ferrets in which correlated retinal activity has been disrupted by intraocular injections of an immunotoxin that targets cholinergic ACs, segregated retinogeniculate projections are formed normally (Huberman *et al.*, 2003). Complete blockade of retinal activity by pharmacological treatment of the newborn ferret retina prevents segregation of ocular projections, in line with what has been reported by other studies cited above. Thus, unlike the studies on $\beta 2^{-/-}$ mice, the work on the immunotoxin-treated ferrets indicates that retinal activity plays a permissive rather than an instructive role in the formation of segregated ocular inputs. A possible means to reconcile these findings on the mutant mouse and immunotoxin-treated ferret has been cogently discussed recently by Grubb and Thompson (2004, and see also Torborg *et al.*, 2004). Collectively, these studies underscore the fact that further work is required to identify the mechanisms responsible for the formation of segregated left and right eye inputs and to establish the role of retinal waves in the formation of the visual system.

13.3.3 Cellular mechanisms underlying the generation and propagation of retinal waves

A number of studies have been concerned with the important issues of how rhythmic bursting is generated in RGCs and what mechanisms lead to the propagation of activity across the retina.

Both acetylcholine and glutamate play an important role in the generation of retinal waves, but their relative contributions are age-dependent. At early stages, nicotinic cholinergic neurotransmission is necessary to generate correlated spontaneous bursting activity in RGCs (Sernagor and Grzywacz, 1996, 1999) and to propagate the waves (Feller *et al.*, 1996; Catsicas *et al.*, 1998; Wong *et al.*, 1998; Bansal *et al.*, 2000; Sernagor *et al.*, 2000; Wong and Wong, 2000; Zhou and Zhao, 2000; Sernagor *et al.*, 2003; Syed *et al.*, 2004b). The cholinergic inputs originate from starburst ACs, the only cholinergic neurons in the retina (Masland and Tauchi, 1986; Vaney, 1990). Indeed, simultaneous patch-clamp recordings from a RGC and a presumed cholinergic AC have demonstrated that these two cell types, are coactivated during spontaneous bursting activity (Zhou, 1998) (Figure 13.4). Using computer-modelling approaches, Feller *et al.* (1997) suggested that uncorrelated spontaneous activity in the ACs spreads laterally to other ACs and converges onto neighbouring RGCs to produce correlated waves. However, this cannot be the sole mechanism for initiating and propagating retinal waves because these ACs also receive synaptic input (Zhou, 1998; Zheng *et al.*, 2004), and because other neurotransmitters, secreted by ACs or by other cell types, regulate the spatiotemporal properties of the waves. In the presence of elevated adenosine, waves become more frequent and they increase in size (Stellwagen *et al.*, 1999). On the other hand GABAergic neurotransmission does not affect the wave spatiotemporal properties at early stages (Stellwagen *et al.*, 1999; Sernagor *et al.*, 2003).

At later stages, wave control switches from acetylcholine to glutamate (Wong *et al.*, 1998; Sernagor *et al.*, 2000; Bansal *et al.*, 2000; Wong and Wong, 2000; Zhou and Zhao, 2000; Syed *et al.*, 2004b). However, both neurotransmitters are required at all times in turtle retina (Sernagor and Grzywacz, 1999; Sernagor *et al.*, 2003). It has been suggested that acetylcholine mediates lateral propagation while glutamate may be involved in local excitability but does not regulate propagation per se (Sernagor and Grzywacz, 1999), so that acetylcholine influences the wave extent, whereas glutamate modulates their speed (Sernagor *et al.*, 2000, 2003). In turtle and ferret, the contribution of glutamate is largely mediated by α-amino-3-hydroxy-5-methyl-4-isoxazolepropionate (AMPA)/kainate rather than by *N*-methyl-D-aspartate (NMDA) receptors (Sernagor and Grzywacz, 1999; Wong *et al.*, 2000; Sernagor *et al.*, 2003), while in chick it appears to be mediated equally by both subtypes (Wong *et al.*, 1998; Sernagor *et al.*, 2000). Glutamate is presumably secreted by BCs but other sources such as RGC collaterals or direct projections from immature rods and cones are possible (see Section 13.2). Other neurotransmitters switch function with development. Nicotinic cholinergic neurotransmission switches to muscarinic in neonatal rabbit (Zhou and Zhao, 2000; Syed *et al.*, 2004b). γ-Aminobutyric acid and glycine, which also participate in wave modulation at later stages (see below), shift from being functionally excitatory to inhibitory (GABA – ferret: Fischer *et al.*, 1998; turtle: Sernagor *et al.*, 2003; rabbit: Zheng *et al.*, 2004; glycine – rabbit: Zhou, 2001).

13.3.4 Gap junctions: do they mediate wave propagation?

Gap junctions represent an attractive mechanism for mediating wave propagation, and we have already seen that they are involved in vertical (Catsicas *et al.*, 1998) and horizontal propagations through the ventricular zone before synaptogenesis (Syed *et al.*, 2004a). Once synaptic contacts have been established in the chick IPL, pharmacological blockade of gap junctions suppresses spontaneous activity (Wong *et al.*, 1998), suggesting some contribution to the wave generation process. Nevertheless, there is no unequivocal demonstration that gap junctions are necessary for wave propagation. Indeed, although tracer-coupling studies in the developing ferret retina indicate that α and γ RGCs demonstrate homologous coupling (Penn *et al.*, 1994), there is no coupling between β cells despite their participation in the waves (Wong *et al.*, 1993; Wong and Oakley, 1996). Moreover, gap junction blockade in turtle does not alter wave properties (Sernagor, unpublished observations).

13.3.5 Age-related changes in wave dynamics

Correlated spontaneous bursting activity and waves occur only during a finite temporal period of development. If such spontaneous activity were to persist once the retina becomes capable of being driven by light, it would interfere with visual responses. It is therefore important to understand what triggers the disappearance of retinal waves.

Both spatial and temporal aspects of wave propagation change with development. At early stages, waves spread relatively fast (at a speed of several micrometres per second) over large retinal areas (reviewed in Sernagor et al., 2001). In a longitudinal study of the changes in retinal wave dynamics in turtle, spanning the last three gestational weeks (gestation takes eight weeks) until a month post-hatching, waves dramatically slow down at embryonic Stage 25 (S25), a week before hatching. Towards hatching they become patches within which RGCs fire in near synchrony, occurring at random both in time and in location (Sernagor et al., 2003; see also Sernagor et al., 2001, Sernagor and Mehta, 2001) (Figure 13.2). These patches become smaller and eventually disappear about one month post-hatching. This gradual restriction in lateral propagation is due to developmental changes in the expression of $GABA_A$ responses (Sernagor et al., 2003). Activity of $GABA_A$ receptors is not involved in the early synaptic network that generates fast and wide-spreading waves. However, at S25, when the waves suddenly slow down and become narrower, they are modulated by GABA acting through type-A receptors. The $GABA_A$ responses at that time are excitatory, as has been reported in many parts of the immature CNS (Ben-Ari, 2002), becoming gradually inhibitory, until there is no correlated spontaneous activity, about one month later. When glutamic acid decarboxylase, the enzyme that synthesizes GABA, is blocked at post-hatching stages with a drug called allylglycine, leading to a decrease in endogenous GABA, spontaneous activity is stronger than in age-matched controls, and it even exhibits propagation (Figure 13.3a), presumably because the synaptic network generating spontaneous activity is relieved from GABAergic inhibition (Figure 13.4). These developmental changes in the dynamics of spontaneous activity coincide with the upregulation in the IPL of the K^+–Cl^- membrane cotransporter KCC2 that extrudes Cl^- from mature cells (Figure 13.3b), thereby causing the equilibrium potential for Cl^- to shift to more hyperpolarized levels, so that $GABA_A$ responses become inhibitory.

Several studies suggest that GABA is important in controlling the disappearance of retinal waves in mammals as well. An elegant recent study, employing dual patch-clamp recordings and Ca^{2+} imaging from pairs of rabbit starburst ACs, shows that these cells make reciprocal GABAergic synapses with each other, and that the GABAergic responses switch from excitatory to inhibitory while the waves disappear (Zheng et al., 2004; see also Zhou, 2001; Syed et al., 2004b) (Figure 13.4). Moreover, GABA shifts from excitatory to inhibitory around P15 to P18 in ferret (Fischer et al., 1998), shortly before waves stop propagating (Wong et al., 1993). In addition, vesicular GABAergic neurotransmission occurs at early stages (P1 to P5) at conventional synapses in the rodent inner retina (Johnson et al., 2003; see Section 13.2), while it has a depolarizing effect on RGCs in mouse (Bahring et al., 1994) and in all cell types in the rabbit (Huang and Redburn, 1996). Imaging of intracellular Cl^- in developing RGCs of the Enhanced Yellow Fluorescent Protein mouse shows that there is a marked increase in the driving force for Cl^- from P1 to P9 (when waves are known to occur) to P30 to P36 (when waves have disappeared) (Sernagor, Mutoh and Knöpfel,

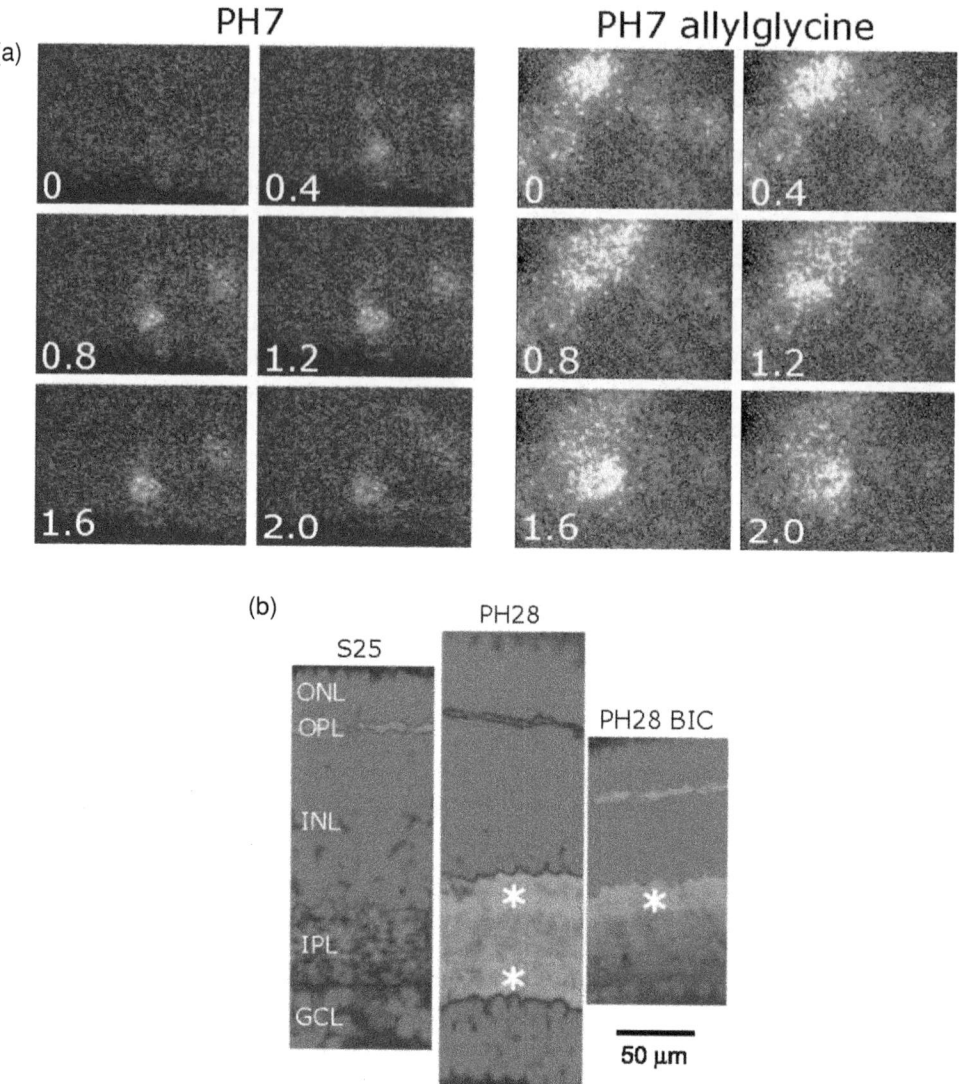

Figure 13.3 The role of GABA in the generation of retinal waves. (a) Time-lapse images (taken every 0.4 s) of spontaneous activity in the turtle retina at seven days post-hatching (PH7). The activity was recorded optically, with RGCs loaded with the Ca^{2+}-sensitive dye, Ca^{2+} green dextran. The background fluorescence is subtracted from the image, so that only increases in fluorescence, associated with neural activity, are seen. In control conditions (left panels), the activity is weak and patchy at that age. Following two days incubation in allylglycine, a drug that blocks the synthesis of GABA, therefore reducing the endogenous GABA levels in the tissue, the activity is much stronger, and even propagates (right panels), indicating that GABA has an inhibitory effect on spontaneous activity at post-hatching stages. (b) Immunofluorescence of KCC2, the transporter that extrudes Cl^- from mature cells in the turtle retina. Cell nuclei are labelled with 4,'6-diamidino-2-phenylindole (DAPI). KCC2 expression is limited to the plexiform layers. At S25, KCC2 expression is relatively weak (this is when GABA is excitatory). At PH21 days, KCC2 levels have increased significantly in the IPL, and there are two sub-bands of stronger expression (white asterisks), presumably corresponding to the

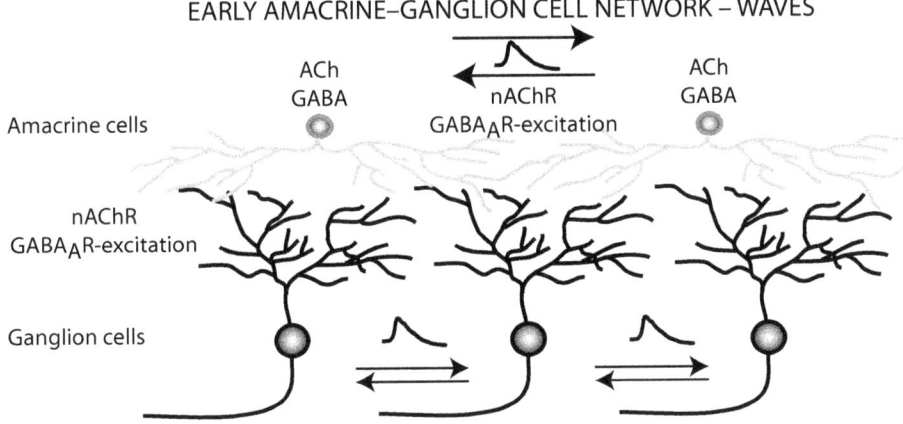

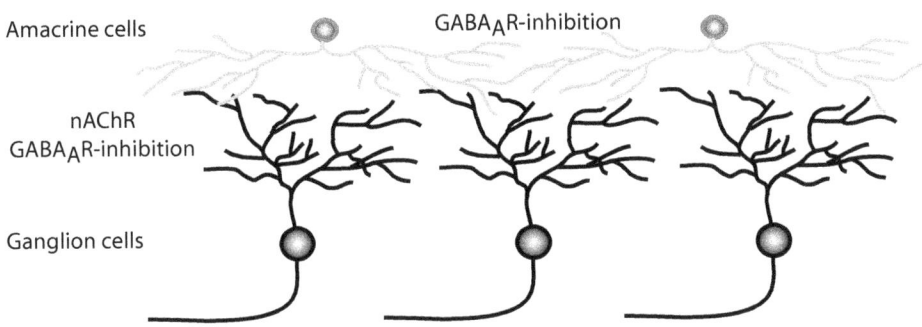

Figure 13.4 Amacrine–ganglion cell interactions during the period of spontaneous waves (upper-panel) and in maturity (lower panel). The diagram is based on findings from rabbit (Zheng et al., 2004) and turtle (Sernagor et al., 2003). During the period of spontaneous waves, ACs make direct connections among themselves as well as with RGCs. These connections are all excitatory: cholinergic nicotinic (nACh) and GABAergic (GABA$_A$). The GABAergic connections are excitatory because of high intracellular Cl$^-$. These connections enable lateral propagation within the amacrine network as well as across the GCL, resulting in propagating waves. Later on, the cholinergic nicotinic connections between ACs withdraw (they remain between ACs and RGCs) and GABA$_A$ responses become inhibitory, resulting in the disappearance of correlated spontaneous activity.

←

Figure 13.3 (cont.) ON and OFF laminas in the IPL. This increase in KCC2 expression suggests that the intracellular concentration of Cl$^-$ must be lower at that age, resulting in inhibitory GABAergic responses (and wave disappearance). When GABAergic activity is chronically blocked from S24 to 28 days PH (PH28 BIC), KCC2 expression is weaker in the IPL than in age-matched controls, with a particular emphasis on the inner part of the IPL, where the ON lamina is now lacking stronger KCC2 expression. There are still waves in these chronic animals, presumably because of weaker cellular extrusion of Cl$^-$ and depolarizing GABAergic responses. GCL, ganglion cell layer; INL, inner nuclear layer; ONL, outer nuclear layer.

unpublished observations), suggesting that GABA-induced Cl⁻ currents are depolarizing during the period of spontaneous waves. Finally, KCC2 expression increases significantly during the second week of postnatal development in rat (Vu *et al.*, 2000).

Apart from changes in the spatial patterns of activity, developmental changes in the temporal firing patterns have been reported in ferret RGCs (Wong and Oakley, 1996). Once eye-specific segregation is complete in the LGN, ON and OFF RGCs develop distinct temporal patterns of spontaneous activity, although both cell types still burst in synchrony. ON cells adopt a much lower burst frequency compared with OFF cells. The emergence of these different ON–OFF rhythms is, once again, related to changes in $GABA_A$ neurotransmission. Indeed, GABA becomes inhibitory at the time when ON and OFF rhythms diversify. At that time, GABA suppresses bursting activity in ON RGCs more effectively than in OFF RGCs (Fischer *et al.*, 1998). In addition to differences in GABAergic circuitry, distinct bursting patterns between ON and OFF cells also rely on differences in intrinsic membrane properties between these neurons (Myhr *et al.*, 2001). At still later stages of development, the three major RGC classes in the ferret retina (α, β and γ) gradually attain distinct spontaneous discharge patterns that are superimposed on their collective waves-like discharges (Liets *et al.*, 2003). Presumably, such class-specific discharges also reflect differences in the intrinsic membrane properties of these neurons.

Since GABA has such a strong impact on age-related changes in spontaneous activity patterns, it is important to understand what factors control developmental changes in $GABA_A$ activity. Sustained $GABA_A$ activity is required for GABA to shift polarity in rodent hippocampal cultures (Ganguly *et al.*, 2001, but see Ludurg *et al.*, 2003). In the turtle retina, chronic blockade of $GABA_A$ receptors in vivo during the period of the switch prevents $GABA_A$ responses from switching polarity, KCC2 expression remains lower (Figure 13.3b) and as a result, strong spontaneous waves still propagate across the retina at one month post-hatching, when normally there is no correlated activity anymore (Leitch *et al.*, 2005).

13.3.6 Why do waves disappear with retinal maturation?

In all species, waves disappear shortly after the onset of visual experience, when RGCs become driven by light. This immediately suggests that exposure to light at birth somehow triggers a mechanism that leads to the disappearance of the waves. In the absence of visual experience, spontaneous bursting activity in turtle RGCs (Sernagor and Grzywacz, 1996) and waves (Sernagor *et al.*, 2003) persist for longer periods post-hatching, suggesting that early visual experience may indeed trigger their disappearance. Interestingly, the activity-enhancing effect of visual deprivation acts via GABA. Indeed, GABA does not switch polarity, KCC2 expression is lower and waves keep propagating following one month of rearing in the dark (Sernagor *et al.*, 2003). This intriguing effect remains to be elucidated. Perhaps it works through a direct effect of light on $GABA_A$ responses, as has been described in adult rat RGCs (Leszkiewicz and Aizenman, 2003).

The situation is different in mammals, however. In the mouse retina, although dark rearing reversibly suppresses the post-eye-opening surge in spontaneous synaptic events (Tian and Copenhagen, 2001), it does not change the period during which RGCs exhibit correlated activity (Demas *et al.*, 2003). Both in control and in dark-reared animals, correlated spontaneous bursting activity disappears by P21. Hence, the effects of dark rearing upon correlated activity vary from species to species, presumably depending on how much they rely on visual experience during early postnatal life. Mammals depend on maternal protection at birth, whereas newly hatched turtles must immediately rely on visual cues for survival. (For example, marine turtles must run to the sea as soon as they hatch to escape predators and start their long migratory journey.) This suggests that the onset of visual experience may be more vital to reptiles than to mammals. It would be interesting to investigate the effect of dark rearing on other mammalian species relying more on vision than rodents. Interestingly, in adult taurine-deficient cats, RGCs within retinal areas depleted of photoreceptors demonstrate bursting behaviour like immature cells (W. R. Levick, personal communication). This suggests that the emergence of the vertical photoreceptor pathways in the maturing retina merely 'hides' the circuitry underlying spontaneous correlated bursting activity and propagating waves.

Whatever the explanation for this discrepancy between species might be, it does not exclude the possibility that a developmental switch in the polarity of $GABA_A$ activity is necessary in all species to induce the disappearance of correlated spontaneous activity. Clearly, many more studies manipulating GABAergic systems are required to reach a better understanding of these issues.

13.4 Concluding remarks

This chapter has reviewed the steps leading to the formation of functional synapses in various retinal layers. It has shown that synapse formation proceeds in a centrifugal manner, from the inner to the outer retina, and it occurs first in horizontal connections within the plexiform layers, followed by vertical connections between layers.

The chapter has highlighted the importance of various proteins involved in synaptic release, as well as extracellular matrix components, in guiding the development of the synaptic release machinery. The use of knockout mice has already clarified important issues on the role of many of these proteins (e.g. Bassoon) during the maturation of neurotransmission processes. It is clear that genetic manipulations are still the key experimental approach that will help us, in the near future, to unravel more details on the complex mechanism of synaptogenesis and various associated diseases. Likewise, the development of new knockout mice with different connexin expressions will shed more light on the precise roles of gap junctions during retinal development.

The chapter has also introduced the earliest form of neural activity, which takes the form of synaptically driven spontaneous rhythmic bursting in ACs and RGCs. These bursts propagate across the retinal surface, enabling synchronization between relatively distant parts of the eye. The possible role played by these waves in guiding the development of retinal projections has been discussed, but it is clear that more work is still required to underpin the precise role of retinal waves in the formation of the visual system. The chapter has also discussed how retinal waves change with development, eventually disappearing in all species during the first postnatal month. Whether visual experience guides wave disappearance is still debatable, and certainly appears to be species-related. There is good evidence that the emergence and maturation of synaptic inhibition is very instrumental in guiding the disappearance of these waves, but more studies manipulating inhibitory systems are still needed to reach a better understanding of these issues.

References

Bahring, R., Standhardt, H., Martelli, E. A. and Grantyn, R. (1994). GABA-activated chloride currents of postnatal mouse retinal ganglion cells are blocked by acetylcholine and acetylcarnitine: how specific are ions channels in immature neurons? *Eur. J. Neurosci.*, **6**, 1089–99.

Bansal, A., Singer, J. H., Hwang, B. J. *et al.* (2000). Mice lacking specific nicotinic acetylcholine receptor subunits exhibit dramatically altered spontaneous activity patterns and reveal a limited role for retinal waves in forming on and off circuits in the inner retina. *J. Neurosci.*, **20**, 7672–81.

Becker, D., Bonness, V. and Mobbs, P. (1998). Cell coupling in the retina: patterns and purpose. *Cell Biol. Int.*, **22**, 781–92.

Becker, D. L., Bonness, V., Catsicas, M. and Mobbs, P. (2002). Changing patterns of ganglion cell coupling and connexin expression during chick retinal development. *J. Neurobiol.*, **52**, 280–93.

Ben-Ari, Y. (2002). Excitatory actions of GABA during development: the nature of the nurture. *Nat. Rev. Neurosci.*, **3**, 728–39.

Brandstätter, J. H., Fletcher, E. L., Garner, C. C., Gundelfinger, E. D. and Wässle, H. (1999). Differential expression of the pre-synaptic cytomatrix protein Bassoon among ribbon synapses in the mammalian retina. *Eur. J. Neurosci.*, **11**, 3683–93.

Buckley, K. and Kelly, R. B. (1985). Identification of a transmembrane glycoprotein specific for secretory vesicles of neural and endocrine cells. *J. Cell Biol.*, **100**, 1284–94.

Catsicas, M., Bonness, V., Becker, D. L. and Mobbs, P. (1998). Spontaneous Ca^{2+} waves and their transmission in the developing chick retina. *Curr. Biol.*, **8**, 283–6.

Crair, M. C. (1999). Neuronal activity during development: permissive or instructive? *Curr. Opin. Neurobiol.*, **9**, 88–93.

Crooks, J., Okada, M. and Hendrickson, A. E. (1995). Quantitative analysis of synaptogenesis in the inner plexiform layer of macaque monkey fovea. *J. Comp. Neurol.*, **360**, 349–62.

Demas, J., Eglen, S. J. and Wong, R. O. L. (2003). Developmental loss of synchronous spontaneous activity in the mouse retina is independent of visual experience. *J. Neurosci.*, **23**, 2851–60.

Dhingra, N. K., Ramamohan, Y. and Raju, T. R. (1997). Developmental expression of synaptophysin, synapsin I and syntaxin in the rat retina. *Brain Res. Dev. Brain Res.*, **102**, 267–73.

Dick, O., Dieck, S. T., Altrock, W. D. et al. (2003). The pre-synaptic active zone protein Bassoon is essential for photoreceptor ribbon synapse formation in the retina. *Neuron*, **37**, 775–86.

Eglen, S. J. (1999). The role of retinal waves and synaptic normalization in retinogeniculate development. *Philos. Trans. R. Soc. London Ser. B*, **354**, 497–506.

Farber, D. B. and Danciger, M. (1997). Identification of genes causing photoreceptor degenerations leading to blindness. *Curr. Opin. Neurobiol.*, **7**, 666–73.

Feller, M. B., Wellis, D. P., Stellwagen, D., Werblin, F. S., Shatz, C. J. (1996). Requirement for cholinergic synaptic transmission in the propagation of spontaneous retinal waves. *Science*, **272**, 1182–7.

Feller, M. B., Butts, D. A., Aaron, H. L., Rokhsar, D. S. and Shatz, C. J. (1997). Dynamic processes shape spatiotemporal properties of retinal waves. *Neuron*, **19**, 293–306.

Fischer, K. F., Lukasiewicz, P. D. and Wong, R. O. L. (1998). Age-dependent and cell class-specific modulation of retinal ganglion cell bursting activity by GABA. *J. Neurosci.*, **18**, 3767–78.

Fisher, L. J. (1979). Development of synaptic arrays in the inner plexiform layer of neonatal mouse retina. *J. Comp. Neurol.*, **187**, 359–72.

Galli, L. and Maffei, L. (1988). Spontaneous impulse activity of rat retinal ganglion cells in prenatal life. *Science*, **24**, 90–1.

Ganguly, K., Schinder, A. F., Wong, S. T. and Poo, M. (2001). GABA itself promotes the developmental switch of neuronal GABAergic responses from excitation to inhibition. *Cell*, **18**, 521–32.

Grubb, M. S. and Thompson, I. D. (2004). The influence of early experience on the development of sensory systems. *Curr. Opin. Neurobiol.*, **14**, 503–12.

Grubb, M. S., Rossi, F. M., Changeux, J.-P. and Thompson, I. D. (2003). Abnormal functional organization in the dorsal lateral geniculate nucleus of mice lacking the β2 subunit of the nicotinic acetylcholine receptor. *Neuron*, **40**, 1161–72.

Grzywacz, N. M. and Sernagor, E. (2000). Spontaneous activity in developing turtle retinal ganglion cells: statistical analysis. *Vis. Neurosci.*, **17**, 229–41.

Horsburgh, G. M. and Sefton, A. J. (1987). Cellular degeneration and synaptogenesis in the developing retina of the rat. *J. Comp. Neurol.*, **263**, 553–66.

Huang, B. O. and Redburn, D. A. (1996). GABA-induced increases in $[Ca^{2+}]_i$ in retinal neurons of postnatal rabbits. *Vis. Neurosci.*, **13**, 441–7.

Huberman, A. D., Wang, G. Y., Liets, L. C. et al. (2003). Eye-specific retinogeniculate segregation independent of normal neuronal activity. *Science*, **300**, 994–8.

Ilia, M. and Jeffery, G. (1996). Delayed neurogenesis in the albino retina; evidence of a role for melanin in regulating the pace of cell generation. *Dev. Brain Res.*, **95**, 176–183.

Johnson, D. A., Mills, S. L., Haberecht, M. F. and Massey, S. C. (2000). Dye coupling in horizontal cells of developing rabbit retina. *Vis. Neurosci.*, **17**, 255–62.

Johnson, J., Tian, N., Caywood, M. S. et al. (2003). Vesicular neurotransmitter transporter expression in developing postnatal rodent retina: GABA and glycine precede glutamate. *J. Neurosci.*, **23**, 518–29.

Johnson, P. T., Williams, R. R., Cusato, K. and Reese, B. E. (1999). Rods and cones project to the inner plexiform layer during development. *J. Comp. Neurol.*, **414**, 1–12.

Kriegstein, K. and Schmitz, F. (2003). The expression pattern and assembly profile of synaptic membrane proteins in ribbon synapses of the developing mouse retina. *Cell Tissue Res.*, **311**, 159–73.

Leitch, E., Coaker, J., Young, C., Mehta, V. and Sernagor, E. (2005). GABA type-A activity controls its own developmental polarity switch in the maturing retina. *J. Neurosci.*, **25**, 4801–5.

Leszkiewicz, D. N. and Aizenman, E. (2003). Reversible modulation of GABA(A) receptor-mediated currents by light is dependent on the redox state of the receptor. *Eur. J. Neurosci.*, **17**, 2077–83.

Libby, R. T., Lavallee, C. R., Balkema, G. W., Brunken, W. J. and Hunter, D. D. (1999). Disruption of laminin β2 chain production causes alterations in morphology and function in the CNS. *J. Neurosci.*, **19**, 9399–411.

Libby, R. T., Champliaud, M. F., Claudepierre, T. *et al.* (2000). Laminin expression in adult and developing retinae: evidence of two novel CNS laminins. *J. Neurosci.*, **20**, 6517–28.

Liets, L. C., Olshausen, B. A., Wang, G. Y. and Chalupa, L. M. (2003). Spontaneous activity of morphologically identified ganglion cells in the developing ferret retina. *J. Neurosci.*, **23**, 7343–50.

Ludwig, A., Li, H. Saarma, M., Kaila, K. and Rivera, C. (2003). Developmental up-regulation of KCC2 in the absence of GABAergic and glutamatergic transmission. *Eur. J. Neurosci.*, **18**, 3199–206.

Maffei, L. and Galli-Resta, L. (1990). Correlation in the discharges of neighboring rat retinal ganglion cells during prenatal life. *Proc. Natl. Acad. Sci. U. S. A.*, **87**, 2861–4.

Masland, R. H. (1977). Maturation of function in the developing rabbit retina. *J. Comp. Neurol.*, **175**, 275–86.

Masland, R. H. and Tauchi, M. (1986). The cholinergic amacrine cells. *Trends Neurosci.*, **9**, 218–23.

Maslim, J. and Stone, J. (1986). Synaptogenesis in the retina of the cat. *Brain Res.*, **373**, 35–48.

McArdle, C. B., Dowling, J. E., Masland, R. H. (1977). Development of outer segments and synapses in the rabbit retina. *J. Comp. Neurol.*, **175**, 253–74.

McLaughlin, T., Torborg, C. L., Feller, M. B. and O'Leary, D. M. (2003). Retinotopic map refinement requires spontaneous retinal waves during a brief critical period of development. *Neuron*, **40**, 1147–60.

Meister, M., Wong, R. O. L., Baylor, D. A. and Shatz, C. J. (1991). Synchronous bursts of action potentials in ganglion cells of the developing mammalian retina. *Science*, **252**, 939–43.

Mooney, R., Penn, A. A., Gallego, R. and Shatz, C. J. (1996). Thalamic relay of spontaneous retinal activity prior to vision. *Neuron*, **17**, 863–74.

Myhr, K. L., Lukasiewicz, P. D. and Wong, R. O. L. (2001). Mechanisms underlying developmental changes in the firing pattern of ON and OFF retinal ganglion cells during refinement of their central projections. *J. Neurosci.*, **21**, 8664–71.

Muir-Robinson, G., Hwang, B. J. and Feller, M. B. (2002). Retinogeniculate axons undergo eye-specific segregation in the absence of eye-specific layers. *J. Neurosci.*, **22**, 5259–64.

Nag, T. C. and Wadhwa, S. (2001). Differential expression of syntaxin-1 and synaptophysin in the developing and adult human retina. *J. Biosci.*, **26**, 179–91.

Nishimura, Y. and Rakic, P. (1985). Development of the rhesus monkey retina. I. Emergence of the inner plexiform layer and its synapses. *J. Comp. Neurol.*, **241**, 420–34.

Nishimura, Y. and Rakic, P. (1987). Development of the rhesus monkey retina. II. A three-dimensional analysis of the sequence of synapse combinations in the inner plexiform layer. *J. Comp. Neurol.*, **262**, 290–313.

O'Donovan, M. J. (1999). The origin of spontaneous activity in developing networks of the vertebrate nervous system. *Curr. Opin. Neurobiol.*, **9**, 94–104.

Pearson, R. A., Dale, N., Llaudet, E. and Mobbs, P. (2005). ATP released via gap junction hemichannels from the pigment epithelium regulates neural retinal progenitor proliferation. *Neuron*, **46**, 731–44.

Penn, A. A. and Shatz, C. J. (1999). Brain waves and brain wiring: the role of endogenous and sensory-driven neural activity in development. *Pediatr. Res.*, **45**, 447–58.

Penn, A. A., Wong, R. O. L. and Shatz, C. J. (1994). Neuronal coupling in the developing mammalian retina. *J. Neurosci.*, **14**, 3805–15.

Penn, A. A., Riquelme, P. A., Feller, M. B. and Shatz, C. J. (1998). Competition in retinogeniculate patterning driven by spontaneous activity. *Science*, **279**, 2108–12.

Rich, K. A., Zhan, Y. and Blanks, J. C. (1997). Migration and synaptogenesis of cone photoreceptors in the developing mouse retina. *J. Comp. Neurol.*, **388**, 47–63.

Robinson, S. R. (1991). Development of the mammalian retina. In *Neuroanatomy of the Visual Pathways and their Development*, ed. B. Dreher and S. R. Robinson. London: Macmillan Press, pp. 69–128.

Rorig, B. and Grantyn, R. (1993). Glutamatergic and GABAergic synaptic currents in ganglion cells from isolated retinae of pigmented rats during postnatal development. *Brain Res. Dev. Brain Res.*, **74**, 98–110.

Rossi, F. M., Pizzorusso, T., Porciatti, V. *et al.* (2001). Requirement of the nicotinic acetylcholine receptor beta 2 subunit for the anatomical and functional development of the visual system. *Proc. Natl. Acad. Sci. U. S. A.*, **98**, 6453–8.

Rothman, J. E. and Warren, G. (1994). Implications of the SNARE hypothesis for intracellular membrane topology and dynamics. *Curr. Biol.*, **4**, 220–33.

Sernagor, E. and Grzywacz, N. M. (1995). Emergence of complex receptive field properties of ganglion cells in the developing turtle retina. *J. Neurophysiol.*, **73**, 1355–64.

Sernagor, E. and Grzywacz, N. M. (1996). Influence of spontaneous activity and visual experience on developing retinal receptive-fields. *Curr. Biol.*, **6**, 1503–8.

Sernagor, E. and Grzywacz, N. M. (1999). Spontaneous activity in developing turtle retinal ganglion cells: pharmacological studies. *J. Neurosci.*, **19**, 3874–87.

Sernagor, E. and Mehta, V. (2001). The role of early neural activity in the maturation of turtle retinal function. *J. Anat.*, **199**, 375–83.

Sernagor, E., Eglen, S. J. and O'Donovan, M. J. (2000). Differential effects of acetylcholine and glutamate blockade on the spatiotemporal dynamics of retinal waves. *J. Neurosci.*, **20**(RC56), 1–6.

Sernagor, E., Eglen, S. J. and Wong, R. O. L. (2001). Development of retinal ganglion cell structure and function. *Prog. Retin. Eye Res.*, **20**, 139–74.

Sernagor, E., Young, C. and Eglen, S. J. (2003). Developmental modulation of retinal wave dynamics: shedding light on the GABA saga. *J. Neurosci.*, **23**, 7621–9.

Sharma, R. K., O'Leary, T. E., Fields, C. M. and Johnson, D. A. (2003). Development of the outer retina in the mouse. *Dev. Brain Res.*, **145**, 93–105.

Shatz, C. J. and Stryker, M. P. (1988). Prenatal tetrodotoxin infusion blocks segregation of retinogeniculate afferents. *Science*, **24**, 87–9.

Sherry, D. M., Wang, M. M., Bates, J. and Frishman, L. J. (2003). Expression of vesicular glutamate transporter 1 in the mouse retina reveals temporal ordering in development of rod vs. cone and ON vs. OFF circuits. *J. Comp. Neurol.*, **465**, 480–98.

Sollner, T., Whiteheart, S. W., Brunner, M. *et al.* (1993). SNAP receptors implicated in vesicle targeting and fusion. *Nature*, **362**, 318–24.

Stellwagen, D. and Shatz, C. J. (2002). An instructive role for retinal waves in the development of retinogeniculate connectivity. *Neuron*, **33**, 357–67.

Stellwagen, D., Shatz, C. J. and Feller, M. B. (1999). Dynamics of retinal waves are controlled by cyclic AMP. *Neuron*, **24**, 673–85.

Syed, M. M., Lee, S., He, S. and Zhou, Z. J. (2004a). Spontaneous waves in the ventricular zone of developing mammalian retina. *J. Neurophysiol.*, **91**, 1999–2009.

Syed, M. M., Lee, S., Zheng, J. and Zhou, Z. J. (2004b). Stage-dependent dynamics and modulation of spontaneous waves in the developing rabbit retina. *J. Physiol.*, **560**, 533–49.

Tian, N. and Copenhagen, D. R. (2001). Visual deprivation alters development of synaptic function in inner retina after eye opening. *Neuron*, **32**, 439–49.

Timpl, R. (1996). Macromolecular organization of basement membranes. *Curr. Opin. Cell Biol.*, **8**, 618–24.

Tootle, J. S. (1993). Early postnatal development of visual function in ganglion cells of the cat retina. *J. Neurophysiol.*, **69**, 1645–60.

Torborg, C., Wang, C. T., Muir-Robinson, G. and Feller, M. B. (2004). L-type calcium channel agonist induces correlated depolarisations in mice lacking the beta2 subunit nAChRs. *Vis. Res.*, **44**, 3347–55.

Vaney, D. I. (1990). The mosaic of amacrine cells in the mammalian retina. *Prog. Retin. Eye Res.*, **9**, 49–100.

Vu, T. Q., Payne, J. A. and Copenhagen, D. R. (2000). Localization and developmental expression patterns of the neuronal K-Cl cotransporter (KCC2) in the rat retina. *J. Neurosci.*, **20**, 1414–23.

Wang, M. M., Janz, R., Belizaire, R., Frishman, L. J. and Sherry, D. M. (2003). Differential distribution and developmental expression of Synaptic Vesicle Protein 2 isoforms in the mouse retina. *J. Comp. Neurol.*, **460**, 106–22.

West Greenlee, M. H., Finley, S. K., Wilson, M. C., Jacobson, C. D. and Sakaguchi, D. S. (1998). Transient, high levels of SNAP-25 expression in cholinergic amacrine cells during postnatal development of the mammalian retina. *J. Comp. Neurol.*, **394**, 374–85.

West Greenlee, M. H., Roosevelt, C. B. and Sakaguchi, D. S. (2001). Differential localization of SNARE complex proteins SNAP-25, Syntaxin, and Vamp during development of the mammalian retina. *J. Comp. Neurol.*, **430**, 306–20.

Wong, R. O. L. (1999). Retinal waves and visual system development. *Annu. Rev. Neurosci.*, **22**, 29–47.

Wong, R. O. L. and Oakley, D. M. (1996). Changing patterns of spontaneous bursting activity of on and off retinal ganglion cells. *Neuron*, **16**, 1087–95.

Wong, R. O. L., Meister, M. and Shatz, C. J. (1993). Transient period of correlated bursting activity during development of the mammalian retina. *Neuron*, **11**, 923–38.

Wong, R. O. L., Chernjavsky, A., Smith, S. J. and Shatz, C. J. (1995). Early functional neural networks in the developing retina. *Nature*, **374**, 716–18.

Wong, W. T. and Wong, R. O. L. (2000). Rapid dendritic movements during synapse formation and rearrangement. *Curr. Opin. Neurobiol.*, **10**, 118–24.

Wong, W. T., Sanes, J. R. and Wong, R. O. L. (1998). Developmentally regulated spontaneous activity in the embryonic chick retina. *J. Neurosci.*, **18**, 8839–52.

Wong, W. T., Myhr, K. L., Miller, E. D. and Wong, R. O. L. (2000). Developmental changes in the neurotransmitter regulation of correlated spontaneous retinal activity. *J. Neurosci.*, **20**, 351–60.

Zheng, J. J., Lee, S. and Zhou, Z. J. (2004). A developmental switch in the excitability and function of the starburst network in the mammalian retina. *Neuron*, **44**, 851–64.

Zhou, Z. J. (1998). Direct participation of starburst amacrine cells in spontaneous rhythmic activities in the developing mammalian retina. *J. Neurosci.*, **18**, 4155–65.

Zhou, Z. J. (2001). A critical role of the strychnine-sensitive glycinergic system in spontaneous retinal waves of the developing rabbit. *J. Neurosci.*, **21**, 5158–68.

Zhou, Z. J. and Zhao, D. (2000). Coordinated transitions in neurotransmitter systems for the initiation and propagation of spontaneous retinal waves. *J. Neurosci.*, **20**, 6570–7.

14

Emergence of light responses

Evelyne Sernagor
University of Newcastle upon Tyne, Newcastle upon Tyne, UK

Leo M. Chalupa
University of California Davis, Davis, USA

14.1 Introduction

Although the newborn retina is highly active, with spontaneous waves propagating across the amacrine and the ganglion cell layers every few minutes (see Chapter 13), at that time it is not yet possible to elicit light responses in retinal ganglion cells (RGCs). This lack of responsiveness to light is due to the immaturity of the vertical synaptic pathway between photoreceptors and RGCs provided by bipolar cells (BCs), despite the fact that lateral connections in the inner retina are already well established (see Chapter 13). Moreover, rod and cone opsins are not yet functional at birth. In mouse for example, ultraviolet cone opsin does not appear until postnatal day (P)1, rod opsin until P5 and green cone opsin until P7 (Tarttelin *et al.*, 2003). Hence, RGCs become visually responsive only shortly before eye opening (around P10 in rabbit; Masland, 1977; Dacheux and Miller, 1981a,b; P7 to P10 in cat; Tootle, 1993; P12 in mouse; Sekaran *et al.*, 2005). Humans and other primates, on the other hand, are born with their eyes open and although primate vision is poor at birth a newborn human infant is capable of tracking visual stimuli (Teller, 1997).

This chapter reviews the earliest light responses that can be detected in the developing retina. New studies show that the newborn retina is actually not insensitive to light and this will be considered in the first part of the chapter. Photosensitive, melanopsin-expressing RGCs are functional from the day of birth, providing information on levels of illumination to the suprachiasmatic nucleus of the hypothalamus, where our circadian rhythms are generated and controlled.

The second part of the chapter will review the emergence of conventional image-forming light responses, resulting from synaptic processing through the photoreceptors/BCs/RGCs vertical pathway. We will consider the properties of these early responses to light and discuss the nature of receptive field properties of maturing retinal cells, mainly RGCs.

A central question in developmental neurobiology is whether synaptic wiring is preprogrammed through cellular genetic profiles, or whether it is prone to change with experience. We will address these issues here as well. In particular, we will consider whether early

Retinal Development, ed. Evelyne Sernagor, Stephen Eglen, Bill Harris and Rachel Wong.
Published by Cambridge University Press. © Cambridge University Press 2006.

spontaneous activity and visual experience can shape maturing functional visual properties of retinal neurons.

14.2 Neonatal intrinsic photosensitivity in ganglion cells

Recent studies are telling us that the idea that the neonatal retina of non-primates is insensitive to light is a misconception (reviewed in Sernagor, 2005). We now know that there are responses to light at birth in some RGCs of probably all species (Sekaran et al., 2005). These responses originate in a small subset of RGCs that are photosensitive (Berson et al., 2002) because they contain an opsin-like protein called melanopsin (Provencio et al., 1998). In adults, these intrinsically photosensitive RGCs (ipRGCs) convey information on levels of illumination, or irradiance rather than on image formation. They project via the retinohypothalamic tract to the suprachiasmatic nucleus of the hypothalamus, where circadian rhythms are generated, and influence pineal melatonin levels (reviewed in Berson, 2003) (see Figure 14.1). They also project to the brain area that uses information on irradiance to control pupil constriction. Intrinsically sensitive RGCs are not photosensitive in melanopsin knockout mice and as a result these animals have impaired non-image-forming vision (Berson, 2003).

Since ipRGCs are directly activated by light (although they do receive retinal synaptic contacts (Belenky et al., 2003)) whilst other RGCs respond to light through several synaptic relays, there is no logical reason to believe that light responses emerge concomitantly in these two retinal pathways. Melanopsin is already expressed halfway through gestation in mouse and at 8.6 weeks post-conception in humans (Tarttelin et al., 2003). It is therefore reasonable to assume that melanopsin-expressing RGCs become functional long before other RGCs. Furthermore, in neonatal rodents light induces an increase in the expression of the immediate early gene c-*fos* (a marker of neural activity) in the suprachiasmatic nucleus (Leard et al., 1994; Weaver and Reppert, 1995; Munoz Lamosas et al., 2000; Hannibal and Fahrenkrug, 2004; Sekaran et al., 2005), suggesting that photo-entrainment of the circadian clock begins soon after the eye starts experiencing light at birth, long before image-forming vision is even possible.

Using immunocytochemistry and *in situ* hybridization in rat, Fahrenkrug et al. (2004) reported that melanopsin is expressed from embryonic day (E)18 (when cellular differentiation begins) in cells of the inner neuroblast layer, with increasing expression and migration towards the ganglion cell layer (GCL) around birth. These melanopsin-positive embryonic cells also express pituitary adenylate cyclase-activating polypeptide (PACAP), a neurotransmitter found exclusively in RGCs that form the retinohypothalamic tract (and therefore project to the suprachiasmatic nucleus) in adult. Light induces c-*fos* expression from the day of birth both in these melanopsin/PACAP-expressing RGCs and in the suprachiasmatic nucleus (Hannibal and Fahrenkrug, 2004).

Using Ca^{2+} imaging, a new study has recently provided the first direct physiological evidence that mouse melanopsin-expressing ipRGCs respond to light from birth (Sekaran et al., 2005) (Figure 14.1). Like in adult ipRGCs (Berson et al., 2002) these responses persist

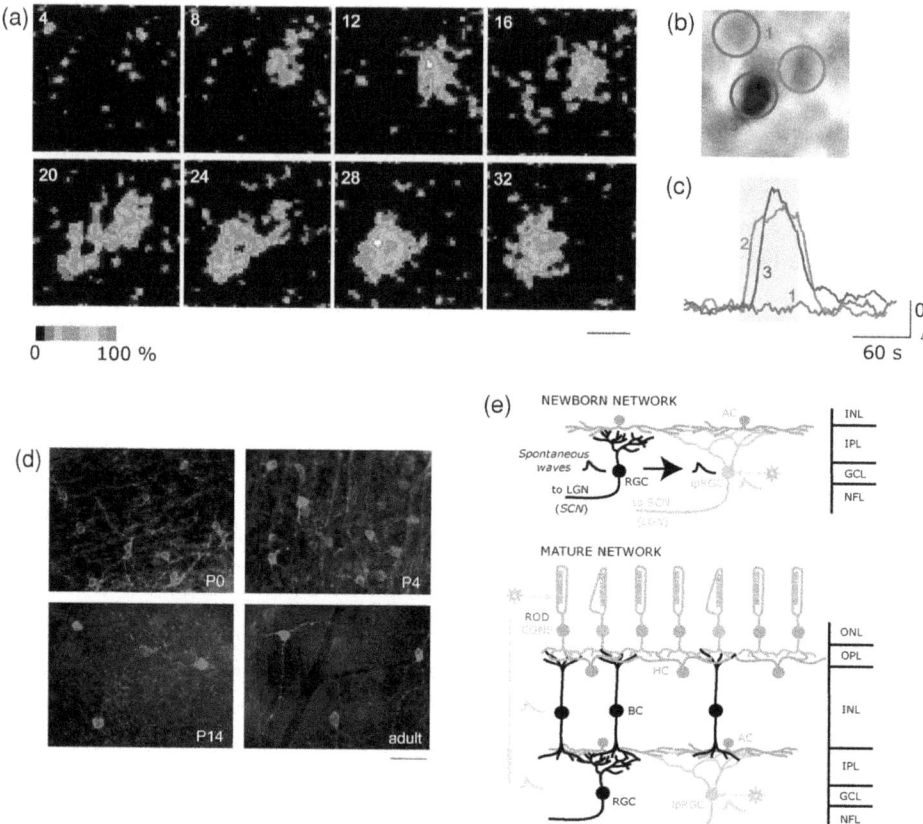

Figure 14.1 Intrinsic photosensitivity in neonatal retinal RGCs. (a) Optical recordings (with the Ca^{2+}-sensitive dye fura-2 AM) of responses to light (470 nm) in neonatal ipRGCs. Responses are shown every 4 s. Scale bar: 10 μm. (b) Illustrates the three cells whose activity is shown in (a).(c) Traces of the change in fluorescence in the three cells illustrated in (a) and (b). Cell 1 is not active, whereas Cells 2 and 3 exhibit clear light-driven responses. (d) Melanopsin-expressing RGCs at different stages of development. There are significantly more melanopsin-positive cells at P0 to P4 than at P14 or in adult. Scale bar: 50 μm. (e) Diagram summarizing mammalian retinal networks involved in the generation of neural responses in RGCs at birth (upper panel) and at maturity (lower panel). At birth, ipRGCs (in blue) respond to light (indicated by the sun symbol, the light response is illustrated by the yellow trace). Intrinsically photosensitive RGCs and other RGCs (in black) as well as amacrine cells (ACs) participate in spontaneously generated waves (illustrated by the black traces). At maturity, the rod/cone BC (in black) pathway is functional, and responses to light propagate vertically through the retinal layers to the RGCs (and to a lesser extent to ipRGCs). Spontaneous waves have disappeared. AC, amacrine cell; GCL, ganglion cell layer; HC, horizontal cell; INL, inner nuclear layer; IPL, inner plexiform layer; LGN, lateral geniculate nucleus; NFL, nerve fibre layer, comprising the axons of all RGCs, including ipRGCs; ONL, outer nuclear layer; SCN, suprachiasmatic nucleus. Panels (a) to (d) reproduced from Sekaran *et al.*, 2005. Panel (e) reproduced from Sernagor, 2005. With permission from *Curr. Biol.*, Cell Press. For colour version, see Plate 11.

when glutamatergic neurotransmission is blocked, ruling out any possible contribution from the photoreceptor-BC pathway. They are absent in melanopsin knockout mice, confirming that light sensitivity in the GCL at birth originates from melanopsin-expressing ipRGCs. An important aspect of this study is the developmental time course of changes in the number of ipRGCs and in their sensitivity to light. More cells respond to light at P0, 13.7%, than at P4 to P5, 5.4%, and in adults, 2.7%. This developmental decrease in light responsiveness is attributed to a decrease of over 70% in the number of melanopsin-expressing RGCs between P4 and P14, when cell density reaches its minimum, adult level, presumably due to massive RGC death, which peaks between P4 to P6 (Young, 1984). Remarkably, these young ipRGCs are capable of generating sustained responses to low illumination, lasting several minutes. This has significant implications for photo-entrainment of the circadian clock in neonates because these cells make functional connections with the suprachiasmatic nucleus from P0.

The discovery of neonatal retinal photosensitivity raises fundamental issues in brain development. The retina can detect irradiance from birth and transmits the information to the clock, but there is also a progressive decrease in irradiance sensitivity during the first postnatal weeks, while the retina is building up its more conventional role in image-forming vision, which becomes functional at eye opening. We can speculate about the reasons why the retina should need such a strong irradiance detection system at birth. Of course, it may just be an epiphenomenon of RGC death. There is a general overproduction of RGCs at birth, including functional ipRGCs, and a large proportion of these cells simply undergo apoptosis while the retina matures. But it is much more attractive to speculate that there might be some physiological reason for neonatal hyperphotosensitivity. It may help the newborn organism switching fast and efficiently from coordination with the maternal circadian system (Reppert and Schwartz, 1983) to its independent photo-entrainment system. Taking advantage of naturally occurring RGC death, this early hypersensitivity to light would gradually decrease to adult levels, avoiding interference with other photoreceptor systems in maturity. Neonatal photosensitivity perhaps also acts as an immature form of vision because ipRGCs project to image-forming visual areas (Hattar *et al.*, 2002; Dacey *et al.*, 2005). Whilst ipRGCs do not contribute to precise image formation, they may provide global information on other aspects of visual perception such as motion, and this may help neonates escaping predators, for example. The initial over production of ipRGCs may also consolidate the development of retinal projections both to the accessory and to the image-forming visual system. Immature ipRGCs participate in retinal waves (Sekaran *et al.*, 2005) (see Chapter 13) and it is therefore very likely that wiring of ipRGC projections are influenced both by spontaneous and light-driven activity in these immature cells.

14.3 The emergence of light responses in the image-forming pathway

Because it is likely that the basic circuitry underlying the receptive field organization of RGCs is present prior to when these cells can respond to light stimuli, it has been difficult to investigate how connections develop to give rise to specific receptive field properties before vision can be experienced. Although visual stimulation in the isolated retina or in eyecup preparations is possible using the immature retina, the poor quality of the optics in neonates prevents reliable assessment of light responses in vivo (Thorn et al., 1976). Nevertheless, important knowledge concerning the development of light responses and the underlying circuitry responsible for their generation has been gained using several different electrophysiological approaches.

14.3.1 Electroretinograms as an assessment tool to monitor developing light responses

An overall assessment of the development of light responses is possible from electroretinograms (ERGs) (see Chapter 17). Indeed, ERGs provide useful functional physiological information, such as the specific contribution of photoreceptors, outer or inner retinal neurons and glial cells to light responses (Brown and Wiesel, 1961; Dowling, 1987). The initial, downward component, the a-wave, reflects photoreceptor responses. The second, upward component is the b-wave, reflecting the neurotransmission to second-order neurons. The last component, the c-wave, reflects participation of glia. In rabbit, a small a-wave is already present from P6, indicating that photoreceptors are already functional before RGCs become sensitive to light and before eye opening. This initial response triples in amplitude by P9 to P10 (Masland, 1977). From P10, the time of eye opening, there is also a small positive b-wave component in the response, indicating neurotransmission from photoreceptors to BCs. Both the a- and b-wave subsequently increase in amplitude and attain their mature profile a few weeks later (Reuter, 1976).

Abnormal ERGs develop in animals suffering from malformation of synaptic contacts between photoreceptors and second-order neurons. For example, both the amplitude and the slope of the b-wave intensity response function are dramatically attenuated in laminin β2 chain-deficient mutants (see Chapter 13 and Libby et al., 1999), suggesting that neurotransmission between rods and BCs is disrupted in these mice. Electroretinograms in Bassoon-deficient mice are also abnormal. As for laminin β2 chain-deficient mice, it is the b-wave that is most affected (see Chapter 13 and Dick et al., 2003): its amplitude is significantly smaller and the signal develops more slowly than in wild type.

14.3.2 Receptive fields of ganglion cells: development and plasticity

General observations

The initial responses of RGCs to light stimulation are weak, labile and rapidly adapting (around P8 in rabbit – Masland, 1977; around P3 to P4 in cat – Tootle, 1993; two to three weeks before hatching in turtle – Sernagor and Grzywacz, 1995). The responses gain in robustness within a few days, and by that time, several adult features of RGC receptive fields are already apparent. In cat and in rabbit the earliest measurable receptive fields are already concentric in their centre-surround organization (Bowe-Anders *et al.*, 1975; Masland, 1977; Tootle, 1993). Specialized features of RGC receptive fields, such as direction selectivity, are also apparent before eye opening (Masland, 1977).

Centre-surround

The maturation of the surround organization varies with species. In rabbit, before eye opening there are large 'undifferentiated' fields with silent surrounds that can suppress the response to centre stimulation but do not themselves respond to direct light stimulation (Masland, 1977). In the cat, the strength of the antagonistic surround relative to that of the centre does not seem to change with postnatal maturation. Silent inhibitory surrounds, however, are not observed until the third postnatal week in cat (Tootle, 1993).

ON–OFF responses

At maturity, RGCs with dendrites stratifying in the inner portion of the inner plexiform layer (IPL) signal information about increments of light, while those with dendrites stratifying in the outer portion of the IPL signal light decrements (Famiglietti and Kolb, 1976; Nelson *et al.*, 1978; for recent review see Nelson and Kolb, 2004). However, early in development the dendritic processes of RGCs ramify throughout the IPL (Dann *et al.*, 1988; Maslim and Stone, 1988; Ramoa *et al.*, 1988). Several years ago, Bodnarenko and Chalupa discovered that segregation of initially multistratified RGC dendrites into ON and OFF sublaminae of the IPL is dependent on the normal release of glutamate by BCs (Bodnarenko and Chalupa, 1993; Bodnarenko *et al.*, 1995). Treating the postnatal cat retina with 2-amino-4-phosphonobutyric acid (APB) (the glutamate analogue that in the mature retina hyperpolarizes ON-cone BCs and rod BCs, thereby preventing their release of glutamate), resulted in a much higher than normal incidence of RGCs with multistratified dendrites (Figure 14.2).

To explain how BC synaptic activity might regulate the stratification of RGC dendrites, a model was formulated stipulating that ON- and OFF-cone BC axons selectively innervate the multistratified dendrites of immature RGCs (Bodnarenko *et al.*, 1995). In line with this idea, very few RGCs in the neonatal cat retina had been reported to respond with ON–OFF discharges to flashing spots of light (Dubin *et al.*, 1986; Tootle, 1993). These findings were based, however, on extracellular recordings, which do not allow one to relate function to structure. Assessing the light responses of developing RGCs in the ferret retina by means of whole-cell patch-clamp recordings (Figure 14.3), which allow intracellular labelling of the recorded cells, showed that the vast majority of developing RGCs with multistratified dendrites responded to both light onset as well as light offset (Wang *et al.*, 2001). Contrary to the original model, these results demonstrated that immature RGCs with multistratified dendrites are innervated early in development by ON- as well as OFF-cone BCs (Figure 14.4). This suggests that glutamate release by BCs triggers an intrinsic programme in multistratified RGCs leading to the retraction of one or another set of dendritic processes.

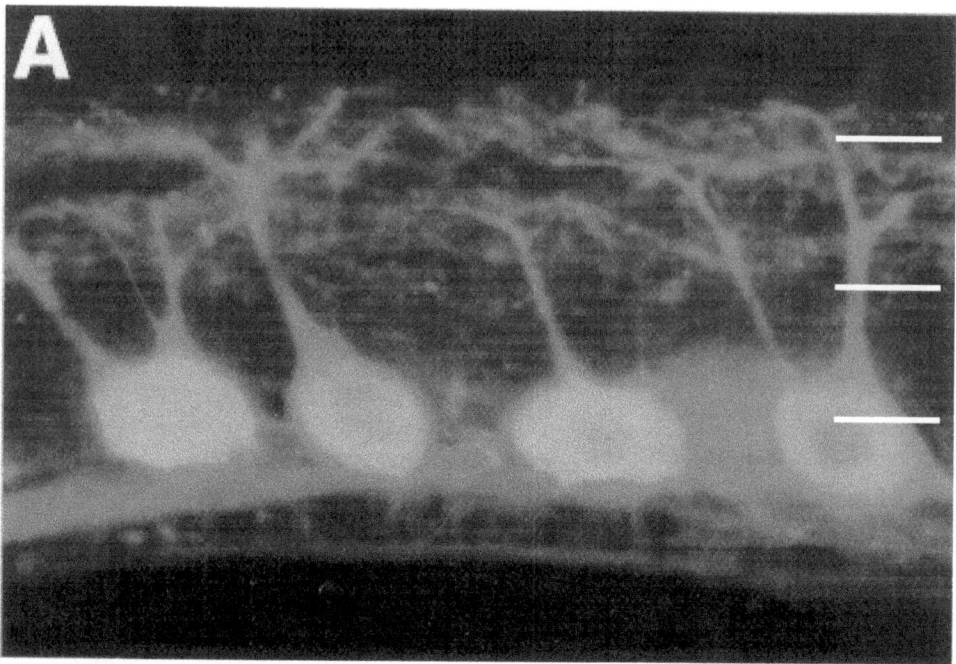

Figure 14.2 Retinal sections showing the stratification patterns of RGC dendrites in the normal cat retina (in panel A) and in a retina treated from birth with APB (in panel B), a drug that blocks the release of glutamate by ON-cone BCs and rod BCs. Note that in A, RGC dendrites terminate within the IPL in one of two strata (boundary denoted by the white lines) either proximal (ON sublayer) or distal (OFF sublayer) to the cell body. This distinct stratification pattern does not develop when glutamate release has been blocked early in development as indicated by the widespread ramification of the RGC dendrites evident in panel B.

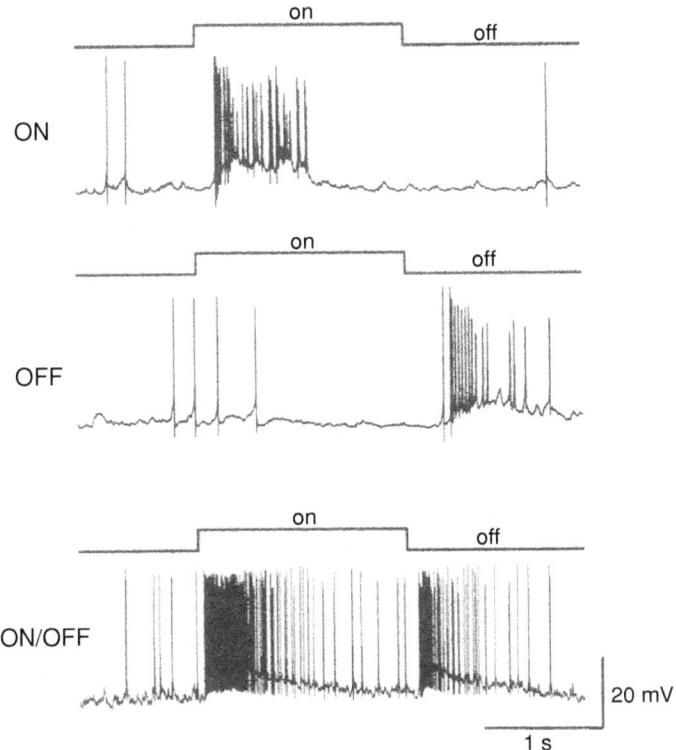

Figure 14.3 Whole-cell patch-clamp recordings from three different RGCs obtained from the neonatal ferret retina to a flashed spot of light. The onset and offset of the light is denoted above each recording. The cell shown on the top responded only to light onset, the one in the middle responded only to light offset; these neurons had their dendrites stratified in either the ON or OFF sublayer of the IPL respectively. By contrast, the cell whose responses are depicted on the bottom responded to both light onset and offset and the dendrites of this RGC were found to span the ON and OFF sublayers of the IPL.

Treating the developing retina with APB perturbs the stratification of both ON- and OFF-RGC dendrites to an approximately equal extent (Bodnarenko and Chalupa, 1993; Bodnarenko et al., 1995; Bisti et al., 1998). It assumes that the effects of APB are basically equivalent in the developing and mature retina. But this is not the case, since both ON and OFF responses can be completely blocked in multistratified RGCs, while only the ON pathway is affected by APB application in the mature retina (Wang et al., 2001). Therefore, we may conclude that the functional circuitry underlying ON–OFF responses in the developing retina is fundamentally different from that found in the mature retina, but the details of these differences remain to be established.

At maturity, the axons of ON-cone and OFF-cone BCs terminate in two distinct strata of the IPL where they synapse onto the dendrites of ON and OFF RGCs. Until recently, nothing was known about the development of BC inputs because it was not feasible to

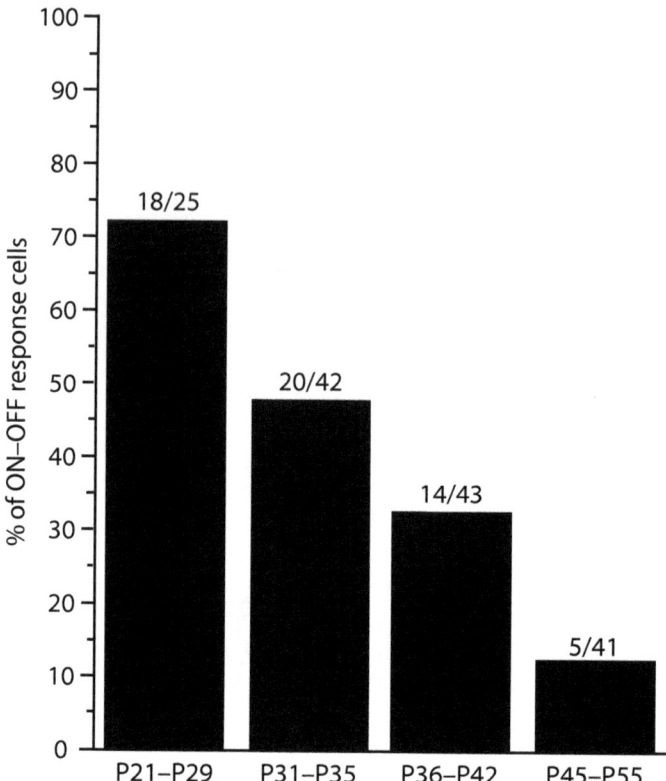

Figure 14.4 The change in the proportion of RGCs manifesting ON–OFF responses during postnatal development of the ferret retina. Note that from postnatal day (P)21 to P29 the majority of RGCs yielded both ON and OFF responses to a flashed spot of light, and that this proportion of cells progressively decreased with maturity, dropping to about 12% by P45 to P55. This functional change parallels a concomitant decrease in the incidence of RGCs with multistratified dendrites.

completely label a large contingent of these retinal interneurons. Fortunately, an antibody against a Ca^{2+}-binding protein, recoverin, which binds to ON- and OFF-cone BCs in the adult retina (Milam et al., 1993; Euler and Wässle, 1995) has been shown to recognize ON- and OFF- cone BCs and their processes in the developing retina. On the basis of recoverin immunostaining it has been inferred that ON- and OFF-cone BCs exhibit a highly specific ingrowth pattern (Miller et al., 1999; Gunhan-Agar et al., 2000). This picture contrasts with the remodelling of RGCs dendrites (discussed above) and with the exuberance and subsequent retraction that has been observed when segregated ON and OFF retinogeniculate pathways are established (Hahm et al., 1991).

What might account for the specific ingrowth patterns of cone BC axons? One possibility is that ON- and OFF- cone BC axons recognize bistratified targets within the developing IPL. This could involve the differential distribution of molecules in the extracellular matrix, as has been shown in the developing cortex (e.g., Pearlman and Sheppard, 1996), and/or cellular processes that might be stratified within the IPL prior to the ingrowth of BC axons. A recent study has shown that the axon arbors of cone BCs still form two distinct strata following depletion of all RGCs by cutting the optic nerve on the day of birth (Gunhan-Agar *et al.*, 2000), suggesting that RGCs are not necessary for the formation of segregated ON and OFF BC inputs. Similar findings have been reported in a zebrafish mutant retina *tak/ath5*, in which RGCs are never born (Kay *et al.*, 2004).

What about the dendrites of starburst amacrine cells (ACs)? At maturity these processes, which can be identified by cholinergic immunostaining, are organized into two strata within the IPL and are innervated selectively by ON- and OFF-cone BCs (Famiglietti *et al.*, 1977; Famiglietti, 1983 a,b; Masland *et al.*, 1984; Tauchi and Masland, 1984; Bloomfield and Miller, 1986; Masland and Tauchi, 1986). Moreover, in both the chick (Layer *et al.*, 1997) and the rat (Koulen, 1997), bistratified cholinergic-positive strata are detectable very early in development. In the rat retina, such bands are already present more than a week before the stratified bipolar axon arbors are formed. To assess the role of ACs in establishing proper innervation of the IPL by BCs, cholinergic neurons can be eliminated with a novel immunotoxin using a saporin-anti-VAChT (vesicular acetylcholine transporter) antibody. Virtually all the retinal cholinergic cells were eliminated after a single intraocular injection of this new immunotoxin as early as the day of birth, more than a week before the stratification of BC axons, and despite that, cone BCs were still found to form their segregated projections (Gunhan *et al.*, 2002).

In summary, there are three retinal cell types with processes that are prominently segregated within the IPL of the mature retina: (1) the dendrites of ON and OFF RGCs; (2) the projections of ON- and OFF-cone BCs; and (3) the processes of cholinergic ACs. The two segregated bands of cholinergic ACs are the first to appear, but the scaffolding provided by these neurons is not essential for the subsequent stratification of cone BCs (Gunhan *et al.*, 2002). Moreover, the stratification of cholinergic processes and cone bipolar projections appears to occur without any obvious refinement, while the dendritic processes of RGCs undergo major structural reorganization to achieve their mature stratified state. Thus, while glutamate release by BCs regulates the stratification of RGC dendrites, the normal ingrowth of cone BCs does not require the presence of RGCs (Gunhan-Agar *et al.*, 2000). In some species such as the cat, the stratification of RGC dendrites commences prior to birth (Bodnarenko *et al.*, 1995), which implies that glutamate release occurs independent of visual input.

Complex receptive fields – insights from a reptilian model

In the last decade, the turtle retina has proven to be an interesting model for studying receptive field development. The reason is that in turtle, as opposed to mammals, embryonic RGCs already have robust light responses. They can be driven by light from embryonic stage 23 (S23) (about three weeks prior to hatching), almost coinciding with the onset of spontaneous bursting activity in these cells (S22) (Sernagor and Grzywacz, 1995). However, at that time, their light responses are immature. Receptive fields are initially small, continuing to expand until about a month post-hatching, when they reach their mature sizes. Surprisingly, RGCs at the early stages respond well to several directions of movement or orientation of the stimulus light edge. This is in contrast to maturity when these cells show a clear preference for a single direction of motion or orientation, or have no preference at all (Sernagor and Grzywacz, 1995). The proportion of RGCs that have circular receptive fields also increases with maturation.

Theoretical studies have suggested that the response to multiple directions or orientations of motion of immature turtle RGC receptive fields is due to polarized and poorly branched dendritic arbors (Burgi and Grzywacz, 1997, 1998). However, concomitant intracellular labelling of embryonic RGCs with Lucifer yellow and mapping of their responses to moving light edges only occasionally reveals a good match between structure and function (Mehta and Sernagor, 2006a). Moreover, large-field RGCs reach peak dendritic proliferation by S25 (following which they undergo pruning), one week prior to hatching (Mehta and Sernagor, 2006b), when their receptive fields still show a high incidence of embryonic irregularities, or 'anisotropies' (Sernagor and Grzywacz, 1995). Thus, rather than being predicted by the shape of dendritic arbors, the responses to multiple directions of movement may arise from immature, sparse sets of excitatory and inhibitory synaptic inputs.

Early neural activity and developing retinal receptive fields

It has generally been assumed that the retina is mature by the time of eye opening (Daw, 1995) and, therefore, few studies have focused on the involvement of visual experience in the maturation of retinal function. For example, one of the earliest studies on that subject showed that visual experience does not affect the development of receptive fields of RGCs in young rabbits (Daw and Wyatt, 1974). Shortly after eye opening, rabbits were reared in an environment with unidirectionally moving stimuli. ON or OFF responses as well as directional selectivity developed normally. It is, however, difficult to interpret these results, because rabbit RGC receptive fields are fairly mature by the time of eye opening (Masland, 1977) and it is very likely that it is too late at that stage for visual experience to have an impact on developing connections. However, even at that time, these observations could not rule out that electrical activity per se may guide developing retinal receptive fields.

More recent studies demonstrate a regulatory role for visual experience in the development of mammalian RGC responses to light. Indeed, dark rearing from early postnatal life transiently suppresses the developmental increase in peak amplitude and time-to-peak light-driven responses normally observed in mouse RGCs after eye opening (Tian and Copenhagen, 2001). Electroretinograms from these animals reveal that visual deprivation reversibly enhances synaptic function in the outer retina and weakens it in the inner retina, suggesting that synaptic development and plasticity does continue after eye opening in the mammalian retina.

In turtle, where RGCs become sensitive to light much earlier (see Complex receptive fields – insights from a reptilian model, p. 297), experimental manipulations that modify spontaneous activity in vivo affect the development of RGC receptive field properties.

Dark rearing from hatching prolongs the period of spontaneous activity and retinal waves (Sernagor et al., 2003) and results in larger receptive fields (Sernagor and Grzywacz, 1996). On the other hand, when cholinergic transmission is chronically blocked from embryonic stages by in vivo application of curare, receptive field areas remain small (Sernagor et al., 2001). Moreover, exposure to curare from the day of hatching prevents the dark-induced expansion of receptive fields (Sernagor and Grzywacz, 1996). These observations suggest that it is the mere presence of spontaneous activity rather than visual experience that really matters for the expansion of receptive field areas. Both curare treatment and dark rearing reduce the amount of mature RGCs with circular receptive fields, while the incidence of circular receptive fields in dark-reared turtles was similar to that of either dark- or light-reared turtles whose retinas had been exposed to curare from hatching (Sernagor et al., 2001). Finally, when waves are chronically made larger and stronger (by increasing the wave cellular recruitment and spatial extent) by enhancing cholinergic activity through the blockade of anticholinesterase with neostigmine from S21 or by blocking γ-aminobutyric acid A receptors with bicuculline from S24, the degree of receptive field anisotropies is also lower (Mehta and Sernagor, unpublished results). Taken together, these observations indicate that retinal waves may be important in guiding the maturation of RGC receptive fields.

The role of retinal waves in the stratification of retinal RGC dendrites has also been examined (Bansal et al., 2000). Recall, as discussed in Chapter 13, that $\beta 2^{-/-}$ mice exhibit no correlated activity, although individual RGCs do manifest spikes. In these animals stratified dendrites are formed, but this developmental process is delayed compared with that in wild-type mice. This indicates that retinal waves are not essential for the formation of ON and OFF stratified retinal pathways.

As discussed above, Tian and Copenhagen (2003) demonstrated that visual experience is required for the maturation of ON and OFF responses in mouse RGCs, showing that light *is* important for the maturation of retinal function in mammals. In that study, dark rearing prevents the normal developmental loss of RGCs responding to both ON and OFF light stimuli. Concomitantly, light deprivation prevents the normal course of age-related loss of bistratified RGCs (see Section 14.3.2).

These observations in mouse suggest that manipulating early neural activity can alter the organization of dendritic arbors of RGCs. Modifying early activity changes dendritic organization in turtle RGCs as well. Dark rearing leads to abnormally high dendritic proliferation (there is an increase both in branch number and length), whereas curare treatment leads just to the opposite (Sernagor et al., 2001; Sernagor and Mehta, 2001; Mehta and Sernagor, 2006b). These structural changes may explain, in part, the enhancement and reduction in receptive field size in dark-reared and curare-treated animals, respectively.

14.4 Concluding remarks

This chapter has reviewed the earliest light responses that can be detected in the developing vertebrate retina.

It is now established that ipRGCs are present and functional from the day of birth in mammals, long before vision is possible through the image-forming pathway. This significant discovery is undoubtedly going to change our perception of how important early visual experience is for the maturation of the visual system and circadian rhythmicity.

The chapter has also reviewed the early light responses in the image-forming pathway, from photoreceptors to RGCs, with an emphasis on the development of ON–OFF responses and RGC complex receptive field properties.

Collectively, the available evidence on the development of retinal ON and OFF pathways would seem to suggest that retinal interneurons follow a set of developmental rules distinct from those obeyed by RGCs (cf. Chalupa and Gunhan, 2004). The morphological properties of dendrites (as well as axon terminals) of RGCs undergo considerable reshaping during the course of normal development and they appear to be susceptible to various types of environmental manipulation. By comparison, the processes of cholinergic ACs and cone BCs appear more rigidly pre-programmed. It remains to be seen whether this generalization gains support as we continue to accumulate new information on the development of retinal circuitry.

Studies on the role of early activity in guiding the development of light responses suggest that the maturation of visual function is highly plastic in early postnatal life. Early neural activity, either in the form of spontaneous waves, or in the form of visual experience can influence the development and refinement of retinal circuitry, ultimately influencing how retinal neurons will process visual information once the retina has reached maturity. Moreover, both forms of early activity interact, making the system even more prone to developmental plasticity. In future studies, to reach a better understanding of the specific contribution provided by early activity in shaping retinal function, it will be important to design experiments where distinct aspects of the wave dynamics and visual processing are specifically targeted.

References

Bansal, A., Singer, J. H., Hwang, B. J. *et al.* (2000). Mice lacking specific nicotinic acetylcholine receptor subunits exhibit dramatically altered spontaneous activity patterns and reveal a limited role for retinal waves in forming on and off circuits in the inner retina. *J. Neurosci.*, **20**, 7672–81.

Belenky, M. A., Smeraski, C. A., Provencio, I., Sollars, P. J. and Pickard, G. E. (2003). Melanopsin retinal ganglion cells receive bipolar and amacrine cell synapses. *J. Comp. Neurol.*, **460**, 380–93.

Berson, D. M. (2003). Strange vision: ganglion cells as circadian photoreceptors. *Trends Neurosci.*, **26**, 314–20.

Berson, D. M., Dunn, F. A. and Takao, M. (2002). Phototransduction by retinal ganglion cells that set the circadian clock. *Science*, **295**, 1065–70.

Bisti, S., Gargini, C. and Chalupa, L. M. (1998). Blockade of glutamate-mediated activity in the developing retina perturbs the functional segregation of ON and OFF pathways. *J. Neurosci.*, **18**, 5019–25.

Bloomfield, S. A., Miller, R. F. (1986). A functional organization of ON and OFF pathways in the rabbit retina. *J. Neurosci.*, **6**, 1–13.

Bodnarenko, S. R. and Chalupa, L. M. (1993). Stratification of ON and OFF ganglion cell dendrites depends on glutamate-mediated afferent activity in the developing retina. *Nature*, **364**, 144–6.

Bodnarenko, S. R., Jeyarasasingam, G. and Chalupa, L. M. (1995). Development and regulation of dendritic stratification in retinal ganglion cells by glutamate-mediated afferent activity. *J. Neurosci.*, **15**, 7037–45.

Bowe-Anders, C., Miller, R. F. and Dacheux, R. F. (1975). Developmental characteristics of receptive organization in the isolated retina eye-cup of the rabbit. *Brain Res.*, **87**, 61–5.

Brown, K. T. and Wiesel, T. N. (1961). Localization of origins of electroretinogram components by intra-retinal recording in the intact cat eye. *J. Physiol.*, **158**, 257–80.

Burgi, P. Y. and Grzywacz, N. M. (1997). Possible roles of spontaneous waves and dendritic growth for retinal receptive field development. *Neural Comput.*, **9**, 533–53.

Burgi, P. Y. and Grzywacz, N. M. (1998). A biophysical model for the developmental time course of retinal orientation selectivity. *Vis. Res.*, **38**, 2787–800.

Chalupa, L. M. and Gunhan, E. (2004). Development of On and Off retinal pathways and retinogeniculate projections. *Prog. Retin. Eye Res.*, **23**, 31–51.

Dacey, D. M., Liao, H.-S., Peterson, B. *et al.* (2005). Melanopsin-expressing ganglion cells in primate retina signal color and irradiance and project to the LGN. *Nature*, **433**, 749–54.

Dacheux, R. F. and Miller, R. F. (1981a). An intracellular electrophysiological study of the ontogeny of functional synapses in the rabbit retina. I. Receptors, horizontal, and bipolar cells. *J. Comp. Neurol.*, **198**, 307–26.

Dacheux, R. F. and Miller, R. F. (1981b). An intracellular electrophysiological study of the ontogeny of functional synapses in the rabbit retina. II. Amacrine cells. *J. Comp. Neurol.*, **198**, 327–34.

Dann, J. F., Buhl, E. H. and Peichl, L. (1988). Postnatal dendritic maturation of alpha and beta ganglion cells in cat retina. *J. Neurosci.*, **8**, 1485–99.

Daw, N. W. (1995). *Visual Development*. New York: Plenum Press.

Daw, N. W. and Wyatt, H. J. (1974). Raising rabbits in a moving visual environment: an attempt to modify directional sensitivity in the retina. *J. Physiol.*, **240**, 309–30.

Dick, O., Dieck, S. T., Altrock, W. D. *et al.* (2003). The pre-synaptic active zone protein Bassoon is essential for photoreceptor ribbon synapse formation in the retina. *Neuron*, **37**, 775–86.

Dowling, J. E. (1987). *The Retina; an Approachable Part of the Brain*. Cambridge, MA: Harvard University Press.

Dubin, M. W., Stark, L. A. and Archer, S. M. (1986). A role for action-potential activity in the development of neuronal connections in the kitten retinogeniculate pathway. *J. Neurosci.*, **6**, 1021–36.

Euler, T. and Wässle, H. (1995). Immunocytochemical identification of cone bipolar cells in the rat retina. *J. Comp. Neurol.*, **361**, 461–78.

Fahrenkrug, J., Nielsen, H. S. and Hannibal, J. (2004). Expression of melanopsin during development of the rat retina. *NeuroReport*, **15**, 781–4.

Famiglietti, E. V. Jr. (1983a). 'Starburst' amacrine cells and cholinergic neurons: mirror-symmetric on and off amacrine cells of rabbit retina. *Brain Res.*, **261**, 138–44.

Famiglietti, E. V. Jr. (1983b). On and off pathways through amacrine cells in mammalian retina: the synaptic connections of 'starburst' amacrine cells. *Vis. Res.*, **23**, 1265–79.

Famiglietti, E. V. Jr. and Kolb, H. (1976). Structural basis for ON-and OFF-center responses in retinal ganglion cells. *Science*, **194**, 193–5.

Famiglietti, E. V. Jr, Kaneko, A. and Tachibana, M. (1977). Neuronal architecture of on and off pathways to ganglion cells in carp retina. *Science*, **198**, 1267–9.

Gunhan, E., Choudary, P. V., Landerholm, T. E. and Chalupa, L. M. (2002). Depletion of cholinergic amacrine cells by a novel immunotoxin does not perturb the formation of segregated on and off cone bipolar cell projections. *J. Neurosci.*, **22**, 2265–73.

Gunhan-Agar, E., Kahn, D. and Chalupa, L. M. (2000). Segregation of on and off bipolar cell axonal arbors in the absence of retinal ganglion cells. *J. Neurosci.*, **20**, 306–14.

Hahm, J. O., Langdon, R. B. and Sur, M. (1991). Disruption of retinogeniculate afferent segregation by antagonists to NMDA receptors. *Nature*, **351**, 568–70.

Hannibal, J. and Fahrenkrug, J. (2004). Melanopsin containing retinal ganglion cells are light responsive from birth. *NeuroReport*, **15**, 2317–20.

Hattar, S., Liao, H. W., Takao, M., Berson, D. M. and Yau, K. W. (2002). Melanopsin-containing retinal ganglion cells: architecture, projections, and intrinsic photosensitivity. *Science*, **295**, 1065–70.

Kay, J. N., Roeser, T., Mumm, J. S. *et al.* (2004). Transient requirement for ganglion cells during assembly of retinal synaptic layers. *Development*, **131**, 1331–42.

Koulen, P. (1997). Vesicular acetylcholine transporter (VAChT): a cellular marker in rat retinal development. *NeuroReport*, **8**, 2845–8.

Layer, P. G., Berger, J. and Kinkl, N. (1997). Cholinesterases precede "ON-OFF" channel dichotomy in the embryonic chick retina before onset of synaptogenesis. *Cell Tissue Res.*, **288**, 407–16.

Leard, L. E., Macdonald, E. S., Heller, H. C. and Kilduff, T. S. (1994). Ontogeny of photic-induced c-fos mRNA expression in rat suprachiasmatic nuclei. *NeuroReport*, **5**, 2683–7.

Libby, R. T., Lavallee, C. R., Balkema, G. W., Brunken, W. J. and Hunter, D. D. (1999). Disruption of laminin β2 chain production causes alterations in morphology and function in the CNS. *J. Neurosci.*, **19**, 9399–411.

Masland, R. H. (1977). Maturation of function in the developing rabbit retina. *J. Comp. Neurol.*, **175**, 275–86.

Masland, R. H. and Tauchi, M. (1986). The cholinergic amacrine cells. *Trends Neurosci.*, **9**, 218–23.

Masland, R. H., Mills, J. W. and Hayden, S. A. (1984). Acetylcholine-synthesising amacrine cells: identification and selective staining by using radioautography and fluorescent markers. *Proc. R. Soc. London B. Biol. Sci.*, **223**, 79–100.

Maslim, J. and Stone, J. (1988). Time course of stratification of the dendritic fields of ganglion cells in the retina of the cat. *Brain Res. Dev. Brain Res.*, **44**, 87–93.

Mehta, V. and Sernagor, E. (2006a). Receptive-field structure function correlates in developing turtle retinal ganglion cells. *Eur. J. Neurosci.*, In press.

Mehta, V. and Sernagor, E. (2006b). Early neural activity and dendritic growth in turtle retinal ganglion cells. *Eur. J. Neurosci.*, In press.

Milam, A. H., Dacey, D. M. and Dizhoor, A. M. (1993). Recoverin immunoreactivity in mammalian cone bipolar cells. *Vis. Neurosci.*, **10**, 1–12.

Miller, E. D., Tran, M. I., Wong, G. K., Oakley, D. M. and Wong, R. O. (1999). Morphological differentiation of bipolar cells in the ferret retina. *Vis. Neurosci.*, **16**, 1133–44.

Munoz Llamosas, M., Huerta, J. J., Cernuda-Cernuda, R. and Garcia-Fernandez, J. M. (2000). Ontogeny of a photic response in the retina and suprachiasmatic nucleus in the mouse. *Brain Res. Dev. Brain Res.*, **120**, 1–6.

Nelson, R. and Kolb, B. (2004). ON and OFF pathways in the vertebrate retina and visual system. In *The Visual Neurosciences*, ed. L. M. Chalupa and J. S. Werner. Cambridge, MA: MIT Press, pp. 260–78.

Nelson, R., Famiglietti, E. V. Jr and Kolb, H. (1978). Intracellular staining reveals different levels of stratification for on- and off-center ganglion cells in cat retina. *J. Neurophysiol.*, **41**, 472–83.

Pearlman, A. L. and Sheppard, A. M. (1996). Extracellular matrix in early cortical development. *Prog. Brain Res.*, **108**, 117–34.

Provencio, I., Jiang, G., De Grip, W. J., Hayes, W. P. and Rollag, M. D. (1998). Melanopsin: an opsin in melanophores, brain, and eye. *Proc. Natl. Acad. Sci. U. S. A.*, **95**, 340–5.

Ramoa, A. S., Campbell, G. and Shatz, C. J. (1988). Dendritic growth and re-modeling of cat retinal ganglion cells during fetal and postnatal development. *J. Neurosci.*, **8**, 4239–61.

Reppert, S. M. and Schwartz, W. J. (1983). Maternal coordination of the fetal biological clock in utero. *Science*, **220**, 969–71.

Reuter, J. H. (1976). The development of the electroretinogram in normal and light-deprived rabbits. *Pflügers Arch.*, **363**, 7–13.

Sekaran, S., Lupi, D., Jones, S. L. et al. (2005). Melanopsin dependent photoreception provides earliest light detection in the mammalian retina. *Curr. Biol*, **15**, 1099–107.

Sernagor, E. (2005). Retinal development: second sight comes first. *Curr. Biol.*, **15**, R556–9.

Sernagor, E. and Grzywacz, N. M. (1995). Emergence of complex receptive field properties of ganglion cells in the developing turtle retina. *J. Neurophysiol.*, **73**, 1355–64.

Sernagor, E. and Grzywacz, N. M. (1996). Influence of spontaneous activity and visual experience on developing retinal receptive-fields. *Curr. Biol.*, **6**, 1503–8.

Sernagor, E. and Mehta, V. (2001). The role of early neural activity in the maturation of turtle retinal function. *J. Anat.*, **199**, 375–83.

Sernagor, E., Eglen, S. J. and Wong, R. O. L. (2001). Development of retinal ganglion cell structure and function. *Prog. Retin. Eye Res.*, **20**, 139–74.

Sernagor, E., Young, C. and Eglen, S. J. (2003). Developmental modulation of retinal wave dynamics: shedding light on the GABA saga. *J. Neurosci.*, **23**, 7621–9.

Tarttelin, E. E., Bellingham, J., Bibb, L. C. et al. (2003). Expression of *opsin* genes early in ocular development of humans and mice. *Exp. Eye Res.*, **76**, 393–6.

Tauchi, M. and Masland, R. H. (1984). The shape and arrangement of the cholinergic neurons in the rabbit retina. *Proc. R. Soc. London B. Biol. Sci.*, **223**, 101–19.

Teller, D. Y. (1997). First glances: the vision of infants. The Friedenwald lecture. *Invest. Ophthalmol. Vis. Sci.*, **38**, 2183–203.

Thorn, F., Gollender, M. and Erikson, P. (1976). The development of the kitten's visual optics. *Vis. Res.*, **16**, 1145–9.

Tian, N. and Copenhagen, D. R. (2001). Visual deprivation alters development of synaptic function in inner retina after eye-opening. *Neuron*, **32**, 439–49.

Tian, N. and Copenhagen, D. R. (2003). Visual stimulation is required for refinement of On and Off pathways in postnatal retina. *Neuron*, **39**, 85–96.

Tootle, J. S. (1993). Early postnatal development of visual function in ganglion cells of the cat retina. *J. Neurophysiol.*, **69**, 1645–60.

Wang, G. Y., Liets, L. C. and Chalupa, L. M. (2001). Unique functional properties of on and off pathways in the developing mammalian retina. *J. Neurosci.*, **21**, 4310–7.

Weaver, D. R. and Reppert, S. M. (1995). Definition of the developmental transition from dopaminergic to photic regulation of *c-fos* gene expression in the rat suprachiasmatic nucleus. *Brain Res. Mol. Brain Res.*, **33**, 136–48.

Young, R. W. (1984). Cell death during differentiation of the retina in the mouse. *J. Comp. Neurol.*, **229**, 362–73.

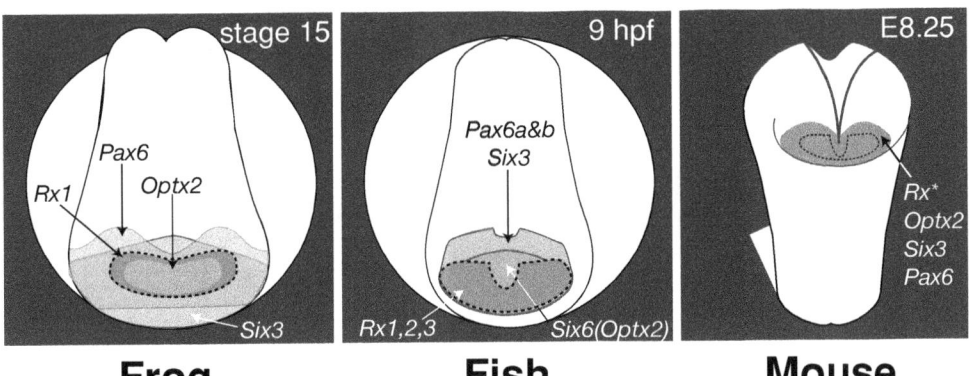

Plate 1

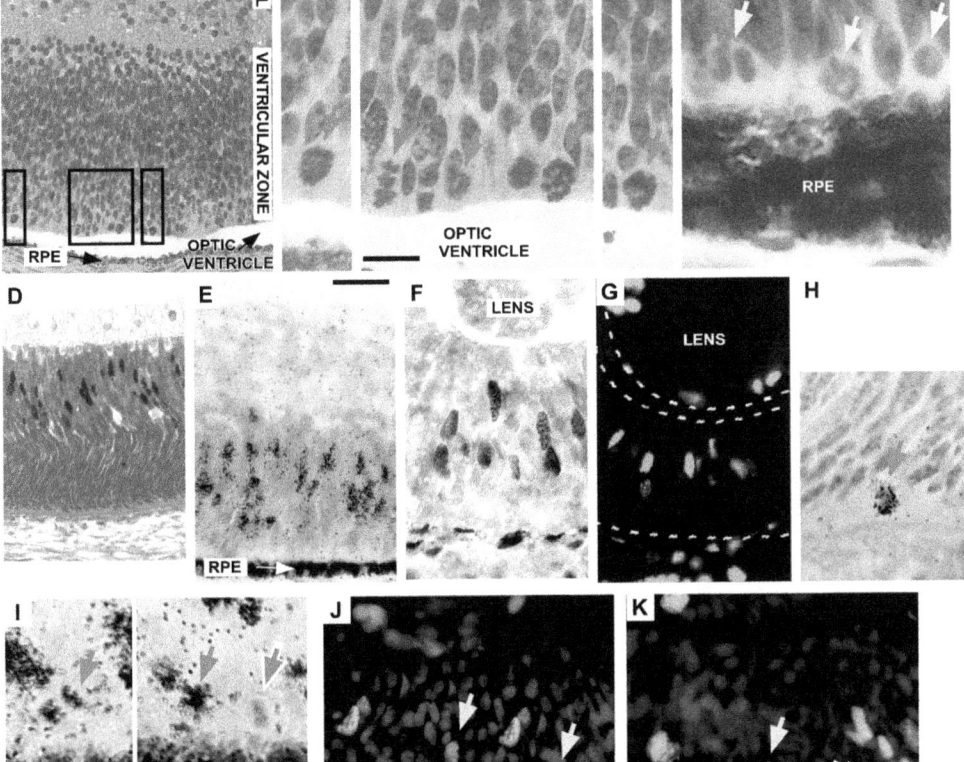

Plate 2

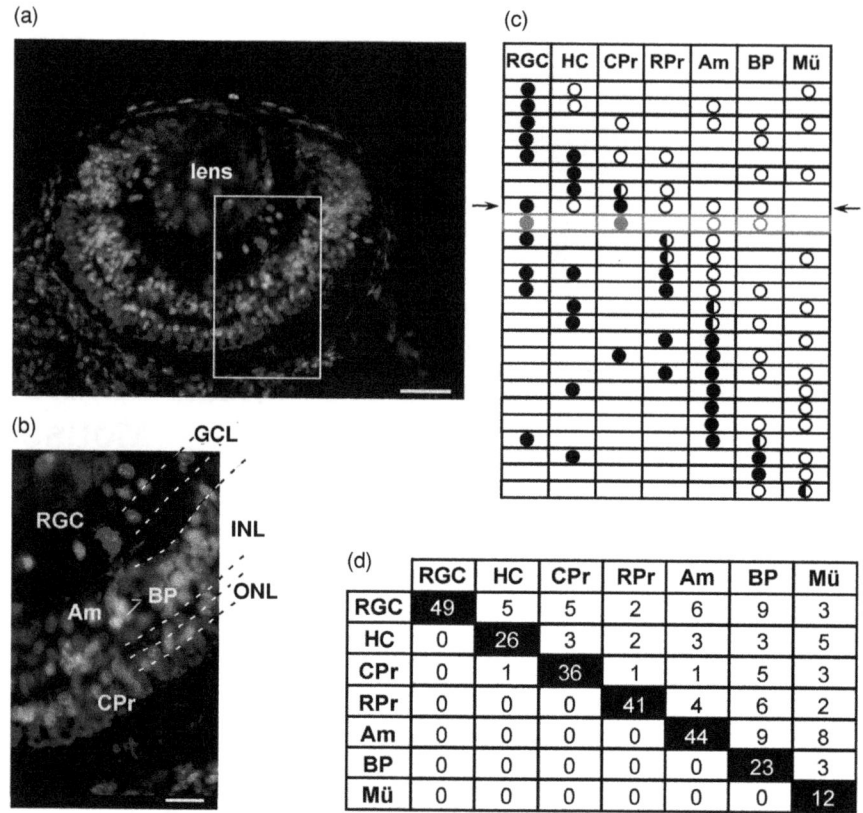

Plate 3

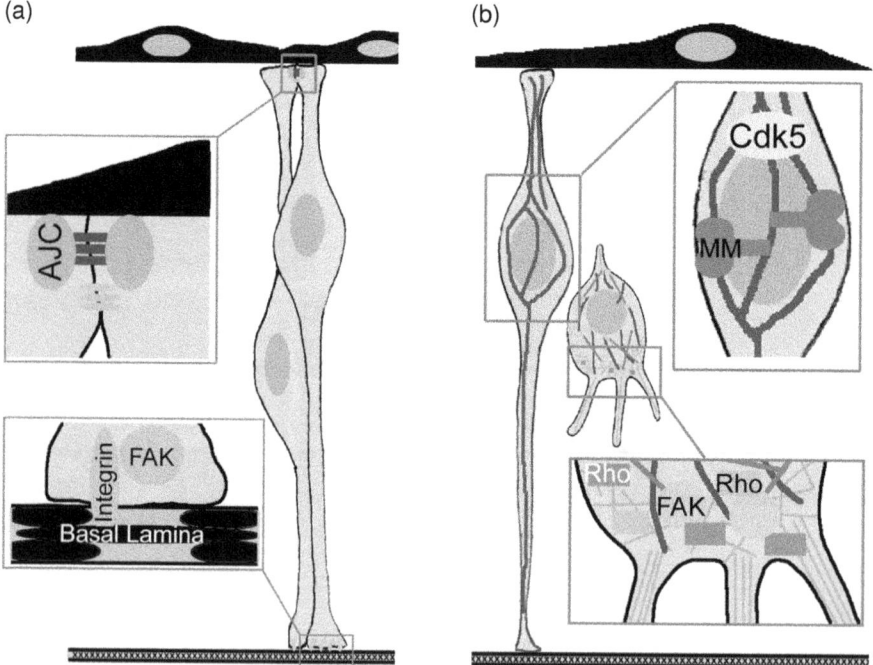

Plate 4

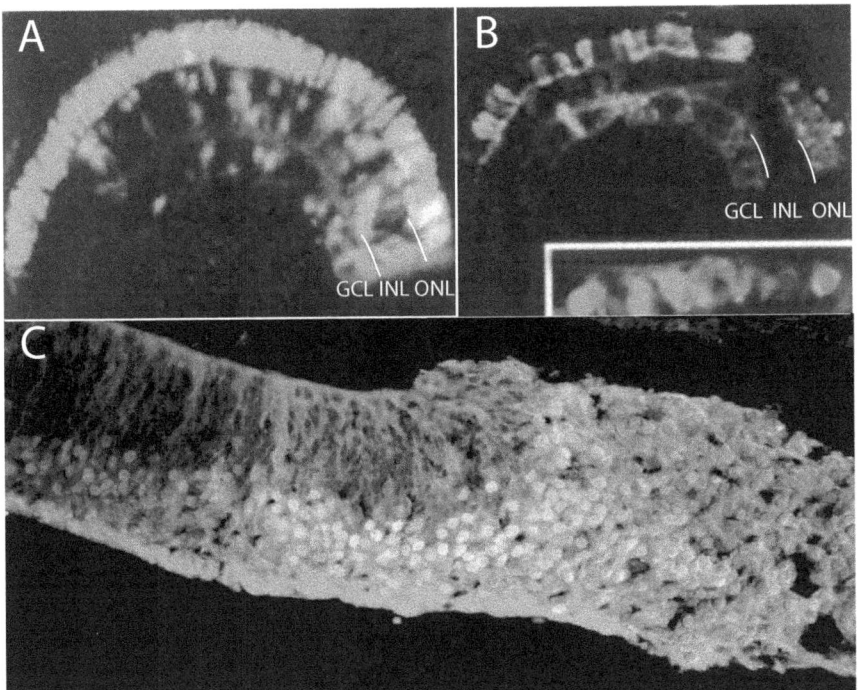

Plate 5

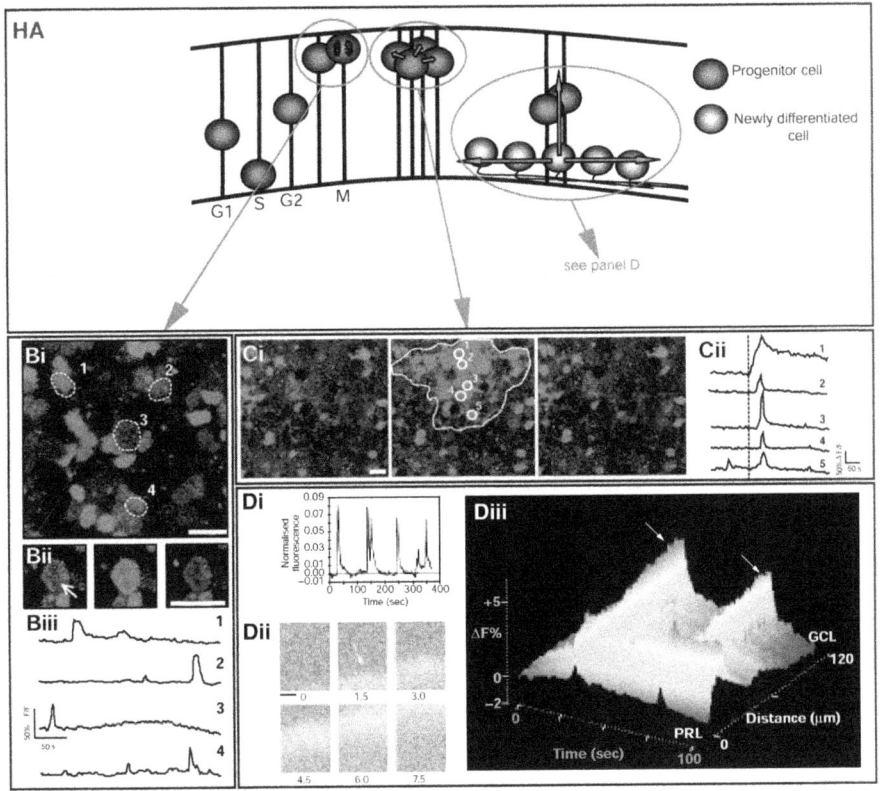

Plate 6

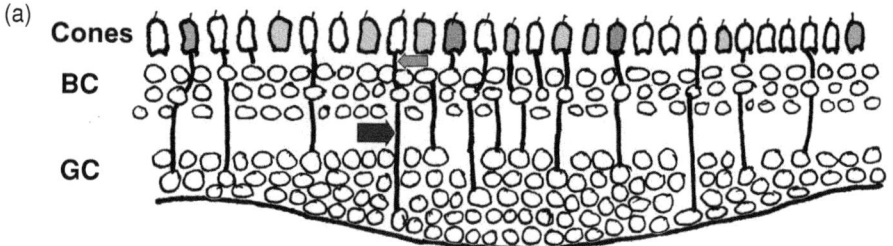

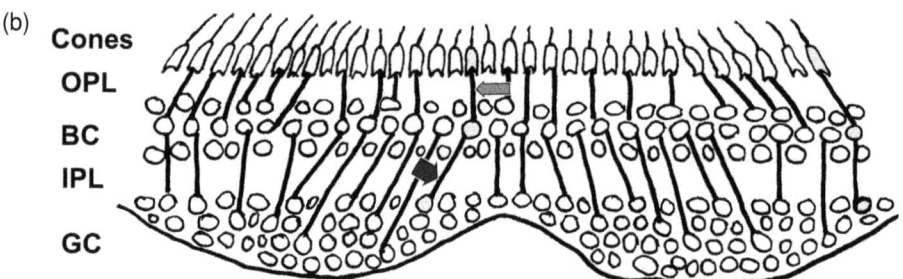

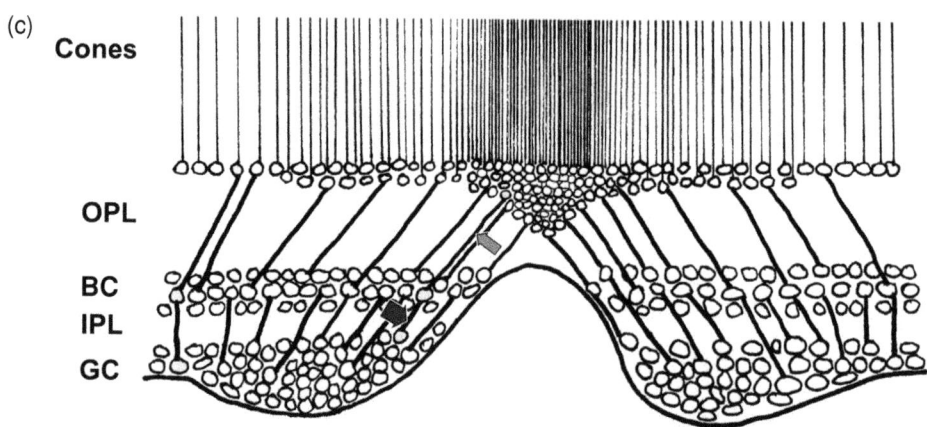

Plate 7

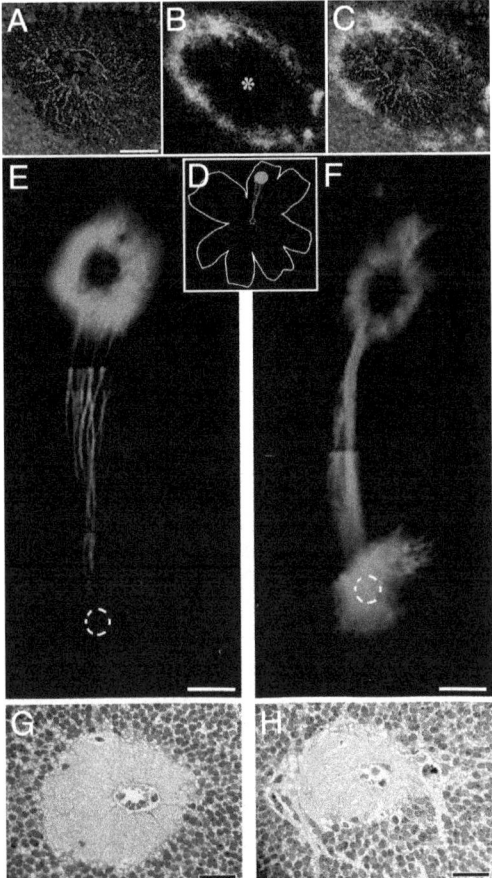

Plate 8

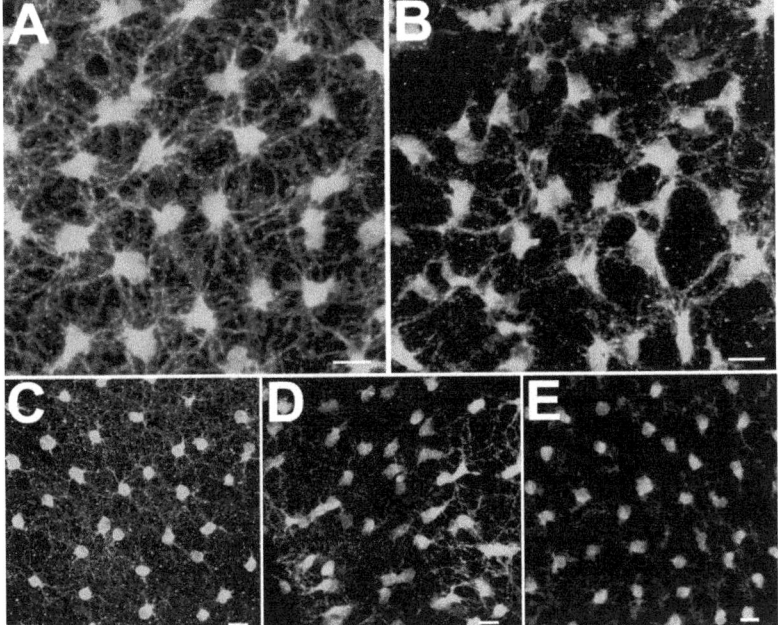

Plate 9

(a) Neighbour interactions determine shape of dendritic territories

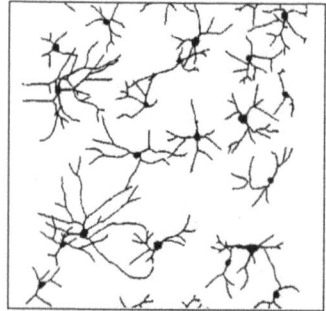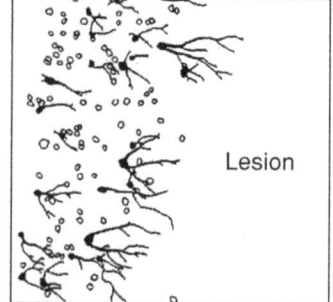

(b) Neighbours may interact via dendro-dendritic contacts

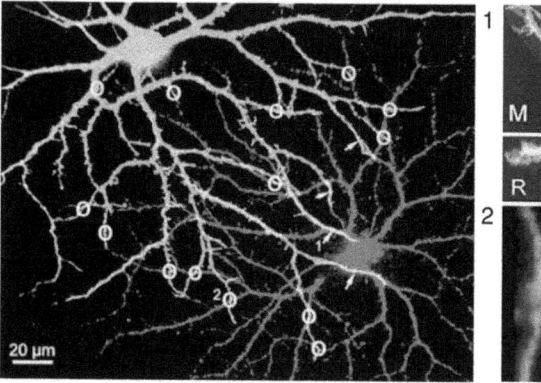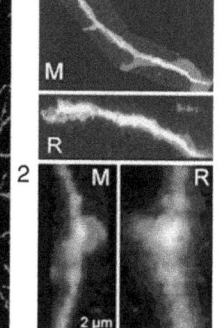

(c) Area of dendritic fields is intrinsically predefined

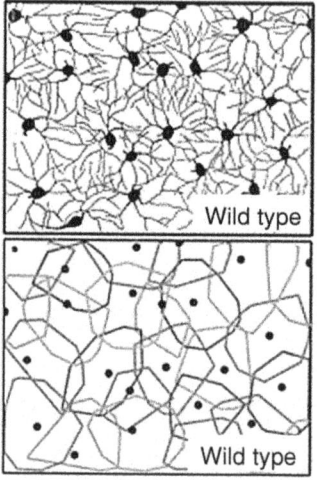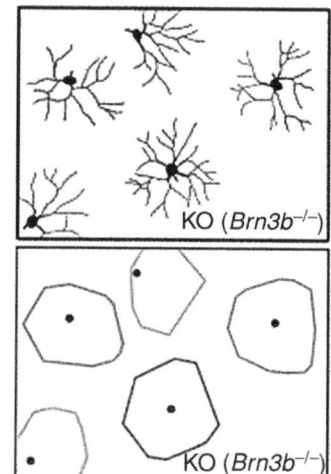

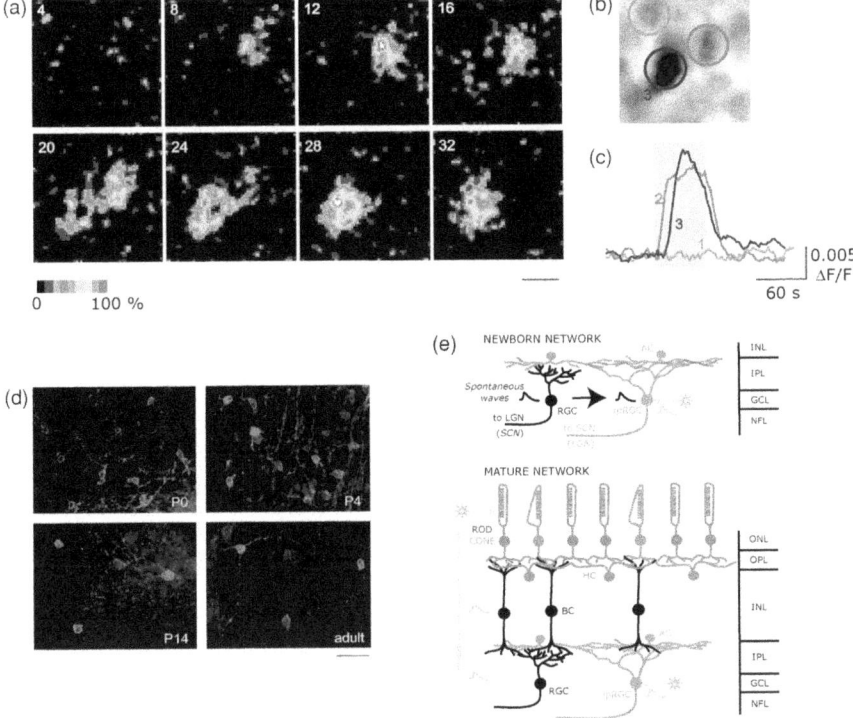

Plate 11

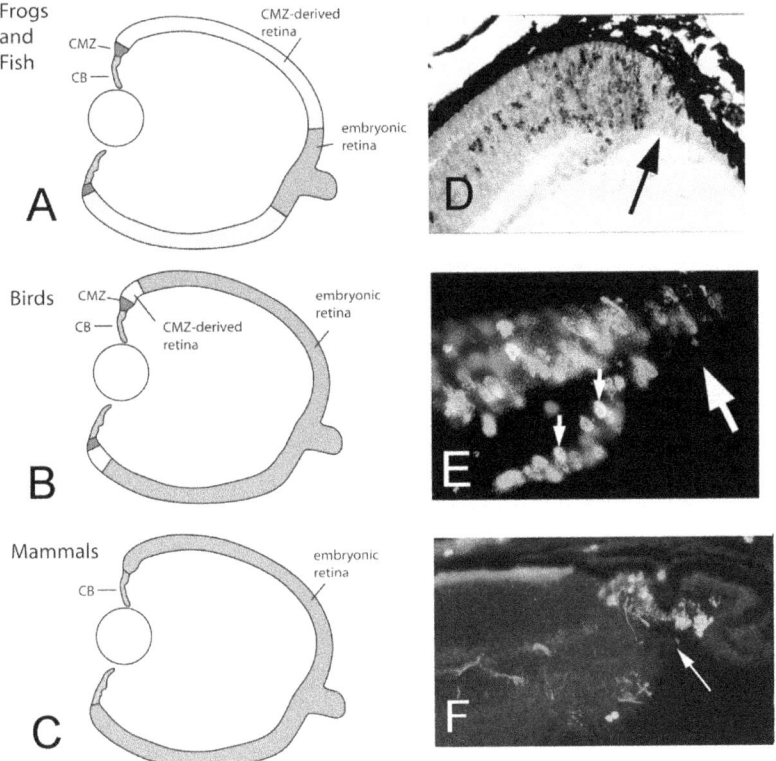

Plate 12

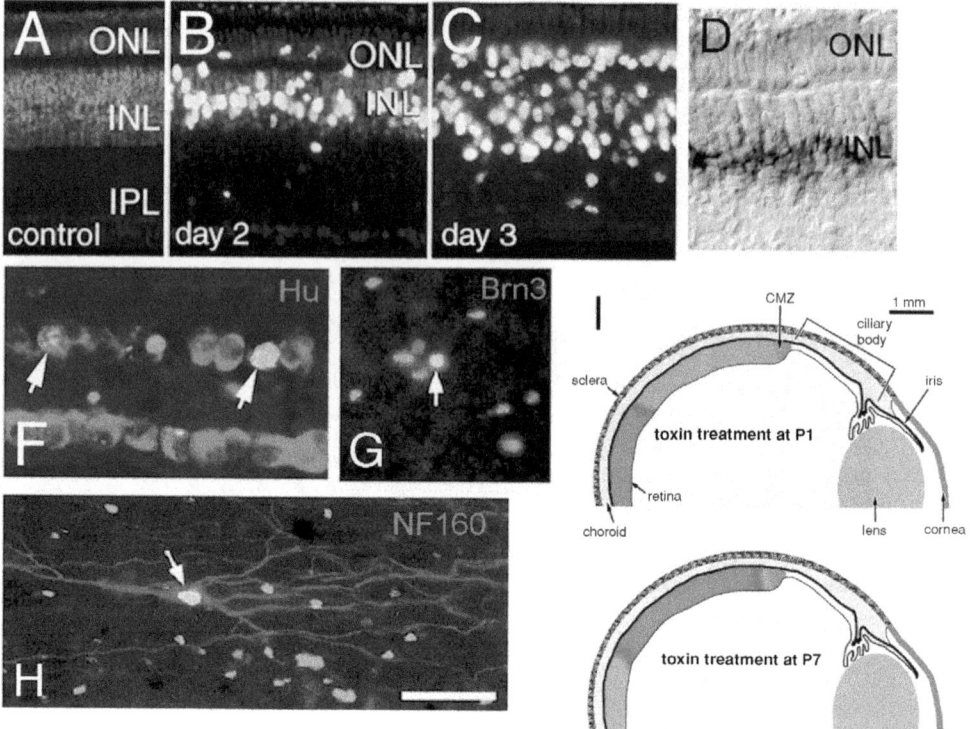

Plate 13

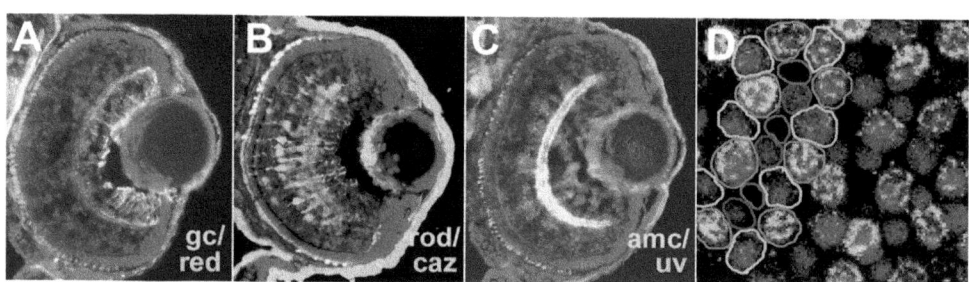

Plate 14

New perspectives

15

Regeneration: transdifferentiation and stem cells

Jennie Leigh Close and Thomas A. Reh

University of Washington, School of Medicine, Seattle, USA

15.1 Introduction

The study of regeneration in the vertebrate began with the pioneering experiments of Claude Bonnet in 1781. He found that if part of the eye of an adult newt (*Triturus cristatus*) was removed, a smaller, but complete, eye was regenerated within a few months. All of the ocular tissues, including the cornea, lens and retina, were capable of regenerating. Subsequent work by biologists, working primarily in the 1800s and early 1900s, characterized many critical features of the regeneration process in the eye. The molecular basis for this remarkable process is still not understood. However, recent progress in eye development research has spurred new lines of investigation into this question. In this review, we briefly discuss highlights of historical work and then focus on recent experiments in a variety of species that illustrates the complexities of the questions being investigated today.

15.2 A brief history of retinal regeneration

One of the first questions that arose historically concerning retinal regeneration in newts was the nature of the cells that provided the regenerated tissue. Early studies argued that a ring of cells at the peripheral retinal margin, what is now most commonly called the ciliary margin zone (CMZ), was the primary source of regenerated retina (Colucci, 1891 (cited in Keefe, 1973d); Fujita, 1913). Later studies confirmed the CMZ as a source of regeneration, but also demonstrated that the retinal pigmented epithelium (RPE) could regenerate neural retina in the posterior eye (Wachs, 1914, 1920). In the 1930s and 1940s, Leon Stone and his colleagues confirmed the earlier studies and, in a now classic experiment, isolated RPE from a newt and transplanted it into the vitreous of another animal. The transplanted pigmental epithelial sheet gave rise to new retina in the vitreous, clearly demonstrating that the RPE can serve as a source for retinal regeneration (Stone, 1950; Stone and Steinitz, 1957).

Stone and his colleagues also established several other features of retinal regeneration: they determined how many times the retina could regenerate in a single animal (four!) and they demonstrated that the regenerated retina can function in simple behavioural tasks

Retinal Development, ed. Evelyne Sernagor, Stephen Eglen, Bill Harris and Rachel Wong.
Published by Cambridge University Press. © Cambridge University Press 2006.

(Stone and Farthing, 1942). They were even able to exchange eyes between adult animals, with recovery of function. Studies in the 1960s and 1970s applied the technologies of electron microscopy and ^{3}H-thymidine autoradiography to characterize the regeneration process further (Hendrickson, 1964; Keefe, 1973a,b,c,d). These studies confirmed that both the CMZ and the RPE contributed to the regenerated retina, the former to the anterior retina and the latter to the posterior retina. Moreover, by this time it was recognized that the CMZ represents a zone of continued neurogenesis in adult amphibians and fish (Gaze and Watson, 1968; Hollyfield, 1968), and thus the regeneration from this zone represents an increase in normal retinal growth.

Studies of retinal regeneration in fish have shown that there is a source for genesis of new retinal neurons within the adult retina (Raymond and Hitchcock, 1997), in addition to the CMZ and the RPE. When small regions of gold fish retina are excised, or more widespread damage is induced by neurotoxins, new retinal neurons are generated. The regenerated neurons do not come from either the CMZ or the RPE, but rather arise from a stem-like cell within the retinal parenchyma. Recent work in the post-hatch chick retina is also consistent with an intrinsic source for retinal regeneration (Fischer and Reh, 2001; see Section 15.5). Taken together, the studies indicate that there are many different sources of retinal regeneration in various species (Figure 15.1). All of these sources share several common features: (1) although they are very different tissues histologically, they are all derived from the neural tube; (2) they all have the capacity to re-enter the cell cycle, while the retinal neurons are incapable of proliferation; and (3) they all express several genes typically present in retinal progenitors at some stage during the regeneration process. In this chapter, we will highlight the common features of the process with an eye towards understanding the essential elements of retinal regeneration.

15.3 Regeneration from the ciliary margin zone

As noted above, the CMZ was initially thought to be the only source for retinal regeneration. In addition, this is the one source of neurogenesis in the adult retina common to fish, amphibians and birds. The CMZ is in many ways similar to other regions of persistent progenitors in the CNS, like the subventricular zone and the hippocampal progenitor zone. The CMZ of fish and amphibians allows the retinal growth to keep pace with the overall growth of the eye and the animal. The CMZ of some fish and urodele amphibians continues to generate new retina throughout life, while neurogenesis at the CMZ of anuran amphibians slows considerably at metamorphosis.

The CMZ contributes to the regeneration of the anterior retina in cold-blooded vertebrates. In adult urodeles and larval anurans, proliferation of CMZ cells is up-regulated after retinal damage. Following retinal destruction through devascularization or surgical removal in newts, the CMZ generates a considerably greater amount of new retina than it would in the undamaged eye, and, as described above, early investigators believed that all of the regeneration of the retina was derived from the CMZ. In newts, Keefe reported that the

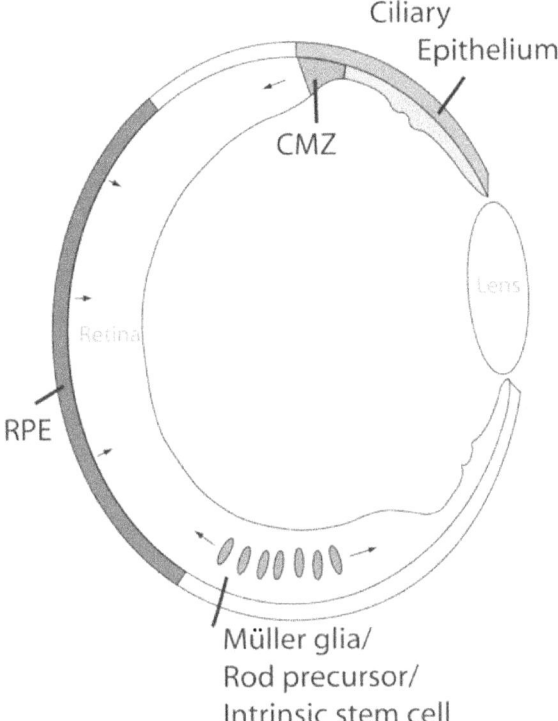

Figure 15.1 There are many different sources of retinal regeneration in various species. In amphibians, the retinal pigmented epithelium (RPE) is the primary source of new retinal progenitors or stem cells. The proliferating retinal progenitors of the ciliary margin zone (CMZ) generate new neurons in response to damage of the anterior retina in amphibians, fish and birds to a limited extent. The primary source for retinal regeneration in fish is an intraretinal source, likely either the rod progenitor or an intrinsic stem cell; however, in both fish and birds, there is evidence for Müller glial-mediated regeneration of retinal neurons. The various parts of the anterior eye, including both the pigmented and non-pigmented epithelia of the ciliary body, and the iris, can generate neurons of various types in mammals and birds, though this has not been shown in fish and amphibians.

number of ^{3}H-thymidine-labelled cells in the CMZ (pars ciliaris retinae) increased approximately tenfold by 20 days after retinal damage (Keefe, 1973a). Similarly, in *Rana* tadpoles, kainic acid neurotoxic lesions of the retina caused the number of ^{3}H-thymidine-labelled cells to more than double within two or three weeks following damage (Reh, 1987).

Goldfish and zebrafish also display an excellent capacity for retinal regeneration after different types of damage. However, most studies have focused on regeneration from a source within the retinal parenchyma. Nevertheless, evidence indicates that the CMZ survives neurotoxic damage, like ouabain, and that the anterior retina is regenerated from the CMZ in goldfish (Stenkamp et al., 2001). In addition, neurotoxic doses of 6-hydroxydopamine increase the width of the CMZ in fish by 50% (Negishi et al., 1988).

Does the CMZ contribute to retinal regeneration in birds and mammals? In the past few years, our lab has found evidence for a CMZ in the retina of post-hatch chicks and quails up to one year of age (Fischer and Reh, 2000; Kubota *et al.*, 2002). The chicken CMZ can be stimulated to regenerate new retinal neurons after damage as the fish and amphibian CMZs do, but only if exogenous growth factors are injected into the vitreous after neurotoxin treatment (Fischer *et al.*, 2002; Fischer and Reh, 2002). While the normal mammalian retina does not continue to generate new neurons at the retinal margin in adult animals, a recent analysis of a mouse mutant has provided evidence that the mammalian retina may be capable of regeneration (Fischer *et al.*, 2002; Fischer and Reh, 2002) at the margin (Figure 15.2). In mice with a single functioning allele of the Shh receptor, *patched*, a CMZ-like zone forms at the retinal margin (Moshiri and Reh, 2004). Cells labelled with 5'-Bromo-2'-deoxy-uridine (BrdU) can be found for weeks after the normal cessation of retinal neurogenesis. Moreover, the proliferation of the progenitors at the retinal marginal zone is increased significantly when the *patched* mutant mouse is crossed onto a retinal degeneration background.

The CMZ of persistent progenitors or retinal stem cells that exists in cold-blooded vertebrates is thus a key source of retinal regeneration that has progressively receded in homeothermic vertebrates like chickens and mice. While the CMZ can regenerate (and generate) all types of retinal neurons in frogs and fish, only a few cell types have been demonstrated to be regenerated from this zone in birds (Fischer and Reh, 2000; Fischer *et al.*, 2002), and even fewer have been found in mice (Moshiri and Reh, 2004). Thus, there is a limitation in both the quantity of neurogenesis at the retinal margin in warm-blooded vertebrates, as well as in the regeneration potential of the proliferating precursor cells and/or their local microenvironment. Regeneration from zones of persistent progenitors in other areas of the CNS, like the subventricular zone or the hippocampal progenitor zone, is similarly limited, with primarily granule neurons generated from both regions (Doetsch *et al*, 1999). Pyramidal cells in the hippocampus are not replaced by the hippocampal progenitors after lesions (Nakatomi *et al.*, 2002), and, in the song bird, only those neurons normally generated in the adult are capable of being regenerated (Scharff *et al.*, 2000). It should also be emphasized that, even though nearly all animals examined demonstrate some capacity for regeneration from the stem/progenitor cells at the retinal margin, and it is possible to stimulate proliferation and regeneration from these cells with intraocular growth factor injections, the regeneration is relatively local and confined to a few hundred microns of the marginal zone. There is no evidence for long-distance migration from this region to lesions in the central retina.

15.4 Regeneration from the retinal pigmented epithelium

As noted above, historically, the second cell type recognized as providing a source for retinal regeneration in the newt is the RPE. The RPE is morphologically very distinct from neural retina: it is a monolayer of cuboidal cells without any evidence for neurons in normal animals, and it provides critical functions for the rod and cone photoreceptors, including

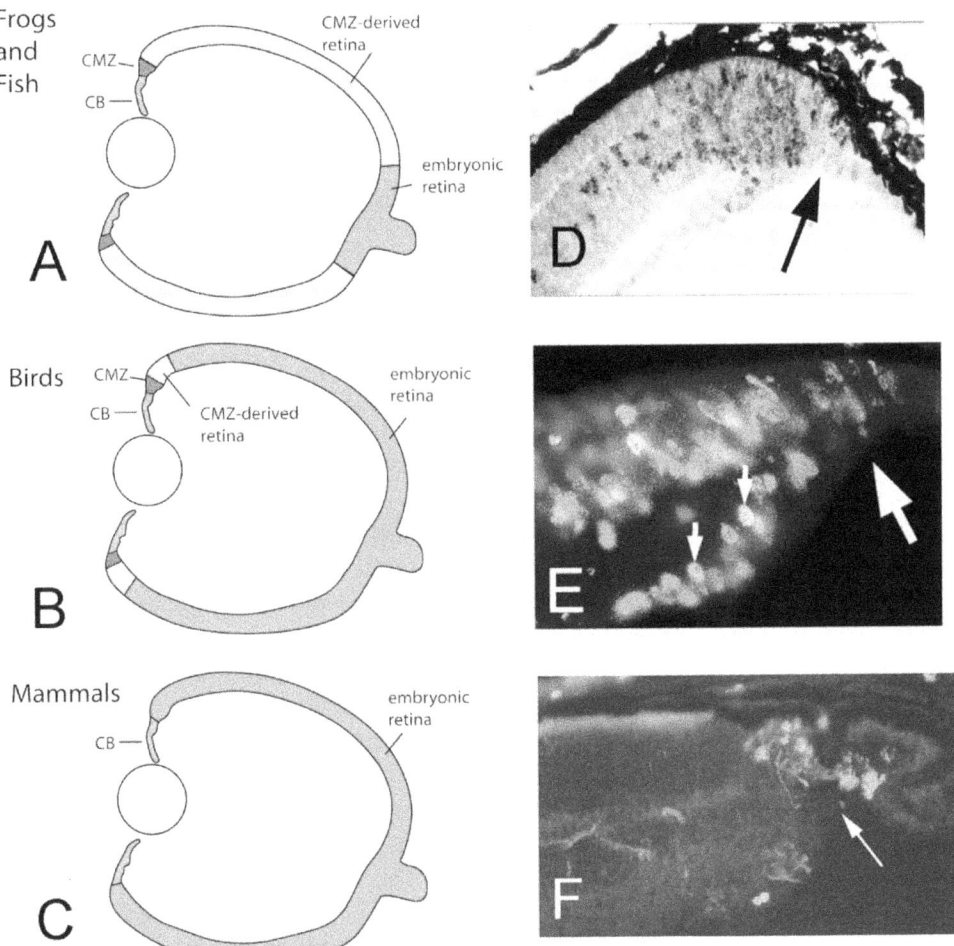

Figure 15.2 The ciliary margin zone (CMZ) is a region of retinal progenitors that persists into adulthood in fish, amphibians (A,D) and, to a more limited extent, birds (B,E). (C) Although the CMZ is not present in normal mice, in mice with a single functioning allele of the Sonic hedgehog receptor, *patched* (F), a CMZ-like zone forms at the retinal margin (modified from Moshiri and Reh, 2004). 5'-Bromo-2'-deoxy-uridine-labelled cells are green in E and F, while the silver grains in D show [^{3}H]-thymidine incorporation. The large arrow points to the retinal margin in D–F. CB, ciliary body. For colour version, see Plate 12.

outer segment phagocytosis and visual pigment regeneration. The common embryological origin of these tissues as the two layers of the optic cup, an evagination of the neural tube, belies their apparent morphological and physiological differences.

Retinal regeneration from the RPE in amphibians has been most extensively studied in newts, salamanders and axolotls, though anuran (frog) tadpoles are also able to regenerate retina from the RPE. The most common experimental design is to remove the retina, leaving the RPE intact. The RPE subsequently loses pigmentation, proliferates and generates two new epithelial layers, a pigmented layer and a non-pigmented layer. The non-pigmented layer begins to express genes typical of retinal progenitor cells and undergoes extensive cell division to produce sufficient new neurons for the new retina (Reh and Nagy, 1987, 1989). Thus, retinal regeneration occurs in two phases. In the first phase, the RPE cells dedifferentiate to become retinal progenitors. The second phase is much like normal development of the retina and follows a similar time course and developmental programme. In vitro experiments have confirmed that the RPE is the source of neural retinal tissue; RPE cells can dedifferentiate in vitro and generate new retinal neurons (Reh et al., 1987). The demonstration of RPE-cell dedifferentiation has been facilitated by the cells' pigmentation, which provides an intrinsic marker. The regeneration of retina from the RPE was therefore one of the first well-recognized examples of 'transdifferentiation' (Okada, 1981). However, it should be noted that this process involves extensive cell proliferation, and a direct conversion between a RPE cell and a retinal neuron is not typically observed.

The embryonic chick eye is capable of a similar form of RPE transdifferentiation (Coulombre and Coulombre, 1965; Park and Hollenberg, 1989; Pittack et al., 1991). Removal of the retina from a chick embryo within the first three to four days of incubation causes the RPE of the chick to undergo a transdifferentiation into neural retinal progenitors, very similar to that observed in the amphibian. The retina that forms from the RPE is laminated like normal retina, and contains relatively normal ratios of the different retinal cell types. Although the RPE can give rise to new retina in amphibians, embryonic chicks and embryonic mammals, there is an important difference in the process in amphibians that is critical for functional regeneration: the RPE generates normally oriented retina in the amphibian, but generates retina of inverted polarity in embryonic chicks and mammals. The reason for this difference is shown in Figure 15.3. The neural tube is an epithelium, with a basal surface and an apical surface. The involution of the optic vesicle that allows optic cup formation leads to the retinal and pigmented epithelia lying adjacent to one another, but with opposite polarities; i.e. their apical surfaces are adjacent. Ultimately, as the retina develops, the photoreceptor outer segments form at the apical surface of the retinal epithelium, while the RPE microvilli form at the apical surface of the RPE. During retinal regeneration in the amphibian, one of the first stages in the process is when the RPE cells detach from their basement membrane (Bruch's membrane) and round up (Keefe, 1973a,b,c,d; Reh and Nagy, 1987). The rounded-up RPE cells have apparently lost their polarity, but become repolarized when they make contact with remnants of the vitreal basement membrane. The progenitor cells that are produced by the RPE thus have the normal retinal polarity. By contrast, in embryonic chicks and mammals, there is direct conversion of the RPE cell layer into neural retina, without the cells ever detaching from Bruch's membrane (Coulombre and Coulombre, 1965; Park and Hollenberg, 1989; Pittack et al., 1991). As a result, the regenerated retina retains the polarity of the original RPE and thus the retina is inverted from its 'normal' orientation. In addition to the inverted polarity, the regenerated retina of chicks and mammals has another obvious problem: in regions where the RPE is converted to retina,

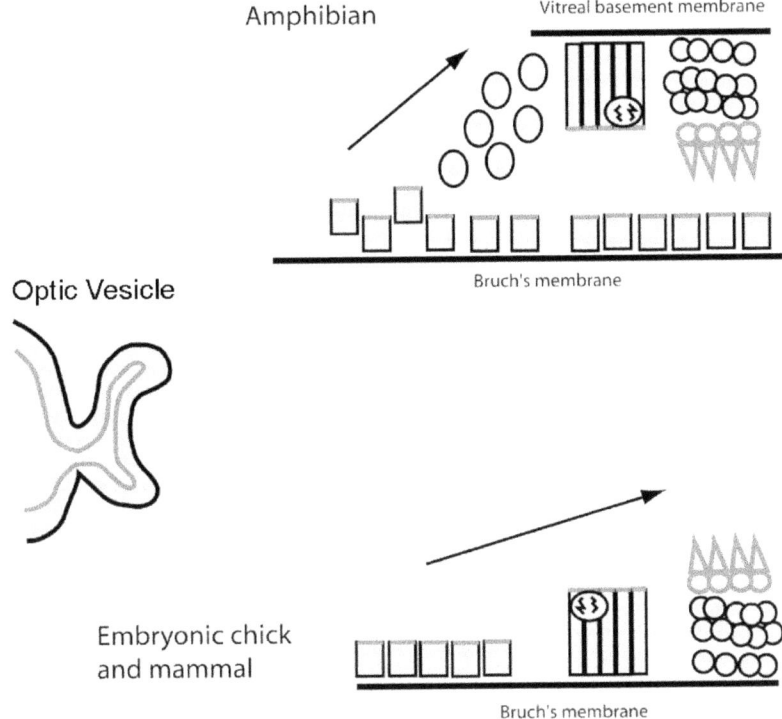

Figure 15.3 Retinal regeneration from the pigmented epithelium in amphibians and chick embryos results in an oppositely oriented retina. The optic vesicle is shown for orientation, with a basal surface and an apical surface. During retinal regeneration in the amphibian, one of the first stages in the process is when the RPE cells detach from their basement membrane (Bruch's membrane), migrate into the vitreous and become repolarized when they make contact with remnants of the vitreal basement membrane. The progenitor cells that are produced by the RPE thus have the normal retinal polarity. By contrast, in embryonic chicks and mammals, there is direct conversion of the RPE cell layer into neural retina, without the cells ever detaching from Bruch's membrane. As a result, the regenerated retina retains the polarity of the original RPE and thus the retina is inverted from its 'normal' orientation.

no RPE remains. Thus, while the RPE's ability to convert to neural retina in embryonic chicks and mammals indicates that some parts of the regeneration process of amphibians are retained in higher vertebrates, it is clear that there are important differences that need to be understood if the process is to be stimulated in the mammalian retina.

Although our understanding of the molecular events critical for regeneration are far from complete, several key events in the process appear to mirror aspects of normal development (Moshiri et al., 2004). Signalling molecules, including fibroblast growth factors (FGFs), bone morphogenetic proteins and hedgehogs, have been shown to be critical signalling molecules in both retinal development and regeneration. Coulombre first noted that the transdifferentiation of the RPE in chick embryos required that a small amount of neural retinal tissue remain in the eye (Coulombre and Coulombre, 1965). In *Xenopus* tadpoles and chick embryos, FGFs can replace the piece of neural retina and stimulate the process of regeneration from the RPE alone (Park and Hollenberg, 1989; Pittack *et al.*, 1991; Sakaguchi *et al.*, 1997). Fibroblast growth factor-1, FGF-2, or FGF-8 can all induce retinogenesis in the RPE (Vogel-Hopker *et al.*, 2000). By contrast, Sonic hedgehog (Shh) can inhibit regeneration from the RPE, and antagonists to Shh can promote more effective regeneration in chick embryos (Spence *et al.*, 2004). Thus, the same critical factors that regulate the fate decision between the RPE and the retina during development are also critical for the transition from RPE to retinal progenitors that occurs during regeneration.

15.5 Regeneration from intrinsic retinal sources

Cells within the retina itself can also provide a source of regeneration. The fish retina has rod precursor cells in the outer nuclear layer (Johns and Fernald, 1981) and quiescent stem cells in the inner nuclear layer throughout the retinal circumference (Julian *et al.*, 1998; Hitchcock *et al.*, 2004). When a small patch of central retina is surgically removed in a fish, the border of the excised retina forms a blastema, which proliferates to replace the removed retina (see Raymond and Hitchcock, 1997, for review). The source of the blastema is not the CMZ, which is located too far peripherally to the excision site. However, there are several other possible sources for the regenerating cells. While the RPE has also been ruled out in fish (Knight and Raymond, 1995), one or more of the following sources may be involved: (1) the intrinsic rod progenitors of the outer nuclear layer (ONL), which normally only generate rod photoreceptors; (2) a normally quiescent 'stem' cell, located in the retina (Otteson and Hitchcock, 2003); and/or (3) a subpopulation of the Müller glial cells. Since each of these cell types has the capacity to re-enter the cell cycle, the blastema may receive a contribution from them all.

The possibility that one of the intrinsic sources of regeneration might be the Müller glia was first raised by Braisted *et al.* (1994) in the goldfish. In these experiments, laser damage caused a proliferative response in the Müller cells and a concomitant replacement of the damaged cone photoreceptors. Recent evidence from a variety of different experimental systems indicates that cells once considered glial cells can act as neuronal progenitors in both the developing and mature mammalian brain (Doetsch *et al.*, 1999; Alvarez-Buylla *et al.*, 2000, 2001; Laywell *et al.*, 2000; Malatesta *et al.*, 2000; Noctor *et al.*, 2001, 2002; Heins *et al.*, 2002). Radial glia, whose primary function was once considered to provide the guidance scaffold for migrating neurons, produce many, if not all, of the neurons in the brain (Anthony *et al.*, 2004). In the brain of the mature mouse, Alvarez-Buylla and colleagues have established that glial fibrillary acidic protein-expressing cells are the progenitors/stem cells of the granule neurons of both the olfactory bulb and the hippocampal dentate gyrus (Merkle *et al.*, 2004).

Several lines of evidence support a close relationship between the Müller glia and retinal progenitors. Examination of multi-cell clones derived from late-stage retinal progenitors indicates that Müller glia and later-born retinal neurons are derived from a common precursor (Turner and Cepko, 1987). Notch signalling, known to be important for maintaining neuronal progenitors in an undifferentiated state, is also critical for promoting the Müller glial fate (Dorsky *et al.*, 1997; Henrique *et al.*, 1997; Furukawa *et al.*, 2000; Hojo *et al.*, 2000; Satow *et al.*, 2001). Recent gene-expression profiling studies also demonstrate a large degree of overlap in expressed genes between Müller glia and late retinal progenitors (Blackshaw *et al.*, 2003). Müller glia of humans and rodents alike express *Nestin*, a neural progenitor marker, both in vitro and in vivo after retinal damage (Cattaneo and McKay, 1990; Close *et al.*, 2000; Walcott and Provis, 2003).

In the post-hatch chick retina, several types of neurotoxic injury can stimulate a regenerative process from the Müller glia (Fischer and Reh, 2001). Following intraocular injections of *N*-methyl-D-aspartate (NMDA) in post-hatch chicks, which causes a rapid degeneration of most of the amacrine cells, and subsequent injections of BrdU to assay for cell proliferation, Fischer and Reh (2001) reported a burst of cell proliferation in the inner nuclear layer (INL), 48 hours after the toxin injection. The cells in the INL that re-entered the cycle were determined to be Müller glia by double-labelling with antibodies specific for Müller cells, like one against glutamine synthetase (GS). However, by three days post-toxin treatment, the majority of the BrdU-labelled cells also expressed several genes normally present in retinal progenitors (Figure 15.4), including chicken achaete-scute homologue 1 (*Cash1*), *Pax6* and *Chx10* (Fischer and Reh, 2001). When the fate of these dividing cells was analysed several days later, most of the BrDU-positive cells appeared to maintain the progenitor-like phenotype, continuing to coexpress *Pax6* and *Chx10*; approximately 20% of the BrdU-labelled cells went on to differentiate as GS-positive glia. However, a small percentage of the BrdU-labelled cells also differentiated into neurons, expressing neuronal markers characteristic of amacrine cells and bipolar cells (Figure 15.4).

The regenerative process in birds can also be stimulated with other types of neurotoxic damage (Fischer and Reh, 2001). Following treatment with either kainic acid or colchicine, Müller cells re-entered the cell cycle in similar proportion and timing to that observed after NMDA injection. However, unlike the NMDA-treated retinas, when injection of colchicine or kainic acid was combined with FGF-2 and insulin treatment, BrdU-labelled cells were later observed that were also positive for the ganglion cell markers neurofilament and Brn3.0 (Figure 15.4 G,H) (Fischer *et al.*, 2002). Thus, it appears that loss of a particular type of neuron stimulates its replacement in the post-hatch chick retina, albeit at a low frequency.

These studies show that the Müller glia of the vertebrate retina do retain some of the essential characteristics of neuronal progenitor cells: first, the ability to undergo cell division; second, the ability to express the markers of multipotent progenitor cells; and finally, the ability to generate progenitors, neurons and glia. However, there are several key differences between the regeneration observed in fish and chickens. First, only a few of the different types of retinal cells – amacrine cells, bipolar cells and ganglion cells – regenerate in the chick retina, whereas in the fish retina, all cells types are regenerated. Second, in the chick retina, most of the Müller glia that re-enter the cell cycle remain in an undifferentiated state, as *Pax6/Chx10*-expressing cells in the inner and outer nuclear layers, while in the fish, it appears that all the progeny of the proliferation, regardless of their source, differentiate into retinal neurons and glia; a large undifferentiated pool of cells has not been reported to

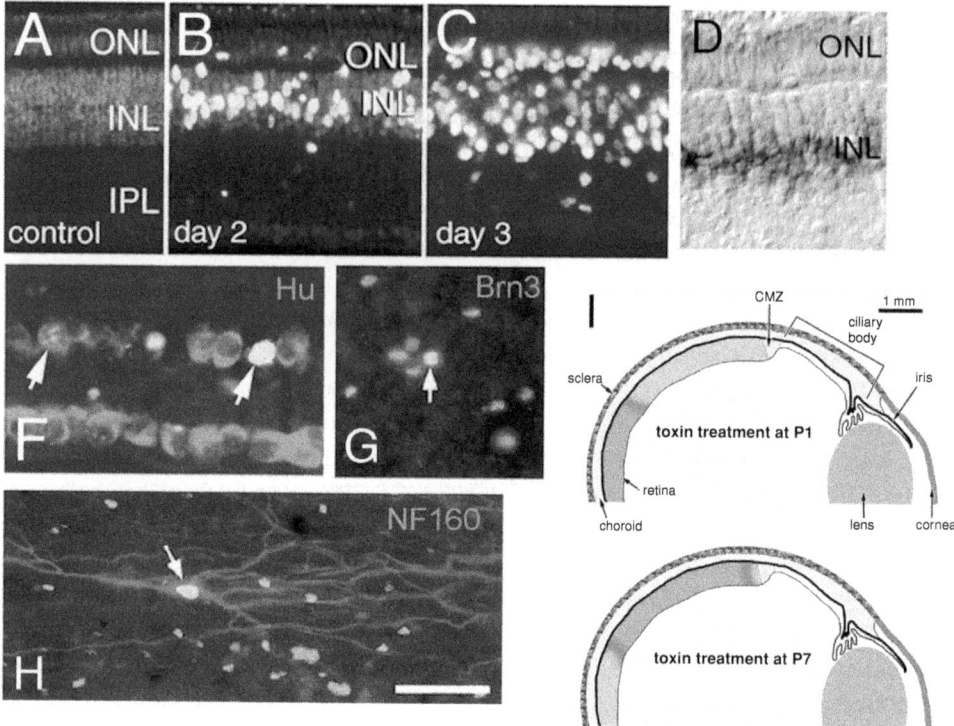

Figure 15.4 Regeneration in the chick retina following neurotoxic damage. (A–C) Injection of NMDA in post-hatch chicks causes a rapid degeneration of many of the amacrine cells, followed by a burst of cell proliferation in the inner nuclear layer, 48 hours after the toxin injection. (D) By three days post-toxin treatment, the majority of the BrdU-labelled cells also expressed several genes normally present in retinal progenitors, including chicken achaete-scute homologue 1 (*Cash1*). When the fate of these dividing cells was analysed several days later, a small percentage of the BrdU-labelled cells also differentiated into neurons, expressing neuronal markers characteristic of amacrine cells (F) and when treated with FGF, ganglion cells (G,H). BrdU is shown in green in (F–H), and double-labelled with the antibody shown on the panel. (I) Diagram showing the area of retina that shows regeneration at either postnatal day (P)1 or P7. INL, inner nuclear layer; IPL, inner plexiform layer; ONL, outer nuclear layer, (Modified from Fischer and Reh, 2001.) For colour version, see Plate 13.

remain after damage. Third, while the fish retina regenerates throughout its entire extent, the regeneration in the chick retina is regionally localized (Figure 15.4 I). In the new post-hatch chick, the central retina shows the regenerative response, but this moves peripherally as the animal grows, such that by postnatal day 30 regeneration only occurs in the peripheral retina (Figure 15.4 I).

The limited regeneration of new neurons in the chick still stands in contrast to an even more limited regenerative response in the mammalian retina. As noted above, regeneration in either the fish or avian retina begins with the re-entry of Müller glia into the cell cycle and/or the up-regulation of proliferation of a more slowly cycling intrinsic precursor/stem cells. In rats and mice, by contrast, retinal progenitor proliferation peaks around the day of birth and declines until approximately the end of the first postnatal week (Sidman, 1960; Young, 1985). After this time, there is little evidence for renewed proliferation of either progenitors or Müller glia in the mammalian retina. Even in cases of severe retinal degeneration, the Müller glia show only a low level of proliferation (Nork et al., 1986, 1987; Robison et al., 1990; Sueishi et al., 1996; Dyer and Cepko, 2000; Fariss et al., 2000; Moshiri and Reh, 2004). Thus, there appears to be a fundamental difference in the ability of different cells in the mammalian retina to re-enter the cell cycle after damage. This lack of Müller proliferation stands in contrast to observations from many studies that mammalian Müller glia grow and proliferate well in vitro (Sarthy, 1985). Recent evidence from our lab indicates that transforming growth factor (TGF)-β2 produced by postnatal retinal neurons inhibits Müller glial proliferation and may be responsible for this difference between mammals and other vertebrates (Close et al., 2005).

A key difference noted above between the fish retinal response to injury and that of the chick is the fate of the newly generated cells following injury. In the fish, nearly all of the newly generated cells differentiate into appropriate numbers of neurons and glia. By contrast, in the bird retina, most of the newly generated cells remain in an undifferentiated state for several weeks, while only a small number differentiate as retinal neurons. While growth factor treatment can change the types of neurons that differentiate, we have not found a treatment condition that substantially increases the rate of differentiation after neurotoxic injury. Moreover, transfection of Müller glia with the proneural basic helix-loop-helix transcription factor NeuroD1 promotes the expression of certain neuronal genes, such as the photoreceptor marker visinin, in vitro (Fischer et al., 2004), but fails to activate a repertoire of gene expression that generates a full complement of retinal neurons. Instead, the majority of the progeny of Müller glia in the post-hatch chick retina that arise following neurotoxin treatment do not differentiate into neurons either in vivo or in vitro.

Despite the similarities between retinal progenitors and Müller glia noted above, there are clear differences as well. First, although the cell bodies of both Müller glia and retinal progenitors are situated in the retinal INL, and their processes span the width of the retina, Müller glia have much more complex branching of their vitreal processes. Also, like astrocytes, Müller glia buffer intercellular ion concentrations and recycle neurotransmitters released by their neuronal neighbours (Newman and Reichenbach, 1996). Therefore, Müller glia express high levels of genes involved in these support and homeostatic functions, such as *Glast* and *GS*; progenitor cells do not. The work from the fish and bird retinas indicate that these cells may play important roles in the regenerative processes, but much more research into the limits on their proliferation and phenotypic plasticity is necessary before we understand the limits to intrinsic regeneration in the mammalian retina.

15.6 Neogenesis of neurons from the ciliary epithelium?

Although the CMZ is a specialized region for the genesis of new retinal neurons in frogs, fish and birds, as noted above, this zone is not found in the mammalian retina. Nevertheless, there have been several recent reports demonstrating that the anterior eye of mammals contains cells that can at least express neuronal antigens, if not differentiate into functional neurons. There are several regions of the anterior eye that have been implicated as a source of 'retinal stem cells' including the ciliary body and the iris (Ahmad *et al.*, 2000; Tropepe *et al.*, 2000). Both the iris and the ciliary epithelium of the ciliary body are derived from the anterior rim of the optic cup and have the same two-layered structure, a pigmented layer and a non-pigmented layer. The ciliary epithelium can be further subdivided into the more proximal pars plana and the distal, folded pars plicata.

The cells of the anterior eye undergo considerable phenotypic plasticity in vitro. When the mammalian ciliary body and iris are dissociated and maintained in vitro, some of the pigmented cells undergo transdifferentiation, similar to the transdifferentiation of the RPE of amphibians, and give rise to unpigmented spheres (Tropepe *et al.*, 2000). When placed under differentiation conditions, these anterior eye cells express neuronal antigens, and some retinal-specific antigens, and so they have been labelled retinal stem cells. Due to the small size of the mouse eye, however, it has been difficult to identify the specific region from which these pigmented cells are derived, since this area contains pigmented cells from the iris, ciliary epithelium and the neural crest-derived choroids. Experiments using explant cultures of iris have shown that this region provides at least one source for neural precursors. Haruta *et al.* (2001) found that some cells within the mammalian iris, either the pigmented or non-pigmented epithelial tissue, have the capacity for generating neuronal-like cells in vitro. Monolayer cultures of the iris from adult rats contained cells that express neurofilament immunoreactivity and had a distinctive morphology characteristic of neurons. Although they did not demonstrate that these neurons were generated *de novo*, when they expressed the gene *Crx* in these cells with a retroviral construct, they found that iris-derived cells could express *rhodopsin*, and *recoverin*, two genes characteristic of photoreceptors. Moreover, the infected cells assumed the small round morphology characteristic of photoreceptors. More recently, Akagi *et al.* (2004) have found that the cells of the iris and the ciliary epithelium of adult rats could be grown as spheres. In contrast to the report by Tropepe *et al.* (2000), however, they found that these cells did not express photoreceptor-specific genes unless transfected with either *Otx2* or *Crx*. The human eye also has the potential to express neuronal antigens in the pigmented cells of the anterior eye. Coles *et al.* (2004) reported that cells from the human iris and ciliary epithelium also express neuronal markers when dissociated and cultured, though the *rhodopsin*- and *rom1*-expressing cells do not take on the characteristic morphology of cultured rod photoreceptors, as they appear to after transfection with *Crx* or *Otx2*. Taken together, these results suggest that the iris and ciliary epithelium of the adult mammalian retina will grow in dissociated cell cultures and can respond to specific transcription factors to transcribe photoreceptor-specific genes.

Do the cells from these most-anterior eye tissues normally continue to regenerate? In vivo reports also indicate that the cells of the ciliary epithelium are capable of giving rise to neuronal cells. Intraocular injections of insulin or FGF-2 into the eye of post-hatch chickens results in the production of neuronal cells in both the pars plana and the pars plicata of the ciliary epithelium (Fischer and Reh, 2003). These cells are immunoreactive for a variety

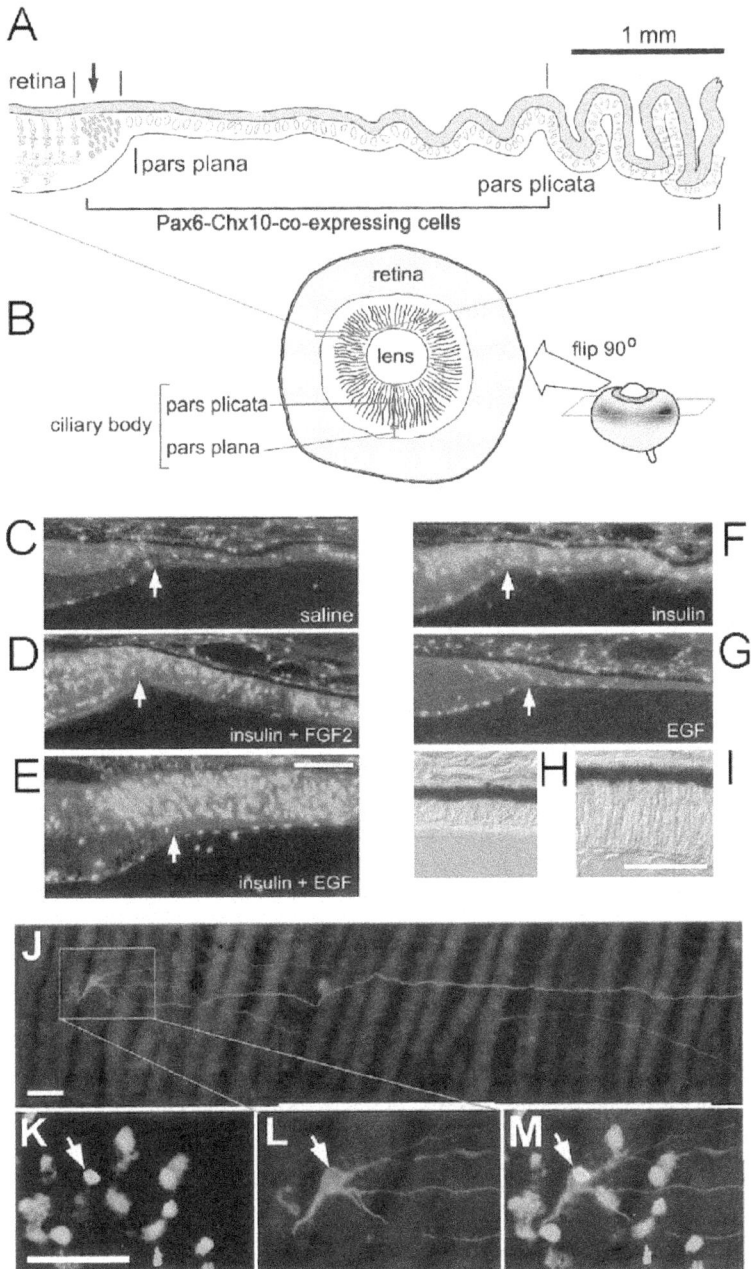

Figure 15.5 The ciliary epithelium is capable of generating neuronal cells. (A,B) Diagram showing the relationships among the anterior eye structures. (C–I) Intraocular injections of insulin, FGF-2 or epidermal growth factor (EGF) into the eye of post-hatch chickens results in the stimulation of proliferation in both the pars plana and the pars plicata of the ciliary epithelium. The combination of insulin and EGF causes the greatest amount of proliferation; however, neuronal differentiation of the newly generated cells was greatest in the FGF + insulin condition. (J–M) Example of a newly generated (BrdU) neuron (NF160) in the ciliary epithelium after intraocular injection of insulin and FGF. The cells typically have complex morphologies, and many have axons that extend for hundreds of microns through the ciliary epithelium. (Modified from Fischer and Reh, 2003.)

of neuron-specific antigens, including Hu, Islet1, RA4, β3-tubulin and calretinin. The cells typically have complex morphologies, and many have axons that extend for hundreds of microns through the ciliary epithelium (Figure 15.5). In the monkey, there is also evidence for neuronal differentiation in the ciliary epithelium of older adults, though it is not clear whether this is due to existing cells expressing neuronal antigens, or through *de novo* generation of new neurons, as occurs in the avian eye (Fischer *et al.*, 2001). Together with the in vitro studies, these results support the possibility that the ciliary epithelium and the iris contain cells capable of differentiating towards a neuronal lineage when stimulated appropriately. One reason for this may be that the cells of both the ciliary epithelium (particularly the pars plana) are less differentiated, and express genes associated with developing retina. It may also be that because the ciliary epithelium is immediately adjacent to the CMZ, that some of the same factors that maintain the immaturity of this region of the eye in lower vertebrates extend into the ciliary epithelium.

15.7 Concluding remarks

In this review, we have attempted to show the diversity of potential sources of retinal regeneration. It is interesting that so many different strategies appear to have been used to repair retinal damage, particularly in cold-blooded vertebrates, yet all of these sources exist in only a rudimentary fashion in homeothermic vertebrates. Nevertheless, vestiges of nearly all of the potential sources of regeneration appear to persist in birds and mammals, and progress is being made towards stimulating the process of regeneration in these species. Future work in this area will likely concentrate on understanding the molecular basis of the limitations on regeneration in higher vertebrates, as well as on developing a better understanding of the mechanisms by which the different sources of retinal regeneration converge on a common outcome.

Acknowledgements

Thanks to Paige Etter and Melissa Lee Phillips for her helpful comments on the manuscript. The authors acknowledge support from National Institutes of Health grant EY13475.

References

Ahmad, I., Tang, L. and Pham, H. (2000). Identification of neural progenitors in the adult mammalian eye. *Biochem. Biophys. Res. Commun.*, **270**, 517–21.

Akagi, T., Mandai, M., Ooto, S. *et al.* (2004). *Otx2* homeobox gene induces photoreceptor-specific phenotypes in cells derived from adult iris and ciliary tissue. *Invest. Ophthalmol. Vis. Sci.*, **45**, 4570–5.

Alvarez-Buylla, A., Herrera, D. G. and Wichterle, H. (2000). The subventricular zone: source of neuronal precursors for brain repair. *Prog. Brain Res.*, **127**, 1–11.

Alvarez-Buylla, A., Garcia-Verdugo, J. M. and Tramontin, A. D. (2001). A unified hypothesis on the lineage of neural stem cells. *Nat. Rev. Neurosci.*, **2**, 287–93.

Anthony, T. E., Klein, C., Fishell, G. and Heintz, N. (2004). Radial glia serve as neuronal progenitors in all regions of the central nervous system. *Neuron*, **41**, 881–90.
Blackshaw, S., Kuo, W. P., Park, P. J. et al. (2003). MicroSAGE is highly representative and reproducible but reveals major differences in gene expression among samples obtained from similar tissues. *Genome Biol.*, **4**, R17.
Braisted, J. E., Essman, T. F. and Raymond, P. A. (1994). Selective regeneration of photoreceptors in goldfish retina. *Development*, **120**, 2409–19.
Cattaneo, E. and McKay, R. (1990). Proliferation and differentiation of neuronal stem cells regulated by nerve growth factor. *Nature*, **347**, 762–5.
Close, J., Fischer, A. J., Roberts, M. and Reh, T. A. (2000). Damage induced expression of progenitor and neural cell markers in adult rodent retinal Muller glia. *Society for Neuroscience Poster*, 601.16.
Close, J. L., Gumuscu, B. and Reh, T. A. (2005). Retinal neurons regulate proliferation of postnatal progenitors and Müller glia in the rat retina via TGFβ signaling. *Development*, **132**(13), 3015–2.
Coles, B. L., Angenieux, B., Inoue, T. et al. (2004). Facile isolation and the characterization of human retinal stem cells. *Proc. Natl. Acad. Sci. U. S. A.*, **101**, 15 772–7.
Colucci, V. S. (1891). Sulla rigenerazione parziale dell'occhio nei Tritoni. Istogenesi e siluppo. *Mem. della Roy. Acad. della Science dell'Inst. di Bologna*, **Ser V 1**, 593–629. (cited in Keefe, 1973a)
Coulombre, J. L. and Coulombre, A. J. (1965). Regeneration of neural retina from the pigmented epithelium in the chick embryo. *Dev. Biol.*, **12**, 79–92.
Doetsch, F., Caille, I., Lim, D. A., Garcia-Verdugo, J. M. and Alvarez-Buylla, A. (1999). Subventricular zone astrocytes are neural stem cells in the adult mammalian brain. *Cell*, **97**, 703–16.
Dorsky, R. I., Chang, W. S., Rapaport, D. H. and Harris, W. A. (1997). Regulation of neuronal diversity in the Xenopus retina by Delta signaling. *Nature*, **385**, 67–70.
Dyer, M. A. and Cepko, C. L. (2000). Control of Müller glial cell proliferation and activation following retinal injury. *Nat. Neurosci.*, **3**, 873–80.
Fariss, R. N., Li, Z. Y. and Milam, A. H. (2000). Abnormalities in rod photoreceptors, amacrine cells, and horizontal cells in human retinae with retinitis pigmentosa. *Am. J. Ophthalmol.*, **129**, 215–23.
Fischer, A. J. and Reh, T. A. (2000). Identification of a proliferating marginal zone of retinal progenitors in postnatal chickens. *Dev. Biol.*, **220**, 197–210.
Fischer, A. J. and Reh, T. A. (2001). Müller glia are a potential source of neural regeneration in the postnatal chicken retina. *Nat. Neurosci.*, **4**, 247–52.
Fischer, A. J. and Reh, T. A. (2002). Exogenous growth factors stimulate the regeneration of ganglion cells in the chicken retina. *Dev. Biol.*, **251**, 367–79.
Fischer, A. J. and Reh, T. A. (2003). Growth factors induce neurogenesis in the ciliary body. *Dev. Biol.*, **259**, 225–40.
Fischer, A. J., Hendrickson, A. and Reh, T. A. (2001). Immunocytochemical characterization of cysts in the peripheral retina and pars plana of the adult primate. *Invest. Ophthalmol. Vis. Sci.*, **42**, 3256–63.
Fischer, A. J., Dierks, B. D. and Reh, T. A. (2002). Exogenous growth factors induce the production of ganglion cells at the retinal margin. *Development*, **129**, 2283–91.
Fischer, A. J., Wang, S. Z. and Reh, T. A. (2004). NeuroD induces the expression of visinin and calretinin by proliferating cells derived from toxin-damaged chicken retina. *Dev. Dyn.*, **229**, 555–63.

Fujita, H. (1913). Regenerations prozess der Netzhaut des Tritons und des Frosches. *Arch. vergl. Ophthalmol.*, **3**, 356–68.
Furukawa, T., Mukherjee, S., Bao, Z. Z., Morrow, E. M. and Cepko, C. L. (2000). rax, Hes1, and notch1 promote the formation of Müller glia by postnatal retinal progenitor cells. *Neuron*, **26**, 383–94.
Gaze, R. and Watson, W. E. (1968). *Growth of the Nervous System*. London: Ciba Foundation Symposium.
Haruta, M., Kosaka, M., Kanegae, Y. et al. (2001). Induction of photoreceptor-specific phenotypes in adult mammalian iris tissue. *Nat. Neurosci.*, **4**, 1163–4.
Heins, N., Malatesta, P., Cecconi, F. et al. (2002). Glial cells generate neurons: the role of the transcription factor *Pax6*. *Nat. Neurosci.*, **5**, 308–15.
Henrique, D., Hirsinger, E., Adam, J. et al. (1997). Maintenance of neuroepithelial progenitor cells by Delta-Notch signaling in the embryonic chick retina. *Curr. Biol.*, **7**, 661–70.
Hendrickson, A. E. (1964). Regeneration of the retina of the newt: an electron microscopic study. Unpublished Ph.D. thesis, University of Washington, Seattle.
Hitchcock, P., Ochocinska, M., Sieh, A. and Otteson, D. (2004). Persistent and injury-induced neurogenesis in the vertebrate retina. *Prog. Retin. Eye Res.*, **23**, 183–94.
Hojo, M., Ohtsuka, T., Hashimoto, N. et al. (2000). Glial cell fate specification modulated by the bHLH gene *Hes5* in mouse retina. *Development*, **127**, 2515–22.
Hollyfield, J. G. (1968). Differential addition of cells to the retina in *Rana pipiens* tadpoles. *Dev. Biol.*, **18**, 163–79.
Johns, P. R. and Fernald, R. D. (1981). Genesis of rods in teleost fish retina. *Nature*, **293**, 141–2.
Julian, D., Ennis, K. and Korenbrot, J. I. (1998). Birth and fate of proliferative cells in the inner nuclear layer of the mature fish retina. *J. Comp. Neurol.*, **394**, 271–82.
Keefe, J. R. (1973a). An analysis of urodelian retinal regeneration. I. Studies of the cellular source of retinal regeneration in *Notophthalmus viridescens* utilizing 3H-thymidine and colchicine. *J. Exp. Zool.*, **184**, 185–206.
Keefe, J. R. (1973b). An analysis of urodelian retinal regeneration. II. Ultrastructural features of retinal regeneration in *Notophthalmus viridescens*. *J. Exp. Zool.*, **184**, 207–32.
Keefe, J. R. (1973c). An analysis of urodelian retinal regeneration. 3. Degradation of extruded malanin granules in *Notophthalmus viridescens*. *J. Exp. Zool.*, **184**, 233–8.
Keefe, J. R. (1973d). An analysis of urodelian retinal regeneration. IV. Studies of the cellular source of retinal regeneration in *Triturus cristatus carnifex* using 3H-thymidine. *J. Exp. Zool.*, **184**, 239–58.
Knight, J. K. and Raymond, P. A. (1995). Retinal pigmented epithelium does not transdifferentiate in adult goldfish. *J. Neurobiol.*, **27**, 447–56.
Kubota, R., Hokoc, J. N., Moshiri, A., McGuire, C. and Reh, T. A. (2002). A comparative study of neurogenesis in the retinal ciliary marginal zone of homeothermic vertebrates. *Brain Res. Dev. Brain Res.*, **134**, 31–41.
Laywell, E. D., Rakic, P., Kukekov, V. G., Holland, E. C. and Steindler, D. A. (2000). Identification of a multipotent astrocytic stem cell in the immature and adult mouse brain. *Proc. Natl. Acad. Sci. U. S. A.*, **97**, 13 883–8.
Malatesta, P., Hartfuss, E. and Gotz, M. (2000). Isolation of radial glial cells by fluorescent-activated cell sorting reveals a neuronal lineage. *Development*, **127**, 5253–63.

Merkle, F. T., Tramontin, A. D., Garcia-Verdugo, J. M. and Alvarez-Buylla, A. (2004). Radial glia give rise to adult neural stem cells in the subventricular zone. *Proc. Natl. Acad. Sci. U. S. A.*, **101**, 17528–32.

Moshiri, A. and Reh, T. A. (2004). Persistent progenitors at the retinal margin of ptc+/− mice. *J. Neurosci.*, **24**, 229–37.

Moshiri, A., Close, J. and Reh, T. A. (2004). Retinal stem cells and regeneration. *Int. J. Dev. Biol.*, **48**, 1003–14.

Nakatomi, H., Kuriu, T., Okabe, S. (2002). Regeneration of hippocampal pyramidal neurons after ischemic brain injury by recruitment of endogenous neural progenitors. *Cell*, **110**, 429–41.

Negishi, K., Teranishi, T., Kato, S. and Nakamura, Y. (1988). Immunohistochemical and autoradiographic studies on retinal regeneration in teleost fish. *Neurosci. Res.*, **Suppl. 8**, S43–57.

Newman, E. and Reichenbach, A. (1996). The Müller cell: a functional element of the retina. *Trends Neurosci.*, **19**, 307–12.

Noctor, S. C., Flint, A. C., Weissman, T. A., Dammerman, R. S. and Kriegstein, A. R. (2001). Neurons derived from radial glial cells establish radial units in neocortex. *Nature*, **409**, 714–20.

Noctor, S. C., Flint, A. C., Weissman, T. A. *et al.* (2002). Dividing precursor cells of the embryonic cortical ventricular zone have morphological and molecular characteristics of radial glia. *J. Neurosci.*, **22**, 3161–73.

Nork, T. M., Ghobrial, M. W., Peyman, G. A. and Tso, M. O. (1986). Massive retinal gliosis. A reactive proliferation of Müller cells. *Arch. Ophthalmol.*, **104**, 1383–9.

Nork, T. M., Wallow, I. H., Sramek, S. J. and Anderson, G. (1987). Müller's cell involvement in proliferative diabetic retinopathy. *Arch. Ophthalmol.*, **105**, 1424–9.

Okada, T. S. (1981). Phenotypic expression of embryonic neural retinal cells in cell culture. *Vis. Res.*, **21**, 83–6.

Otteson, D. C. and Hitchcock, P. F. (2003). Stem cells in the teleost retina: persistent neurogenesis and injury-induced regeneration. *Vis. Res.*, **43**, 927–36.

Park, C. M. and Hollenberg, M. J. (1989). Basic fibroblast growth factor induces retinal regeneration *in vivo*. *Dev. Biol.*, **134**, 201–5.

Pittack, C., Jones, M. and Reh, T. A. (1991). Basic fibroblast growth factor induces retinal pigment epithelium to generate neural retina in vitro. *Development*, **113**, 577–88.

Raymond, P. A. and Hitchcock, P. F. (1997). Retinal regeneration: common principles but a diversity of mechanisms. *Adv. Neurol.*, **72**, 171–84.

Reh, T. A. (1987). Cell-specific regulation of neuronal production in the larval frog retina. *J. Neurosci.*, **7**, 3317–24.

Reh, T. A. and Nagy, T. (1987). A possible role for the vascular membrane in retinal regeneration in *Rana catesbienna* tadpoles. *Dev. Biol.*, **122**, 471–82.

Reh, T. A. and Nagy, T. (1989). Characterization of Rana germinal neuroepithelial cells in normal and regenerating retina. *Neurosci. Res.*, **Suppl 10**, S151–61.

Reh, T. A., Nagy, T. and Gretton, H. (1987). Retinal pigmented epitheial cells induced to transdifferentiate to neurons by laminin. *Nature*, **330**, 68–71.

Robison, W. G., Jr, Tillis, T. N., Laver, N. and Kinoshita, J. H. (1990). Diabetes-related histopathologies of the rat retina prevented with an aldose reductase inhibitor. *Exp. Eye Res.*, **50**, 355–66.

Sakaquchi, D. S., Janick, L. M. and Reh, T. A. (1997). Basic fibroblast growth factor (FGF2) induced transdifferentiation of retinal pigment epitheliums: generation of retinal neurons and glial. *Dev. Dyn.*, **209**, 387–98.

Sarthy, P. V. (1985). Establishment of Müller cell cultures from adult rat retina. *Brain Res.*, **337**, 138–41.

Satow, T., Bae, S. K., Inoue, T. *et al.* (2001). The basic helix-loop-helix gene *hesr2* promotes gliogenesis in mouse retina. *J. Neurosci.*, **21**, 1265–73.

Scharff, C., Kirn, J. R., Grossman, M., Macklis, J. D. and Nottebohm, F. (2000). Targeted neuronal death affects neuronal replacement and vocal behavior in adult songbirds. *Neuron*, **25**, 481–92.

Sidman, R. L. (1960). The structure of the eye. In *Seventh International Congress of Anatomists*, ed. G. K. Smelser. New York, NY: Academic Press, pp. 487–505.

Spence, J. R., Madhavan, M., Ewing, J. D. *et al.* (2004). The hedgehog pathway is a modulator of retina regeneration. *Development*, **131**, 4607–21.

Stenkamp, D. L., Powers, M. K., Carney, L. H. and Cameron, D. A. (2001). Evidence for two distinct mechanisms of neurogenesis and cellular pattern formation in regenerated goldfish retinae. *J. Comp. Neurol.*, **431**, 363–81.

Stone, L. S. (1950). Neural retina degeneration followed by regeneration from surviving retinal pigment cells in grafted adult salamander eyes. *Anat. Rec.*, **106**, 89–109.

Stone, L. S. and Steinitz, H. (1957). Regeneration of neural retina and lens from retina pigment cell grafts in adult newts *J. Exp. Zool.*, **135**, 301–17.

Stone, L. S. and Farthing, T. E. (1942). Return of vision four times in the same adult salamander eye (*Triturus viridescens*) repeatedly transplanted. *J. Exp. Zool.*, **91**, 265–85.

Sueishi, K., Hata, Y., Murata, T. *et al.* (1996). Endothelial and glial cell interaction in diabetic retinopathy via the function of vascular endothelial growth factor (VEGF). *Pol. J. Pharmacol.*, **48**, 307–16.

Tropepe, V., Coles, B. L., Chiasson, B. J. *et al.* (2000). Retinal stem cells in the adult mammalian eye. *Science*, **287**, 2032–6.

Turner, D. L. and Cepko, C. L. (1987). A common progenitor for neurons and glia persists in rat retina late in development. *Nature*, **328**, 131–6.

Vogel-Hopker, A., Momose, T., Rohrer, H. *et al.* (2000). Multiple functions of fibroblast growth factor-8 (FGF-8) in chick eye development. *Mech. Dev.*, **94**, 25–36.

Wachs, H. (1914). Neue Versuche zue Wolffschen Lensenregeneration. *Arch. f. Entw. Mech. d. Org.*, **39**, 384–451.

Wachs, H. (1920). Restitution des Auges nach Extirpation von Retina und Linse bein Tritonen. *Arch. f. Entw. Mech. d. Org.*, **46**, 328–90.

Walcott, J. C. and Provis, J. M. (2003). Müller cells express the neuronal progenitor cell marker nestin in both differentiated and undifferentiated human foetal retina. *Clin. Exp. Ophthalmol.*, **31**, 246–9.

Young, R. W. (1985). Cell proliferation during postnatal development of the retina in the mouse. *Brain Res.*, **353**, 229–39.

16

Genomics

Seth Blackshaw

Johns Hopkins University School of Medicine, Baltimore, USA

16.1 Introduction

Many distinct processes occur during the course of retinal development. These range from regulation of mitosis and cell fate specification to axon outgrowth and targeting, dendritogenesis and terminal differentiation of different cell types. Since all of these events require changes in gene expression, it follows that global analysis of changes in transcription during development should reveal the identity of many of the genes that mediate these processes. This has been the logic underlying genomic studies of the developing retina, which have so far been undertaken by a number of groups.

The retina has many features that make it well suited to genomic studies. In both invertebrates and vertebrates, the major cell subtypes in the retina are easily distinguished by both molecular and morphological criteria. Compared with other parts of the nervous system, the number of distinct retinal cell subtypes is quite limited and, in both rodents and flies, photoreceptors make up the majority of retinal cells. The birth order of each major cell type is known, and in vertebrates these generation times are distinct and only partially overlapping. Cell types are readily identified by spatial position, which renders *in situ* hybridization-based verification of primary expression data relatively straightforward. Interpretation of expression data in model organisms is also aided by previous work that has already identified large numbers of genes that are selectively expressed in specific cell types of the mature and differentiating retina. Finally, a wealth of mutations that disrupt different aspects of retinal development are available. These can be used to identify genes that are direct or indirect targets of the mutated gene. In the case of mutants that disrupt the development of relatively rare cell types such as ganglion cells, expression profiling of mutant and wild-type animals can help identify novel genes that are selectively expressed in those cell types.

Since expression profiling approaches typically generate enormous quantities of data, the design of experiments and the secondary verification of expression data become key. Considerable experimental and biological variability is often seen among expression profiling experiments of all techniques, so multiple replicates of each experiment are highly

Retinal Development, ed. Evelyne Sernagor, Stephen Eglen, Bill Harris and Rachel Wong.
Published by Cambridge University Press. © Cambridge University Press 2006.

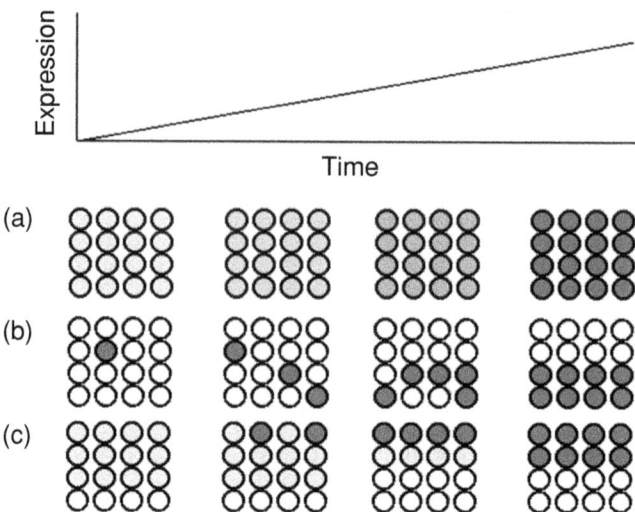

Figure 16.1 Identical changes in gene expression in any heterogeneous tissue can result from very different expression patterns at the cellular level. In the hypothetical case shown here, a gene is observed to steadily increase in expression over time. Three possible cellular expression patterns of the gene in question are shown. (a) The expression level of the gene increases equally in all cells. (b) The number of cells expressing the gene increases, but the level of the transcript in each cell that does express it remains the same. (c) A complex pattern, in which both the number of cells expressing the gene and the level of expression within expressing cells changes over time.

desirable (Pritchard *et al.*, 2001; Blackshaw *et al.*, 2003). Time-course experiments aimed at identifying pathways of genes that regulate specific processes, such as changes in the developmental competence of mitotic progenitors or photoreceptor maturation, must bear in mind the fact that genes with identical temporal expression profiles can have completely non-overlapping cellular expression patterns. See Figure 16.1 for a schematic demonstration of this fact. The considerable cellular complexity of the developing retina prevents one from concluding anything definitive from expression profiling data of retinal tissue – cellular expression patterns of genes of interest must be first verified by *in situ* hybridization or immunohistochemistry. In cases where expression profiling of mutant animals is undertaken to identify direct targets of the gene in question, care must be taken to conduct initial studies at the very onset of the expression of that gene, rather than at the time at which the mutant phenotype is observed. Finally, as the aims of a specific experiment will likely fit best with a specific technique of expression profiling, the experimental technique must be carefully thought out ahead of time.

16.2 Different expression profiling techniques – their pros and cons

High-throughput expression profiling techniques fall into three basic categories, each with their own unique advantages and disadvantages. These are hybridization-based approaches,

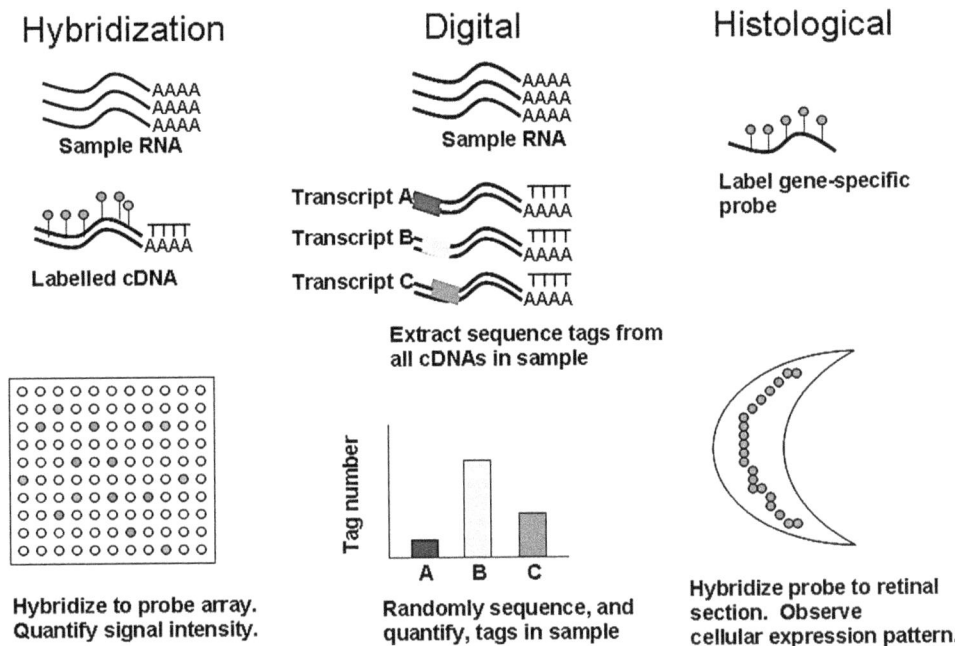

Figure 16.2 The three main high-throughput approaches to expression profiling in the developing retina are outlined here. Hybridization-based methods directly detect transcript levels in cDNA preparations. Digital methods randomly sample and count sequence tags derived from individual transcripts present in cDNA preparations. Histological methods measure the cellular expression patterns of individual genes in tissue sections.

which are typified by microarrays; digital, or sequence-based, approaches exemplified by serial analysis of gene expression (SAGE); and histological approaches such as high-throughput *in situ* hybridization. Although new variations of each of these approaches will undoubtedly emerge over the next decade, it is likely that all high-throughput approaches to expression profiling will remain as part one of these three broad categories. See Figure 16.2 for a schema illustrating how each of these techniques works, and Table 16.1 for a summary of their relative advantages and disadvantages.

Hybridization-based approaches are conceptually simple – labelled nucleic acids from a sample of interest are hybridized onto a printed array of DNA probes corresponding to the transcripts one is seeking to detect. This printed array is then washed, and quantified following scanning – and operate very much like a reverse Northern blot. Microarray hybridization is easy to initiate and comparatively inexpensive. It is the method of choice for analysis of gene expression in very small amounts of starting material – even as little as a single cell (Tietjen *et al.*, 2003; Gustincich *et al.*, 2004). Oligonucleotide arrays covering the full set of predicted mouse and human genes are now commercially available, although recent studies have indicated that a significant fraction of the probes on these arrays may

Table 16.1. *Pros and cons of different expression profiling techniques*

Method	Advantages	Disadvantages	Cost
Microarray (general points)	Rapid, low cost, good for very small samples	Many genes may be missed	Depends on platform used
cDNA microarrays	Enriched in retinal genes (if retinal libraries are used to array), useful for non-model organisms	Production inefficient, cross-hybridization with homologous genes	~$50/array (parts only)
Oligonucleotide microarrays	Commercially available, good coverage, specific probes	Lower sensitivity than cDNA arrays, many probes work poorly, often can only use one probe/array	~$400/array
SAGE	Comprehensive, can be compared with other SAGE data, allows gene discovery (21 bp tags)	Tag ambiguities (conventional SAGE), requires library construction, relatively expensive, poor detection of rare transcripts	~$10 000/library (60 000 21 bp tags)
High-throughput *in situ* hybridization	Rapid, scaleable, allows preselection of genes of interest (transcription factors, etc.)	Poor coverage	~$5–10/probe if cDNA template available/~$20–40/probe if not

not efficiently or selectively detect expression of the target gene (Mecham *et al.*, 2004; Irizarry *et al.*, 2005). Microarray probes can also be generated from cDNA clones. The sensitivity of cDNA arrays also often exceeds that of oligonucleotide arrays, although the amplification and printing of cDNA clones, however, is often slow and subject to problems of quality control. Both oligonucleotide and cDNA arrays have been used successfully to examine gene expression in mature retinas in a variety of different genetic backgrounds and disease conditions (Livesey *et al.*, 2000; Kennan *et al.*, 2002; Yoshida *et al.*, 2002; Chowers *et al.*, 2003), and microarrays remain the most widely used expression profiling technology in both retinal research and in other fields. An array, however, can only detect genes for which it has a matching probe and no vertebrate array yet available completely covers all expressed transcripts, though coverage continues to improve with time.

Digital expression profiling approaches such as SAGE offer an alternative complementary approach to microarray analysis (Velculescu *et al.*, 1995). This is conceptually very similar to taking an opinion poll – a random sample of all expressed genes in the cell or tissue of interest is extracted, and a statistical snapshot of the levels of gene-specific tags is obtained. Serial analysis of gene expression libraries are generated by extracting a short (14 to 21 bp) tag from a defined point near the 3' end of each cDNA in the sample of interest. These tags are then concatenated and subcloned, and individual clones are then sequenced. The more accurate a picture of gene expression in the sample that is desired, the more clones are sequenced – in principal there is no limit to sensitivity. Though SAGE sequencing is relatively expensive, the availability of commercial high-throughput sequencing services has made generation of SAGE data relatively quick. Recent breakthroughs in sequencing technology may result in considerably lower costs in the near future, however (Margulies *et al.*, 2005; Shendure *et al.*, 2005). Other than cost and speed, the main drawbacks of SAGE are its ability to detect low-abundance genes only at very high tag numbers and the ambiguous tag to gene matches that are often seen with short 14 bp tags. Massively parallel signature sequencing (MPSS) (Brenner *et al.*, 2000) and similar technologies offer a conceptually similar tag profiling approach that does not involve a cloning step and can rapidly identify more than one million tags per sample. Approaches of this sort may effectively replace SAGE in the future, although they are still quite expensive.

Histological approaches to expression profiling, such as high-throughput *in situ* hybridization, can be used either on their own or as a secondary screen to examine the cellular expression patterns of genes that are identified as differentially expressed using SAGE or microarray-based approaches. These can be done in an unbiased fashion – by testing cDNAs randomly picked from libraries – or in directed studies, where identified cDNAs from a clone collection are selected and tested based on functional criteria. For vertebrates, the relative ease of performing whole-mount *in situ* hybridization has led to major efforts along these lines to test cellular patterns of gene expression in early embryos (Gawantka *et al.*, 1998; Neidhardt *et al.*, 2000), although the data from such efforts have not usually been made fully available to the research community. Some useful public resources do exist, however. In the case of *Drosophila*, a large database of useful *in situ* images already exists for the early developing retina, as a systematic effort has been undertaken to generate embryonic expression patterns of every gene in the genome (accessible at www.fruitfly.org) (Tomancak *et al.*, 2002). In the case of mice, the efforts of the Gensat consortium to generate bacterial artificial chromosome (BAC) lines that faithfully duplicate the expression of large numbers of endogenous genes (www.gensat.org), has generated hundreds of lines that have been examined at various points in development and which in some cases examine gene expression in the retina as well as the brain (Gong *et al.*, 2003).

The main advantages of high-throughput *in situ* hybridization studies are the ease of initiating studies and the low cost of directed candidate-based approaches. If one is mainly interested in the 20 genes known or suspected to regulate a given signal transduction pathway, there is no need to resort to more expensive expression profiling methods. Moreover, since the cellular expression pattern of genes found to be of interest by microarray or SAGE studies needs to be tested before conclusions can be drawn about that gene's site or mechanism of action, *in situ* hybridization-based studies effectively skip ahead a step in this analytic process. The main disadvantage of these studies is, of course, that they are far from comprehensive – given the fact that mammals possess <35 000 protein-coding genes and produce an unknown but far larger number of distinct mRNAs (Cheng *et al.*, 2005), most efforts will only end up sampling a few per cent of all transcribed genes.

Data generated by any one of these methods of expression profiling will typically only partially overlap with data obtained from another technique. Serial analysis of gene expression to the level of 50 to 100 000 tags will detect virtually all of the moderate to high-abundance transcripts in a sample, but will miss most of the low-abundance transcripts, although detection will get progressively more sensitive as more tags are sequenced. Oligonucleotide-based arrays have similar overall sensitivity to SAGE (Evans *et al.*, 2002), although the fact that many of the probes on these arrays do not work well limits the comprehensiveness of even 'full-genome' arrays. Complementary DNA arrays, while not comprehensive, can often be very effective at detecting low-abundance transcripts. *In situ*-based screens do not make any presuppositions about transcript abundance, and can potentially detect transcripts that are highly expressed in small subsets of cells that other methods will miss.

The method of choice for a given experiment will vary depending on the design of the experiment and on how much information is needed. If one is very specifically interested in the role of a specific category of genes – such as homeodomain-containing transcription factors – in retinal development, an *in situ*-based screen is the most appropriate approach. If a large number of samples or conditions are being tested, or if using very small amounts of starting material (i.e. <1 μg of total RNA), microarrays are the method of choice. Commercially available microarrays are also very useful if the objective of a study is simply to obtain a limited number of differentially expressed genes to use as markers of developmental stages or cell types. Custom-produced small microarrays, containing probes for a limited number of genes of interest, can be a useful and inexpensive alternative to commercially available larger arrays when conducting very large numbers of experiments that are aimed at identifying mechanistic changes in response to gain or loss of function of specific genes of interest. Digital approaches, such as SAGE or MPSS, are perhaps the method of choice if one is seeking a comprehensive readout of gene expression in a limited number of samples, or if work is aimed primarily at identifying unannotated transcripts.

16.3 A survey of genomic studies of vertebrate retinal development

Work on the genomics of vertebrate retinal development has focused on a variety of topics – some broad, others narrow in scope – and have used microarrays, SAGE and high-throughput *in situ* hybridization. See Table 16.2 for a summary of the studies published to date. One microarray-based study focused on the identification of progenitor-enriched genes in retina and other regions of the CNS. For this work, murine retinal cells were labelled with Hoescht 33042, a vital dye that labels nucleic acids, and fluorescence-activated cell sorting analysis

Table 16.2. *Summary of retinal developmental genomics studies*

Microarray studies	Species	Technique used	Time-points/ conditions tested	Number of probes used	Genes subject to secondary verification
(Livesey et al., 2004)	Mouse	cDNA array	E15/E18/P0 4N vs. 2N cells FACS sorted cells	8000	11 (ISH)
(Diaz et al., 2003)	Mouse	cDNA array	E14 dorsal/ nasal/ventral/ temporal retina	19 000	14 (ISH)
(Dorrell et al., 2004)	Mouse	Affymetrix array	P0/P4/P7/P10/ P14/P21 retina	11 000	6 (qRT-PCR)
(Hackam et al., 2003)	Chick	cDNA array	E18 retina vs. brain+liver	5000	15 (ISH)
(Zhang et al., 2004)	Mouse	cDNA array	P5.5 wt vs. $Rb1^{-/-}$	9000	0
(Mu et al., 2004)	Mouse	cDNA array	E14/16/18 wt vs. $Brn3b^{-/-}$	15 000	18 (qRT-PCR)/11 (ISH)
(Yoshida et al., 2004) (Yu et al., 2004a)	Mouse	cDNA array+ Affymetrix array	P2/10/21/60 wt vs. $Nrl^{-/-}$	6 500 (cDNA), 12 000 (Affymetrix)	76 (qRT-PCR)
SAGE studies				*Tags sequenced*	
(Blackshaw et al., 2004)	Mouse	SAGE (14 bp tag)	E12/14/16/18/ P0/P2/P4/ P6/P10/P50	750 000	1051 (ISH)
High- throughput ISH studies				*Probes tested*	
(Thut et al., 2001)	Mouse	DIG cRNA probes	E12/13/14/16/ 18/P2	1035	NA

DIG, digoxygenin-UTP; FACS, fluorescence-activated cell sorting; ISH, *in situ* hybridization; NA, not applicable; RT-PCR, quantitative RT-PCR; wt, wild type.

was used to separate 4N cells (i.e. mitotic cells between late S- and early M-phase) from 2N cells (i.e. cells that are either postmitotic or in phase G1 of the cell cycle) at embryonic day (E)16, E18 and postnatal day (P)0. Gene expression in these sorted cell populations was then examined by cDNA microarray (Livesey et al., 2004). A number of genes known to be expressed in mitotic retinal progenitors, including *Pax6*, *Chx10*, *Otx2*, *cyclinD1* and *cdk4*, were detected as progenitor-enriched, along with components of the transforming growth factor-β (TGF-β) and Wnt signalling pathway. Using a relatively relaxed selection criterion (i.e. >1.3-fold higher expression in at least one of three replicate hybridizations), roughly 800 genes were found to be enriched in 4N cells from retina, and of these roughly 600 were also enriched in 4N cortical and cerebellar progenitors. Many of these encoded genes are involved in regulation of protein synthesis, protein degradation, or microtubule function – very much in line with the SAGE data for *Drosophila* retinal progenitors. The small handful of genes found to be enriched exclusively in retinal progenitors included known genes such as *RORβ* and *Six3*, along with regulators of growth factor signalling such as *Grb10* and cytoskeletal proteins such as the smooth muscle-enriched gene *transgelin 2*. *In situ* hybridization analysis of 11 genes enriched in 4N retinal cells revealed that 10 were expressed in the outer neuroblastic layer (ONBL) of embryonic day (E)14 retina and either weakly or not expressed in the inner neuroblastic layer. Three genes – *Otx2*, *NOPE* and *SFRP2* were more strongly expressed in subsets of cells in the ONBL.

Other studies have looked specifically at genes that are dynamically expressed during postnatal development. In one study, a time course of gene expression in the postnatal mouse retina was examined using Affymetrix chips, with samples tested from postnatal day (P)0, P4, P7, P10, P14 and P21 (Dorrell *et al.*, 2004). Roughly half of all genes detected showed dynamic expression, with 14% clearly showing a constant increase through development, 21% showing a steady decrease, and 7% showing a peak or trough of expression at P10, near the middle of the time course. Genes that showed a constant increase in expression included known phototransduction genes, while genes that showed a constant decrease were enriched for markers of mitotic progenitors. Genes that showed intermediate peaks included genes implicated in vasculogenesis, which peaks during the second week of life. Quantitative RT-PCR analysis confirmed the temporal expression profile of 11 of these genes. Many genes that showed a late onset of expression mapped to chromosomal intervals containing uncloned Mendelian photoreceptor dystrophy gene, although this study did not confirm the cellular expression pattern of any differentially expressed transcripts.

Another study aimed to identify retina-enriched genes. In this study 5000 cDNAs from a chick retinal cDNA library were spotted onto a nylon filter. Embryonic day 18 chick retinal cDNA was radiolabelled and hybridized, as were E18 brain and liver cDNA, and 272 clones enriched in retinal cDNA were identified (Hackam *et al.*, 2003). Developmental *in situ* hybridization analysis of 15 cDNAs revealed a variety of patterns, ranging from strong and selective expression in developing ganglion cells (*stathmin-like 2*) to expression only in mature photoreceptors (*Gtγ2*).

Another study examined spatial rather than differences in gene expression in the developing retina. In this study, the E14 mouse retina was dissected into four quadrants, representing nasal, temporal, anterior and posterior poles of the retina and each possible two-way hybridization was performed against a 19 000 spot cDNA microarray derived from the RIKEN consortium mouse cDNA collection (Diaz *et al.*, 2003), and 367 genes were detected as differentially expressed. A number of genes known to show dorsoventral or naso-temporal gradients in developing retina, such as *ephrin* family members, were also found to do so by microarray, and a total of five genes, including the progenitor-enriched genes *μ-crystallin* and *Id3*, not previously reported to be spatially restricted in retina. However, the false-positive rate was high, and most of the cDNAs tested showed no differences in spatial expression.

Three studies looked specifically at genes that showed differential expression in mutants known to influence retinal differentiation. One study examined a time course of expression in mutant animals that show defects in ganglion cell development. In this study ~15 000 unique cDNAs derived from an E14 mouse retinal library were spotted onto glass slides, and both retinal cDNA from E14, E16 and E18 wild-type and $Brn3b^{-/-}$ mice were hybridized to the array (Mu *et al.*, 2004). Eighty-seven distinct genes were strongly and reproducibly upregulated or downregulated at one of the three time-points. The cellular expression patterns of 23 of these were determined by *in situ* hybridization or obtained from the literature and these, strikingly, fell into two categories. Sixteen genes were selectively expressed in retinal ganglion cells, including transcription factors, such as *Brn3a*, *Irx2* and *Olf-1*, which may represent direct targets of *Brn3b*, and markers of terminal differentiation such as neurofilament genes. Two of the ganglion cell-enriched genes encoded secreted factors – the TGF-β family members *GDF* and *Shh*. However, five of the genes were selectively expressed in mitotic progenitor cells, including *Ptc2* and *Gli1*, two direct targets of *Shh* signalling. Whether *cyclinD1*, *Dlx1* and *Dlx2* – the other three progenitor-enriched genes in this category – also are directly regulated by *Shh* or *GDF8* signalling remains to be determined.

Another study looked at mutant animals at a single time point. In this study, gene expression in retinas explanted at E13.5 from wild-type and $Rb1^{-/-}$ mice and maintained in culture for 14 days were compared on a cDNA microarray (Zhang *et al.*, 2004). Though *Rb1* has been studied primarily in the context of its role as a negative regulator of cell cycle progression, microarray studies revealed a dramatic downregulation in rod- and cone-specific genes, a fact that was confirmed using immunohistochemical techniques, and an upregulation of genes that regulate mitotic progression – consistent with *Rb*'s role as a negative regulator of mitosis.

A third set of studies analysed gene expression at P2, P10, P21 and 2 months of age in wild-type mice and mice mutant for *Nrl* (Yoshida *et al.*, 2004; Yu *et al.*, 2004a), a transcription factor that plays a dual function both in promoting rod photoreceptor development and in preventing these cells from differentiating into cone photoreceptors (Mears *et al.*, 2001). Very few genes showed significant differences in expression at P2. However, by P10 a large set of known rod-specific genes were downregulated in $Nrl^{-/-}$ animals, while a roughly equal number of genes were upregulated in these animals. A number of these genes mapped to chromosomal intervals containing uncloned Mendelian photoreceptor dystrophy genes. The great majority of these transcripts remained differentially expressed at two months of age, and are enriched in photoreceptors of the adult retina (Supplementary Table 16.1). Decreases in expression of components of the bone morphogenetic protein (BMP) signalling pathway were observed, and chromosome immunoprecipitation (ChIP) analysis was used to demonstrate decreased Smad4 binding to the rhodopsin promoter in $Nrl^{-/-}$ animals. While many known cone-specific transcripts are found in the set of up-regulated genes (*Opn1sw*, *Gnat2*), at least one subset of these genes are expressed either in mitotic progenitors (*Fgf15*), Müller glia (*Fabp7*, *ApoE*, *Gdc*), or amacrine cells (*Dfy*). As is the case with the *Brn3b* mutant animals, disruption of development of a specific cell type clearly produces a variety of cell non-autonomous effects on gene expression. The mechanism by which this occurs, and the consequences for the development of other cells in the retina, has yet to be explored.

One published study has used high-throughput *in situ* hybridization as an initial screen to identify genes that show dynamic cellular expression patterns during retinal development. In this study (Thut *et al.*, 2001), 1035 individual IMAGE consortium cDNAs were examined against a series of retinal sections from E12, E13, E14, E16, E18 and P2 mice. Though these cDNAs were selected on the basis of known or putative roles in regulating development, only 17 genes tested showed clear cell-specific expression in the developing neuroretina, 5 of which had been previously reported to be expressed during retinal development. *Fgf15* was confirmed as a prominent marker of retinal progenitors, while *Ptmb10* and *thyroid hormone receptor alpha* were strongly expressed in the inner neuroblastic layer. Six additional genes were selectively expressed at the ciliary margin, including *Tgfbli14* and *Ptmb4*, and it was shown that expression of these two genes could be induced in E10 to E13 neuroretina by direct contact with cocultured chick lens.

So far, only one study has combined both expression profiling with large-scale *in situ* hybridization analysis of genes that were observed to show differential expression in developing retina. In this final study, SAGE libraries were constructed from mouse retinas at two-day intervals from E12.5 to P6.5, with libraries also being made from P10.5 and P50 retina (Blackshaw *et al.*, 2004). A total of 1051 genes that showed dynamic expression during development by either visual inspection or cluster analysis were examined by *in situ* hybridization of a panel of retinal sections that recapitulated the time course of the SAGE libraries. A molecular atlas of gene expression in the developing and mature retina was thereby constructed, along with a taxonomic classification of developmental gene expression patterns.

Genes were identified that label both temporal and spatial subsets of mitotic progenitor cells, thus identifying potential molecular mediators of changes in the developmental competence of progenitors and demonstrating that there is considerable heterogeneity among progenitors. For each developing and mature major retinal cell type, genes selectively expressed in that cell type were identified. The gene expression profiles of retinal Müller glia and mitotic progenitor cells were found to be highly similar, suggesting that Müller glia might serve to produce multiple retinal cell types under the right conditions. Strong and transient expression of many metabolic enzymes was observed in immature Müller glia, though the significance of this finding is unclear. In addition, multiple transcripts that did not appear to encode open reading frames (ORFs) that were evolutionarily conserved or <100 amino acids in length ('non-coding RNAs') were found to be dynamically and specifically expressed in developing and mature retinal cell types. Finally, many photoreceptor-enriched genes that mapped to chromosomal intervals containing uncloned retinal disease genes were identified.

16.4 Comparing the results of the mouse studies

Four studies have surveyed a large fraction of the mouse retinal transcriptome at similar time-points, and thus fruitful comparisons can be made. Serial analysis of gene expression tags are detected that match the majority of the 800-odd transcripts enriched in 4N cells of the developing retina, with 125 genes being present at >0.1% of tags in at least one SAGE library (see Supplementary Table 16.2). The temporal expression profile of 84% of these genes was generally falling over time, with 13% basically unchanging and only 3% rising through development – just as would be expected for progenitor-enriched genes. Twenty-two of these genes were tested by *in situ* hybridization in the SAGE study, and the great majority were found to be progenitor-enriched, with some showing broad expression throughout mitosis and some showing restricted expression in subsets of cells. When only genes for which probes are present on the array and genes sufficiently abundant to be detected by SAGE are considered, agreement between the two data sets is quite good. A caveat must be made that it is usually the more abundant and dynamically expressed genes in any primary data set that get examined by *in situ* hybridization, so the very high true-positive rate seen here probably represents an overestimate when considering the entire data set.

The time course of gene expression in the postnatal retina as measured by both SAGE and Affymetrix chips could also be compared. Eight hundred and seventy-five of the 5249 genes detected by the arrays in the postnatal retina were also present at >0.1% of tags in at least one SAGE library (see Supplementary Table 16.3). Here again, there is reasonable correlation between the strongly expressed genes in the two data sets, particularly for genes that show steadily increasing expression through development. Many of these turn out to be photoreceptor-enriched genes, although these also include genes selectively expressed in every other major types of retinal cell. Combining these results from the array data set obtained from $Nrl^{-/-}$ animals reveals a core set of 41 genes that are very likely to be rod-enriched, even in the absence of cellular expression data, on the basis of showing both late onset of expression by both SAGE and Affymetrix chips as well as reduced expression in $Nrl^{-/-}$ animals. This includes a total of ten genes whose cellular expression pattern in the retina has not yet been previously analysed (see Supplementary Table 16.3). As a general rule, however, the more highly expressed a gene, the more likely the SAGE and array data are to correlate with one another, and low-abundance transcripts often show little or no correlation between the two data sets.

16.5 What have genomics studies told us about mechanisms of retina development?

So far, these genomic studies have left us at a tantalizing but somewhat frustrating point. They have given us an extensive but incomplete parts list for the assembly of a mature retina – potentially the starting point of 1000 projects – and have given us a vastly expanded list of cell-specific markers. Table 16.3 contains a partial list for the mouse retina. However, these

Table 16.3. *Cell-specific markers identified in mouse retina using genomic approaches*

Developmental stage	Rods	Horizontal cells	Bipolar cells	Müller glia	Amacrine cells	Ganglion cells	Pan-progenitor	Subset of progenitors
Early	Cst, RNCR3	Borg4	Gli5	Fabp7	Plcl3, Lmo4	Gdf8	Dlx1, Dlx2, Hmga1	Sox4, Eya2, sFrp2
Intermediate	PIAS3, Cdgap, Cpx2	Sal3	Zf-1, Lhx4	BC016235, crym, BC064011	Dusp26, Robo3	KIAA0133	Edr	Lhx2, Edr
Late	ERR2	Sept4	Zfh4	Clu	6330527O06Rik			

studies have not yet given us much coherent data about mechanisms of retinal development. The fact that no method of expression profiling gives a fully comprehensive picture of gene expression has limited our ability to extract patterns from the data. None of these studies has yet moved to functional analysis of any of the developmentally dynamic genes identified in these studies, so we still have little clue what mechanistic role these differentially expressed genes play in retinal development.

In studies where expression profiling was combined with large-scale *in situ* hybridization of differentially regulated genes, the sheer number of the cellular expression patterns examined has allowed some general conclusions to be drawn. These studies have also reinforced some previously anticipated details regarding retinal development. Dynamic waves of expression of selected genes are observed in mitotic progenitors, as would be predicted on the basis of the changing developmental competence of retinal progenitors (Cepko *et al.*, 1996). A great diversity of expression patterns is observed for amacrine-enriched genes, as might be expected from the considerable morphological diversity of amacrine cells (MacNeil and Masland, 1998).

These studies have also identified some general, unanticipated themes in retinal development. These include the considerable heterogeneity of mitotic progenitors, the similarity of mitotic progenitors and Müller glia, the dramatically elevated expression of various metabolic enzymes in newly formed glia, and the prominent and dynamic expression of mRNA-like molecules that do not encode proteins in developing retina. Genomic studies have also yielded a potential wealth of clinically important data. Nearly half of all cloned Mendelian photoreceptor dystrophy genes are highly and selectively expressed in either developing or mature photoreceptors (Blackshaw *et al.*, 2001; Katsanis *et al.*, 2002; Pacione *et al.*, 2003; Yu *et al.*, 2004b). As a result, the genomic studies aimed at identifying photoreceptor-enriched genes have uncovered a wealth of potential candidate genes for inherited photoreceptor dystrophies that have not yet been cloned. The results of one SAGE-based study were used to pinpoint the photoreceptor-enriched *IMPDH1* gene as being mutated in RP10 patients (Bowne *et al.*, 2002). Other genes that were later implicated in retinal disease or photoreceptor survival, including *RDH12* (Haeseleer *et al.*, 2002; Janecke *et al.*, 2004), *BBS5* (Li *et al.*, 2004), and *rod-derived cone survival factor* (Leveillard *et al.*, 2004), were also prospectively identified as photoreceptor-enriched genes using genomics approaches (although the approaches used to clone these genes did not directly use this genomic data). The catalogue of disease genes identified using genomic approaches is likely to grow considerably in the years ahead.

It remains the case, however, that for most genomic expression studies, the cellular expression patterns of differentially expressed genes have not been examined. Without this data, it is generally not yet possible to draw broader mechanistic conclusions about retinal development from these genomic data. The situation is somewhat clearer for studies that examine mutant animals, such as those performed on mice deficient for *Nrl*, *Rb1*, and *Brn3b*. In these cases, clear sets of cell-specific genes are disregulated. Given that the mutated genes are transcription factors, it's likely that at least some of the differentially expressed genes will indeed represent direct targets of those factors. Though some progress has been made on this topic, particularly in the demonstration that the BMP/Smad pathway modulates expression of rod-specific genes (Yu *et al.*, 2004a), on the whole this awaits further experimental investigation.

16.6 Future directions

There is clearly an essential need to broaden and deepen the body of genomic data on retinal development, and to systematically characterize the cellular expression patterns of genes already identified in previous studies. However, beyond descriptive anatomical studies lies the promise of functional genomics. Every one of the studies described here has generated many genes that can be fruitfully investigated by conventional reverse genetic and biochemical approaches. Nonetheless, the availability of collections of both expressible full-length cDNAs and small interfering RNAs that cover a large fraction of the *Drosophila* and mouse genome (Berns *et al.*, 2004; Boutros *et al.*, 2004; Paddison *et al.*, 2004; Zheng *et al.*, 2004), along with means of delivering them to the developing retina in vivo through electroporation or viral transduction, implies that medium-throughput functional examination of large numbers of genes identified in these studies may soon be practical.

However, perhaps the main outstanding challenge in developmental genomics is not insufficient data, or a lack of means to examine the function of genes of interest, but rather the poor accessibility of most of the data to the research community. Several factors have limited the ability of the community to access and formulate hypotheses from the data produced by these studies. First, direct comparison of data obtained by different studies is made difficult by the lack a common identifier for individual genes. Second, the great majority of expression data in these studies is usually contained in the supplementary data of publications, and is neither deposited in publicly available databases nor visible to search engines such as PubMed or Google. Expression data for a given gene can thus only be found by combing each paper for the specific reference.

An agreement on the part of the research community to adopt a common intuitive vocabulary for gene identification, and to deposit published expression data in public databases, would go a long way to addressing these problems and greatly increase the usefulness of this data to the broader research community.

References

Berns, K., Hijmans, E. M., Mullenders, J. *et al.* (2004). A large-scale RNAi screen in human cells identifies new components of the p53 pathway. *Nature*, **428**, 431–7.

Blackshaw, S., Fraioli, R. E., Furukawa, T. and Cepko, C. L. (2001). Comprehensive analysis of photoreceptor gene expression and the identification of candidate retinal disease genes. *Cell*, **107**, 579–89.

Blackshaw, S., Kuo, W. P., Park, P. J. *et al.* (2003). MicroSAGE is highly representative and reproducible but reveals major differences in gene expression among samples obtained from similar tissues. *Genome Biol.*, **4**, R17.

Blackshaw, S., Harpavat, S., Trimarchi, J. *et al.* (2004). Genomic analysis of mouse retinal development. *PLoS Biol*, **2**, E247.

Boutros, M., Kiger, A. A., Armknecht, S. *et al.* (2004). Genome-wide RNAi analysis of growth and viability in Drosophila cells. *Science*, **303**, 832–5.

Bowne, S. J., Sullivan, L. S., Blanton, S. H. *et al.* (2002). Mutations in the inosine monophosphate dehydrogenase 1 gene (IMPDH1) cause the RP10 form of autosomal dominant retinitis pigmentosa. *Hum. Mol. Genet.*, **11**, 559–68.

Brenner, S., Johnson, M., Bridgham, J. *et al.* (2000). Gene expression analysis by massively parallel signature sequencing (MPSS) on microbead arrays. *Nat. Biotechnol.*, **18**, 630–4.

Cepko, C. L., Austin, C. P., Yang, X., Alexiades, M. and Ezzeddine, D. (1996). Cell fate determination in the vertebrate retina. *Proc. Natl. Acad. Sci. U. S. A.*, **93**, 589–95.

Cheng, J., Kapranov, P., Drenkow, J. *et al.* (2005). Transcriptional maps of 10 human chromosomes at 5-nucleotide resolution. *Science*, **308**, 1149–54.

Chowers, I., Liu, D., Farkas, R. H. *et al.* (2003). Gene expression variation in the adult human retina. *Hum. Mol. Genet.*, **12**, 2881–93.

Diaz, E., Yang, Y. H., Ferreira, T. *et al.* (2003). Analysis of gene expression in the developing mouse retina. *Proc. Natl. Acad. Sci. U. S. A.*, **100**, 5491–6.

Dorrell, M. I., Aguilar, E., Weber, C. and Friedlander, M. (2004). Global gene expression analysis of the developing postnatal mouse retina. *Invest. Ophthalmol. Vis. Sci.*, **45**, 1009–19.

Evans, S. J., Datson, N. A., Kabbaj, M. (2002). Evaluation of Affymetrix Gene Chip sensitivity in rat hippocampal tissue using SAGE analysis. Serial Analysis of Gene Expression. *Eur. J. Neurosci.*, **16**, 409–13.

Gawantka, V., Pollet, N., Delius, H. *et al.* (1998). Gene expression screening in Xenopus identifies molecular pathways, predicts gene function and provides a global view of embryonic patterning. *Mech. Dev.*, **77**, 95–141.

Gong, S., Zheng, C., Doughty, M. L. *et al.* (2003). A gene expression atlas of the central nervous system based on bacterial artificial chromosomes. *Nature*, **425**, 917–25.

Gustincich, S., Contini, M., Gariboldi, M. *et al.* (2004). Gene discovery in genetically labeled single dopaminergic neurons of the retina. *Proc. Natl. Acad. Sci. U. S. A.*, **101**, 5069–74.

Hackam, A. S., Bradford, R. L., Bakhru, R. N. *et al.* (2003). Gene discovery in the embryonic chick retina. *Mol. Vis.*, **9**, 262–76.

Haeseleer, F., Jang, G. F., Imanishi, Y. *et al.* (2002). Dual-substrate specificity short chain retinol dehydrogenases from the vertebrate retina. *J. Biol. Chem.*, **277**, 45 537–46.

Irizarry, R. A., Warren, D., Spencer, F. *et al.* (2005). Multiple-laboratory comparison of microarray platforms. *Nat. Methods*, **2**, 345–50.

Janecke, A. R., Thompson, D. A., Utermann, G. *et al.* (2004). Mutations in RDH12 encoding a photoreceptor cell retinol dehydrogenase cause childhood-onset severe retinal dystrophy. *Nat. Genet.*, **36**, 850–4.

Katsanis, N., Worley, K. C., Gonzalez, G., Ansley, S. J. and Lupski, J. R. (2002). A computational/functional genomics approach for the enrichment of the retinal transcriptome and the identification of positional candidate retinopathy genes. *Proc. Natl. Acad. Sci. U. S. A.*, **99**, 14 326–31.

Kennan, A., Aherne, A., Palfi, A. *et al.* (2002). Identification of an IMPDH1 mutation in autosomal dominant retinitis pigmentosa (RP10) revealed following comparative microarray analysis of transcripts derived from retinae of wild-type and Rho(−/−) mice. *Hum. Mol. Genet.*, **11**, 547–57.

Leveillard, T., Mohand-Said, S., Lorentz, O. et al. (2004). Identification and characterization of rod-derived cone viability factor. *Nat. Genet.*, **36**, 755–9.

Li, J. B., Gerdes, J. M., Haycraft, C. J. et al. (2004). Comparative genomics identifies a flagellar and basal body proteome that includes the BBS5 human disease gene. *Cell*, **117**, 541–52.

Livesey, F. J., Furukawa, T., Steffen, M. A., Church, G. M. and Cepko, C. L. (2000). Microarray analysis of the transcriptional network controlled by the photoreceptor homeobox gene *Crx*. *Curr. Biol.*, **10**, 301–10.

Livesey, F. J., Young, T. L. and Cepko, C. L. (2004). An analysis of the gene expression program of mammalian neural progenitor cells. *Proc. Natl. Acad. Sci. U. S. A.*, **101**, 1374–9.

MacNeil, M. A. and Masland, R. H. (1998). Extreme diversity among amacrine cells: implications for function. *Neuron*, **20**, 971–82.

Margulies, M., Egholm, M., Altman, W. E. et al. (2005). Genome sequencing in microfabricated high-density picolitre reactors. *Nature*, **437**, 376–80.

Mears, A. J., Kondo, M. and Swain, P. K. (2001). Nrl is required for rod photoreceptor development. *Nat. Genet.*, **29**, 447–52.

Mecham, B. H., Klus, G. T., Strovel, J. et al. (2004). Sequence-matched probes produce increased cross-platform consistency and more reproducible biological results in microarray-based gene expression measurements. *Nucleic Acids Res.*, **32**, e74.

Mu, X., Beremand, P. D., Zhao, S. et al. (2004). Discrete gene sets depend on POU domain transcription factor Brn3b/Brn-3.2/POU4f2 for their expression in the mouse embryonic retina. *Development*, **131**, 1197–210.

Neidhardt, L., Gasca, S., Wertz, K. et al. (2000). Large-scale screen for genes controlling mammalian embryogenesis, using high-throughput gene expression analysis in mouse embryos. *Mech. Dev.*, **98**, 77–94.

Pacione, L. R., Szego, M. J., Ikeda, S., Nishina, P. M. and McInnes, R. R. (2003). Progress toward understanding the genetic and biochemical mechanisms of inherited photoreceptor degenerations. *Annu. Rev. Neurosci.*, **26**, 657–700.

Paddison, P. J., Silva, J. M., Conklin, D. S. et al. (2004). A resource for large-scale RNA-interference-based screens in mammals. *Nature*, **428**, 427–31.

Pritchard, C. C., Hsu, L., Delrow, J. and Nelson, P. S. (2001). Project normal: defining normal variance in mouse gene expression. *Proc. Natl. Acad. Sci. U. S. A.*, **98**, 13 266–71.

Shendure, J., Porreca, G. J., Reppas, N. B. et al. (2005). Accurate multiplex polony sequencing of an evolved bacterial genome. *Science*, **309(5741)**, 1728–32.

Thut, C. J., Rountree, R. B., Hwa, M. and Kingsley, D. M. (2001). A large-scale in situ screen provides molecular evidence for the induction of eye anterior segment structures by the developing lens. *Dev. Biol.*, **231**, 63–76.

Tietjen, I., Rihel, J. M., Cao, Y. et al. (2003). Single-cell transcriptional analysis of neuronal progenitors. *Neuron*, **38**, 161–75.

Tomancak, P., Beaton, A., Weiszmann, R. et al. (2002). Systematic determination of patterns of gene expression during Drosophila embryogenesis. *Genome Biol*, **3**, RESEARCH0088.

Velculescu, V. E., Zhang, L., Vogelstein, B. and Kinzler, K. W. (1995). Serial analysis of gene expression. *Science*, **270**, 484–7.

Yoshida, S., Yashar, B. M., Hiriyanna, S. and Swaroop, A. (2002). Microarray analysis of gene expression in the aging human retina. *Invest. Ophthalmol. Vis. Sci.*, **43**, 2554–60.

Yoshida, S., Mears, A. J., Friedman, J. S. *et al.* (2004). Expression profiling of the developing and mature Nrl−/− mouse retina: identification of retinal disease candidates and transcriptional regulatory targets of Nrl. *Hum. Mol. Genet.*, **13**, 1487–503.

Yu, J., He, S., Friedman, J. S. *et al.* (2004a). Altered expression of genes of the Bmp/Smad and Wnt/calcium signaling pathways in the cone-only Nrl−/− mouse retina, revealed by gene profiling using custom cDNA microarrays. *J. Biol. Chem.*, **279**, 42 211–20.

Yu, J., Mears, A. J., Yoshida, S. *et al.* (2004b). From disease genes to cellular pathways: a progress report. *Novartis Found. Symp.*, **255**, 147–60; discussion 160–4, 177–8.

Zhang, J., Gray, J., Wu, L. *et al.* (2004). Rb regulates proliferation and rod photoreceptor development in the mouse retina. *Nat. Genet.*, **36**, 351–60.

Zheng, L., Liu, J., Batalov, S. *et al.* (2004). An approach to genomewide screens of expressed small interfering RNAs in mammalian cells. *Proc. Natl. Acad. Sci. U. S. A.*, **101**, 135–40.

17

Zebrafish models of retinal development and disease

James M. Fadool
Florida State University, Tallahassee, USA

John E. Dowling
Harvard University, Cambridge, USA

17.1 Introduction

The zebrafish (*Danio rerio*; *Brachydanio rerio* in older literature) has become a powerful model system to study genetic mechanisms of vertebrate development and disease. Much of the current success can be traced back to the pioneering work of George Streisinger and colleagues at the University of Oregon. Like many of his peers, Streisinger had an acclaimed research programme on phage genetics but sought a eukaryotic system to expand further the known roles of genes in biological processes. Whereas Seymour Benzer focused his efforts on *Drosophila* and Sydney Brenner (Brenner, 1974) adopted the nematode worm, Streisinger, a fish hobbiest, turned his efforts towards the zebrafish (Streisinger *et al.*, 1981; Chakrabarti *et al.*, 1983; Walker and Streisinger, 1983; Grunwald and Streisinger, 1992). Streisinger first recognized many of the oft-cited advantages for the use of zebrafish as a genetic model (Mullins and Nusslein-Volhard, 1993; Driever *et al.*, 1994; Solnica-Krezel *et al.*, 1994). Zebrafish, small freshwater teleosts, are easily adapted to the laboratory setting and can be maintained in a relatively small space. The fish typically reach sexual maturity in 3 to 4 months, and a breeding pair of fish can produce >200 fertilized eggs per mating. Fertilization is external, and the egg and embryo are transparent, facilitating visual identification of morphogenetic movements and organogenesis with a standard dissecting microscope. Development is rapid; by 24 hours post-fertilization (hpf) all of the major organ systems have formed and spontaneous muscle flexures soon begin. Prior to 48 hpf the first behavioural responses can be observed, and by 3 days post-fertilization (dpf) a free-swimming larva that actively feeds upon small prey has emerged. Many of the methods in use today, including gamma ray and chemical mutatgenesis, haploid screens and diploidization, transgenesis and forward and reverse genetic approaches, have underpinned its rapid success for experimental and genetic manipulations of the visual system.

17.2 Mutagenesis screens

Forward genetic screens represent an unbiased approach to uncover novel genes or novel gene functions. An organism is mutagenized with a chemical, radiation or a DNA mutagen,

Retinal Development, ed. Evelyne Sernagor, Stephen Eglen, Bill Harris and Rachel Wong.
Published by Cambridge University Press. © Cambridge University Press 2006.

and the appearance of an interesting phenotype is sought in subsequent generations. The mutated gene leading to the phenotype is isolated, cloned and sequenced. Not only can the function of the mutated gene be elucidated by this method, but also fundamental cellular or behavioural processes can be studied in the absence of the specific gene product. Following the pioneering work at the University of Oregon, two laboratories developed methods for efficient and large-scale chemical mutagenesis of zebrafish for the expressed purpose of identifying recessive mutations affecting embryonic development (Mullins and Nusslein-Volhard, 1993; Driever et al., 1994; Solnica-Krezel et al., 1994). Both screens used the alkylating agent N-ethyl-N-nitrosourea (ENU) to induce point mutations in zebrafish spermatagonia. The effectiveness of ENU mutagenesis typically generates more than one mutant phenotype per genome. Recessive mutations are then recovered in a traditional third-generation screen. The high rate of mutagenesis combined with morphological analysis, enabled the isolation of thousands of mutations affecting hundreds of loci essential to development of the vertebrate embryo, including the eye and visual system (Brockerhoff et al., 1995; Baier et al., 1996; Driever et al., 1996; Haffter et al., 1996; Malicki et al., 1996). These methods subsequently have been adopted for several small-scale highly focused screens to identify mutations specific to affecting the development and function of the retina (Fadool, et al., 1997; Li and Dowling, 1997; Vihtelic, et al., 2001; Perkins et al., 2002).

Another major advance in zebrafish forward genetic screens occurred with the application of a pseudotype retrovirus vector for insertional mutagenesis (Lin et al., 1994; Gaiano et al., 1996; Golling et al., 2002; Amsterdam et al., 2004). First developed as a vector for gene therapy and genetic studies, the engineered virus can infect a wide range of organisms and efficiently integrate into the genome. In zebrafish, transformation rates are approaching 100%, with most founders transmitting on average 10 proviral inserts to progeny. One in 80 inserts results in an embryonic lethal mutation, and in a large-scale screen hundreds of insertional mutations were recovered over a several year period, although only a fraction of these resulted in specific developmental phenotypes. It is estimated that the 315 saved mutants reflect 25% of the genes essential for the development of many different embryonic structures and organs including the eye (Allende et al., 1996; Becker et al., 1998; Amsterdam et al., 2004; Gross et al., 2005). Comparisons to other species, namely Saccharomyces cervesiae and Caernorhabditis elegans, revealed that 77% and 72% respectively of the essential fish genes are evolutionarily essential in the other species. One clear advantage of insertional mutagenesis is that the proviral insert acts as a molecular tag that facilitates the rapid cloning of the mutated gene as compared with the laborious effort required for positional cloning of ENU-induced mutations (Gaiano et al., 1996). However, the use of the retrovirus techniques in modestly sized screens or to isolate specific phenotypes may be limited.

Although embryonic stem cells and targeted mutagenesis have not been developed for zebrafish, alternative reverse genetic and gene knock-down methods in zebrafish have enabled the analysis of phenotypes of known genes. Targeting induced local lesions in genomes (TILLING), a method originally developed for plant mutagenesis screens, has been used successfully to identify genetic lesions in specific genes of interest (McCallum *et al.*, 2000; Wienholds *et al.*, 2002, 2003; Henikoff *et al.*, 2004; Till *et al.*, 2004). Although providing the opportunity to screen for mutations in virtually any gene, TILLING and similar approaches are very labour intensive, require large-scale ENU mutagenesis combined with PCR-based assays to identify fish carrying lesions in the genes of interest, followed by the recovery of the alleles in subsequent generations. The early reports also provide evidence that many of the mutations are silent or that the amino acid substitutions do not alter the function of the gene product.

The more common strategy is to determine the function of a known gene by decreasing the level of its expression through injection of modified antisense oligonucleotides (morpholinos) to block translation of the desired mRNA (Nasevicius and Ekker, 2000). More recently, peptide nucleic acids have been evaluated as an alternative to morpholinos. Peptide nucleic acids demonstrate stringent hybridization properties, are resistant to most peptidases and nucleases (Urtishak *et al.*, 2003; Wickstrom *et al.*, 2004a,b). Although not truly genetic methods, these knock-down approaches allow investigators to test the role of a specific gene in a given process, confirm that a mutant phenotype can be phenocopied by blocking expression of the suspected gene, or used to inhibit functions of several genes simultaneously without the time-constraints imposed by interbreeding of heterozygous carriers of different mutations. Morpholino-modified oligonucleotides are designed to be complimentary to the translation initiation sequence or putative intron – exon splice-sites of the target gene. When injected into the one-cell-stage embryo, the modified oligonucleotides hybridize to the target sequence and inhibit translation of the mRNA or splicing of the preRNA respectively. These approaches have been used successfully in several studies to address specifically early patterning and cellular differentiation in the neural retina, although several limitations have been identified (Malicki, 2000; Gregg *et al.*, 2003; Tsujikawa and Malicki, 2004a; Van Epps *et al.*, 2004). For example, morpholinos typically inhibit protein expression for only 2 to 3 days. Interestingly, peptide nucleic acids targeted to the *dharma* (*bozozok*) gene effectively phenocopied the genetic mutation while morpholinos did not. Most recently, the potential use of short interfering RNAs (siRNAs) has been demonstrated in zebrafish, offering yet a third possibility for altering the expression of the target gene (Boonanuntanasarn *et al.*, 2003; Dodd *et al.*, 2004). Taken together, the knock-down strategies provide necessary and viable alternatives to genetic methods for investigating developmental processes by manipulating specific gene products.

17.3 Retinal anatomy

Many of the aforesaid techniques have been successfully applied to the zebrafish for the systematic analysis of retinal development and physiology. Like most classes of extant vertebrates, the zebrafish retina is composed of seven major cell types, six neurons and a single glial cell, the Müller cell (Figure 17.1). However, this simplified view grossly understates the true diversity of neuronal types that contribute to the complex circuitry of the vertebrate retina (Kolb *et al.*, 2001; Masland, 2001). Unlike rodent models, the zebrafish is diurnal and its retina contains a large number of diverse cone subtypes in addition to rods

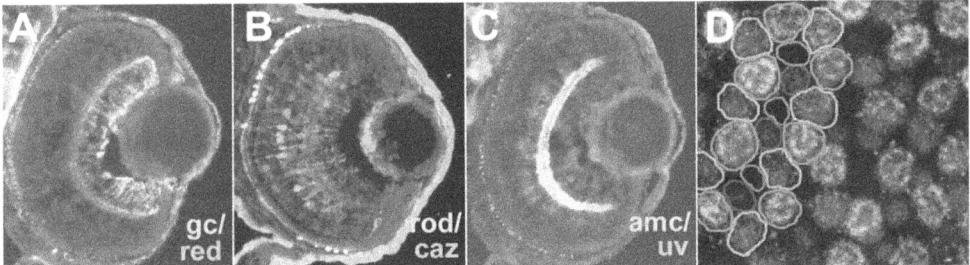

Figure 17.1 Histology of the zebrafish retina. (A–C) Fluorescent double immunolabelling of specific cell types in transverse section of the eye of a zebrafish larvae and DAPI (4′,6-diamidino-2-phenylindole) counterstaining reveal the archetypical laminar arrangement of the retina. (A) Labelling with the Zn-8 monoclonal antibody specific for ganglion cells (gc) and the red-cone opsin. (B) Labelling for the rod photoreceptors with the 1D1 antibody and Müller glia with anti-CAZ. (C) Labelling of amacrine cells (amc) with the 5E11 antibody and co-labelling for rhodopsin. (D) Mosaic organization of the larval retina revealed by labelling red and green cones with the zrp1 antibody and DAPI counterstaining of nuclei. The red and green cones can be distinguished by the greater fluorescent labelling of the red cones, and the positions of the blue and ultravoilet cones can be discerned by labelling with DAPI. The identity of the cone subtypes are diagrammatically represented and colour-coded. For colour version, see Plate 14.

(Branchek, 1984; Branchek and Bremiller, 1984; Larison and Bremiller, 1990; Raymond *et al.*, 1993; Raymond *et al.*, 1995; Schmitt and Dowling, 1996). The cones are subdivided into four classes based upon spectral sensitivity and morphology (Raymond *et al.*, 1993; Robinson *et al.*, 1993). The cone photoreceptors are tiered within the outer nuclear layer with the red- and green-sensitive cones paired as distinct long double cones. The red cone is the slightly longer principal member, whereas the green cone is the shorter accessory member. The blue-sensitive cones are long single cones while the ultraviolet (UV)-sensitive cones are short single cones. The rod cell bodies are located vitreal to the cone nuclei, and in the light-adapted retina, the thin rod inner and outer segments project beyond the cones to interdigitate with the apical microvilli of the pigment epithelium (Burnside, 2001).

The major classes of interneurons, the horizontal, bipolar and amacrine cells, can also be subdivided into numerous subpopulations based upon morphological, immunohistochemical and physiological profiles. Using a recently developed method to randomly label cells with lipophilic dyes, referred to as DiOlistics, Connaughton and colleagues have provided a detailed classification of the retinal interneurons based upon morphology, synaptic location and terminal arborization (Gan *et al.*, 2000; Connaughton *et al.*, 2004). The ability of the lipophilic dyes to freely diffuse within the plasma membrane of the labelled neurons provides a method reminiscent of the Golgi staining so elegantly utilized by Cajal. From the fluorescent images, the horizontal cells were classified into three major subtypes, the HA-1, HA-2 and HB, and whole-cell recording demonstrates many properties similar to other teleost horizontal cells (McMahon, 1994). Amacrine cells were categorized into seven morphological subtypes, and though these morphological classifications are in basic agreement with previous reports using immunolabelling and metabolic signatures, a greater diversity

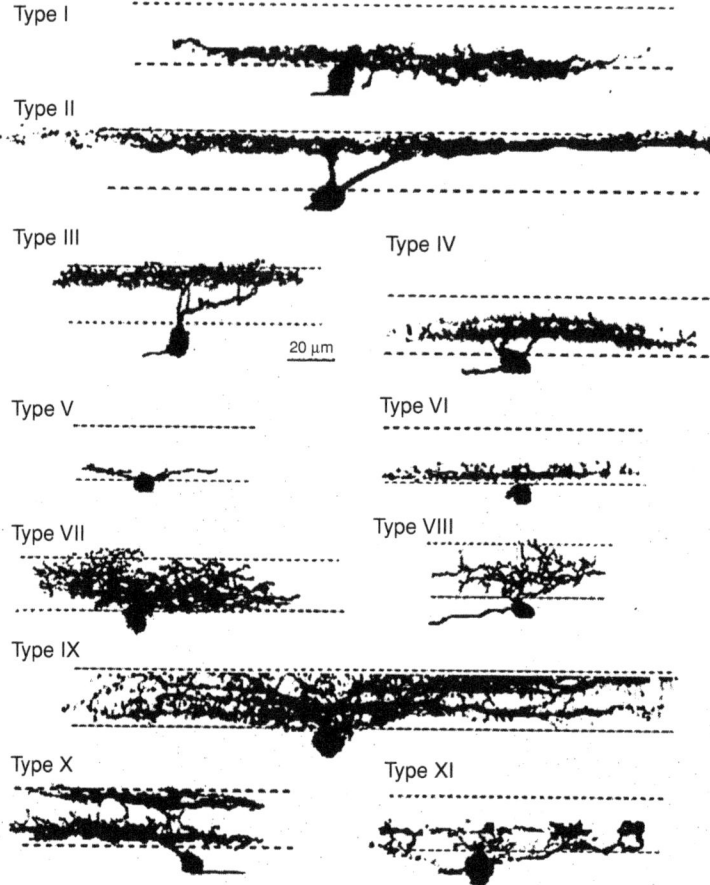

Figure 17.2 Summary diagram of the 11 morphological types of ganglion cell found in the zebrafish retina shown in radial view at the same magnification. The dotted lines mark the boundary of the inner nuclear layer and ganglion cell layer, respectively. Types I and II denote wide-field ganglion cells, while Types III, IV, V and VI denote narrow-field ganglion cells. Types VII and VIII are diffusely branching ganglion cell types, while Types IX, X and XI are bistratified or multistratified ganglion cell types. (Modified from Mangrum et al., 2002; reprinted with permission of the publisher.)

of the amacrine cell subtypes will likely be identified using additional methods (Marc and Cameron, 2001; Yazulla and Studholme, 2001). Bipolar cells have been best characterized in the zebrafish using diolistic labelling or a combination of physiological recordings followed by backfilling with fluorescent dye (Connaughton and Nelson, 2000; Connaughton et al., 2004). The majority of the bipolar cells demonstrate ON or OFF responses consistent with their terminals being in sublamina b of the inner plexiform layer or sublamina a respectively. The remaining are multistratified with terminals in both sublaminae. Ganglion cells have been characterized by fluorescent labelling in flat mounted retinas in combination with confocal image reconstruction (Figure 17.2; Mangrum et al., 2002). Eleven morphological ganglion cell subtypes have been identified including wide-field and narrow-field ganglion cells as well as unistratified, multistratified and diffusely branching types. These anatomical studies provide a backdrop for both physiological and genetic studies of the circuitry of the zebrafish retina as well as unlocking developmental mechanisms generating the diversity of the neuronal subtypes (Baier and Copenhagen, 2000; Wong et al., 2004).

The well-characterized laminar organization of the retina is complemented by the non-random or mosaic organization of the neuronal populations within each of the layers (Wässle and Riemann, 1978; Cameron and Carney, 2000; Rockhill et al., 2000; Fadool, 2003). Gaps in the distribution of cells or random clustering would result in under-representation or over-sampling of information in those regions of the visual field. In the fish retina, this arrangement is most evident in the outer nuclear layer where the position of each cone subtype is precisely arranged relative to the others (Robinson et al., 1993; Fadool, 2003) resulting in a highly ordered crystalline-like mosaic (Figure 17.1). In adult zebrafish, the mosaic is composed of columns of alternating blue- and UV-sensitive single cones that alternate in turn with columns of red- and green-sensitive double cones. The parallel columns are aligned such that in a horizontal row, the green-sensitive members of the double cones flank the short single cones, whereas the long single cones flank the red-sensitive member of the double cone. In the larval retina, the basic rules governing photoreceptor cell patterns are observed, with green photoreceptors alternating with red cones and blues alternating with UV cones, however, the mosaic pattern is far less precise than that of the adult (Figure 17.1D).

Although the necessity for a mosaic organization is well recognized, little is known about the mechanisms underlying this organization. The conservation between the fly and vertebrates of numerous aspects of neurogenesis has lead to the speculation that lateral inhibition plays a role in mosaic formation in the vertebrate retina analogous to the role of lateral inhibition during specification and patterning of photoreceptor cells in the *Drosophila* ommatidia (Cagan and Ready, 1989; Baker et al., 1990; Raymond et al., 1995; Schmitt and Dowling, 1996). Although lateral inhibition plays a role in the specification of early versus late neuronal cell types in the vertebrate retina, a role in mosaic patterning has yet to be demonstrated (Austin et al., 1995; Ahmad et al., 1997; Waid and McLoon, 1998). Further, there are conflicting views from mammals on the role of lateral cellular migration and pruning of dendritic processes during formation of the mosaics of interneurons and ganglion cells (Cook and Chalupa, 2000; Eglen et al., 2000; Galli-Resta, 2002; Lin et al., 2004; Novelli et al., 2004). With the genetic tools afforded by the zebrafish, in combination with its highly ordered arrangements of neurons, it should be possible to distinguish between these hypotheses and dissect the gene networks involved in the mosaic patterning of the neurons.

17.4 Retinal development

During zebrafish development, eye and lens morphogenesis, retinal histology and the expression of transcription factors exhibit a great deal of consistency with other vertebrates. During neurulation, expression of the transcription factors *Six3a* and *Pax6* in the anterior neural plate specify the ocular tissues (Loosli et al., 1998, 1999, 2003; Nornes et al., 1998; Seo et al., 1998; Wargelius et al., 2003). Through subsequent morphogenetic movements and inductive interactions, zebrafish eyes develop from bilateral paddle-shaped masses of cells that evaginate from the forebrain (Schmitt and Dowling, 1994). Disruption of the *chokh/rx3* genes results in a failure of the retinal progenitor cells to evaginate leading to an eyeless phenotype (Loosli et al., 2003). Twenty-two hours post fertilization, invagination of the central portions of this eye-mass and formation of the optic lumen contribute to the formation of an optic cup (Table 17.1; Schmitt and Dowling, 1994, 1999). The inner layer continues to proliferate and produces the neural retina, whereas the outer layer gives rise to the retinal pigment epithelium (RPE), likely through the action of *mitf* expression (Lister et al., 2001). The positioning of the optic stalk is regulated by the expression of the *Pax2* and *Pax6* genes (Macdonald et al., 1995).

Birthdating studies have described an orderly process of neurogenesis. Similar to that observed in other species, the first cells to exit the cell cycle differentiate into ganglion cells. Neurogenesis then follows in an approximate inner to outer retinal order (Hu and Easter, 1999). The first postmitotic cells and differentiation of ganglion cells are first identifiable between 28 hpf and 32 hpf, in the ventral patch, a region of precocious neural development in the ventral nasal retina (Kljavin, 1987; Schmitt and Dowling, 1994, 1996; Burrill and Easter, 1995; Hu and Easter, 1999). Differentiation then spreads dorsally around to the ventral temporal retina in a wave-like manner reminiscent of the movement of the morphogenetic furrow in *Drosophila* (Schmitt and Dowling, 1996). Also similar to *Drosophila*, the wave of differentiation is associated with a wave of *Sonic hedgehog* expression by the differentiating cells (Neumann and Nuesslein-Volhard, 2000). The similarities do not end there. The specification of ganglion cells requires the expression of the basic helix-loop-helix transcription factor *atonal5*, a homologue of the *Drosophila* gene *atonal*, a gene required for the specification of the R8 photoreceptor. *atonal5* is required for ganglion cell specification in many vertebrate species (Brown et al., 1998; Kay et al., 2001; Liu et al., 2001). In the absence of *ath5* in the *lakritz* mutant, neuroblasts fail to be specified as ganglion cells and remain in the cell cycle giving rise to later born cell types (Kay et al., 2001). However, expression of *Sonic hedgehog* by amacrine cells appears to mediate specification of the other retinal neurons (Shkumatava et al., 2004). As we have alluded to, the differentiation of the ganglion cells is followed very closely by the differentiation of amacrine cells, interneurons and retinal lamination. By 48 hpf, lamination has spread across most of the retina (Schmitt and Dowling, 1999). Cytochemically, Müller glia are amongst the last to express the mature phenotype in many species including zebrafish (Peterson, R. E. et al., 2001). Although cells of the inner nuclear layer are postmitotic by 48 hpf, glutamine synthetase and carbonic anhydrase, two markers of functional Müller glia cells are not detectable until approximately 72 and 96 hpf, respectively.

Rhodopsin and red-cone opsin are initially detected in the ventral patch by 50 hpf, preceding the expression of the other cone opsins (Kljavin, 1987; Larison and Bremiller, 1990; Raymond et al., 1995; Schmitt and Dowling, 1996). The expression of the cone opsins

Table 17.1. *Comparison of retinal development in the zebrafish between 32 and 74 hpf with that of retinotectal projections and behaviour*[a]

Hours post-fertilization	Outer retina		Inner retina			Other features	Retinotectal projections	Behaviour/retinal responses
	Photoreceptor	OPL	INL	IPL	Ganglion cell layer			
32					Differentiation in ventronasal patch	Pigmentation within pigment epithelium	Optic axons exit the retina	Weak touch response
36						Optic stalk attenuation commences	Optic axons reach optic chiasm	
40					Ganglion cell birth-days complete/ differentiation in ventrotemporal region	Optic stalk attenuation complete	Optic axons reach tectum	Vigorous touch responses
50	Inner segment formation/opsin expression in ventronasal patch	Horizontal cell processes with zonulae adherentes	Differentiation of amacrine and horizontal cells	Amacrine and ganglion cell processes/ presumptive synaptic zones	Displaced amacrine cells in intraplexiform location	GABA and Trk immunoreac- tivity in optic nerve		
55	Outer segments in ventronasal patch							
60	Outer segments dorsal/nasal to optic nerve	Rod and cone synaptic terminals	Differentiation of bipolar cells	Bipolar cell processes/ conventional synapses	Displaced amacrine cells in ganglion cell layer	GABA and Trk immunoreac- tivity in OPL, INL, IPL		Hatching begins

(*cont.*)

Table 17.1. (cont.)

Hours post-fertilization	Outer retina			Inner retina			Other features	Retinotectal projections	Behaviour/retinal responses
	Photoreceptor	OPL	INL	IPL	Ganglion cell layer				
65		Synaptic ribbons (dyads and triads) in ventronasal patch							
70	Rhodopsin expression in dorsal retina	Synaptic ribbons in other regions		Bipolar cell ribbon synapses in ventronasal patch		Extraocular eye muscle maturation		Arborization within nine fields in tectum	First visual-evoked responses
72						Emmetropization of eye and lens		Arborization within tenth field in tectum	First electroretinographic responses
74	Specialized area in temporal retina first evident			Bipolar cell ribbon synapses in other regions/four sublaminae					Optokinetic responses

[a] Schmitt and Dowling (1999).

GABA, γ-aminobutyric acid; INL, inner nuclear layer; IPL, inner plexiform layer; OPL, outer plexiform layer.

then spreads in a wave-like manner into the dorsal and temporal retina. Interestingly, the UV-sensitive cones are the first cones to mature in zebrafish, whereas the red–green double cones are the last to mature. Between 72 to 96 hpf, most major classes of cells can be identified in the central retina by morphological or cytochemical criteria. During this period of differentiation, the first behavioural responses can be elicited, coincident with the appearance of the outer segments and synaptic ribbons (Easter and Nicola, 1996, 1997; Schmitt and Dowling, 1996, 1999). In teleosts, differentiation of rods follows a developmental programme distinct from that of the cones (reviewed in Johns, 1982; Raymond and Rivlin, 1987; Otteson et al., 2001; Otteson and Hitchcock, 2003). In contrast to the regular spread of cone opsin expression into the dorsal and temporal retina, cells expressing rhodopsin are sporadically distributed across the retina (Raymond et al., 1995; Schmitt and Dowling, 1996; Fadool, 2003). The first detectable rod responses in electroretinogram (ERG) recordings appear between 15 and 18 dpf (Bilotta et al., 2001).

17.5 Retinal mutants

The zebrafish has proven a powerful tool for the genetic analysis of visual system development and function. The large-scale genetic screens, and many other smaller screens, have recovered numerous loci with discrete functions in cellular specification and morphogenesis, retinal lamination, axonal guidance and photoreceptor cell function (Brockerhoff et al., 1995; Allende et al., 1996; Baier et al., 1996; Karlstrom et al., 1996; Malicki et al., 1996; Trowe et al., 1996; Fadool et al., 1997; Li and Dowling, 1997; Neuhauss et al., 1999; Doerre and Malicki, 2001, 2002; Vihtelic et al., 2001; Holzschuh et al., 2003; Jensen and Westerfield, 2004). Three types of assays have been utilized to identify mutations affecting the visual system – morphology, fluorescent labelling and visually evoked behaviours, with additional screens having been proposed or initiated. The clarity of the early embryo and relatively large size of the eye and lens make screening for morphological defects relatively straightforward. In the large-scale screen in Boston, mutations at 36 loci affecting various aspects of eye development were identified (Malicki et al., 1996). These were classified into seven categories based on phenotype, including alterations in retinal patterning, photoreceptor cell survival, eye shape and size, and pigmentation. In an ongoing, multifaceted screen for morphological and behavioural defects of the visual system, 17 mutations leading to morphological defects were initially reported, including defects of the anterior chamber, altered retinal lamination and several demonstrating diminished cell proliferation at the retinal margin (Brockerhoff et al., 1995; Fadool et al., 1997; Li and Dowling, 1997; Perkins et al., 2002; Kainz et al., 2003). Though many of the morphological mutants fell into several of the categories described in the large-scale screen, the latter demonstrated that a single lab could successfully conduct a highly focused multifaceted screen directed at a single organ system (Fadool et al., 1997). In total, the morphological screens have provided significant inroads into the genetic pathways or cellular functions essential to fundamental processes of retinal development. For example, from several different labs mutations have been isolated resulting in altered retinal lamination and formation of photoreceptor rosettes but otherwise with normal neuronal differentiation (Malicki et al., 1996, 2003; Jensen et al., 2001; Masai et al., 2003). In the mutant embryos, mitotic activity was not restricted to the apical margin but rather was distributed across the width of the neuroepithelium. Subsequent cloning revealed that several of the mutations disrupted genes involved in epithelial junctional complexes, demonstrating the importance of maintaining epithelial polarity for the appropriate radial arrangement of neuroprogenitors and lamination of the retina (Wei and Malicki, 2002; Jensen and Westerfield, 2004; Wei et al., 2004).

The *young* (*yng*) mutant embryos display a variety of defects including a failure of retinal cell differentiation (Link et al., 2000). Surprisingly, the normal expression pattern of *ath5* in *yng* mutant larvae was observed, as was the wave of expression of the signalling molecule *Sonic hedgehog*. However, mitogen-activated protein kinase (MAPK) and *Brn3.2*, a transcription factor necessary for ganglion cell differentiation, were severely hindered in expression suggesting a role for the *yng* gene product in the transition from cellular specification to neuronal differentiation (Gregg et al., 2003; DeCarvalho and Fadool, unpublished observations). Positional cloning of the *yng* mutation identified an essential role for the *brahma-related* (*brg1*) chromatin-remodelling complex in mediating retinal cell differentiation (Gregg et al., 2003). Brg1 is a helicase associated with a large megadalton chromatin-remodelling complex implicated in development, cell proliferation and tumorigenesis. The mutation was partially rescued by injection of a bacterial artificial chromosome (BAC) encompassing the entire *brg1* gene sequence and was phenocopied with morpholinos confirming that the *brg1* is the gene disrupted in *yng* mutant embryos. A similar retinal phenotype was also identified in larvae mutant for *baf53* (hi550), a factor known to interact with Brg1, supporting a novel role for chromatin remodelling in the differentiation of cells in the vertebrate retina (Gregg et al., 2003; Amsterdam et al., 2004). Interestingly, mutation of *snf2h*, an ATPase gene of a different chromatin-remodelling complex, did not affect retinal cell differentiation, providing evidence for tissue-specific roles for different chromatin-remodelling complexes. In support of this conclusion, expression of several associated factors in the mouse also demonstrated neural-specific expression patterns during development (Olave et al., 2002; Seo et al., 2005). However, the targeted disruption of *Brg1* in the mouse led to lethality in the preimplantation embryo, precluding analysis of a potential role in neural development (Bultman et al., 2000). This latter result raises an important issue; if *brg1* is essential for early stages of murine development, then how did the zebrafish embryos survive? In zebrafish, transcripts for many genes involved in early development such as *brg1* are expressed maternally (Gregg et al., 2003; Dosch et al., 2004; Wagner et al., 2004). These maternal stores likely permit embryos to progress through the early stages of development thereby uncovering the novel function of the gene in retinal development. For more detailed descriptions of the nature of other genes and their roles in the early stages of ocular development, the reader is referred to two recent publications on the topic (Malicki, 2000; Malicki et al., 2002). We shall focus our discussion on the many advantages offered by the zebrafish as a behavioural model to uncover novel gene functions or reveal fundamental processes in retinal physiology.

17.6 Behavioural screens for alterations in retinal development and function

The development of behavioural assays to detect visual system deficits in zebrafish may hold the greatest potential to contribute to our understanding of retinal function and visual system processing. Zebrafish larvae and adults are highly visual animals. The first visually evoked startle responses are observed 3 dpf. By 4 dpf, many larvae demonstrate an optokinetic reflex (OKR) in response to moving objects, and by 5 dpf, >95% of zebrafish larvae display smooth pursuit and saccade eye movements in response to illuminated rotating stripes (Brockerhoff et al., 1995; Easter and Nicola, 1996, 1997). The basic function of the OKR is to keep an object stably positioned on the retina while moving through the environment. The robustness of the OKR, the ability to screen young larvae and the potential to vary the assay to detect multiple types of visual system defects led Brockerhoff and colleagues (Brockerhoff et al., 1995) to use the OKR as a robust method to identify recessive mutations affecting the visual system in otherwise normal appearing larvae (Brockerhoff et al., 1995, 2003; Taylor et al., 2004). The assay is rapid, the responses from several larvae can be obtained simultaneously and an entire clutch can be assayed in minutes. To conduct the assay, larvae are immobilized in a petri dish containing methylcellulose and placed on a stationary pedestal in the middle of a rotating drum. Rotating the drum elicits eye movements in the direction of the rotation of the stripe. Although the rate of isolating mutations affecting the OKR in otherwise normal appearing larvae is several-fold less than the frequency of morphological mutants, the benefits are apparent. By varying the stimulus from bright to dim white light or using a long-wavelength illumination, subtle defects affecting specific aspects of photoreceptor function, single photoreceptor cell types, synaptic activity or biochemical pathways have been isolated (Brockerhoff et al., 1995, 1997, 2003; Allwardt et al., 2001; Kainz et al., 2003). One potential drawback of any behavioural screen is isolating the origin of the defect to the region of the CNS of interest, in our case, the retina. Therefore, recording of the ERG is routinely applied as a secondary screen to distinguish between a retinal defect versus alteration in another tissue necessary for the OKR such as the neuromuscular junction. Once an interesting defect is confirmed as retinal in origin, positional cloning and a candidate gene approach are used to identify the mutated gene. The assay was also used to screen a collection of 450 mutants previously identified by morphological criteria, of which a total of 25 displayed visual system impairment (Neuhauss et al., 1999). And others continue to refine the paradigm to evaluate motion detection, colour discrimination and higher order processes.

The potential of the assay to identify larvae with subtle defects is well illustrated by the identification of a red-blind mutant, *partial optokinetic response b* (*pob*) (Brockerhoff et al., 1995, 1997). In most respects, *pob* mutant larvae are morphologically indistinguishable from their wild-type siblings, and they demonstrate robust eye movements in response to moving black and white stripes illuminated with white light. However, they do not move their eyes when the stripes are illuminated with red light. This difference in response to red versus white light strongly suggested a retinal rather than a central deficit. In the secondary screen, the ERG confirmed the retinal nature of the defect; *pob* mutant larvae showed markedly attenuated responses to red light compared to wild-type larvae (Figure 17.3).

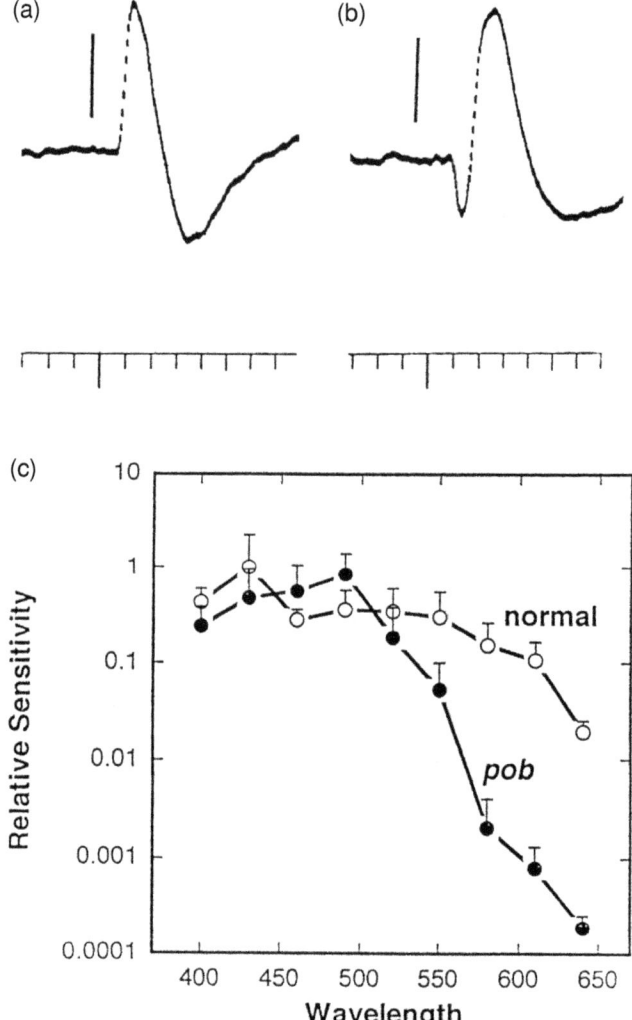

Figure 17.3 Electroretinograms (ERGs) from 6-day-old OKR+ (a) and *pob* mutant larvae (b). The responses were elicited with a short 0.01 s flashes of green (520 nm) light at the same intensity. In (a), only the b-wave is evident but both the a- and b-waves are present in (b). The vertical bars represent 50 μV (a) or 100 μV (b). Time markers = 0.1 ms. (c) Comparison of the spectral sensitivities of *pob* and normal sibling larvae. Spectral sensitivities were determined by ERG analysis. The inverse of the number of photons required to generate a threshold (20 μV) b-wave response was calculated at each wavelength and normalized to the sensitivity of the normal larvae at 430 nm. Note the *pob* mutant larvae are about 2 log units less sensitive to red light. (From Brockerhoff *et al.*, 1997; reprinted with permission of the publisher.)

In situ hybridization for the cone opsins and cell counts demonstrated a selective loss of the red cones in the retina of the *pob* mutant larvae although the probe for the red-cone opsin labelled small cone profiles near the margin suggesting the red cones initially begin to differentiate and then die (Brockerhoff *et al.*, 1997). Surprisingly, cloning of the locus demonstrated that *pob* encodes a widely distributed 30-kDa protein of unknown function (Taylor *et al.*, 2005), but based upon highly conserved sequence homology, the authors proposed that the protein product plays a role in protein sorting and/or trafficking essential to red cone function.

Two other mutations, *nrc* and *nof* illustrate the value of the forward genetic screen as a means to develop greater understanding of cellular physiology through the analysis of retinal function in the absence of specific genes products (Brockerhoff *et al.*, 1995, 2003). Neither *nrc* nor *nof* mutant larvae demonstrate an OKR, and the ERGs are consistent with photoreceptor-specific defects. Curiously, the ERG of *nrc* mutant larvae demonstrated an odd oscillatory wave, similar to that observed in individuals affected by Duchenne's muscular dystrophy. Ultrastructural analysis of the cone terminals of *nrc* mutant larvae showed a lack of proper invaginating synapse development, with free floating ribbons and fewer synaptic vesicles than observed in wild-type larvae, consistent with the altered ERG (Figure 17.4; Allwardt *et al.*, 2001). However, no alterations of ribbon synapses in the inner nuclear layer were observed. Positional cloning of the *nrc* locus revealed a premature stop codon in synaptojanin1, a phosphotidyl phosphatase previously implicated in clathrin-mediated endocytosis and actin cytoskeleton rearrangements (Van Epps *et al.*, 2004). Although synaptojanin1 had previously been cloned in mammalian models, identification of the *nrc* mutation revealed a novel role for phosphotide metabolism in cone photoreceptor synapse organization and function.

In *nof* mutant larvae, ERG recordings also suggested a photoreceptor origin to the visual deficit (Brockerhoff *et al.*, 2003). Positional cloning identified a premature stop codon in the alpha subunit of cone transducin in *nof* mutant larvae, and the behavioural effect could be phenocopied by morpholinos. The study demonstrated first of all that transducin is not essential for normal cone development. In the absence of any obvious ultrastructural changes in the photoreceptor cells, whole-cell electrical recording was used to investigate the cellular physiology of cones in the absence of transducin-mediated phototransduction. The dark currents for *nof* and wild-type cones differed by less than 30%, and as anticipated, no light-induced changes in current were detected with moderate intensity stimulation. However, photoresponses could be elicited in cones isolated from *nof* mutant when stimulated with a step increase in bright light that bleached a few per cent of the visual pigment per second. The response demonstrated a slow onset, on the order of $\sim$1 second compared with 0.1 to 0.2 seconds for wild-type cones, and the low response amplitude suggested a mechanism different from the canonical transducin-mediated phototransduction. The response of *nof* cones was attenuated by preloading the cones with the membrane permeant form of the Ca^{2+} chelator BAPTA (1,2-*bis* (*o*-aminophenoxy)ethane-N,N,N′,N′-tetraacetic acid), suggesting a role for Ca^{2+} in the observed currents. Imaging the responses with the fluorescent Ca^{2+} indicator Fluo-4 provided proof that the observed currents in *nof* cones were mediated by

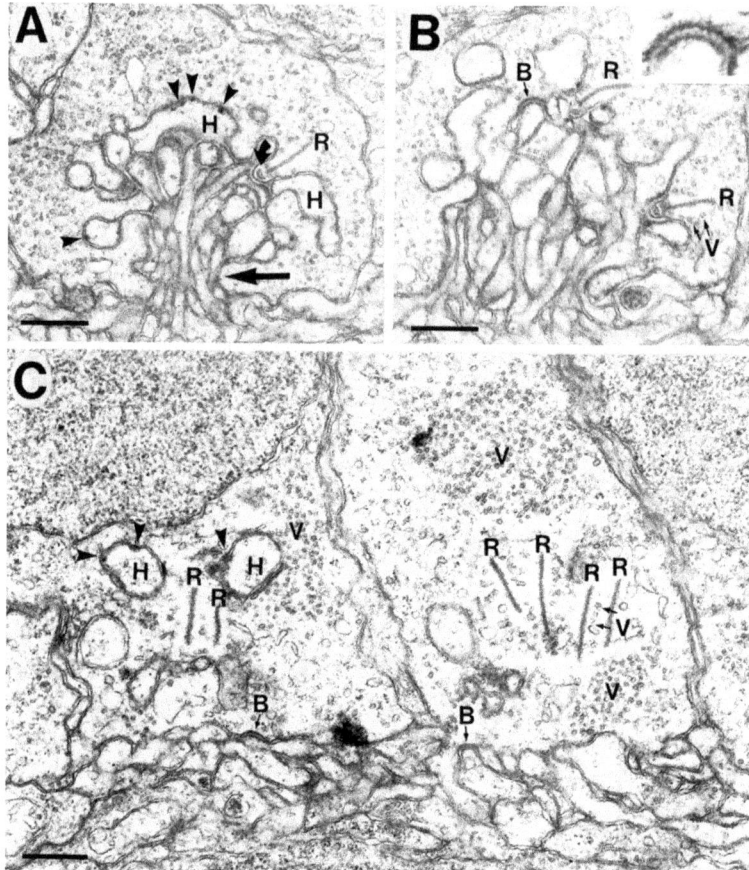

Figure 17.4 Electron micrographs of cone terminals in wild-type (A and B) and *nrc* mutant larva (C). (A) In the wild-type retina, bipolar and horizontal cell processes invaginate the pedicle in a tight bundle (*arrow*). Horizontal cell processes (*H*) are easily recognized by their large size, electron-lucent cytoplasm and characteristic densities (*small arrowheads*). Synaptic ribbons (*R*) are associated with the presynaptic membrane via an arciform density (*curved arrow*). (B) Basal contacts (*B*) are found in wild-type cones between the ribbon synapses. *Inset*, Under high power, the basal contacts show fluffy cytoplasmic material on both sides of the junction. Synaptic vesicles (*V*) surround the synaptic ribbons (*R*). (C) In the *nrc* retina, synaptic ribbons (*R*) in most of the pedicles appear to be floating in the cytoplasm, unassociated with an arciform density and the presynaptic membrane. Few postsynaptic processes invaginate the pedicles. Many of these processes have small densities (*arrowheads*) suggesting they are horizontal cell processes. Basal contacts are made onto bipolar cells at the base of the pedicle (*B*). Synaptic vesicles (*V*) often clump and fail to distribute evenly in the pedicle. However, they surround synaptic ribbons as they do in wild-type pedicles (*small arrows*). Scale bar, 0.5 μm. (From Allwardt *et al.*, 2001; reprinted with permission of the publisher.)

a transducin-independent increase in cytosolic Ca^{2+} following light stimulation. Calcium changes have been recorded in other photoreceptors exposed to light; however, the role of the observed changes in photoreceptor cell physiology is not fully understood (Matthews and Fain, 2001, 2002). Further biochemical and physiological studies should help resolve the source of the Ca^{2+} pool and elucidate the role of the Ca^{2+} release in mediating the changes in whole-cell current in wild-type and mutant photoreceptors.

Behavioural analysis of mutant larvae has also revealed aspects of the circuitry underlying the elicited eye movements. The *belladonna* mutation (*bel*), so named for a pigmentation defect of the eye resulting in the appearance of a dilated pupil, also was found to display a misrouting of ganglion cell axons (Karlstrom *et al.*, 1996; Trowe *et al.*, 1996; Neuhauss *et al.*, 1999; Rick *et al.*, 2000). Whereas in wild-type larvae contralateral projections from ganglion cells to the tectum are the norm, in *bel* mutants ganglion cell axons project to the ipsilateral tectum (Figure 17.5A and B). The phenotype in *bel* mutant larvae ranges from relatively mild, displaying few altered projections, to fully penetrant with only ipsilateral projections. Analysis of the OKR in *bel* mutant larvae revealed two interesting properties. First and foremost, in response to the moving stripes, the eyes of the mutant larvae moved in the direction opposite to the direction of the stimulus; for example, in response to stripes sweeping across the right eye in a nasal to temporal direction, the eye moved in a temporal to nasal direction. Second, for *bel* mutant larvae demonstrating reverse eye movements, eye velocity was independent of stimulus velocity. The movement of the stripes initiated eye movement but did not influence the rate of the pursuit, and, in contrast to wild-type larvae, the amplitude of the movement of the stimulated eye was less than the amplitude of movement of the opposite eye. Although the optic tectum does not mediate the OKR in zebrafish, the level of misrouting to the tectum correlated well with the altered behaviour and may therefore reflect the degree of misrouting of ganglion cell projects to other nuclei, including pretectal nuclei thought to be involved in mediating the OKR (Roeser and Baier, 2003). To explain these behaviours, the authors proposed the following: in the wild-type fish, visual stimulation of one eye drives movement of that eye through projections to a contralateral OKR-mediated nucleus and integrator nucleus that ultimately controls the ipsilateral motor nuclei and the ocular muscles of the stimulated eye (Figure 17.5C and D; Rick *et al.*, 2000). In this model, the neural basis of the altered behaviour in *bel* mutants can be attributed to the singular defect in the projection of ganglion cell axons to the ipsilateral OKR-mediated nucleus. The ipsilateral projections innervate the ipsilateral OKR-mediated nucleus and integrator nucleus, but the output neurons from the integrator nucleus still cross the midline and subsequently drive the motor nucleus of the unstimulated eye.

17.7 Behavioural mutants as models of human disease

Heritable diseases are among the leading causes of blindness in developed countries. *Retinitis pigmentosa* (RP) and allied dystrophies represent a heterogeneous collection of diseases

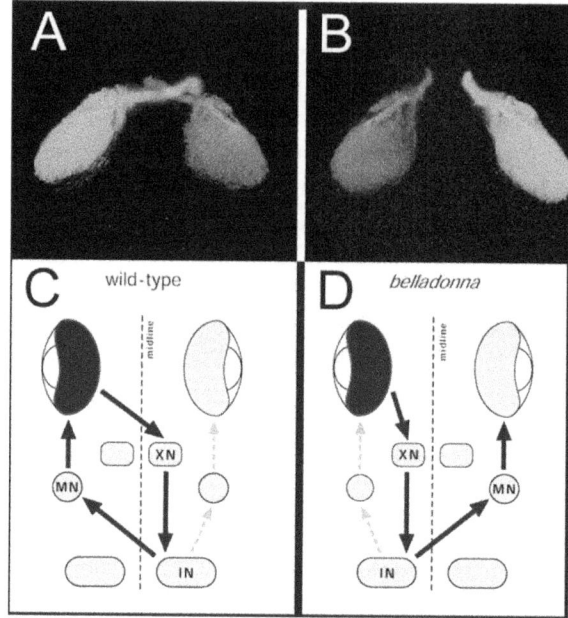

Figure 17.5 Projection defect of retinal ganglion cells in *bel* mutant larvae revealed by injection of DiI (left eye, black) and DiO (right eye, grey) into either eye. (A) Wild-type larvae have a complete contralateral projection with the optic nerves crossing at the chiasm. (B) *bel* mutant larva demonstrating complete ipsilateral projection with no formation of the optic chiasm. (C and D) Model of reversal in *bel* mutant. (C) In wild-type larvae, the visual stimulus is perceived in the stimulated eye (black) and transferred across the midline into the OKR-mediating nucleus (XN). This nucleus connects to an integrator nucleus (IN), which in turn controls the motor nucleus (MN) after crossing the midline. The IN also controls the movement of the unstimulated eye, albeit less robustly. (D) In *bel* mutant larvae the only defect is that the initial connections do not cross the midline but instead innervate the ipsilateral OKR-mediating nucleus. The result is the stimulated eye (black) drives the movement of the unstimulated eye. (Modified from Rick *et al.*, 2000; reprinted with permission of the publisher.)

that affect the function and survival of the photoreceptor cells of the retina, in many cases leaving the second-order neurons intact (Berson *et al.*, 2002; Rivolta *et al.*, 2002). Patients suffering from RP lose their peripheral vision in adolescence or as young adults and become completely blind between 30 and 60 years of age. Approximately 40% of the cases of RP demonstrate an autosomal dominant form of inheritance. By comparison, Leber congenital amuraurosis has a much lower incidence with an autosomal recessive mode of inheritance and typically presents at birth. Since the seminal identification in 1989 of the first locus associated with an inherited photoreceptor cell degeneration, and during the following year (McWilliam *et al.*, 1989; Dryja *et al.*, 1990a,b), the subsequent determination of mutations in the *rhodopsin* (*RHO*) gene responsible for autosomal dominant RP, over 150 loci and 100 genes have been associated with photoreceptor cell dystrophies (a comprehensive and updated list can be found at Retnet (http://www.sph.uth.tmc.edu/Retnet/disease.htm)). It is not surprising that many of the initial discoveries were genes associated with the unique processes of photoreceptor cells, such as phototransduction, photoreceptor cell structure or cellular interactions unique to the photoreceptor cells and the retinal pigment epithelium (RPE). For example, mutations in *RHO* are the single most prevalent alterations leading to RP (Berson et al., 2002; Rivolta et al., 2002). But these findings led to new questions, such as how mutations in genes exclusively expressed by rod photoreceptors result in the progressive yet irreversible loss of cones (Papermaster, 1995). Unforeseen were the significant numbers of mutations in genes with very diverse functions and widespread patterns of expression that led to loss of vision. These genes include mitochondrial-specific factors, RNA splicing components or metabolic proteins. It has been postulated that the effects may be associated with the unique metabolic and structural features of the photoreceptor cells, which render them hypersensitive to mutations in these additional genes; however, this requires additional support.

Several recessive mutations affecting visual function underscore the power of the zebrafish as a genetic model of human congenital defects. The *noa* locus was identified based on the absence of the OKR (Brockerhoff et al., 1995). Recent cloning of the *noa* gene product demonstrated a deficiency for dihydrolipoamide S-acetyltransferase, and the neurological phenotype of *noa* mutant fish displays several characteristics similar to pyruvate dehydrogenase deficiency (Taylor et al., 2004). Furthermore, rescue of the severe effects of the mutation was accomplished by dietary supplementation, providing a novel model for understanding this human disease. Similarly, two genes essential for the function of photoreceptor cells and associated with retinal distrophy have recently been identified. The mutation in the zebrafish orthologue of the human *choroideremia* gene, which encodes the Rab escort protein-1 and is responsible for a human disease marked by slow-onset degeneration of rod photoreceptors and retinal pigment epithelial cells, was recently identified (Starr et al., 2004), as was the *oval* gene product, an intraflagellar transport protein locus, known to be associated with proper function of the connecting cilium in photoreceptor cells (Tsujikawa and Malicki, 2004). Therefore, the zebrafish has the potential of providing additional models for diseases of the visual system.

It is anticipated that in the forthcoming years, many more candidates for disease-causing genes will be identified in zebrafish genetic screens. Unfortunately, the majority of the published mutations affecting the zebrafish are larval lethal or require extraordinary measures to maintain the mutant fish beyond larval stages (Brockerhoff et al., 1995; Malicki et al., 1996; Fadool et al., 1997). Therefore, a systematic screen of adult and late-larval-stage zebrafish for both dominant and recessive mutations is necessary to generate more representative models of retinal dystrophies. The visually mediated escape response was developed as an assay to quantify visual sensitivity as a potential tool to detect retinal dystrophies in adult zebrafish (Li and Dowling, 1997). To assay the escape response, free-swimming adult zebrafish are placed in a small circular dish. On a rotating drum located outside of the dish is a single black spot, simulating a threatening object. Upon encountering the rotating black spot, the zebrafish takes refuge behind a single post located in the middle of the dish. The direction of the rotation can be altered, and the intensity of illumination can be easily controlled with neutral density filters. Using this simple paradigm, visual threshold, circadian control and dark adaptation were evaluated in a screen of 245 adult F1 generation zebrafish of mutagenized adults (Figure 17.6). Seven dominant mutations (*night blind a, nba; night blind b, nbb, etc.* ....) affecting visual sensitivity (Li and Dowling, 1997) were identified. In subsequent generations, the onsets of the dominant phenotypes were found to vary from several months to greater than two years suggesting a situation similar to late-onset retinal dystrophies in humans. The severity of phenotypes also varied. Fish heterozygous for the *nba* mutation demonstrate slow progressive photoreceptor cell degeneration with loss of both rods and cones in patches across the retina, with a corresponding alteration in the ERG. By comparison, *nbc* heterozygous fish demonstrated changes ranging from the slow progressive loss of rod and cone photoreceptor outer segments, to others demonstrating only alterations in rods, and still others displaying no obvious morphological phenotype (Maaswinkel *et al.*, 2003). However, the ERGs of *nbc* mutant and wild-type fish were similar, making the origin of the behavioural deficit unknown. Unexpectedly, larvae homozygous for either mutation displayed severe and widespread neural degeneration, suggesting the affected genes are not photoreceptor cell specific (Li and Dowling, 1997; Maaswinkel *et al.*, 2003). Interestingly, in a subsequent off-the-shelf screen, several heterozygous adults for previously identified recessive mutations displayed an adult phenotype – that is, the fish were night-blind (Darland and Li, personal communication). Taken as a whole, the genetic screens of zebrafish and the continued development of novel screening strategies should provide a rich resource for investigating fundamental processes of visual system development and physiology.

17.8 Chemical genetics

The combination of external fertilization and clarity of the embryo that has propelled zebrafish as a genetic model of vertebrate development likewise enables chemical screens to identify agents that specifically alter retinal development and nervous system function. In one of the early chemical screens, Hyatt and colleagues looked for compounds that altered development of the eyes and discovered a novel role for retinoic acid (RA) in visual system development (Hyatt *et al.*, 1992). Retinoic acid is a potent morphogen and its importance during neural development is well documented (Ross *et al.*, 2000; Maden and Holder, 1992; Hyatt and Dowling, 1997). At high concentrations it displays teratogenic effects, and its absence can lead to visual impairment among other congenital defects. Application of RA during early neurulation of zebrafish resulted in an apparent duplication of the neural retina (Hyatt *et al.*, 1992). The retinas of the treated larvae had an expanded ventral retina, producing two concave surfaces that in some cases were associated with duplicated lenses. Endogenous RA is synthesized from retinaldehyde by a dehydrogenase. In the zebrafish and mouse, expression of a specific dehydrogenase in the ventral retina potentially leads to a gradient of RA across the neural retina suggesting a necessary function during patterning of the dorsoventral axis (Marsh-Armstrong *et al.*, 1994; Hyatt *et al.*, 1996b). Subsequently it was demonstrated that inhibition of the dehydrogenase leads to a lack of ventral retinal structures in a stage-specific manner. The application of RA at later stages of development

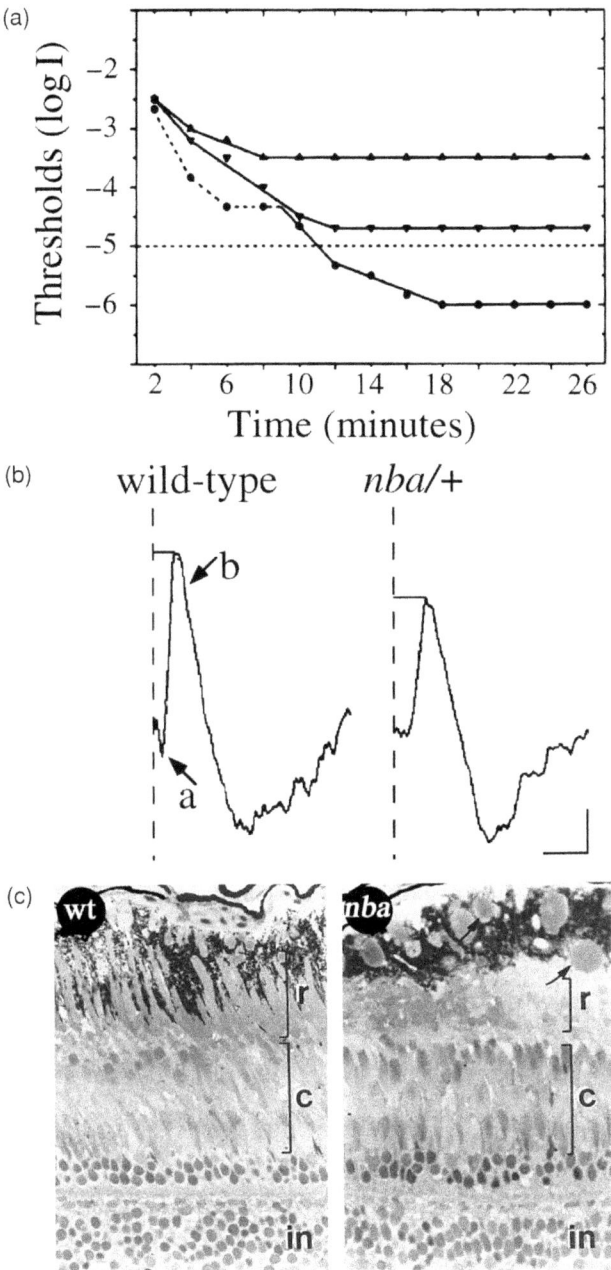

Figure 17.6 (a) Dark adaptation curves for wild-type (circles) and two *nba* fish (triangles) determined by behavioural testing. The biphasic curve for the wild-type larva reflects cone dark adaptation (dashed line) and the second phase reflects rod adaptation (solid line). (b) Full-field ERGs of wild-type (left) and *nba* (right) fish to white light stimuli. a, a-wave; b, b-wave. Calibration bars (right lower) signify 0.2 s horizontally and 50 µV vertically. (c) Histological sections showing the photoreceptor layer of 13-month-old wild-type (wt) and *nba* retina. Note the thinning of the rod outer segments (r) in the *nba* retina and the accumulation of lipid droplets in the RPE (arrow). c, cones; in, inner nuclear layer. (From Li and Dowling, 1997; reprinted with permission of the publisher, copyright © 1993–2005 by The National Academy of Sciences of the United States of America, all rights reserved.)

promoted rod differentiation while it inhibited cone maturation consistent with other models on the role of RA in photoreceptor cell development (Hyatt et al., 1996a; Levine et al., 2000).

These studies highlighted the potential use of the zebrafish for large-scale chemical genetic screens for small molecules that perturb specific aspects of organogenesis, pattern formation and neural genesis (Peterson, R. T. et al., 2000, 2001). However, unlike the example of the RA pathway screen, the chemical screens do not necessarily target a known biochemical pathway; rather, like forward genetic screens, they take an unbiased approach to identify compounds that when applied to developing vertebrate embryos yield a specific developmental phenotype. Similar to the design of cell culture assays, zebrafish embryos are arrayed into microtitre plates and exposed to chemical agents by adding the dissolved compounds into embryo medium. In this way, tens of thousands of small molecules previously arrayed into microtitre dishes can be systematically screened for effects upon discrete aspects of development. The in vivo model has several clear advantages over other culture assays. First, cellular and tissue interactions not present in vitro are maintained in vivo thus expanding the assay to detect alterations in tissue induction, cell migration and morphogenesis. Second, compounds may be applied to the embryos at any stage of development, thereby revealing the timing of gene action and limiting the pitfalls associated with the loss of function mutations displaying an earlier developmental phenotype that may otherwise obscure later functions of the gene.

The ability to screen large numbers of compounds led to a novel application of the chemical genetic screen to identify compounds with the potential to suppress the lethal phenotype of a genetic mutation (Peterson et al., 2004). Two of the 5000 compounds tested rescued the embryonic vascular defect associated with the mutation *gridlock* (*grl* affects *hey2*), and following the early treatment, the rescued mutants remained viable into adulthood. Thus there exists the potential for retinal-specific mutations in zebrafish to be models for identifying novel therapeutic agents to circumvent a pathway disrupted by a mutation or stimulate a compensatory pathway to alleviate a congenital defect even in the absence of identifying the mutated gene.

17.9 Concluding remarks

Even with the wealth of information gained by the analysis of the existing mutations in zebrafish, additional novel screens are necessary to reveal mutations not detected by current assays. Just as the OKR offered a clear advantage over morphological screens for detecting some types of visual deficits in otherwise normal larvae, other well thought out assays can uncover additional phenotypes. For example, it is probable that mutations specifically affecting rods were not detected in the previous larval screens. Based upon the ERG and visual-evoked behaviours, rod function usually cannot be detected prior to 21 dpf (Saszik et al., 1999; Bilotta et al., 2001) whereas most screens are conducted at 5 dpf. To detect changes in the rods, alternative approaches needed to be developed. One such method was the development of transgenic lines of zebrafish demonstrating rod-specific expression of a protein chimera between the enhanced green fluorescent protein and the C-terminal sequence of opsin as a reporter (Perkins et al., 2002). This permits screening live zebrafish larvae for changes in the number and spacing of rod photoreceptors as well as the vectorial sorting of opsin to the outer segment. Others have demonstrated the utility of *in situ* antibody labelling or histology-based screens to detect changes in specific cells of the neural retina (Morris and Fadool, 2005). Although somewhat more labour intensive than a transgenic screen, the latter offers the potential to label with multiple probes to thereby detect simultaneously changes in numerous cell types. With the growing number of transgenic lines demonstrating retinal-specific expression of fluorescent reporter genes, the development of more sophisticated behavioural assays and the rapidly advancing cloning techniques, the analysis of new or existing mutations should continue to uncover a wealth of information on the development and physiology of the retina.

17.10 Acknowledgements

The authors wish to thank Stephan Neuhauss for images used in this paper. The work from the authors' laboratories was supported by grants National Institutes of Health grants EY00811 and EY00824 to J. E. D. and EY13020 to J. M. F.

References

Ahmad, I., Dooley, C. M. and Polk, D. L. (1997). Delta-1 is a regulator of neurogenesis in the vertebrate retina. *Dev. Biol.* **185**, 92–103.

Allende, M. L., Amsterdam, A., Becker, T. *et al.* (1996). Insertional mutagenesis in zebrafish identifies two novel genes, *pescadillo* and *dead eye*, essential for embryonic development. *Genes. Dev.*, **10**, 3141–55.

Allwardt, B. A., Lall, A. B., Brockerhoff, S. E. and Dowling, J. E. (2001). Synapse formation is arrested in retinal photoreceptors of the zebrafish *nrc* mutant. *J. Neurosci.*, **21**, 2330–42.

Amsterdam, A., Nissen, R. M., Sun, Z. *et al.* (2004). Identification of 315 genes essential for early zebrafish development. *Proc. Natl. Acad. Sci. U. S. A.*, **101**, 12 792–7.

Austin, C. P., Feldman, D. E., Ida, J. A., Jr and Cepko C. L. (1995). Vertebrate retinal ganglion cells are selected from competent progenitors by the action of Notch. *Development*, **121**, 3637–50.

Baier, H. and Copenhagen, D. (2000). Combining physiology and genetics in the zebrafish retina. *J. Physiol.*, **524** Pt 1:1.

Baier, H., Klostermann, S., Trowe, T. *et al.* (1996). Genetic dissection of the retinotectal projection. *Development*, **123**, 415–25.

Baker, N. E., Mlodzik, M. and Rubin, G. M. (1990). Spacing differentiation in the developing *Drosophila* eye: a fibrinogen-related lateral inhibitor encoded by scabrous. *Science*, **250**, 1370–7.

Becker, T. S., Burgess, S. M., Amsterdam, A. H., Allende, M. L. and Hopkins, N. (1998). *not really finished* is crucial for development of the zebrafish outer retina and encodes a transcription factor highly homologous to human Nuclear Respiratory Factor-1 and avian Initiation Binding Repressor. *Development*, **125**, 4369–78.

Berson, E. L., Rosner, B., Weigel-DiFranco, C., Dryja, T. P. and Sandberg, M. A. (2002). Disease progression in patients with dominant *retinitis pigmentosa* and rhodopsin mutations. *Invest. Ophthalmol. Vis. Sci.*, **43**, 3027–36.

Bilotta, J., Saszik, S. and Sutherland, S. E. (2001). Rod contributions to the electroretinogram of the dark-adapted developing zebrafish. *Dev. Dyn.*, **222**, 564–70.

Boonanuntanasarn, S., Yoshizaki, G. and Takeuchi, T. (2003). Specific gene silencing using small interfering RNAs in fish embryos. *Biochem. Biophys. Res. Commun.*, **310**, 1089–95.

Branchek, T. (1984). The development of photoreceptors in the zebrafish, *Brachydanio rerio*. II. Function. *J. Comp. Neurol.*, **224**, 116–22.

Branchek, T. and Bremiller, R. (1984). The development of photoreceptors in the zebrafish, *Brachydanio rerio*. I. Structure. *J. Comp. Neurol.*, **224**, 107–15.

Brenner, S. (1974). The genetics of *Caenorhabditis elegans*. *Genetics*, **77**, 71–94.

Brockerhoff, S. E., Hurley, J. B., Janssen-Bienhold, U. *et al.* (1995). A behavioral screen for isolating zebrafish mutants with visual system defects. *Proc. Natl. Acad. Sci. U. S. A.*, **92**, 10 545–9.

Brockerhoff, S. E., Hurley, J. B., Niemi, G. A. and Dowling, J. E. (1997). A new form of inherited red-blindness identified in zebrafish. *J. Neurosci.*, **17**, 4236–42.

Brockerhoff, S. E., Rieke, F., Matthews, H. R. *et al.* (2003). Light stimulates a transducin-independent increase of cytoplasmic Ca^{2+} and suppression of current in cones from the zebrafish mutant nof. *J. Neurosci.*, **23**, 470–80.

Brown, N. L., Kanekar, S., Vetter, M. L. *et al.* (1998). Math5 encodes a murine basic helix-loop-helix transcription factor expressed during early stages of retinal neurogenesis. *Development*, **125**, 4821–33.

Bultman, S., Gebuhr, T., Yee, D. *et al.* (2000). A Brg1 null mutation in the mouse reveals functional differences among mammalian SWI/SNF complexes. *Mol. Cell*, **6**, 1287–95.

Burnside, B. (2001). Light and circadian regulation of retinomotor movement. *Prog. Brain Res.*, **131**, 477–85.

Burrill, J. D., Easter, S. S., Jr. (1995). The first retinal axons and their microenvironment in zebrafish: cryptic pioneers and the pretract. *J. Neurosci.*, **15**, 2935–47.

Cagan, R. L. and Ready, D. F. (1989). Notch is required for successive cell decisions in the developing Drosophila retina. *Genes Dev.*, **3**, 1099–112.

Cameron, D. A. and Carney, L. H. (2000). Cell mosaic patterns in the native and regenerated inner retina of zebrafish: implications for retinal assembly. *J. Comp. Neurol.*, **416**, 356–67.

Chakrabarti, S., Streisinger, G., Singer, F. and Walker, C. (1983). Frequency of *gamma*-ray induced specific locus and recessive lethal mutations in mature germ cells of the zebrafish, *Brachydanio rerio*. *Genetics*, **103**, 109–36.

Connaughton, V. P. and Nelson, R. (2000). Axonal stratification patterns and glutamate-gated conductance mechanisms in zebrafish retinal bipolar cells. *J. Physiol.*, **524** Pt 1, 135–46.

Connaughton, V. P., Graham, D. and Nelson, R. (2004). Identification and morphological classification of horizontal, bipolar, and amacrine cells within the zebrafish retina. *J. Comp. Neurol.*, **477**, 371–85.

Cook, J. E. and Chalupa, L. M. (2000). Retinal mosaics: new insights into an old concept. *Trends Neurosci.*, **23**, 26–34.

Dodd, A., Chambers, S. P. and Love, D. R. (2004). Short interfering RNA-mediated gene targeting in the zebrafish. *FEBS Lett.*, **561**, 89–93.

Doerre, G. and Malicki, J. (2001). A mutation of early photoreceptor development, *mikre oko*, reveals cell–cell interactions involved in the survival and differentiation of zebrafish photoreceptors. *J. Neurosci.*, **21**, 6745–57.

Doerre, G. and Malicki, J. (2002). Genetic analysis of photoreceptor cell development in the zebrafish retina. *Mech. Dev.*, **110**, 125–38.

Dosch, R., Wagner, D. S., Mintzer, K. A. *et al.* (2004). Maternal control of vertebrate development before the midblastula transition: mutants from the zebrafish I. *Dev. Cell*, **6**, 771–80.

Driever, W., Stemple, D., Schier, A. and Solnica-Krezel, L. (1994). Zebrafish: genetic tools for studying vertebrate development. *Trends Genet.*, **10**, 152–59.

Driever, W., Solnica-Krezel, L., Schier, A. F. *et al.* (1996). A genetic screen for mutations affecting embryogenesis in zebrafish. *Development*, **123**, 37–46.

Dryja, T. P., McGee, T. L., Hahn, L. B. *et al.* (1990a). Mutations within the *rhodopsin* gene in patients with autosomal dominant retinitis pigmentosa. *New Engl. J. Med.*, **323**, 1302–7.

Dryja, T. P., McGee, T. L., Reichel, E. *et al.* (1990b). A point mutation of the *rhodopsin* gene in one form of retinitis pigmentosa. *Nature*, **343**, 364–6.

Easter, S. S., Jr and Nicola, G. N. (1996). The development of vision in the zebrafish (*Danio rerio*). *Dev. Biol.*, **180**, 646–63.

Easter, S. S., Jr and Nicola, G. N. (1997). The development of eye movements in the zebrafish (*Danio rerio*). *Dev. Psychobiol.*, **31**, 267–76.

Eglen, S. J., van Ooyen, A. and Willshaw, D. J. (2000). Lateral cell movement driven by dendritic interactions is sufficient to form retinal mosaics. *Network: Comput. Neural Syst.*, **11**, 103–18.

Fadool, J. M. (2003). Development of a rod photoreceptor mosaic revealed in transgenic zebrafish. *Dev. Biol.*, **258**, 277–90.

Fadool, J. M., Brockerhoff, S. E., Hyatt, G. A. and Dowling, J. E. (1997). Mutations affecting eye morphology in the developing zebrafish (*Danio rerio*). *Dev. Genet.*, **20**, 288–95.

Gaiano, N., Amsterdam, A., Kawakami, K. *et al.* (1996). Insertional mutagenesis and rapid cloning of essential genes in zebrafish. *Nature*, **383**, 829–32.

Galli-Resta, L. (2002). Putting neurons in the right places: local interactions in the genesis of retinal architecture. *Trends Neurosci.*, **25**, 638–43.

Gan, W. B., Grutzendler, J., Wong, W. T., Wong, R. O. and Lichtman, J. W. (2000). Multicolor 'DiOlistic' labeling of the nervous system using lipophilic dye combinations. *Neuron*, **27**, 219–25.

Golling, G., Amsterdam, A., Sun, Z. *et al.* (2002). Insertional mutagenesis in zebrafish rapidly identifies genes essential for early vertebrate development. *Nat. Genet.*, **31**, 135–40.

Gregg, R. G., Willer, G. B., Fadool, J. M., Dowling, J. E. and Link, B. A. (2003). Positional cloning of the *young* mutation identifies an essential role for the Brahma chromatin re-modeling complex in mediating retinal cell differentiation. *Proc. Natl. Acad. Sci. U. S. A.*, **100**, 6535–40.

Gross, J. M., Perkins, B. D., Amsterdam, A. *et al.* (2005). Identification of zebrafish insertional mutants with defects in visual system development and function. *Genetics*, **170**, 245–61.

Grunwald, D. J. and Streisinger, G. (1992). Induction of recessive lethal and specific locus mutations in the zebrafish with ethyl nitrosourea. *Genet. Res.*, **59**, 103–16.

Haffter, P., Granato, M., Brand, M. *et al.* (1996). The identification of genes with unique and essential functions in the development of the zebrafish, *Danio rerio*. *Development*, **123**, 1–36.

Henikoff, S., Till, B. J. and Comai, L. (2004). TILLING. Traditional mutagenesis meets functional genomics. *Plant Physiol.*, **135**, 630–6.

Holzschuh, J., Hauptmann, G. and Driever, W. (2003). Genetic analysis of the roles of Hh, FGF8, and nodal signaling during catecholaminergic system development in the zebrafish brain. *J. Neurosci.*, **23**, 5507–19.

Hu, M. and Easter, S. S. (1999). Retinal neurogenesis: the formation of the initial central patch of post-mitotic cells. *Dev. Biol.*, **207**, 309–21.

Hyatt, G. A. and Dowling, J. E. (1997). Retinoic acid. A key molecule for eye and photoreceptor development. *Invest. Ophthalmol. Vis. Sci.*, **38**, 1471–5.

Hyatt, G. A., Schmitt, E. A., Marsh-Armstrong, N. R. and Dowling, J. E. (1992). Retinoic acid-induced duplication of the zebrafish retina. *Proc. Natl. Acad. Sci. U. S. A.*, **89**, 8293–7.

Hyatt, G. A., Schmitt, E. A., Fadool, J. M. and Dowling, J. E. (1996a). Retinoic acid alters photoreceptor development *in vivo*. *Proc. Natl. Acad. Sci. U. S. A.*, **93**, 13 298–303.

Hyatt, G. A., Schmitt, E. A., Marsh-Armstrong, N. *et al.* (1996b). Retinoic acid establishes ventral retinal characteristics. *Development*, **122**, 195–204.

Jensen, A. M. and Westerfield, M. (2004). Zebrafish *mosaic eyes* is a novel FERM protein required for retinal lamination and retinal pigmented epithelial tight junction formation. *Curr. Biol.*, **14**, 711–17.

Jensen, A. M., Walker, C. and Westerfield, M. (2001). *mosaic eyes*: a zebrafish gene required in pigmented epithelium for apical localization of retinal cell division and lamination. *Development*, **128**, 95–105.

Johns, P. R. (1982). Formation of photoreceptors in larval and adult goldfish. *J. Neurosci.*, **2**, 178–98.

Kainz, P. M., Adolph, A. R., Wong, K. Y. and Dowling, J. E. (2003). *Lazy eyes* zebrafish mutation affects Müller glial cells, compromising photoreceptor function and causing partial blindness. *J. Comp. Neurol.*, **463**, 265–80.

Karlstrom, R. O., Trowe, T., Klostermann, S. *et al.* (1996). Zebrafish mutations affecting retinotectal axon pathfinding. *Development*, **123**, 427–38.

Kay, J. N., Finger-Baier, K. C., Roeser, T., Staub, W. and Baier, H. (2001). Retinal ganglion cell genesis requires *lakritz*, a zebrafish *atonal* homolog. *Neuron*, **30**, 725–736.

Kljavin, I. J. (1987). Early development of photoreceptors in the ventral retina of the zebrafish embryo. *J. Comp. Neurol.*, **260**, 461–71.

Kolb, H., Nelson, R., Ahnelt, P. and Cuenca, N. (2001). Cellular organization of the vertebrate retina. *Prog. Brain Res.*, **131**, 3–26.

Larison, K. D. and Bremiller, R. (1990). Early onset of phenotype and cell patterning in the embryonic zebrafish retina. *Development*, **109**, 567–76.

Levine, E. M., Fuhrmann, S. and Reh, T. A. (2000). Soluble factors and the development of rod photoreceptors. *Cell Mol. Life Sci.*, **57**, 224–34.

Li, L. and Dowling, J. E. (1997). A dominant form of inherited retinal degeneration caused by a non-photoreceptor cell-specific mutation. *Proc. Natl. Acad. Sci. U. S. A.*, **94**, 11645–50.

Lin, B., Wang, S. W. and Masland, R. H. (2004). Retinal ganglion cell type, size, and spacing can be specified independent of homotypic dendritic contacts. *Neuron*, **43**, 475–85.

Lin, S., Gaiano, N., Culp, P. et al. (1994). Integration and germ-line transmission of a pseudotyped retroviral vector in zebrafish. *Science*, **265**, 666–9.

Link, B. A., Fadool, J. M., Malicki, J. and Dowling, J. E. (2000). The zebrafish *young* mutation acts non-cell-autonomously to uncouple differentiation from specification for all retinal cells. *Development*, **127**, 2177–88.

Lister, J. A., Close, J. and Raible, D. W. (2001). Duplicate *mitf* genes in zebrafish: complementary expression and conservation of melanogenic potential. *Dev. Biol.*, **237**, 333–44.

Liu, Y., Shen, Y., Rest, J. S., Raymond, P. A. and Zack, D. J. (2001). Isolation and characterization of a zebrafish homologue of the cone rod homeobox gene. *Invest. Ophthalmol. Vis. Sci.*, **42**, 481–7.

Loosli, F., Koster, R. W., Carl, M., Krone, A. and Wittbrodt, J. (1998). *Six3*, a medaka homologue of the *Drosophila* homeobox gene *sine oculis* is expressed in the anterior embryonic shield and the developing eye. *Mech. Dev.*, **74**, 159–64.

Loosli, F., Winkler, S. and Wittbrodt, J. (1999). *Six3* over-expression initiates the formation of ectopic retina. *Genes Dev.*, **13**, 649–54.

Loosli, F., Staub, W., Finger-Baier, K. C. et al. (2003). Loss of eyes in zebrafish caused by mutation of *chokh/rx3*. *EMBO Rep.*, **4**, 894–9.

Maaswinkel, H., Ren, J. Q. and Li, L. (2003). Slow-progressing photoreceptor cell degeneration in *night blindness c* mutant zebrafish. *J. Neurocytol.*, **32**, 1107–16.

Macdonald, R., Barth, K. A., Xu, Q. et al. (1995). Midline signaling is required for *Pax* gene regulation and patterning of the eyes. *Development*, **121**, 3267–78.

Maden, M. and Holder, N. (1992). Retinoic acid and development of the central nervous system. *BioEssays*, **14**, 431–8.

Malicki, J. (2000). Harnessing the power of forward genetics–analysis of neuronal diversity and patterning in the zebrafish retina. *Trends Neurosci.*, **23**, 531–41.

Malicki, J., Neuhauss, S. C., Schier, A. F. et al. (1996). Mutations affecting development of the zebrafish retina. *Development*, **123**, 263–73.

Malicki, J., Jo, H., Wei, X., Hsiung, M. and Pujic, Z. (2002). Analysis of gene function in the zebrafish retina. *Methods*, **28**, 427–38.

Malicki, J., Jo, H. and Pujic, Z. (2003). Zebrafish, N-cadherin, encoded by the *glass onion* locus, plays an essential role in retinal patterning. *Dev. Biol.*, **259**, 95–108.

Mangrum, W. I., Dowling, J. E. and Cohen, E. D. (2002). A morphological classification of ganglion cells in the zebrafish retina. *Vis. Neurosci.*, **19**, 767–79.

Marc, R. E. and Cameron, D. (2001). A molecular phenotype atlas of the zebrafish retina. *J. Neurocytol.*, **30**, 593–654.

Marsh-Armstrong, N., McCaffery, P., Gilbert, W., Dowling, J. E. and Drager, U. C. (1994). Retinoic acid is necessary for development of the ventral retina in zebrafish. *Proc. Natl. Acad. Sci. U. S. A.*, **91**, 7286–90.

Masai, I., Lele, Z., Yamaguchi, M. et al. (2003). N-cadherin mediates retinal lamination, maintenance of forebrain compartments and patterning of retinal neurites. *Development*, **130**, 2479–94.

Masland, R. H. (2001). The fundamental plan of the retina. *Nat. Neurosci.*, **4**, 877–86.

Matthews, H. R. and Fain, G. L. (2001). A light-dependent increase in free Ca^{2+} concentration in the salamander rod outer segment. *J. Physiol.*, **532**, 305–21.

Matthews, H. R. and Fain, G. L. (2002). Time course and magnitude of the calcium release induced by bright light in salamander rods. *J. Physiol.*, **542**, 829–41.

McCallum, C. M., Comai, L., Greene, E. A. and Henikoff, S. (2000). Targeted screening for induced mutations. *Nat. Biotechnol.*, **18**, 455–7.

McMahon, D. G. (1994). Modulation of electrical synaptic transmission in zebrafish retinal horizontal cells. *J. Neurosci.*, **14**, 1722–34.

McWilliam, P., Farrar, G. J., Kenna, P. *et al.* (1989). Autosomal dominant retinitis pigmentosa (ADRP): localization of an ADRP gene to the long arm of chromosome 3. *Genomics*, **5**, 619–622.

Morris, A. C. and Fadool, J. M. (2005). Studying rod photoreceptor development in zebrafish. *Physiol. Behav.*, **86**, 306–13.

Mullins, M. C. and Nusslein-Volhard, C. (1993). Mutational approaches to studying embryonic pattern formation in the zebrafish. *Curr. Opin. Genet. Dev.*, **3**, 648–54.

Nasevicius, A. and Ekker, S. C. (2000). Effective targeted gene 'knockdown' in zebrafish. *Nat. Genet.*, **26**, 216–20.

Neuhauss, S. C., Biehlmaier, O., Seeliger, M. W. *et al.* (1999). Genetic disorders of vision revealed by a behavioral screen of 400 essential loci in zebrafish. *J. Neurosci.*, **19**, 8603–15.

Neumann, C. J. and Nuesslein-Volhard, C. (2000). Patterning of the zebrafish retina by a wave of sonic hedgehog activity. *Science*, **289**, 2137–39.

Nornes, S., Clarkson, M., Mikkola, I. *et al.* (1998). Zebrafish contains two *pax6* genes involved in eye development. *Mech. Dev.*, **77**, 185–96.

Novelli, E., Resta, V. and Galli-Resta, L. (2004). Mechanisms controlling the formation of retinal mosaics. *Prog. Brain Res.*, **147**, 141–53.

Olave, I., Wang, W., Xue, Y., Kuo, A. and Crabtree, G. R. (2002). Identification of a polymorphic, neuron-specific chromatin remodeling complex. *Genes Dev.*, **16**, 2509–17.

Otteson, D. C. and Hitchcock, P. F. (2003). Stem cells in the teleost retina: persistent neurogenesis and injury-induced regeneration. *Vis. Res.*, **43**, 927–36.

Otteson, D. C., D'Costa, A. R. and Hitchcock, P. F. (2001). Putative stem cells and the lineage of rod photoreceptors in the mature retina of the goldfish. *Dev. Biol.*, **232**, 62–76.

Papermaster, D. S. (1995). Necessary but insufficient. *Nat. Med.*, **1**, 874–75.

Perkins, B. D., Kainz, P. M., O'Malley, D. M. and Dowling, J. E. (2002). Transgenic expression of a GFP-rhodopsin COOH-terminal fusion protein in zebrafish rod photoreceptors. *Vis. Neurosci.*, **19**, 257R–64R.

Peterson, R. E., Fadool, J. M., McClintock, J. and Linser, P. J. (2001). Müller cell differentiation in the zebrafish neural retina: evidence of distinct early and late stages in cell maturation. *J. Comp. Neurol.*, **429**, 530–40.

Peterson, R. T., Link, B. A., Dowling, J. E. and Schreiber, S. L. (2000). Small molecule developmental screens reveal the logic and timing of vertebrate development. *Proc. Natl. Acad. Sci. U. S. A.*, **97**, 12 965–9.

Peterson, R. T., Mably, J. D., Chen, J. N. and Fishman, M. C. (2001). Convergence of distinct pathways to heart patterning revealed by the small molecule concentramide and the mutation heart-and-soul. *Curr. Biol.*, **11**, 1481–91.

Peterson, R. T., Shaw, S. Y., Peterson, T. A. *et al.* (2004). Chemical suppression of a genetic mutation in a zebrafish model of aortic coarctation. *Nat. Biotechnol.*, **22**, 595–9.

Raymond, P. A. and Rivlin, P. K. (1987). Germinal cells in the goldfish retina that produce rod photoreceptors. *Dev. Biol.*, **122**, 120–38.

Raymond, P. A., Barthel, L. K., Rounsifer, M. E., Sullivan, S. A. and Knight, J. K. (1993). Expression of rod and cone visual pigments in goldfish and zebrafish: a rhodopsin-like gene is expressed in cones. *Neuron*, **10**, 1161–74.

Raymond, P. A., Barthel, L. K. and Curran, G. A. (1995). Developmental patterning of rod and cone photoreceptors in embryonic zebrafish. *J. Comp. Neurol.*, **359**, 537–50.

Rick, J. M., Horschke, I. and Neuhauss, S. C. (2000). Optokinetic behavior is reversed in achiasmatic mutant zebrafish larvae. *Curr. Biol.*, **10**, 595–8.

Rivolta, C., Sharon, D., DeAngelis, M. M. and Dryja, T. P. (2002). Retinitis pigmentosa and allied diseases: numerous diseases, genes, and inheritance patterns. *Hum. Mol. Genet.*, **11**, 1219–27.

Robinson, J., Schmitt, E. A., Harosi, F. I., Reece, R. J. and Dowling, J. E. (1993). Zebrafish ultraviolet visual pigment: absorption spectrum, sequence, and localization. *Proc. Natl. Acad. Sci. U. S. A.*, **90**, 6009–12.

Rockhill, R. L., Euler, T. and Masland, R. H. (2000). Spatial order within but not between types of retinal neurons. *Proc. Natl. Acad. Sci. U. S. A.*, **97**, 2303–7.

Roeser, T. and Baier, H. (2003). Visuomotor behaviors in larval zebrafish after GFP-guided laser ablation of the optic tectum. *J. Neurosci.*, **23**, 3726–34.

Ross, S. A., McCaffery, P. J., Drager, U. C. and De, Luca L. M. (2000). Retinoids in embryonal development. *Physiol. Rev.*, **80**, 1021–54.

Saszik, S., Bilotta, J. and Givin, C. M. (1999). ERG assessment of zebrafish retinal development. *Vis. Neurosci.*, **16**, 881–8.

Schmitt, E. A. and Dowling, J. E. (1994). Early eye morphogenesis in the zebrafish, *Brachydanio rerio*. *J. Comp. Neurol.*, **344**, 532–42.

Schmitt, E. A. and Dowling, J. E. (1996). Comparison of topographical patterns of ganglion and photoreceptor cell differentiation in the retina of the zebrafish, *Danio rerio*. *J. Comp. Neurol.*, **371**, 222–34.

Schmitt, E. A. and Dowling, J. E. (1999). Early retinal development in the zebrafish, *Danio rerio*: light and electron microscopic analyses. *J. Comp. Neurol.*, **404**, 515–36.

Seo, H. C., Drivenes, Ellingsen, S. and Fjose, A. (1998). Expression of two zebrafish homologs of the murine *Six3* gene demarcates the initial eye primordia. *Mech. Dev.*, **73**, 45–57.

Seo, S., Richardson, G. A. and Kroll, K. L. (2005). The SWI/SNF chromatin re-modeling protein Brg1 is required for vertebrate neurogenesis and mediates transactivation of Ngn and NeuroD. *Development*, **132**, 105–15.

Shkumatava, A., Fischer, S., Muller, F., Strahle, U. and Neumann, C. J. (2004). Sonic hedgehog, secreted by amacrine cells, acts as a short-range signal to direct differentiation and lamination in the zebrafish retina. *Development*, **131**, 3849–58.

Solnica-Krezel, L., Schier, A. F. and Driever, W. (1994). Efficient recovery of ENU-induced mutations from the zebrafish germline. *Genetics*, **136**, 1401–20.

Starr, C. J., Kappler, J. A., Chan, D. K., Kollmar, R. and Hudspeth, A. J. (2004). Mutation of the zebrafish *choroideremia* gene encoding Rab escort protein 1 devastates hair cells. *Proc. Natl. Acad. Sci. U. S. A.*, **101**, 2572–7.

Streisinger, G., Walker, C., Dower, N., Knauber, D. and Singer, F. (1981). Production of clones of homozygous diploid zebra fish (*Brachydanio rerio*). *Nature*, **291**, 293–6.

Taylor, M. R., Hurley, J. B., Van Epps, H. A. and Brockerhoff, S. E. (2004). A zebrafish model for pyruvate dehydrogenase deficiency: rescue of neurological dysfunction and embryonic lethality using a ketogenic diet. *Proc. Natl. Acad. Sci. U. S. A.*, **101**, 4584–9.

Taylor, M. R., Kikkawa, S., Diez-Juan, A. *et al.* (2005). The zebrafish *pob* gene encodes a novel protein required for survival of red cone photoreceptor cells. *Genetics*, **170**, 263–73.

Till, B. J., Reynolds, S. H., Weil, C. et al. (2004). Discovery of induced point mutations in maize genes by TILLING. *BMC Plant Biol.*, **4**, 12.

Trowe, T., Klostermann, S., Baier, H. et al. (1996). Mutations disrupting the ordering and topographic mapping of axons in the retinotectal projection of the zebrafish, *Danio rerio*. *Development*, **123**, 439–50.

Tsujikawa, M. and Malicki, J. (2004). Intraflagellar transport genes are essential for differentiation and survival of vertebrate sensory neurons. *Neuron*, **42**, 703–16.

Urtishak, K. A., Choob, M., Tian, X. et al. (2003). Targeted gene knockdown in zebrafish using negatively charged peptide nucleic acid mimics. *Dev. Dyn.*, **228**, 405–13.

Van Epps, H. A., Hayashi, M., Lucast, L. et al. (2004). The zebrafish nrc mutant reveals a role for the polyphosphoinositide phosphatase synaptojanin 1 in cone photoreceptor ribbon anchoring. *J. Neurosci.*, **24**, 8641–50.

Vihtelic, T. S., Yamamoto, Y., Sweeney, M. T., Jeffery, W. R. and Hyde, D. R. (2001). Arrested differentiation and epithelial cell degeneration in zebrafish lens mutants. *Dev. Dyn.*, **222**, 625–36.

Wagner, D. S., Dosch, R., Mintzer, K. A., Wiemelt, A. P. and Mullins, M. C. (2004). Maternal control of development at the midblastula transition and beyond: mutants from the zebrafish II. *Dev. Cell.*, **6**, 781–90.

Waid, D. K. and McLoon, S. C. (1998). Ganglion cells influence the fate of dividing retinal cells in culture. *Development*, **125**, 1059–66.

Walker, C. and Streisinger, G. (1983). Induction of mutations by *gamma*-rays in pregonial germ cells of zebrafish embryos. *Genetics*, **103**, 125–36.

Wargelius, A., Seo, H. C., Austbo, L. and Fjose, A. (2003). Retinal expression of zebrafish six3.1 and its regulation by *Pax6*. *Biochem. Biophys. Res. Commun.*, **309**, 475–81.

Wässle, H. and Riemann, H. J. (1978). The mosaic of nerve cells in the mammalian retina. *Proc. R. Soc. London B Biol. Sci.*, **200**, 441–61.

Wei, X. and Malicki, J. (2002). *nagie oko*, encoding a MAGUK-family protein, is essential for cellular patterning of the retina. *Nat. Genet.*, **31**, 150–7.

Wei, X., Cheng, Y., Luo, Y. et al. (2004). The zebrafish *Pard3* ortholog is required for separation of the eye fields and retinal lamination. *Dev. Biol.*, **269**, 286–301.

Wickstrom, E., Choob, M., Urtishak, K. A. et al. (2004a). Sequence specificity of alternating hydroyprolyl/phosphono peptide nucleic acids against zebrafish embryo mRNAs. *J. Drug Target.*, **12**, 363–72.

Wickstrom, E., Urtishak, K. A., Choob, M. et al. (2004b). Downregulation of gene expression with negatively charged peptide nucleic acids (PNAs) in zebrafish embryos. *Methods Cell Biol.*, **77**, 137–58.

Wienholds, E., Schulte-Merker, S., Walderich, B. and Plasterk, R. H. (2002). Target-selected inactivation of the zebrafish *rag1* gene. *Science*, **297**, 99–102.

Wienholds, E., van Eeden, F., Kosters, M. et al. (2003). Efficient target-selected mutagenesis in zebrafish. *Genome Res.*, **13**, 2700–7.

Wong, K. Y., Gray, J., Hayward, C. J. C., Adolph, A. R. and Dowling, J. E. (2004). Glutamatergic mechanisms in the outer retina of larval zebrafish: analysis of electroretinogram b- and d-waves using a novel preparation. *Zebrafish*, **1**, 121–31.

Yazulla, S. and Studholme, K. M. (2001). Neurochemical anatomy of the zebrafish retina as determined by immunocytochemistry. *J. Neurocytol.*, **30**, 551–92.

Index

ACh (acetylcholine) 99
   mechanisms of release in development 99–100, 101
   modes of action 102
   role in retinal wave generation 275–6, 279
   sources in development 101–2
AChRs (ACh receptors), in the developing retina 103–4, 105
amacrine cells 1–2, 3, 30
   AII type 2–3
   cholinergic amacrine cells 212, 274–6, 279
   displaced 44
   early synapse formation 266–7, 268
   evidence for programmed cell death 212
   factors which shape arborization 245, 246
   identification 44
   morphological development 245, 246
   neuroblast migrations 60, 63–4
   size 44
   spontaneous bursting in immature cells 272–5, 279
   starburst amacrine cells 297
   subtypes 30, 44–5
   timing of genesis 44–5
   see also retinal waves
ANB (anterior neural border) 12–13
anterior neural plate, separation of eye field 19–20, 21
apical polarity complex, role in retinal cell migration 67, 68
apoptosis 130–1, 136, 208
   evidence from pyknotic profiles 213–14
   molecular mechanisms 208
   process 213–14
   see also cell death; PCD (programmed cell death)
arcuate fibres 159
astrocytes in the retina 172–3
   origin of 173
   role in vascular development 182–3, 184
asymmetrical cell division, RPCs 32–3, 35–8
Ath5 gene, bHLH proneural gene 79–80, 81

ath5 gene, zebrafish 348
atonal5 gene, zebrafish 348
ATP (adenosine triphosphate) 99
   e-ATP (extracellular ATP) 199
   mechanisms of release in development 99–100, 101
   modes of action 102
   receptors in the developing retina 104–5, 106
   receptors on developing Müller cells 182
   sources in development 102
auto correlation plot, for retinal mosaics 201–2, 203
autophagy, form of cell death 216–17
autoradiography 31–3, 34–5
axon guidance/pathfinding *see* neurite field; neurite growth; RGC axons
axon pathways, optic nerve 150
axonal regeneration 150
axonal transport 150
Axonin 1 (IgCAM), role in axon guidance 155–6, 157

basal lamina *see* ILM (inner limiting membrane; retinal basal lamina)
Bassoon 270, 292
bax pro-apoptotic gene 215
Bcl-2 gene family 214, 216–17
BDNF (brain-derived neurotrophic factor) 115
   and programmed cell death 116–17
   receptor binding 115
   role in cell death and survival 219
behavioural screens, for zebrafish mutants 353–4, 356, 357, 358
bel (*belladonna*) zebrafish mutant 357, 358
bFGF (basic fibroblast growth factor), production by Müller cells 176
bHLH proneural genes
   activation by *Pax6* 82
   Ath5 79–80, 81
   generation of multiple cell fates 79–80, 81
   glial cell fate regulation 80
bHLH transcription factors, in the Notch pathway 78–9
bicuculline 299

bipolar cells 1–2, 3, 30
  evidence for programmed cell death 212
  factors which shape arborization 245, 246
  midget bipolar cells 126
  morphological development 245, 246
  recoverin immunostaining 293–6
  ribbon synapse formation 266–7, 268, 269, 270
  role in IPL ON and OFF sublamination 293–4, 295, 296, 297
blindness, heritable causes 357–9
blood supply to the eye 1, 2 see also retinal vasculature
BMP (bone morphogenetic protein) inhibition
  and eye formation 9–11
  role in neural induction and patterning 11–14
BrdU (5′-Bromo-2′-deoxy-uridine) 33, 34–5
brg1 (Brahma related) gene function, zebrafish 352

c-fos gene, light-induced expression 289
Ca^{2+} role in the CNS 103, 112
Ca^{2+}-stimulated synaptic vesicle exocytosis 269
$[Ca^{2+}]_i$
  role in development 103, 112
  modulation by mAChRs 103–4, 105
  modulation by neurotransmitters 103, 112
$[Ca^{2+}]_i$ local waves 112, 113
$[Ca^{2+}]_i$ propagating waves 112, 114
$[Ca^{2+}]_i$ transients 103, 112
  modulation by neurotransmitters 111–12, 113
  role in cell cycle regulation 111–12, 113
cadherin 2 (CDH2; N-cadherin), role in retinal cell migration 68
caecal period 128–9, 130
caspase 3-mediated apoptosis 214, 216–17
catecholamine receptors, on developing Müller cells 182
β-catenin 12
cDNA microarrays 327–8
cell adhesion molecules
  role in retinal cell migration 66, 68
  signal transduction 66
cell autonomous factors, in cell fate determination 42, 47–51
cell birth order in the retina 34–5, 39, 40–2, 47
  amacrine cells 44–5
  horizontal cells 34–5, 45
  Müller glia 47
  photoreceptors 42, 45–7
  retinal ganglion cells 43
cell birth order within clones 42, 47–51
cell birthday 39, 40–1, 42
cell–cell interactions, retinogenic zone 8–10, 11
cell cycle in the retina 31–3, 34–5
  and cell fate determination 87–9, 90
  effects of bHLH proneural genes 79–80, 81
  exit on differentiation 87–8
  impact of cell fate determinants 51–2, 90

modulated timing 35, 37
timing 33–5, 36–7
cell death 115
  apoptosis in the early retina 130–1, 136
  apoptotic properties of NGF 116–17
  in retinal mosaic development 197–9
  programmed phases in the retina 116–17
  role of e-ATP 199
  see also apoptosis; PCD (programmed cell death)
cell differentiation in the retina 75, 128–9, 130
  and NT-3 115–16
  combinatorial action of transcription factors 82–3
  direction of progress 128–9, 130
  exit from the cell cycle 87–8
  influence of $[Ca^{2+}]_i$ levels 103, 112
  influence of neurotransmitters 109–10, 111–12, 113, 114–15
  starting point 128–9, 130
cell displacement
  foveal cone packing 141–2, 143, 144
  foveal pit formation 132, 138–9, 140
  to form the fovea 127, 128
cell division
  asymmetrical 32–3, 35–8
  cytoplasmic contents division 38–9, 40
  symmetrical 32–3, 35–8
  terminal (RPCs) 32–3, 35–8
cell division mode, and plane of cleavage 38–9, 40
cell-extrinsic cues, retinal progenitor regulation 76–9
cell fate determination in the vertebrate retina 30–1
  action of homeobox transcription factors 81
  and cell cycle progression 87–9, 90
  cell autonomous and non-autonomous factors 42, 47–51, 52
  compared with Drosophila eye and CNS 90–1
  compared with vertebrate spinal cord 91
  competence model 83, 84–5
  future investigations 91–2
  gene interactions 51–2
  impact on cell cycle 90
  in retinal mosaic development 195–6
  models 51–2
  progenitor heterogeneity (mosaic) model 84–5, 87
  role of p27Xic1 cell cycle inhibitor 88, 89
  roles of cell cycle components 88–9
  stochastic choice model 84–5, 87
  study of the retina 75
  timing of 90
cell interactions, in retinal mosaic development 194–5
cell-intrinsic factors
  multiple interactions between 82–3
  retinal progenitor regulation 76, 79–81, 83
cell lineage 42, 47–51

Index

cell migration
  cortical cell studies 60, 70–1
  glial-guided migration 60, 61–2
  in retinal mosaic development 196–7
  in the retina 59–61
  mechanisms 60, 61–4
  retinal and cortical comparisons 60, 70–1
  somal translocation 60, 62–3
  time-lapse imaging 60, 70
  unconstrained migration 60, 63–4
cell migration signalling 66–7, 69
  apical polarity complex 67, 68
  cell adhesion molecules 66, 68
  directed cell motility 67, 69
  extracellular guidance cues 66–7, 68
  intracellular mechanisms 67, 69
  intracellular signalling molecules 67, 69
  mechanisms of somal translocation 67, 69
  retinal basal lamina (ILM) 66–7, 68
  role of cadherin 2 (CDH2; N-cadherin) 68
  role of integrins 68
  role of RhoGTPases 69
cell migration trajectories 64–5
  bi directional 64–5
  radial and tangential 64, 65
  role of the ILM 64–5
  uni directional 64–5
cell non-autonomous factors, in cell fate determination 42, 47–51
cell signalling 30–1 *see also specific processes*
cell-specific markers, gene expression profiling 328, 335, 336
cellular polarity, effects on retinal lamination 242–4
cerberus gene mis expression 11
Cerberus inhibition of signalling cascades 9–11
chemical genetic screens 360–2
chemical synapses 266–7, 271
chemokines, roles in cell death and survival 219–20
cholinergic amacrine cells 212
  mediation of retinal waves 274–6, 279
cholinergic neurotransmitters, regulation of progenitor cell cycle 109–10, 111–12, 113
chromosome duplication (S-phase) 31–3, 34–5
ciliary body, development of 1, 2
ciliary epithelium (mammals), possible neogenesis of neurons 317–19, 320
circadian rhythm generation 289–91
clonal genesis 42, 47–51
CMZ (ciliary margin zone)
  continued neurogenesis in fish and amphibians 308–9, 310, 311
  persistent progenitor/retinal stem cells 308–10, 311
  source for retinal regeneration 307–9, 310, 311
CNS axon pathways 150
coloboma 152–4

color vision *see* fovea
competence model, retinal cell fate determination 83, 84–5
computer models
  d_{min} model 194–5, 199
  retinal mosaic formation 194–5, 199
cone bipolar cells, ON and OFF types 2–3
cone opsin expression 135–6, 137–40, 288
cone photoreceptors 1–2, 3, 30
  cone dominated or pure cone retinas 42, 45–6
  cone packing in the fovea 127, 128, 141–2, 143, 144
  development of inner and outer segments 141–2, 144
  development of synaptic pedicle 141–2, 144
  evidence for programmed cell death 212
  Henle fibre (axon) development 141–2, 143, 144
  maturation time 46–7
  morphological changes during foveal development 141–2, 143, 144
  morphological development 245, 246
  multiple types 46
  ON and OFF cones 293–4, 295, 296, 297
  specialized subcircuits 2–3
  subtype arrangement in zebrafish retina 345, 347
  timing of genesis 42, 45–6
  see also photoreceptors
connexins 271–2
contrast sensitivity in the retina 243, 249–50
cornea, development of 1, 2
cortical cell migration 60, 70–1
  glial-guided migration 60, 61–2
  somal translocation 63
  unconstrained migration 63–4
cross-correlation plot, for retinal mosaics 201–2, 203
cross-correlation studies, retinal mosaic development 195
CSPG (chondroitin-sulphate proteoglycans), role in orientation of axons 154–5
curare 299
cyc (*cyclops*) gene 21
cyclopia 151
cyclopic mutations 20–1
cytokines, roles in cell death and survival 219–20

dark rearing
  effects on retinal waves 280–1
  effects on RGC receptive fields 298–9
DCC guidance molecule 159–60, 161, 163–5
De Morsier's syndrome 162–3
decussation patterns of RGCs 224, 225–6
Delaunay triangulation 203–4
dendrites
  field size 245–8, 249
  interactions in retinal mosaic development 199–200, 201
  see also neurite field; neurite growth
dendritic filopodial motility 250–1, 252
dendritic nets 199–200, 201

dendro-dendritic contacts among RGCs 247–8, 249
developmental competence, RPCs 42, 47–51
DHA (docosahexaenoic acid), trafficking by Müller cells 175–6
dkk1 (dickkopf1), Wnt inhibition 12–13
dkk1 (dickkopf1) gene, effects on forebrain patterning 12–13
differentiation *see* cell differentiation
digital gene expression profiling 326–7, 328, 329
displaced amacrine cells 44
DM-GRASP (IgCAM), role in axon guidance 155–6, 157
dopamine, role in cell death and survival 220–1
dopamine receptors, on developing Müller cells 182
dopaminergic amacrine cells 212
Drosophila
   eye and CNS cell fate determination 90–1
   eye formation genes 16–17
   gene expression database 329
Dsh (Dishevelled) intracellular protein 12

e-ATP (extracellular adenosine triphosphate), role in cell death 199
ECM (extracellular matrix)
   influence on synapse formation 270–1
   modulation of developmental cell death 221
ectopic eyes
   induction in *Drosophila* 16–17
   induction with EFTFs 16–17
   response to light 16–17
EFTFs (eye field transcription factors) 8
   eye field specification 14–15, 17, 19
   genome duplicates and homologues 18
   midline repression to form eye primordia 19–20
electrical synapses 271–2
embryogenesis, development of the eye 1, 2
ENU mutagenesis 343
Eph proteins, role in RGC axon pathfinding 157–8, 159
ephrin proteins, role in RGC axon pathfinding 157–8, 159
ERGs (electroretinograms)
   a-, b- and c-wave components 292
   to monitor developing light responses 292
extracellular matrix *see* ECM
extrinsic cues, retinal progenitor regulation 76–9
eye, development 1, 2
eye field (eye anlage)
   origins 8
   specification 8–9, 10, 11, 14–15, 17, 19
   see also EFTFs (eye field transcription factors); eye primordia formation
eye field formation
   and Wnt signalling 12–13
   fate-mapping experiments 14, 15
   transplantation experiments 14
eye field patterning
   role of Wnt signalling 11–14
   signalling systems 12

eye formation, and neural induction 9–11
eye formation genes 14–15, 17, 19
   Drosophila 16–17
eye growth, retinal stretch 132, 144–5
eye opening, light responsiveness 288
eye primordia formation
   displacement of neural plate cells 20–1
   genes and signalling pathways 20–1
   reprogramming of midline cells 19–20
   separation of single eye field 19–20, 21
   see also eye field
eye regeneration, history of study 307–8, 309
eye-specific inputs, segregation in the lateral geniculate nucleus 274–5

FGFs (fibroblast growth factors), neural induction and patterning 11–14
forebrain patterning, and Wnt activity 11–14
forkhead (foxb1.2) gene 20–1
fovea
   avoidance by RGC axons 159
   cell types not found in 130–1, 136
   colour vision 128
   development 127, 128
   midget pathway 126, 127, 128
   starting point for cell differentiation 128–9, 130
   structure 127, 128
   synapse formation 131–2, 134, 135, 138–9
   visual acuity 128
fovea centralis
   early studies 126
   structure 127, 128
   see also fovea
foveal avascular zone (FAZ) 127, 128, 132, 138–9, 140
foveal cone mosaic 132, 137–9, 140
   cone packing 141–2, 143, 144
foveal depression formation 132, 134, 137–9, 140
   cell displacement 132, 138–9, 140
   pit stage 132, 138–9, 140
   pre-pit stage 134, 137–9, 140
foveal rim 127, 128
foveal slope 127, 128
Fz•LRP (Frizzled•low density lipoprotein receptor-related protein) complex 12

GABA (γ-aminobutyric acid) 99
   mechanisms of release in development 99–100, 101
   modes of action 102
   role in retinal wave propagation 273, 276–9, 280–1
   sources in development 102
GABA receptors
   in the developing retina 105, 106–7
   on developing Müller cells 182
$GABA_A$, role in retinal wave propagation 273, 276–9, 280–1

GABAergic neurotransmission 268–9
ganglion cells *see* retinal ganglion cells (RGCs)
gap junctions 271–2
  mediation of developmental cell death 222
  role in retinal wave propagation 276
gap-junction hemichannels, neurotransmitter release 101
GCL (ganglion cell layer) formation 59–61
gene expression profiling 325–6
  cDNA microarrays 327–8
  cell-specific markers 328, 335, 336
  choice of appropriate technique 330
  digital or sequence-based approaches 326–7, 328, 329
  Drosophila database 329
  high-throughput *in situ* hybridization 326–7, 328, 329–30
  histological approaches 326–7, 328, 329–30
  hybridization-based approaches 326–8
  microarray hybridization 326–8
  mouse database 329
  mouse studies comparison 328, 335, 336
  MPSS (massively parallel signature sequencing) 329
  oligonucleotide arrays 327–8
  SAGE (serial analysis of gene expression) 326–7, 328, 329
  sensitivity of different techniques 330
  survey of retinal development studies 328, 330–4, 335, 336
  techniques 326–7, 328, 330
gene knockdown methods 343–4
genes, expression during retinal development 325
genomic studies
  clinically important data 337
  disease genes identification 337
  expression profiling 325–6
  future directions 338
  mechanisms of retinal development 335–8
  suitability of retinal development for study 325
  survey of retinal development studies 328, 330–4, 335, 336
GLAST (glutamate aspartate transporter), expression in Müller cells 177
glaucoma 165, 166
  secondary 181
Gli gene, target for *Shh* 161
glial cells
  as neuronal progenitors 314–16, 317
  astrocytes in the retina 172–3
  effects on neuronal process outgrowth 174
  glutamatergic neurotransmission 177–8, 179
  in the optic nerve 150, 162
  migration and differentiation 163, 164
  migration guidance molecules 163–5
  role in blood vessel formation 172–3
  role in retinal vascular development 182–3, 184
  signalling for 51–2
  see also microglia; Müller glial cells; radial glia
glial-guided cell migration 60, 61–2
glial–neuronal interactions, mediating molecules 174–5, 177
glutamate 99
  levels in early retinal development 177
  mechanisms of release in development 99–100, 101
  modes of action 102
  role in cell death and survival 220
  role in retinal wave generation 275–6
  sources in development 102
glutamate receptors
  in the developing retina 105, 107–9
  on developing Müller cells 181–2
glutamatergic neurotransmission
  by glial cells 177–8, 179
  in early development 268–9
glutamine synthetase, expression in Müller cells 177–8
glycine receptors, on developing Müller cells 182
glycinergic neurotransmission 268–9
GSK-3β•APC•Axin (glycogen synthase kinase-3β•Adenomatous Polyposis Coli•Axin protein complex) 12
growth factors *see* neurotrophins

hedgehog signalling system 21 *see also Shh (Sonic hedgehog)*
Henle fibres 127, 128, 141–2, 143, 144
heritable causes of blindness 357–9
HesX1 mutations, link to septo-optic dysplasia 162–3
heterotypic cell interactions, in retinal mosaic development 194–5
high-throughput *in situ* hybridization 326–7, 328, 329–30
histogenesis in the retina 76
histological gene expression profiling 326–7, 328, 329–30
homeobox transcription factors, effects on retinal progenitors 81
homotypic cell interactions, in retinal mosaic development 194–5
horizontal cells 1–2, 3, 30
  cell survival mechanisms 212
  factors which shape arborization 245, 246
  maturation time 34–5, 45
  morphological development 245, 246
  naturally occurring cell death 212
  timing of genesis 34–5, 45
human retina
  cone packing in the foveal cone mosaic 141–2, 143, 144
  cone types 127–8
  developmental milestones 128–9, 130–1, 136
  foveal depression formation 134, 137–9, 140
  opsin expression in photoreceptors 135–6, 137–40
  retinal growth patterns 144–5
  synapse formation in the fovea 131–4, 135, 138–9

human retinal disease
   heritable causes of blindness 357–9
   zebrafish mutants as models 357–60, 361
   see also specific conditions
hybridization-based gene expression profiling 326–8

IgCAMs (immunoglobulin superfamily cell adhesion molecules), role in axon guidance 155–6, 157
IGFs (insulin growth factors), Wnt inhibition 12–13
ILM (inner limiting membrane; retinal basal lamina)
   formation 59–61, 66–7, 68
   influence on cell migration trajectories 64–5
   role in retinal cell migration 66–7, 68
INL (inner nuclear layer), evidence for programmed cell death 211–13
integrins, role in cell migration signalling 68
interkinetic nuclear migration 31–3, 34–5, 59–61
interneurons
   factors which shape arborization 245, 246
   recoverin immunostaining 293–6
interphase 31–3, 34–5
intrinsic factors
   multiple interactions between 82–3
   retinal progenitor regulation 76, 79–81, 83
IPL (inner plexiform layer) 2–3
   synapse formation 266–7, 268
IPL structural lamination
   dendritic filopodial motility 250–1, 252
   early synaptogenic events 250–1, 252
   mechanisms 246, 249–51, 253, 258
   ON and OFF regions of the IPL 243, 249–50, 251–5, 293–4, 295, 296, 297
   possible sublamination strategies 252–3, 255
   role of molecular guidance cues 255–6
   role of neuronal activity 251, 256–8
   sublaminae of the IPL 243, 249–50
   segregation of cell processes in sublaminae 293–4, 295, 296, 297
ipRGCs (intrinsically photosensitive RGCs) 289–90, 291
iris, development of 1, 2
irradiance (illumination level) detection 289–90, 291

K$^+$ channels
   in Müller cells 179–80, 181
   in retinal pathology 181
K$^+$ spatial buffering in the retina 179–80, 181
KCC2 (K$^+$–Cl$^-$ membrane co-transporter) 277–9, 280
kynurenic acid 179

L- / M-cones 135–6, 137–40
L- / M-opsin expression 135–6, 137–40
L1 (IgCAM), role in axon guidance 155–6, 157
lacritz mutant, zebrafish 348

lamina cribosa, lack of myelination 165
laminar organization of the retina 2–3, 30, 193
   factors affecting 242–5
   formation 59–61
   role of Müller cells 173–4
   zebrafish retina 345, 347
   see also IPL structural lamination
laminins 270–1, 292
lateral geniculate nucleus 274–5
lateral migration, in retinal mosaic development 196–7
Leber congenital amauraurosis 357–9
LEF1(lymphoid enhancer binding factor 1) 12
lens development 1, 2
lens placode 1, 2
Lhx2 gene, EFTF expression 14–15, 17, 19
light onset and offset responses 293–4, 295, 296, 297
light responsiveness
   and eye opening 288
   circadian rhythm generation 289–91
   electroretinograms to monitor development 292
   emergence in image-forming pathway 291–4, 295, 296, 299
   ipRGCs (intrinsically photosensitive RGCs) 289–90, 291
   irradiance (illumination level) detection 289–90, 291
   neonatal retina (non-primates) 288
   receptive field development and plasticity 292–4, 295, 296, 299
   turtle embryonic RGCs 297–8

M-phase (mitotic division) 31–3, 34–5, 59–61
macaque retina
   cone packing in the foveal cone mosaic 141–2, 143, 144
   cone types 127–8
   developmental milestones 128–9, 130–1, 136
   foveal depression formation 132, 134, 137–40
   opsin expression in photoreceptors 135–6, 137–40
   retinal growth patterns 132, 144–5
   synapse formation in the fovea 131–2, 134, 135
mAChRs (ACh receptors), in the developing retina 103–4, 105
macrophages, role in developmental cell death 222
macula lutea ('yellow spot') 126, 127, 128
maternal determinants of retinogenic zone 8–10, 11
mbl (*masterblind*) gene, in zebrafish embryos 12
melanoma 181
melanopsin-expressing ipRGCs 289–90, 291
mGluRs (metabotropic glutamate receptors) in the developing retina 105, 107–9
microarray hybridization 326–8
microglia, role in developmental cell death 222
micromechanical hypothesis, for retinal mosaic development 199–200, 201
microtubules, support for axonal and dendritic processes 199–200, 201

midget bipolar cells 126, 127, 128
midget ganglion cells 126, 127, 128
midget pathway in the fovea 126, 127, 128
mitotic figures 31–3, 34–5
mitotic spindle orientation, and mode of cell division 38–9, 40
modulated cell cycle timing 35, 37
morpholinos 344
morphological screening, for zebrafish mutants 351–2
mouse
    cell cycle timing 33–5, 37
    gene expression database 329
    gene expression profiling studies 328, 335, 336
MPSS (massively parallel signature sequencing) 329
Müller glial cell development
    ATP receptors 182
    catecholamine receptors 182
    dopamine receptors 182
    GABA receptors 182
    glutamate receptors 181–2
    glycine receptors 182
    neurotransmitter receptors 181–2
    purinergic receptors 182
Müller glial cells 2–3, 30
    as neuronal progenitors 314–16, 317
    as source for retinal regeneration 314–16, 317
    bFGF production 176
    definition of maturity 173
    DHA trafficking 175–6
    differences to retinal progenitors 317
    effects on neuronal process outgrowth 174
    extracellular glutamate regulation 177
    extracellular K^+ regulation 179–80, 181
    functions in the retina 172–3
    glutamate transporter expression 177
    glutamatergic neurotransmission 177–8, 179
    glutamine synthetase expression 177–8
    inwardly rectifying K^+ channels 179–80, 181
    location in the retina 172–3
    nestin expression 130
    neuroprotective trophic factors production 175–6
    origin of 173
    primate retina 127–8
    production of factors toxic to photoreceptors 177
    promoting effect of p27Xic1 88, 89
    rescue of diseased or damaged neurons 175–6
    retinal ganglion activity modulation 172–3
    role in cell migration 60, 61–2
    role in retinal histogenesis 173–5, 177
    role in retinal organization 173–4
    role in retinal vascular development 182–3, 184
    signalling for 51–2
    timing of genesis 47
    see also glial cells; microglia; radial glia

Müller glial cell fate
    effects of Notch signalling 80, 82, 314–15
    effects of proneural genes 80
multipolar migratory cells 60, 63–4
muscarinic neurotransmitters 109–10, 111–12, 113
myelination of the optic nerve 163, 164
    stop signals in the optic nerve head 165
    unmyelinated lamina cribosa 165
    unmyelinated retinal RGC axons 165

nAChRs (ACh receptors), in the developing retina 103–4, 105
nba (*night blind a*) zebrafish mutation series 359–60, 361
NCAM (IgCAM), role in axon guidance 155–6, 157
neonatal retina (non-primates)
    emergence of light responses 291–4, 295, 296, 299
    intrinsically photosensitive ganglion cells (ipRGCs) 289–90, 291
    irradiance (illumination level) detection 289–90, 291
    responsiveness to light 288
    role in circadian rhythm generation 289–90, 291
neostigmine 299
nestin
    expression by Müller cells 130
    neural progenitor marker 314–15
Netrin guidance molecules 159–60, 161, 163–5
neural induction
    and eye formation 9–11
    and patterning 11–14
neural plate, origins of the eye field 8
neurite field
    arborization complexity 247
    complexity and lateral extent 245–8, 249
    contact-mediated inhibition 247–8, 249
    influence of neighbours of the same subtype 245–8, 249
    intrinsic growth restriction 248, 249
    receptive field size 245–8, 249
    RGCs dendro-dendritic contacts 247–8, 249
    size regulation 245–8, 249
neurite growth
    effects of glial cells 174
    effects of intraretinal signalling 245
    effects of neuronal interactions 245
    influence of cell polarity 242–4
    influence of cellular mechanisms 244–5
    modulation by neurotransmitters 114–15
    molecular stimulants of outgrowth 244
    outgrowth initiation 242–5
neurodegenerative diseases 228–9
neuromodulators, effects on programmed cell death 220–1
neuron arbors, organization in relation to functions 242
neuron survival, role of neurotrophins 115
neuron types 1–2, 3
    morphological development 245, 246
    subtypes diversity in zebrafish 344–5, 346, 347

neurons, mispositioning in retinal laminae 224, 226
neuropeptides, role in cell survival 220–1
neurotransmitter receptors (developing retina) 103–5, 109
   ACh receptors (AChRs) 103–4, 105
   ATP receptors 104–5, 106
   GABA receptors 105, 106–7
   glutamate receptors 105, 107–9
   regulatory role 99
neurotransmitters
   adult retina 99
   ionotropic receptors 102
   mechanisms of release in development 99–100, 101
   membrane-bound receptors 102
   metabotropic receptors 102
   modes of action 102
   modulation of $[Ca^{2+}]_i$ during development 103, 112
   modulation of $[Ca^{2+}]_i$ local waves 112, 113
   modulation of $[Ca^{2+}]_i$ propagating waves 112, 114
   modulation of $[Ca^{2+}]_i$ transients in retinal progenitors 111–12, 113
   modulation of PCD 220–1
   progenitor cell cycle regulation 109–10, 111–12, 113
   regulation of differentiation 114–15
   roles in early retinal development 99–100, 105, 110, 112, 115
   roles in retinal wave generation 275–6, 279
   sources in development 101–2
   synaptic vesicular transport 268–9
neurotrophin receptors
   p75 receptor 115
   tyrosine kinase receptors (TrkA, TrkB, TrkC) 115
neurotrophins
   influence on cell death 218–19
   role in early retinal development 99, 115–17
   role in neuron survival 115
NGF (nerve growth factor) 115
   and programmed cell death 116–17
   apoptotic properties 116–17
   receptor binding 115
   role in cell death and survival 218–19
nicotinic receptors, mediation of retinal waves 274–5
nitric oxide, role in cell survival 220
NMDAs (glutamate receptors) 179
   in the developing retina 105, 107–9
noa zebrafish mutant model 359
Nodals
   signalling inhibition 9–11
   signalling mutations 20–1
   role in neural patterning 12
nof zebrafish mutant 355–7
noggin, neural induction 9–11
non-NMDAs (glutamate receptors) 105, 107–9, 179

Notch–Delta pathway 78–9
Notch signalling
   and cell fate 39, 40
   and Müller glial fate 314–15
   gliogenic activity 80, 82
nrc zebrafish mutant 355–6, 357
NT-3 (neurotrophin-3) 115
   and retinal differentiation 115–16
   receptor binding 115
   role in cell death and survival 219
NT-4 (neurotrophin-4), role in cell death and survival 219
NT-4/5 (neurotrophin-4/5) 115
   receptor binding 115
Numb expression, and cell fate 39, 40
numerical matching of interconnecting cells 224, 227–8

ocular domains separation 224, 225
oligodendrocyte precursor cells
   diencephalic origin 163, 164
   long-distance migration 163, 164
   migration guidance molecules 163–5
oligodendrocytes, myelination of RGC axons 163, 164
oligonucleotide arrays 327–8
OLM (outer limiting membrane) formation 59–61
ON- and OFF-centre responding retinal circuits 243, 249–50, 251–5
ONL (outer nuclear layer), evidence for programmed cell death 211–13
OPL (outer plexiform layer) 2–3
   formation 34–5, 45
   synapse formation 266–7, 268
opl (*odd paired-like*) gene 20–1
opsins 288
   expression in human and macaque 135–6, 137–40
   expression in zebrafish retina 348–51
   in cone photoreceptors 46–7
optic chiasm 162
optic cup
   delineation of tissues 153
   development of 1, 2
   developmental abnormalities 153–4
   specification 150–2, 153
optic fissure
   failure to close 152–4
   malformations 153–4
optic nerve 1–2, 3
   cellular organization 150
optic nerve development 1, 2
   axon guidance molecules 161
   developmental abnormalities 153–4, 162–3
   formation by RGC axons 154–5
   glial cells in the optic nerve 162
   growth of RGC axons 161
   myelination 163, 164, 165

optic cup specification 150–2, 153
optic stalk specification 150–2, 153
repositioning of RGC axons 162
studies 150
three phases 150–1
transient optic nerve axons 162
unmyelinated lamina cribosa 165
see also RGC axons
optic nerve glial cells 150
optic nerve head
  expression of axon guidance molecules 161
  malformations 153–4
  stop signals for myelination 165
  targeting by RGC axons 157–8, 159
optic nerve hypoplasia 161, 162–3
optic stalk development
  delineation of tissues 153
  entry of RGC axons 159–60, 161
  transformation into optic nerve 154
optic stalk specification 150–2, 153
optic vesicles 1, 2
Optx2 gene, EFTF expression 14–15, 17, 19
Otx2 gene, anterior neural patterning 16–17
oxygen tension in the retina, effects on cell survival 221

P2 receptors
  in the developing retina 104–5, 106
  on developing Müller cells 182
p27Xic1 cell cycle inhibitor 88, 89
  Müller-cell promoting effect 88, 89
p53 gene, role in apoptosis 215, 216
p75 receptor 115
PACAP-expressing RGCs 289
Pax2 gene, role in optic stalk development 151–2, 153–4
Pax6 gene
  activation of bHLH proneural genes 82
  EFTF expression 14–15, 17, 19
  eye field marker 12–13
  in zebrafish 347–8
  role in optic cup development 153
  role in retinal progenitor regulation 82
PCD (programmed cell death)
  anatomy in the developing retina 209–13
  apoptotic body clearance 222
  cell loss in retinal cells other than RGCs 211–13
  cell loss magnitude determination 208
  cellular redox status effects 221
  definition 208
  ECM effects 221
  evidence from pyknotic profile 210, 211–13
  evidence of cell loss 209–10
  excitotoxicity effects 220
  ganglion cell loss magnitude 210–11
  gap junction mediation 222

identification techniques 208
implications of understanding PCD 228–9
macrophage roles 222
microglial role 222
multiple alternative pathways 208
neural activity effects 218
neuromodulator effects 220–1
neurotransmitter effects 220–1
neurotrophin effects 218–19
optic nerve formation 154
oxygen tension effects 221
proliferating cell loss 213
regulation mechanisms 218–22, 223
retinal ganglion cell axons 154
retinal ganglion cells 209–11
retinal pigment epithelium effects 218
tissue biology 218–22, 223
trophic factor effects 218–20
see also apoptosis; cell death
PCD cellular and molecular biology 213–17
  apoptotic mechanisms 215–17
  apoptotic process 213–14
  autophagy 216–17
  bax pro-apoptotic gene 215
  Bcl-2 family-modulated apoptosis 214, 216–17
  caspase 3-mediated apoptosis 214, 216–17
  genes involved in apoptosis 214, 215–17
  markers for distinct types of cell death 216–17
  multiple pathways of PCD 216–17
  protein synthesis mechanisms 215–17
  pyknotic profiles 213–14
  role of *p53* gene in apoptosis 215, 216
  role of *Rb* family of genes 216
  signature of dying cells 213–14
  transcription factor mechanisms 215–17
PCD in retinal development 116–17, 222–24, 228
  creation of centro-peripheral density gradients 224, 227
  correction of neural system errors 208–9
  elimination of anomalous projections 222–23, 224
  elimination of mispositioned neurons 224, 226
  establishment of ocular domains 224, 225
  formation of regular mosaics 224, 226–27
  numerical matching of interconnecting cells 208–9, 224, 227–8
  refinement of retinotopic mapping 224, 225
  sculpting of decussation patterns 224, 225–6
RPE (pigmented epithelium)
  regeneration in amphibians 310–13, 314
  source for retinal regeneration 307–8, 309, 310–13, 314
  transdifferentiation 312–13, 314
peptide nucleic acids 344
photopigment 46–7
photoreceptor cell dystrophies 337, 357–9

photoreceptor cells 1–2, 3, 30
  maturation in zebrafish retina 348–51
  maturation time 46–7
  timing of genesis 42, 45–7
  see also cones photoreceptors; rod photoreceptors
phototransduction proteins 135–6, 137–40
pineal melatonin levels 289
pob (*partial optokinetic response b*) zebrafish mutant 353–4, 355
prechordal mesoderm, signals for separation of eye field 19
presumptive eye *see* eye field
presumptive neural plate, transformation 11–14
primate retina
  cone types 127–8
  fovea 127, 128
  Müller cells 127–8
  peripheral retina structure 127–8
  specialized central region 127–8
progenitor cells *see* retinal progenitor cells
programmed cell death *see* PCD
proliferating cells, evidence for programmed cell death 213
proliferative vitreoretinopathy 182
proteoglycans, role in RGC axon orientation 154–5
pure-cone area *see* foveal cone mosaic
purinergic neurotransmitters 109–10, 111–12, 113
purinergic receptors, on developing Müller cells 182
pyknotic profiles
  evidence for apoptosis 213–14
  evidence for PCD 210, 211–13

radial glia, and retinal histogenesis 173–4
radial migration, in retinal mosaic development 196–7
rat, cell cycle timing 33–5, 37
Rb (retinoblastoma) gene family, role in apoptosis 216
receptive field
  centre-surround organization 292–3
  complex fields in turtle retina 297–8
  development and plasticity (RGCs) 292–4, 295, 296, 299
  earliest measurable 292–3
  effects of dark rearing 298–9
  field size and visual acuity 245–6
  light responses in turtle embryo 297–8
  role of early neural activity 298–9
  role of retinal waves in development 299
  stratification into ON and OFF sublaminae 293–4, 295, 296, 297
recoverin immunostaining 293–6
redox status, effects on cell survival 221
regeneration *see* retinal regeneration
retina
  development of 1, 2
  early spontaneous activity 268–70
  early spontaneous wave activity 265, 267

  organization in the mature vertebrate 1–3
  specialized subcircuits 2–3
retinal basal lamina *see* ILM (inner limiting membrane; retinal basal lamina)
retinal cells
  centro-peripheral density gradients 224, 227
  displacement to form the fovea 127, 128
  types 30
  spontaneous bursting in immature cells 272–3 *see also* retinal waves
  see also cell differentiation; cell fate determination *and specific cell types*
retinal circuitry development, role of retinal mosaics 204–5
retinal competence 8–9, 10, 11
retinal degenerative diseases 177
retinal detachment 181
retinal development
  growth patterns 132, 144–5
  milestones 128–9, 130–1, 136
  programmed cell death phases 116–17 *see also* PCD
  role of neurotrophins 115–17
retinal development studies
  fluorescent reporters 258–9
  outstanding issues 258
  recently developed techniques 258–9
  survey of gene expression profiling studies 328, 330–4, 335, 336
retinal disease genes, identification 337
retinal dystrophies 228–9, 357–60, 361
retinal ganglion cell (RGC) axons
  arcuate fibres 159
  avoidance of the fovea 159
  axon collaterals 155
  axon guidance molecules 155–6, 157–8, 159–60, 161
  fasciculation as they cross the retina 155–6, 157
  growth cone modulation 155–7
  growth within the optic nerve 161
  intrinsic orientation property of retinal tissue 157
  myelination in the optic nerve 163, 164
  optic nerve formation 154
  optic nerve head orientation 154–5
  optic nerve head targeting 157–8, 159
  optic stalk entry 159–60, 161
  programmed cell death 154
  repositioning in the optic nerve 162
  stop signals for myelination 165
  unmyelinated in the lamina cribosa 165
  unmyelinated in the retina 165
retinal ganglion cells (RGCs) 1–2, 3, 30
  arborization complexity 247
  axons in the optic nerve 150
  cell death regulation mechanisms 218–22, 223
  contralateral and ipsilateral projection 43
  dendritic filopodial motility 250–1, 252

developmental cell death 209
early synapse formation 250–1, 252, 266–7, 268
elimination of anomalous projections 222–3, 224
establishment of ocular domains 224, 225
intrinsically photosensitive RGCs (ipRGCs) 289–90, 291
light responses in turtle embryo 297–8
midget ganglion cells 126
morphological development 245, 246
numerical matching of interconnecting cells 224, 227–8
post mitotic migration 59–61
receptive field development and plasticity 292–4, 295, 296, 299
retinotopic mapping 224, 225
sculpting of decussation patterns 224, 225–6
spontaneous bursting in immature cells 272–5, 279
sublamination into ON and OFF layers 252–5
subtypes 30, 43
timing of genesis 43
use of NGF to regulate numbers 116–17
see also retinal waves
retinal histogenesis
  molecules that mediate glial–neuronal interactions 174–5, 177
  origin of retinal glia 173
  radial glia 173–4
  role of Müller glial cells 173–5, 177
retinal injury 181
retinal lamination *see* laminar organization of the retina
retinal mosaic development 193–7, 200, 201
  cell fate mechanisms 195–6
  computer models 199
  cross-correlation studies 195
  dendritic interactions 199–200, 201
  d_{min} computer model 194–5, 199
  heterotypic cell interactions 194–5
  homotypic cell interactions 194–5
  micromechanical hypothesis 199–200, 201
  radial and lateral migration 196–7
  role of cell death 197–9, 224, 226–7
  spatial independence of cell types 194–5
  zebrafish 345, 347
retinal mosaics 193, 194
  auto correlation plot 201–2, 203
  cell type variation in regularity 204
  cross-correlation plot 201–2, 203
  Delaunay triangulation measures 203–4
  mathematical quantification of regularity 201–2, 203, 204
  RI (regularity index) 201, 202
  role in retinal circuitry development 204–5
  role of dendro-dendritic interactions 247–8, 249
  Voronoi tessellation measures 203–4
retinal neuroepithelium 59–61
retinal pathology
  and cell death mechanisms 228–9

loss of inwardly rectifying K^+ channels 181
see also specific conditions
retinal progenitor cells (RPCs)
  asymmetrical cell division 32–3, 35–8
  cell cycle timing 33–5, 36–7
  developmental competence 42, 47–51
  heterogeneity 75
  multipotency 76
  symmetrical cell division 32–3, 35–8
  terminal cell division 32–3, 35–8
retinal progenitor proliferation 82
  cell cycle regulation by neurotransmitters 109–10, 111–12, 113
retinal progenitor regulation
  bHLH proneural genes 79–80, 81
  cell type-specific feedback 76–7
  competence to differentiate 78–9
  diffusible extracellular signals 77–8
  extrinsic signalling 76–9
  homeobox transcription factors 81
  intrinsic factor interactions 82–3
  intrinsic signalling 76, 79–81, 83
  Notch–Delta pathway 78–9
  Pax6 gene 82
  postmitotic neuron feedback signalling 76–7
retinal projections, refinement 274–5
retinal regeneration
  ciliary epithelium (mammals) 317–19, 320
  common features of sources 308, 309
  differences in fish, birds and mammals 315–16, 317
  from intrinsic retinal sources 308, 309, 314–16, 317
  from the ciliary margin zone 307–9, 310, 311
  from the pigmented epithelium 307–8, 309, 310–13, 314
  history of study 307–8, 309
  Müller glia as source 314–16, 317
  problems with RPE as source 312–13
  signalling molecules 313–14
retinal stretch 132, 144–5
retinal tissue, intrinsic property for axon orientation 157
retinal vasculature
  barrier properties 184
  blood supply to the eye 1, 2
  development 172–3, 182–3, 184
  foveal avascular zone 127, 128
  role of glia in development 182–3, 184
retinal waves 272–5, 279
  age-related changes in dynamics 273, 276–9, 280
  disappearance with retinal maturation 280–1
  effects of dark rearing 280–1
  effects of visual experience 280–1
  generation and propagation mechanisms 275–6, 279
  role in formation of the visual system 274–5
  role in receptive field development 299
  role of cholinergic amacrine cells 274–5

retinal waves (*cont.*)
    role of gap junctions 276
    role of neurotransmitters 275–6, 279
    role of nicotinic receptors 274–5
    spatiotemporal cues about retinal wiring 274–5
retinitis pigmentosa 175–6, 228–9, 357–9
retinoblastoma pathogenesis 228–9
retinogenic blastomeres 8–9, 10, 11
retinogenic zone 8–9, 10, 11
retinohypothalamic tract 289
retinoic acid 12, 360–2
retinotopic mapping 224, 225, 274–5
retroviral mutagenesis techniques 343
RGCs *see* retinal ganglion cells
RhoGTPases, role in retinal cell migration 69
rhythmic bursting activity 272–5, 279
RI (regularity index), for retinal mosaics 201, 202
ribbon synapse formation 266–7, 268, 269, 270
rod bipolar cells 2–3
rod-free zone *see* foveal cone mosaic
rod opsin expression 135–6, 137–40, 288
rod photoreceptors 1–2, 3, 30
    evidence for programmed cell death 212
    maturation time 46–7
    morphological development 245, 246
    rod dominated or pure rod retinas 42, 45–6
    sensitivity to low light 2–3
    timing of genesis 42, 45–6
    see also photoreceptors
RPE (retinal pigment epithelium) 1, 2, 218, 312–13
Rx gene
    EFTF expression 14–15, 17, 19
    role in optic cup development 153
Rx1 gene, eye field marker 12–13
Rx3 gene, eye field marker 12–13

S-cones 135–6, 137–40
S-opsin expression 135–6, 137–40
S-phase (DNA synthesis) 31–3, 34–5, 59–61
SAGE (serial analysis of gene expression) 326–7, 328, 329
sclera, development of 1, 2
secondary glaucoma 181
SFRPs (Secreted Frizzled Related Proteins), Wnt inhibition 12–13
Sema5A, axon guidance molecule 161
Semaphorin guidance molecules 155–7, 161, 163–5
septo-optic dysplasia (SOD) 162–3
sequence-based gene expression profiling 326–7, 328, 329
SFRP1 gene, effects on forebrain patterning 12–13
Shh (*Sonic hedgehog*) gene 21
    delineation of eye tissue regions 153
    expression by RGCs 161
    expression in zebrafish 348
    genetic programmes triggered by 151–2, 153
    Gli target gene 161
    induction of optic stalk 151
    optic cup specification 151–3
    optic stalk specification 151–2, 153
Sidekicks (1 and 2), possible sublamination guidance cues 255–6
siRNAs (short interfering RNAs) 344
Six3 gene
    EFTF expression 14–15, 17, 19
    eye field marker 12–13
Slit1 and 2, axon guidance molecules 157, 161
Smad2 gene 21
Smoothened, transmembrane protein 151
SNAP-25 (synaptosomal-associated protein of 25kDa) 269–70
SNARE complex 269–70
somal translocation 60, 62–3, 67, 69
Sonic hedgehog see *Shh*
spontaneous bursting activity 272–5, 279
starburst amacrine cells 297
suprachiasmatic nucleus, generation of circadian rhythms 289–91
SV2 (synaptic vesicle protein 2) 269
symmetrical cell division, RPCs 32–3, 35–8
synapse formation 266–7, 272
    chemical synapses 266–7, 271
    dyad and triad formation 266–7, 268, 270–1
    early glutamatergic neurotransmission 268–9
    early synapse formation 266–7, 268
    electrical synapses 271–2
    in retinal neurons 265
    in the fovea 131–2, 134, 135, 138–9
    in the IPL and OPL 266–7, 268
    influence of the extracellular matrix 270–1
    markers for synaptic release-associated proteins 268–70
    ribbon synapses 266–7, 268, 269, 270
    role in IPL structural lamination 250–1, 252
    role of visual experience 267
    ultrastructural studies 266–7, 268
    vesicular transporters 268–9
synaptic transmission, steps in the process 268
synaptic vesicular transport 268–9
syntaxin 269–70
system matching 224, 227–8

T cell-specific transcription factor (TCF) 12
TAG-1 (IgCAM), role in axon guidance 155–6, 157
terminal cell division 32–3, 35–8
thymine precursors 31–3, 34–5
TILLING method 343–4
tlc gene, effects on forebrain patterning 12–13
tll gene, EFTF expression 14–15, 17, 19
transcription factors *see* eye field transcription factors (EFTFs)

transdifferentiation in the pigmented epithelium 312–13, 314
tritiated thymidine (^{3}H-TdR) 31–3, 34–5
TrkA (tyrosine kinase receptor) 115
TrkB (tyrosine kinase receptor) 115
TrkC (tyrosine kinase receptor) 115
trophic factors, role in cell death and survival 218–20
TUNEL technique to detect DNA strand breaks 214
turtle retina
    complex receptive fields 297–8
    light responses in embryonic RGCs 297–8

unconstrained cell migration 60, 63–4

VAMP (vesicle-associated membrane protein) 269–70
vascular system *see* retinal vasculature
Vax1 gene, role in optic stalk development 152–3
VEGF (vascular endothelial growth factor) 182–3, 184
vertebrate spinal cord, cell fate determination 91
VGAT 268–9
VGLUT1 268–9
visual acuity, and receptive field size *see also* fovea 245–6
visual disorders
    colobomas 152–4
    glaucoma 165, 166
    optic nerve developmental abnormalities 162–3
    see also specific conditions
visual experience
    effects on retinal waves 280–1
    role in receptive field development 298–9
    role in synapse formation 267
Voronoi tessellation 203–4
VZ (ventricular zone) 31–3, 34–5

Wnt (Wingless-Int) signaling
    and forebrain patterning 11–14
    control of EFTF expression 15, 18–19
    inhibition 9–11, 12–13
    non-canonical 13–15, 18–19, 25
    patterning the eye field 11–14
    role in neural induction and patterning 11–14

Xenopus
    cell cycle timing 33–5, 37
    developmental competence of RPCs 48–9, 50–1
    EFTFs 14–15, 17, 19

yng (*young*) mutant, in zebrafish 352

zebrafish (*Danio rerio*; *Brachydanio rerio*)
    cell cycle timing 33–5, 37
    gene function identification 342–4
    gene identification 342–4
    suitability as a genetic model 342
zebrafish mutagenesis screens 342–4
    ENU mutagenesis 343
    gene knockdown methods 343–4
    recessive mutation identification 342–4
    retroviral mutagenesis techniques 343
zebrafish retinal anatomy 344–5, 346, 347
    cone subtype arrangement 345, 347
    diversity of neuronal subtypes 344–5, 346, 347
    laminar organization 345, 347
    retinal mosaic patterning 345, 347
zebrafish retinal development 347–51
    ath5 gene 348
    atonal5 gene 348
    opsin expression 348–51
    optic cup and optic stalk formation 347–8
    order of neurogenesis 348
    Pax6 gene 347–8
    photoreceptor maturation 348–51
    Sonic hedgehog expression 348
zebrafish retinal mutants 351–2
    as models of human disease 357–60, 361
    behavioral screens 353–4, 356, 357, 358
    bel (*belladonna*) mutant 357, 358
    brg1 (*Brahma-related*) gene function 352
    chemical genetic screens 360–2
    detection of retinal dystrophies 359–60, 361
    effects of retinoic acid 360–2
    escape response assay 359–60, 361
    future assay techniques 362–3
    lacritz mutant 348
    morphological screening 351–2
    nba (*night blind a*) mutation series 359–60, 361
    noa mutant model 359
    nof mutant 355–7
    nrc mutant 355–6, 357
    pob (*partial optokinetic response b*) mutant 353–4, 355
    rod function screens 362–3
    yng (*young*) mutant 352

For EU product safety concerns, contact us at Calle de José Abascal, 56–1°,
28003 Madrid, Spain or eugpsr@cambridge.org.

www.ingramcontent.com/pod-product-compliance
Ingram Content Group UK Ltd.
Pitfield, Milton Keynes, MK11 3LW, UK
UKHW050109230326
469255UK00020B/458